D1345591

The Geology of
MARS

THOMAS A. MUTCH

RAYMOND E. ARVIDSON

JAMES W. HEAD, III

KENNETH L. JONES

R. STEPHEN SAUNDERS

Princeton University Press
Princeton, New Jersey

UNIVERSITY LIBRARY
24 AUG 1981
LANCASTER

Copyright © 1976 by Princeton University Press
Published by Princeton University Press, Princeton and
Guildford, Surrey. All Rights Reserved

Library of Congress Cataloging in Publication Data will
be found on the last printed page of this book

This book has been composed in Linotype Times Roman

Printed in the United States of America
by Princeton University Press, Princeton, New Jersey

Frontispiece. This time-exposure photograph shows an eruption of Stromboli, a conical volcanic island that rises 924 meters above the Mediterranean Sea, 65 kilometers north of Sicily.

The moon is visible directly above the vent. A star-like object appears to the left and just below the moon, framed by trajectories of pyroclastic ejecta. This object is the planet Mars, which on this date was approximately 77 million kilometers from Earth, shortly after its opposition of August 1971.

This photograph was taken by Douglas Hopkins on September 3, 1971, during a NASA-sponsored MIT-based volcanological expedition. Reproduced with permission of T. R. McGetchin. All rights reserved.

80 016383

Table of Contents

Preface

LIKE many projects, the preparation of this book has stretched far beyond our original intentions, in terms of both time and scope. So far back does the planning go, that our original goal was to assemble an atlas of Mariner 6–7 pictures, to be published *before* the Mariner 9 mission. Then, as 1971 drew near, we decided that it might be prudent to wait for Mariner 9 data before applying ink to paper. In retrospect it is difficult to figure out how we ever could have worked ourselves into a position where it was necessary to make this choice.

The Mariner 9 mission, originally planned for 90 days, continued for the better part of a year. The returned pictures were far more numerous than originally anticipated, and their orderly arrangement and analysis turned out to be a frustratingly time-consuming task. Our self-imposed deadlines for preparation of a manuscript inched steadily forward from 1973 to the present time, the spring of 1975. We now look forward to publication of the volume in the spring of 1976, only months before two Viking spacecraft will land on Mars. So, although we have avoided some of the pitfalls of precipitous generalization of Mariner 6–7 data or premature synthesis of Mariner 9 data, we have hardly escaped the criticism of doing the right thing at the wrong time.

Nonetheless, we are not embarrassed. We anticipate that Viking results will add to rather than replace Mariner results. This is particularly true with respect to the Viking chemical analyses of the atmosphere and surface materials, as well as the biological, seismological, and meteorological investigations. Pictures taken from the Viking orbiters will cover large parts of the Martian surface with ground resolution of about 100 m. However, a complete photographic survey of the planet, such as that accomplished by Mariner 9, will not be repeated. Accordingly, the present time appears appropriate for a review of the planet-wide distribution of the more prominent landforms, using Mariner pictures as a data base.

The composition of this book has been dictated by several goals. The first has been to prepare a complete topographic/geologic atlas of Mars. Toward that end we have included a large number of maps which systematically describe the character and distribution of the principal landforms: craters, channels, volcanoes, and faults; also, related properties such as albedo, elevation, and wind streaks. Pictures of all the important topographic features have been included. All photographs have been identified by position and DAS (Data Automation Subsystem) number so that the reader can easily identify the original Mariner product. Additional information regarding the use and availability of Mariner pictures is contained in appendix A.

A second goal has been to examine Mars within a context of interplanetary comparisons. Until very recently, geologists, of necessity, confined their attention to Earth. Then, with the success of the Apollo program, it became possible to make useful comparisons between the Moon and Earth. Within the past several years, receipt of spaceprobe data for Mercury and Mars has greatly enlarged the opportunities for comparative analysis. In the present volume, although the primary focus is on Mars, we have paid liberal attention to possible lunar, Mercurian and terrestrial analogs.

We believe that geological knowledge of Mars has advanced in quantity and certainty to a point where there is a need for a "primer" or "textbook." With that requirement in mind we have tried to give more-or-less equal treatment to all subjects that have a geological base. Although it would be foolish to claim that our bibliography is exhaustive, we *will* assert that we have tried hard to mention every relevant article. With mixed success we have tried to keep the language nontechnical and to develop each subject with few assumptions regarding the reader's background.

All five authors have participated in the operational phase of one or more space missions. This is singularly exciting work. Throughout the volume, and particularly in chapter 2, we have endeavored to communicate that excitement. Readers who find these discussions of space missions out of place in a book on geology can proceed directly from chapter 1 to 3.

One of the more stringent tests of true friendship must be "cooperative" research and publication. Happily, we can report that, in preparing this volume, we have passed the test, but not without several years of internal negotiations and concessions, accompanied by a spiraling exchange of exhortations and ultimatums. All authors have contributed to all chapters, but certain areas of primary responsibility have been defined. The scheme of automated data gathering and processing, described in appendix B, was conceived and implemented by Jones and Arvidson. Jones has also contributed most of the Martian crater analysis in chapter 4 and the thermodynamic discussions of chapter 8. The review of eolian processes and products in chapter 7 has been prepared by Arvidson. Saunders has taken the lead in compiling the physiographic summaries of chapter 3, the geophysical and structural discussions of chapter 6, and the review of Mariner picture format in appendix A. Head has contributed all of the lunar data. Mutch has had the responsibility of melding the separate parts into a cohesive whole.

May, 1975

THOMAS A. MUTCH, *Department of Geological Sciences, Brown University*

RAYMOND E. ARVIDSON, *McDonnell Center for the Space Sciences, Department of Earth and Planetary Sciences, Washington University*

JAMES W. HEAD, III, *Department of Geological Sciences, Brown University*

KENNETH L. JONES, *Department of Geological Sciences, Brown University*

R. STEPHEN SAUNDERS, *Jet Propulsion Laboratory, California Institute of Technology*

Acknowledgments

MANY persons have assisted in the preparation of this book. Margaret A. Power assembled and maintained a Mariner 9 picture library and assisted in the compilation of data for figures. K. Winslow Farrell, Jr., edited many of the computer-generated maps and contributed to the interpretation of erosional landforms in chapter 8. Sven U. Grenander prepared almost all the figures, a demanding exercise in drafting and scientific interpretation. Kathy Travis and Roberta L. Jennings contributed some of the artwork, particularly the airbrush renderings of Mars in chapter 3.

We are grateful to Clark R. Chapman for unselfish help throughout the course of our research, and for a valuable review of the manuscript. Other persons who have generously reviewed parts of the manuscript include William A. Baum, Michael H. Carr, James E. Conel, Edward A. Flinn, Ronald Greeley, William K. Hartmann, John S. King, Richard J. Pike, Roger J. Phillips, Carl Sagan, George D. Sands, S. Ross Taylor, and Joseph Veverka.

Over a period of several years Sarah H. Bosworth, Margaret T. Cummings, and Sybil E. Williams have typed, collated, and distributed numerous versions of the manuscript. Editorial help at the Princeton University Press has been provided by John Hannon, Science Editor; Gail Filion, who edited the manuscript; and Frank Mahood, who was responsible for the final arrangement of text and figures.

We gratefully acknowledge the help of the National Space Science Data Center in providing copies of planetary photographs. Finally, we are indebted to the National Aeronautics and Space Administration. The discipline of planetary geology depends, for its very existence, on NASA's program of planetary exploration.

The Geology of
MARS

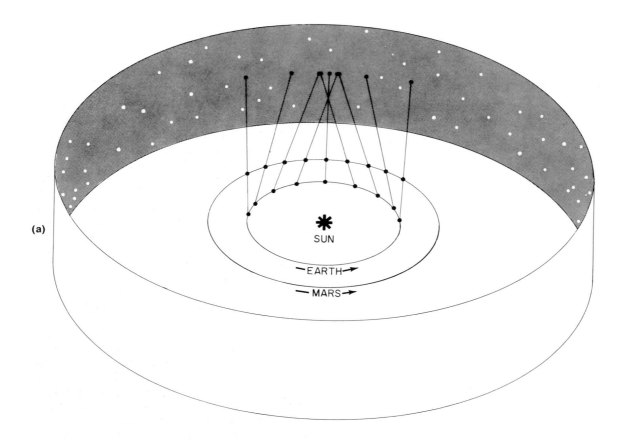

(a)

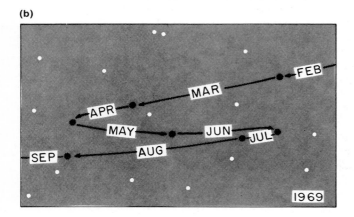

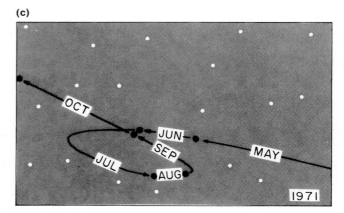

Figure 1.1 (a). The apparent motion of Mars, seen against a fixed star field. When Earth and Mars are close to opposition, Mars, viewed from Earth, appears to reverse its motion relative to the fixed stars. In this schematic drawing the stars are shown close to Earth and Mars. As a result, the relative orientation of rays from Earth and Mars through the star field is distorted. In the actual situation two rays that intersect the star field at the same point will be parallel.

(b). The apparent motion of Mars, viewed from Earth, during the 1969 opposition.

(c). The apparent motion of Mars, viewed from Earth, during the 1971 opposition. Because the orbital planes of Earth and Mars are not coincident, the form of the retrograde loop varies from opposition to opposition, depending on the positions of Earth and Mars in their orbits.

1.

The View from Earth

CLOSETED in window-poor buildings, living in smog-laden cities, and accustomed to judging clear those nights when a few of the brighter stars can be seen directly overhead, we find it difficult to appreciate fully the brilliance of the pre-industrial skies, the immediacy of the sun, moon, and stars, and the benefits—both real and illusory—of arranging one's activities according to the changing celestial patterns.

For our ancestors it was different. The natural world was close at hand and controlled one's daily welfare. Above all, the heavens seemed to hold the clues for understanding the present and predicting the future. Night after night Babylonian astronomers searched the heavens for signs of order, any diagnostic clustering or movement of constellations that, meticulously charted through the nights for many years, might lay bare some small fragment of the future. Of course, there was the sun, setting its daily course from east to west, skimming the horizon of the short winter days and standing nearly overhead during the summer. And there was the moon, growing from a thin sickle to a full luminous circle, then shrinking once more to an arcuate sliver—all this over a regular period of 29½ days. The stars, too, followed regular cycles: nightly risings and settings and longer seasonal periods of appearance and disappearance. However, even though the stars moved steadily through the night sky, they were forever tied to their neighbors. Particular constellations, viewed with considerable artistic license, outlined constant and familiar shapes: the Bull, the Eagle, and the Wolf.

Against the background of fixed stars, five points of bright light with strangely aberrant motion were observed. For many months they would be absent, then they would reappear—dimly at first, but growing in brightness until they rivaled the brightest stars—all the while threading their way through the stationary constellations. Two of the "stars" (actually Mercury and Venus) visible only in the evening and morning, swung back and forth from one side of the sun to the other. The remaining three (Mars, Jupiter, Saturn) usually moved from west to east, but occasionally looped backward before resuming their eastward journey (fig. 1.1). Small wonder that the Greeks called these five bodies "wanderers" or planets.

One of these planets was considered particularly important, not because it was the brightest—Venus held that honor—but because of its peculiar orange-red luster. Naturally enough, the unusual color was considered a portent of war and bloodshed. The Babylonians called the planet Nirgal, master of battles and judge of the dead. The Greek name, Ares, is derived from the verb "to kill." The Romans named it Mars, after the god of war. Even the ancient symbol for the planet Mars (♂) represents a shield and a spear. In the same vein, the National Aeronautic and Space Administration's planned landing of two unmanned spacecraft on Mars in 1976 is termed the Viking mission.

Planetary Orbits

The motion of the planets posed a challenging problem for the early astronomers, most of whom framed their solutions within two preconceptions: first, that the Earth was at the center of the universe and, second, that natural systems favor geometric simplicity. The first historically influential synthesis was by Claudius Ptolemy, a Greek astronomer living in Alexandria in the second century A.D. Ptolemy is generally recalled as the person who mistakenly believed the Earth to be at the center of the solar system. In fact, his scheme was considerably more complicated. The planets, sun, and moon revolved in circular orbits, or epicycles, about a center that itself described a circular orbit, or deferent, about the Earth. The deferents accounted for the general eastward motion of the planets; the epicycles explained their occasional westward retrograde motion. Even this system of multiple circles could not explain all the observations, and it was necessary to make further assumptions about deviations of the deferents from circular paths and variations in the angular velocity of the epicycle center relative to the Earth. To note that Ptolemy's cosmology held sway virtually unchallenged for 1500 years is not, as some historians would have us believe, evidence for a sterile adherence to dogma by the Romans and a general decline in objective analysis during the "dark ages." His mathematical model was durable because it was remarkably successful in predicting the position of a planet at any time. In common with many scientific "laws," it was nothing more than a description that, when repeatedly tested in a variety of situations, was consistent with the natural outcome.

Ptolemy's ideas were challenged by Nicholas Copernicus in 1543 with the publication of *De Revolutionibus Orbium Celestium* (On the Revolutions of the Celestial Spheres). Copernicus was a mathematical astronomer—that is, he was more interested in placing known facts in a correct mathematical context than in making observations himself. His central thesis was that all of the six planets recognized at that time revolve about the sun, following circular paths. Elaborating on this proposition, he was remarkably successful in determining the distances of planets from the sun relative to the Earth-sun distance. His values have errors of less than 10 percent.

Some question exists as to how convinced Copernicus himself was of the reality of his sun-centered configuration. A preface written by a Lutheran cleric responsible for the book's publication notes that the theory provides only a useful mathematical model. Did Copernicus truly believe his thesis; did he deem the disclaimer a convenient foil against possible charges of heresy; or was the preface written in opposition to his own views?

Further understanding of planetary motion awaited the collection of more extensive and precise observations. These were supplied by Tycho Brahe (1546–1601), working with instruments that gave positional information accurate to the limit of visual resolution (the telescope had not yet been invented). Tycho championed an amalgam of Copernican and Ptolemaic ideas. The moon and sun revolved about the Earth, but the other five planets revolved about the sun. The unifying truth imbedded in Tycho's many observations was extracted by Johannes Kepler (1571–1630), a young German astronomer who served as Tycho's assistant during his latter years and continued a detailed analysis of his collected observations following Tycho's death.

At the suggestion of Tycho, Kepler concentrated his attention on determining the orbit of Mars. It was generally known—notwithstanding Tycho's opinions—that it takes Earth 365 days to complete a revolution about the sun. The equivalent figure for Mars is 687 days. If Mars' position is measured on two dates 687 days apart the planet will be at the same orbital point, but Earth will be at two different positions. Knowing the angular position of Mars at these two times one can define the sun-Mars distance in terms of the sun-Earth distance (fig. 1.2). When Kepler first made this calculation, he was distressed to discover that the sun-Mars distance measured for a number of positions in the Martian orbit was not always the same. After numerous false starts stretching out over more than 20 years, he finally derived a solution that accurately explained Tycho's observations. Mars' orbit described an ellipse, that well-known geometric figure in which the sum of distances from two foci to any point on the curve is constant. Like almost anyone who makes an important discovery, Kepler was "lucky." Earth's orbit very closely approximates a circle—which happened to be one of Kepler's initial and necessary assumptions. Finally, he was fortunate in working with observations detailed enough to demonstrate eccentricity, but insufficiently accurate to show the minor perturbations that distort the ellipse and are due to gravitational attraction of neighboring planets.

Kepler is best remembered by students of astronomy for three laws of planetary motion. The first two of these were published in 1609 in a treatise

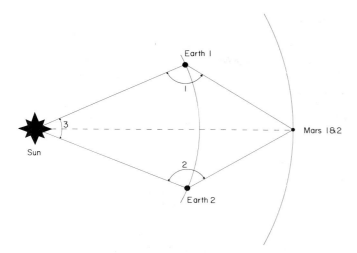

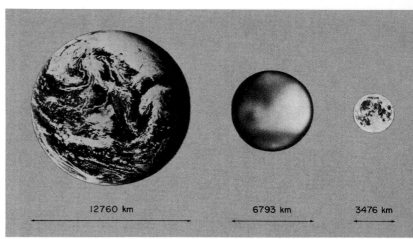

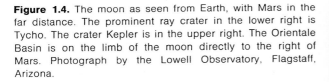

Figure 1.2. Determination of sun-Mars distance. For two observations of Mars taken 687 days apart, Earth occupies orbital positions 1 and 2. Angles 1 and 2 are measured. Angle 3 is calculated, knowing the number of days between observations 1 and 2. Earth's orbit is assumed circular, so distances sun-Earth$_1$ and sun-Earth$_2$ are equal. Knowing three angles and two sides of the quadrangle, the length of the diagonal sun-Mars relative to the Earth-sun distance can be calculated.

Figure 1.3. The relative sizes of Earth, Mars, and the moon. Equatorial diameters are indicated. Photographs of the moon and Mars were taken through Earth-based telescopes. The photograph of Earth was taken from an Apollo spacecraft.

Figure 1.4. The moon as seen from Earth, with Mars in the far distance. The prominent ray crater in the lower right is Tycho. The crater Kepler is in the upper right. The Orientale Basin is on the limb of the moon directly to the right of Mars. Photograph by the Lowell Observatory, Flagstaff, Arizona.

entitled *Motion of Mars*. The first specifies that each planet moves about the sun in an elliptical orbit, with the sun at one focus of the ellipse. The second states that a straight line joining the sun and a planet sweeps out equal areas during equal intervals of time. The third law, appearing in a 1619 volume entitled *The Harmony of the Worlds*, stipulates that the square of a planet's orbital period is proportional to the cube of its mean distance from the sun. This last law is buried in a surplus of more frivolous information, including the musical phrases allegedly sounded by the planets as they move through their orbits. The true significance of Kepler's laws was not appreciated until some eighty years later when Newton demonstrated that the motions of the planets are governed principally by gravitational attraction between them and the sun.

Early Telescopic Observations

The telescope was invented in 1608, and Galileo Galilei was one of the first to use it in study of the planets. He had considerable success delineating the large craters on the moon, but when he turned to Mars he could decipher almost no detail. This exercise has been repeated by many amateur astronomers who are disappointed to discover that the famous red planet is at best resolvable only into a few large bright and dark areas (fig. 1.5). In particular, Galileo was interested in determining whether Mars showed gibbous and full phases analogous to those of the moon, a phenomenon that would be consistent with the planet's revolution about the sun. His own observations were indecisive, but in 1636 Francisco Fontana published a drawing showing a gibbous phase. One might be justified in questioning the reality of Fontana's results. None of the prominent regional variations in brightness is recorded; a dark central circle and a peripheral dark ring were apparently caused by defects in his telescope.

The first informative sketch of Mars was produced by Christian Huygens in 1659 (fig. 1.6). He distinguished a prominent dark marking, probably Syrtis Major. The reliability of his observations is established by the fact that he used the temporal reappearance of this same marking to determine Mars' rotational period as 24 hours, less than one hour removed from the actual value. Huygens also made a crude sketch of the south polar cap, but it was more definitively described by Giacomo Maraldi in 1719 as a white spot located some distance from the rotational pole.

William Herschel, a brilliant English scientist of the late eighteenth century who gave up a distinguished career as a musician in favor of astronomy, extended observation and interpretation far beyond that of his predecessors. He personally constructed

Figure 1.6. Drawing of Mars by Christian Huygens. Telescopic images are inverted, so the south polar cap appears at the top. Dark region probably corresponds to Syrtis Major. From Flammarion (1892).

a series of reflecting telescopes with carefully shaped and surfaced mirrors that produced images of then-unimagined clarity and brightness. He is best remembered for his study of the fixed stars, but his early work centered on the planets. In fact, he discovered a new planet far beyond Saturn. This prestigious discovery brought his work to the attention of King George III, who placed him on the royal payroll. Herschel expressed his gratitude to his patron in the usual way—by naming the new planet "Georgium Sides" (the star of George)—but the designation found little favor on the continent, where it was replaced by the more traditional name Uranus, the mythological father of the Titans and grandfather of Jupiter. Herschel determined the length of the Martian day as 24 hours, 39 minutes, and 22 seconds—only 14 seconds less than the currently accepted value. In making the more difficult measurement of inclination of Mars' rotational axis to the ecliptic, he was in error by about five percent. As on Earth, this inclination is the reason for the day-to-day apparent change in sun position and the succession of the seasons. Herschel correctly associated the changing Martian seasons with the annual growth and shrinkage of the polar caps, which he interpreted as thin deposits of ice and snow. Finally, he detected temporal changes in the patterns of bright and dark regions, a feature he attributed to the passage of clouds.

Herschel's work climaxes the early telescopic studies of Mars. As the eighteenth century drew to a close, a surprisingly Earth-like picture of Mars emerged. Both planets experience a day-night cycle of approximately 24 hours. For both there is a succession of seasons, even though the Martian year, or time of revolution about the sun, is about twice that of Earth. Both planets have polar ice caps that vary seasonally. It was even thought that Mars might have an atmosphere with occasional clouds obscuring the surface.

Seasons

As just mentioned, the seasons are the consequence of a rotational axis inclined to the orbital plane. This inclination, measured from a normal to the orbital plane, is 23.5° for Earth and 25.1° for Mars. Four limiting conditions can be specified. At one point in the planet's orbit, the axis will be pointed toward the sun; at another point it will be tilted away; and at two other points it will be perpendicular to the sun's rays (fig. 1.7). For Earth these conditions occur, respectively, on June 22, December 22, September 23, and March 21. On June 22 the north pole is tilted toward the sun, and a maximum amount of the northern hemisphere is illuminated.

This date is called a solstice, so named for the fact that the sun appears to traverse identical paths on a day-to-day basis. As the planet completes its rotation on June 22, any station north of the equator will remain in sunlight for a maximum period of time. In the southern hemisphere it is the shortest day of the year. On December 22, when the north pole is tilted away from the sun, the situation in the two hemispheres is reversed. On September 23 and March 21 the edge of the illuminated hemisphere passes directly through both poles, and equal areas of both northern and southern hemispheres lie in lightness and dark. Throughout Earth the night and day are equal in length. For that reason, the dates are termed equinoxes.

These four dates are used to define the seasons. For the northern hemisphere, the period between March 21 and June 22 is termed spring. June 22 to September 23 is summer; September 23 to December 22 is autumn; and December 22 to March 21 is winter. Summer starts with the longest day of the year. It is a hot season both because the days are long and because the sun is high in the sky. Clearly the number of sunlight hours are important in determining the amount of solar energy received, but

Figure 1.7. Schematic drawing, showing variable illumination of a planet due to tilt of its rotational axis. In this hypothetical situation, the major axis of the planetary orbit coincides with the line of solstices. At perihelion the northern hemisphere is tilted away from the sun and is experiencing

short days of winter. At the same time in the southern hemisphere summer prevails, an unusually warm season because the planet is at its closest approach to the sun. Earth has approximately this same configuration except that its orbit is more nearly circular.

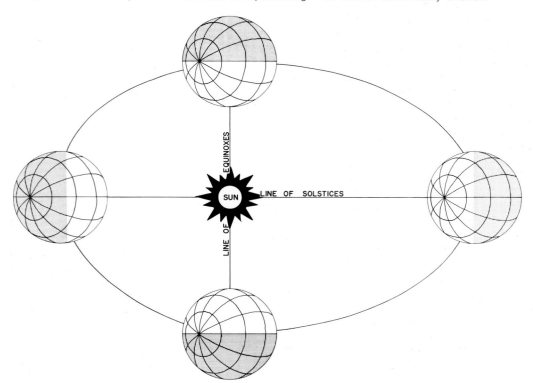

the position of the sun in the sky is also important. When the sun is high, the energy will be spread over much less ground surface than will be the case for a grazing sun. On September 23 and March 21, the noon sun is directly overhead at the equator; between those dates the high sun migrates to a latitude of 23.5° and then back to the equator.

Seasons on Mars can be calculated according to the same relationships just described for Earth. As the planet moves through its orbit, there are two equinoxes and two solstices. The intervening periods correspond to the seasons of spring, summer, autumn, and winter (fig. 1.8). Because the period of Mars' revolution is approximately twice that of Earth, the seasons are twice as long. The eccentricity of Mars' orbit introduces an additional complication. First, the seasons are of unequal duration. For example, spring in the northern hemisphere is 52

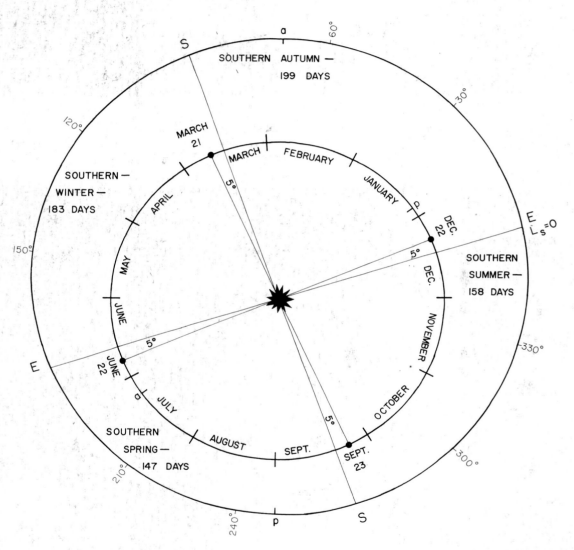

Figure 1.8. Seasons on Earth and Mars. Orbital traces for both planets are drawn to scale. Perihelion (p) and aphelion (a) are shown for the two planets. Equinoxes (E) and solstices (S) are also shown. Note that the lines of equinoxes for the two planets are separated by 85°. Martian seasons are indicated for the southern hemisphere. Aerocentric longitude for the sun (L$_s$) provides a notation for further specifying the season. For example, L$_s$ 315° indicates midsummer in the southern hemisphere. This diagram can be used to predict the Martian season during a particular opposition. For those oppositions that occur close to Martian perihelion, it will be spring or summer in the southern hemisphere. The viewing period from Earth is late summer. For aphelion oppositions, it will be late autumn on Mars and late winter on Earth. The diagram should not be misread to indicate a general temporal correspondence between seasons on the two planets.

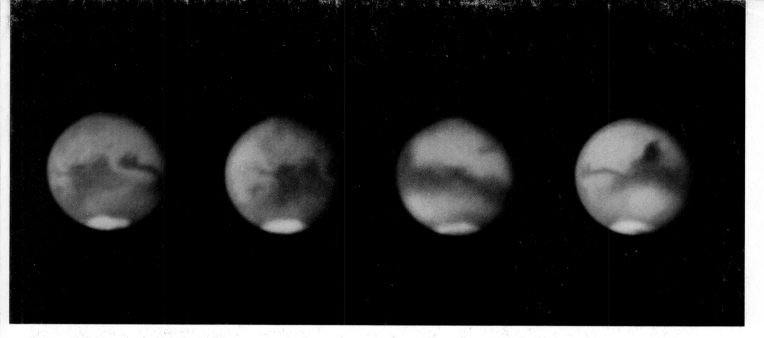

Figure 1.5. A complete circuit of Mars, as viewed from Earth. The pictures were taken during the 1971 opposition when it was summer in the southern hemisphere. The central meridian moves progressively toward the west, reading left to right. The color photographs are composites of three filtered black-and-white images in red, green, and blue light (Jones and Cook, 1974). International Planetary Patrol photographs.

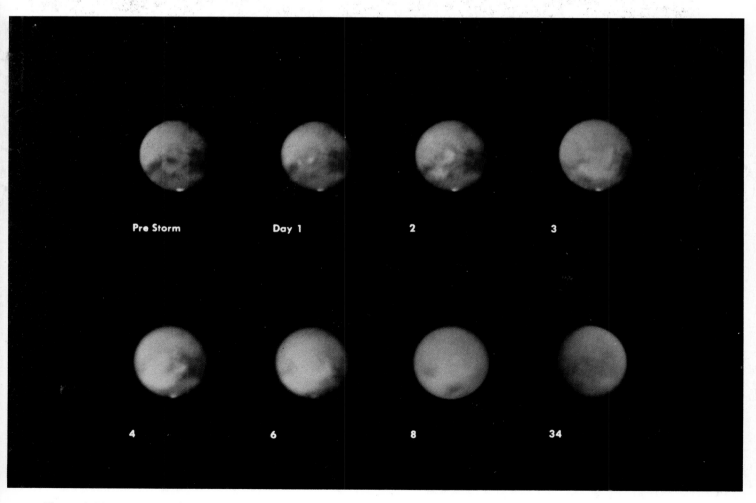

Figure 1.17. A sequence of color pictures illustrating the development of the 1973 dust storm, which originated in the Solis Lacus region. The first image shows the normal appearance of the region; the next six show the buildup of the storm; and the last image shows a partial return to normal (Martin, 1974b). International Planetary Patrol photographs.

This is the first color picture taken on the surface of Mars, approximately 20 hours after the successful landing of Viking 1. The local time is approximately noon; the view is toward the southeast. Orange-red, fine-grained material covers most of the surface. This material is probably limonite (hydrated ferric oxide). Such weathering products form on Earth in the presence of water and an oxidizing environment. Darker, coarser-grained patches occur adjacent to some boulders and in the lower right.

A region of light, apparently fractured, bedrock appears in the upper right, just below the horizon. An isolated dune of fine-grained material appears at the same distance below the horizon at the far left. The sky has a reddish cast, due to scattering and reflection from reddish sediment suspended in the lower atmosphere.

days longer than autumn (table 1.1). An offsetting condition also exists. During the northern spring, Mars passes through its aphelion position. Because it is at its farthest position from the sun it receives relatively less solar radiation, actually 69 percent of that at perihelion. Accordingly, the southern summers are distinguished from the northern ones by being both shorter and hotter.

Table 1.1. Durations of seasons on Earth and Mars. Note that the lengths of solar days for Mars and Earth are not the same.

Season		Earth	Mars	
Northern Hemisphere	Southern Hemisphere	Days	Earth Days	Mars Days
Spring	Autumn	93	199	194
Summer	Winter	93	183	178
Autumn	Spring	90	147	143
Winter	Summer	89	158	154
		365	687	669

Maps of Mars

Throughout the nineteenth century the use of telescopes of increased resolving power led to the construction of progressively more detailed maps of the Martian surface. Wilhelm Beer and Johann Mädler were the first to attempt a complete map of the planet in 1840. It was not particularly successful. Very few of their dark markings can be correlated with specific features. Beer and Mädler complicated an already difficult problem by mapping the planet in two hemispheres as viewed from vantage points directly above the north and the south poles. The view obtained through the telescope is approximately in the equatorial plane, so the polar projection presents a uniquely unfamiliar perspective.

Beer's and Mädler's greatest contribution was in demonstration of techniques. They established a system of latitude and longitude analogous to that of Earth. They arbitrarily defined zero longitude as passing through a small, unusually dark mark. Unfortunately, the establishment of this mapping grid emphasizes the unreliability of their own measurements. Syrtis Major, one of the few markings that can be identified with certainty, is misplaced about 30° relative to the reference mark for 0° longitude.

Much better drawings of Mars were published in 1864 by the Reverend William Dawes. Three years later, Richard Proctor, a fellow Englishman, systematically assigned names to the bright and dark markings (fig. 1.9). Proctor showed a certain favoritism in designating a Dawes Ocean, a Dawes Continent, a Dawes Sea, and a Dawes Forked Bay. He further handicapped his chances for permanent ac-

Figure 1.9. Map of Mars by Richard Proctor. Bright regions are identified as continents. Dark regions are labeled oceans and seas. The Kaiser Sea corresponds to Syrtis Major. From Flammarion (1892).

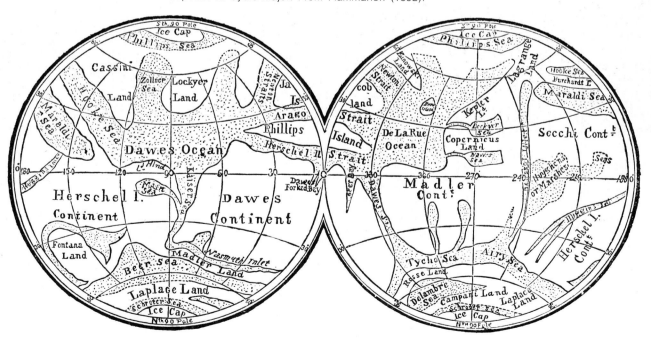

ceptance by relying heavily on the Beer-Mädler depiction of the planet. Predictably, Proctor's efforts were greeted with little enthusiasm by other astronomers.

A completely new terminology was introduced by Giovanni Schiaparelli in 1877. Observing Mars during the favorable oppositions of 1877–1881, when Mars swung particularly close to Earth, Schiaparelli prepared detailed maps (fig. 1.10). Skirting the obvious pitfalls of naming features for actual people, let alone living people, he chose to concoct a romantic blend of Biblical and mythological geography. He was well qualified to do so since he was not only an excellent observational astronomer but also a respected classicist who devoted much of his time to interpretation of Biblical astronomy. His geography was based on the belief that the bright regions were land and the dark areas oceans. Libya (Africa) and Ausonia (Italy) are separated by Tyrrhenum Mare (the Tuscan Sea). Elsewhere the mythical continent of Atlantis is located between Mare Cimmerium and Mare Sirenum, Homer's seas of the Cimmerians and the Sirens. Chryse, the land of gold, is crossed by the Ganges and Indus canals. The bright land of Elysium is bounded by the darker shades of Chaos, Styx, and Cerberus. And so it goes, until more than a hundred markings acquire names rich in history and legend.

Schiaparelli's names were universally accepted and are widely used even today. However, there are difficulties. The first is purely a matter of style. For better or worse most scientists of today have little training in Latin or Greek. Names such as Ogygis Regio and Phaethontis inspire mostly embarrassing uncertainty on the part of the user. Other problems are more substantial. The boundaries for many of Schiaparelli's features are ill defined on telescopic photographs and maps. In an attempt at clarification a committee of the International Astronomical Union recently identified 128 permanent features of special importance. An accompanying map, published in 1958, synthesized many telescopic observations. Comparable Mercator maps recently have been prepared for particular oppositions, utilizing the observations from a half-dozen observatories around the world (fig. 1.11).

When telescopic images are compared with Mariner spacecraft photographs, the lack of detailed correlation between albedo and topography is striking. As discussed further in chapter 3, there is a general tendency for the maria to be heavily cratered and the continents to be relatively featureless. However, the majority of smaller telescopic features have no topographic counterpart. For this reason a new set of names has been assigned to the numerous recognized landmarks: craters, ridges, valleys, mountains, and mesas (fig. 3.1).

Figure 1.10. Map of Mars by Giovanni Schiaparelli. Nomenclature prepared by Schiaparelli has been adopted by almost all succeeding telescopic observers. A network of linear canals is prominently displayed. From Flammarion (1892).

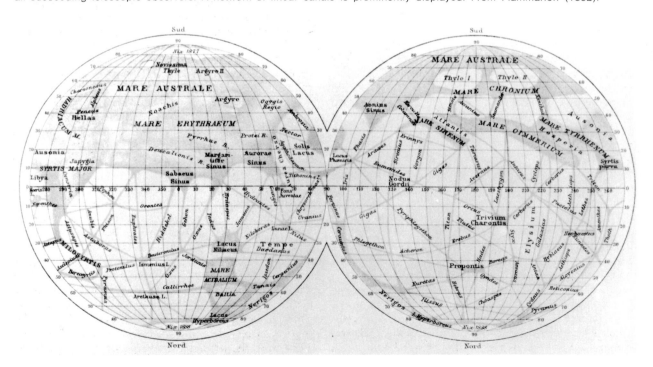

The Violet Layer

When Mars is photographed with a filter that transmits only blue light, the somewhat surprising result is a significant loss of detail that is visible in red light. To explain this phenomenon, early observers postulated an atmospheric layer on Mars that absorbs or scatters blue and violet radiation and thus preferentially reduces surface contrast for the shorter wavelengths of light. Initially, the requirement for such a "violet layer" was countered with the observation that the reflectance properties of bright and dark areas on Mars differ most in red light and converge as wavelengths decrease (see fig. 1.21). Therefore, loss in detail should be expected solely on the basis of surface properties. This proved to be only a partial explanation when Slipher, in 1937, viewing Mars in blue light, observed periods when surface features were clearly delineated. He suggested that such clearings result from temporary dissipation of some atmospheric phenomenon.

Other violet clearings have been observed and recorded since Slipher's initial discovery. Some clearings reportedly reveal selected features; others spread over entire hemispheres. Some last less than a day; others persist for more than a week. Calculations of the magnitude of blue light scattering and absorption by atmospheric hazes or aerosols on Mars suggest that atmospheric layers are probably not sufficient to explain the observations (Van Blerkom, 1971). Pollack and Sagan (1969), in fact, maintain that the blue clearings are illusory, caused by variable viewing conditions of Earth's atmosphere, and by intrinsic differences in Martian surface materials. They point out that clearings are most frequently observed when Syrtis Major is along the central meridian of the Martian disc, as viewed from Earth. Syrtis Major is one of the darkest, most easily recognizable features on Mars. There would be subjective pressure to associate "clearing" with such a favorable orientation of the planet.

An extensive observational survey of the blue clearing phenomenon has been conducted by Boyce and Thompson (Thompson, 1972, 1973a, 1973b; Boyce and Thompson, 1972; Boyce, 1973). They conclude that the apparent contrast changes in blue light can be attributed to changes in reflectance of surface materials caused by varying viewing angles. For instance, the blue clearing of Syrtis Major depends on the phase angle (angle between sun and Earth, measured on Mars). At opposition, when the phase angle is near or at zero, contrast is a maximum. In addition, they find a tendency for clearings to increase through the Martian day, as the local sun angle increases. Although the clearing is most noticeable in blue light, where contrast of Martian materials is low, analogous enhancements take place at all wavelengths. Finally, Boyce and Thompson suggest that the clearings are partly caused by a brightening of bright regions. Dark areas, even though they appear telescopically prominent during a clearing, maintain generally constant reflectivity. They speculate that bright areas are covered with dust and, as solar heating increases during Martian day, atmospheric turbulence is sufficient to suspend the dust. An increase in brightness would then occur by an increase in multiple scattering and reflection of light by the small grains.

White Clouds

White spots that are distinct from the polar caps were first reported by P. A. Secchi in 1858. Several decades later N. Green observed that some of these spots appeared on the limb and remained there while the planet rotated beneath them. He deduced that the spots were detached from the planet and must be clouds. Still later in the nineteenth century, several observers noted that the clouds remained illuminated on the night-time side of the terminator. This is an effect comparable to that experienced by someone in an airplane, traveling during sunset. The ground below is totally dark when the sun, viewed from the plane, is still some distance above the horizon. The apparent difference in sunset time is a function of altitude. In an analogous way, early telescopic observers calculated that the Martian clouds were at heights of 5 to 10 km.

White clouds are photographically distinguished as those that are prominent in blue light. There is some reference in the literature to blue clouds, but the difference between blue and white clouds is uncertain and perhaps nonexistent. Yellow clouds, discussed later, are in a clearly different category and are most easily identified in yellow or red light.

White clouds occur in a number of different situations. Large clouds, 2000 km in diameter, may form at intermediate latitudes and persist for weeks. Smaller clouds occur repeatedly over regions of high elevation (Sagan et al., 1971). They generally are restricted to the lower latitudes, occur in the spring and summer, and form each day in the early afternoon. Favored localities include Nix Olympica, Syrtis Major, Elysium, Hellas, Arcadia, and the Tharsis area. The last is the site of the famed "W" cloud ("M" cloud if you turn Mars right side up—optical systems in telescopes invert images), which was first noted in 1907 and recorded again in 1954 when the distinctive formation appeared daily for a period of more than three weeks (Slipher, 1962). The W cloud was observed in Mariner 6–7 far-encounter pictures (Leovy et al., 1971) and is attributed to the presence of water-ice crystals (Peale,

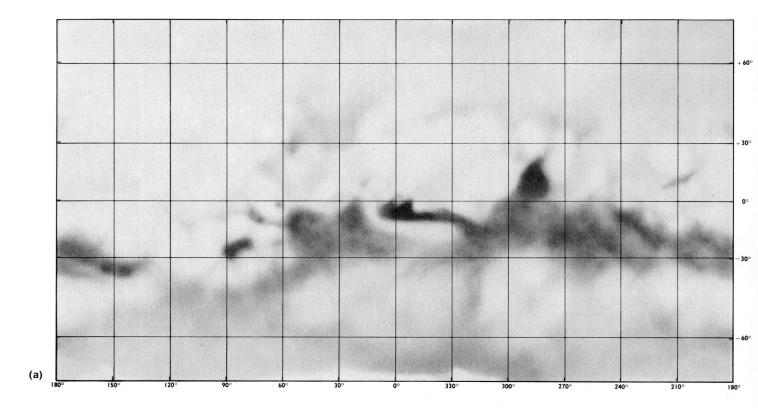

(a)

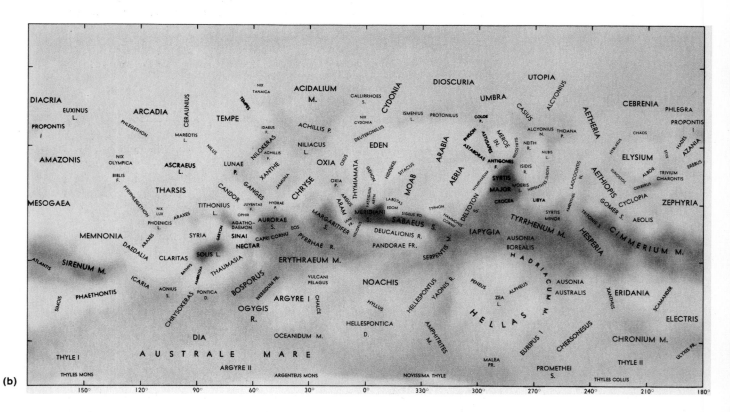

(b)

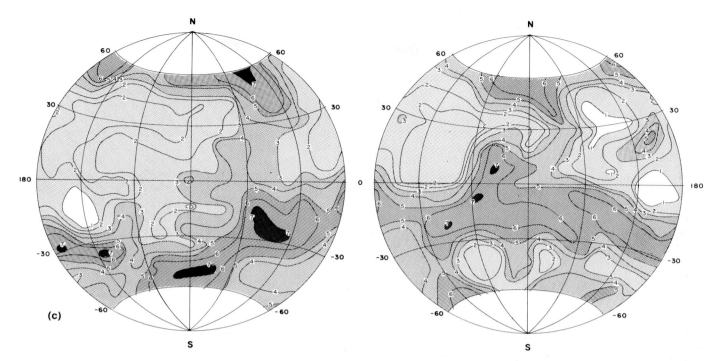

Figure 1.11 (a). Map of Mars showing albedo features visible during 1971 opposition. Prepared by Planetary Research Center, Lowell Observatory, with participation by other observatories contributing to the International Planetary Patrol Program. Compare with telescopic photographs in figure 1.5.

  (b). Nomenclature key for map.

  (c). Visual geometric albedo map of Mars, contoured by DeWys (1968) from data contained in de Vaucouleurs (1967) and obtained during oppositions of 1941 and 1958. Geometric albedo is defined as the ratio of luminance at full phase (phase angle = 0) to the luminance of a perfectly diffusing plane surface perpendicular to the sun. Contours have the following albedo values: 1 = 0.30; 2 = 0.27; 3 = 0.24; 4 = 0.21; 5 = 0.18; 6 = 0.15; 7 = 0.12. Information is plotted on a Lambert equal-area base for east and west hemispheres of Mars.

1973). With receipt of Mariner 9 pictures, a close geographic correlation was revealed between the W cloud and the three volcanoes standing high on the Tharsis ridge (Masursky et al.,1972).

Smith and Smith (1972) have telescopically documented the occurrence of white clouds in the vicinity of Elysium and Nix Olympica. The clouds are most noticeable in Martian midsummer and, in addition, show diurnal variations. They form at local noon and continue to brighten for several hours. The seasonal appearance of the clouds corresponds with high abundance of atmospheric water vapor. The daily changes suggest that water is being cycled between the soil and the lower atmosphere. Relatively warm temperatures during the Martian summer are consistent with the presence of water-ice crystals, which form at temperatures higher than those for CO_2-ice crystals.

Numerous discrete white clouds were delineated in Mariner 9 pictures, particularly for the Tharsis, Amazonis, and Nix Olympica localities (Leovy et al., 1973). The correlation between prior telescopic measurements and spacecraft observations was so precise that it was possible to forecast "scattered clouds" over Olympus Mons and Ascraeus Mons— two prominent volcanoes—for the season of Mariner observations. It seems likely that many of these fixed clouds are orographic in origin, forming as moist air moves up windward slopes, with attendant adiabatic cooling and crystallization to form finely divided ice crystals. (This model was anticipated by de Vaucouleurs, 1954.) Inasmuch as many of the clouds occur around volcanic centers, it is tempting to invoke volcanic degassing (McLaughlin, 1956a), but supporting evidence is lacking. Instead, a small reservoir of adsorbed water is probably stored in the upper few centimeters of the soil and may be cycled daily between soil and atmosphere, as previously described by Pollack et al. (1970a, 1970b), Fanale and Cannon (1971), and Smith and Smith (1972).

Regional hazes are often telescopically observed near the morning terminator (Dollfus, 1961a). These have been interpreted as veils of ice crystals that form during the night, sublimate with the rising sun, and then condense again as the evening air cools during sunset.

The final type of white cloud forms over polar areas during the fall and winter seasons. The polar hood is thick and stable, completely obscuring the surface below during the midwinter months. Each year as the vernal equinox approaches, signaling Martian spring, the hood thins and recedes. By the time the polar hood has disappeared, the underlying cap has already shrunk considerably. It continues to recede throughout the summer (fig. 1.19). Mariner 9 photography amply confirmed the existence of the polar hoods. In order to see the surface above latitudes of 50°N, it was necessary to wait until early June (Martian spring in the northern hemisphere) when the cloud banks dutifully broke up and receded. Several months later, the polar cap was completely cloud-free.

Cloud formations of the north polar hood, photographed by Mariner 9, have been studied by Briggs and Leovy (1974). CO_2 clouds predominate poleward of 60°. Water-ice clouds are prominent between 45° and 55°, where they form spectacular streamers and transverse ripples in the lee of topographic obstacles (fig. 1.12). The twofold division of cloud types is consistent with temperature and

Figure 1.12. Clouds, probably composed of H_2O-ice crystals, formed in the lee of a crater. A bright ring of frost surrounds the crater. Scattered clouds appear in the south. (DAS No. 07903178, rev. 176, A camera, center at 16°W, 56°N.)

composition measurements made with an infrared interferometer spectrometer on board the Mariner 9 spacecraft (Hanel et al., 1972).

Yellow Clouds

The existence of yellow clouds was first noted by H. Falaugergues in 1809. In the following years the phenomenon received relatively little attention, although clouds were observed at some of the more favorable oppositions during the last quarter of the century. Precisely one century after the first observations, E. M. Antoniadi noted that the obscurations preferentially occurred near perihelion—closest approach to the sun. He suggested that increased solar heating during this period provided enough energy for atmospheric winds to raise large amounts of surface dust. Later, analysis of more extensive cloud records by Wells (1966), Capen (1971), and Golitsyn (1973) confirmed that yellow cloud occurrences reach a maximum near perihelion. The more spectacular yellow clouds appear at selected sites in the southern latitudes, Hellas-Noachis for example, and expand to the east and west, eventually covering the whole planet (fig. 1.13).

If yellow clouds have suffered from inattention in the past, the situation was rectified in 1971. On September 22 a bright yellow cloud appeared near Hellespontus (Capen and Martin, 1971). During the next two weeks the cloud belt enlarged to form a continuous girdle at that latitude (figs. 1.14–1.16). A week later the entire planet was obscured—and remained so for two months. Shortly after this yellow shroud enveloped the planet, on November 14, to be exact, Mariner 9 arrived at Mars, only to confirm that the haze cover was depressingly opaque. Very few surface features could be observed because the haze significantly reduced surface contrast. Suddenly there was a frenzied interest in yellow clouds. The obvious question: "When can we expect a clearing?" Review of photographs and literature for previous oppositions indicated that major obscurations were more common than previously appreciated. Once formed, the planet-wide haze commonly persists for two or three months.

In fact, the 1971 storm had been predicted. In an article published in February 1971, C. F. Capen unobtrusively remarked that "A vast atmospheric disturbance could interfere with post-opposition observations of earth-based telescopes as well as the first Mariner orbiter spacecraft mission, which is planned to begin reconnaissance of the planet in November." Quite an understatement, as it turned out.

As discussed in detail in chapter 7, occurrences of yellow clouds establish important constraints for Martian winds. Because the planet has an atmo-

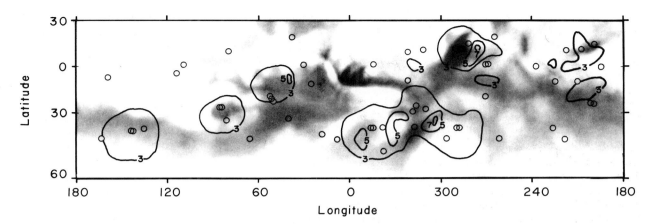

Figure 1.13. Regions in which yellow clouds preferentially form, based on data of Wells (1966) for 58 occurrences. Contours are generated by measuring the number of origins that occur within a 15° radius of a point on the surface. Major concentrations occur over the Noachis-Hellespontus-Hellas region and over Syrtis Major-Isidis. Taken from Arvidson (1972).

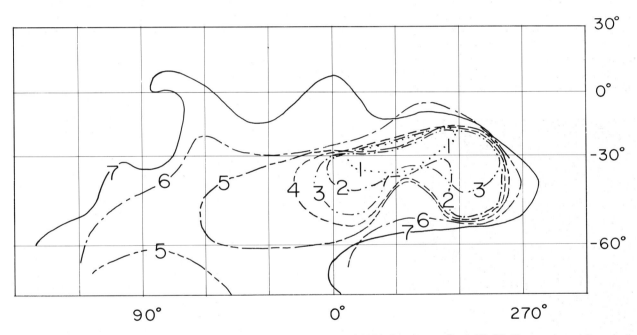

Figure 1.14. Variations in cloud position over the first seven days of 1971 dust storm, Sept. 22–29. The numbered lines indicate the extent of the cloud on particular days. Simplified from Martin (1974a).

spheric density approximately one hundredth that of Earth, very high wind speeds are necessary to move sediment. The calculated required minimum values are 50 m per second (Arvidson, 1972). Available minimum model wind velocities, utilizing topographic information, are 80 m per second (Sagan et al., 1971), and minimum velocities of as much as 110 m/sec have been calculated from Mariner 9 data (Sagan et al., 1974). In such an environment it would be difficult for any spacecraft, manned or unmanned, to survive. Before we land on Mars in 1976, it would be prudent first to check the weather forecasts.

Variable Dark Regions

Nineteenth-century observers of Mars noted that, although the surface of Mars is characterized by bright and dark markings, the shape and size of some markings vary with season and year. In addition, they

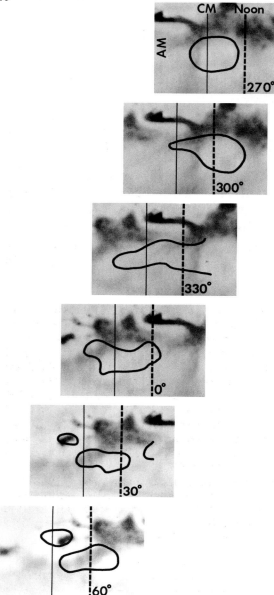

Figure 1.15. Evolution of the 1971 dust storm during the 11th day. Observations are at two-hour intervals. The central meridian is labeled CM. The dotted line indicates the local-noon longitude. Note that the clouds tend to stay left of the noon line, indicating that they are preferentially formed in the morning hours. Simplified from Martin (1974a).

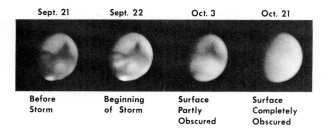

Figure 1.16. The buildup of the 1971 dust storm. The dark triangular region in the upper right is Syrtis Major. The south polar cap is visible in the first two frames. The bright circular area above and to the right of the cap is Hellas. The storm formed just to the left of Hellas. Compare with figure 1.14. International Planetary Patrol photographs.

color is extremely qualitative. Sometimes it is confused with brightness. In other cases, dark regions may appear greenish only because they border regions with reddish hue. The eye tends to "create" the complementary color even when it is not there. In fact, reflectance spectra for all dark regions show greater values in the red than in the blue and green.

A less dramatic if more persuasive interpretation of the variable brightening and darkening is that shifting, wind-driven, dust and sand, perhaps composed of oxidized and hydrated iron-oxide minerals, temporarily cover darker, unoxidized gravel and bedrock. The phenomenon could easily be seasonal since there is observational and theoretical evidence that wind patterns change with season.

The changing patterns of dark areas are both seasonal and secular, regular and erratic. Syrtis Major has remained a dominant feature ever since its detection by the earliest telescopic observers, but it

thought they detected color changes, with many of the dark areas changing from grey to green with the advent of local spring. Naturally enough, the advocates of life on Mars interpreted these observations as evidence of an annual bloom of vegetation.

Both the data and the interpretation are suspect. It is certain enough that the configuration of bright and dark regions changes (fig. 1.18), but beyond that there is little agreement. Visually determined

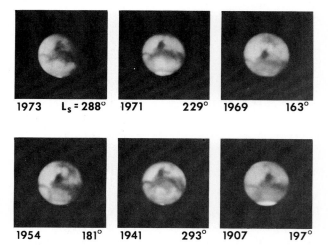

Figure 1.18. Changes in the Syrtis Major region over a period of 64 years. Photographs taken at the Lowell Observatory, Flagstaff, Arizona. For definition of L_s refer to figure 1.8.

displays remarkable seasonal variations. Solis Lacus has repeatedly changed in appearance between oppositions, but nonetheless is an apparently permanent feature. Some regions develop dark areas where none previously had been recognized. Slipher (1962) describes such a famous occurrence east of Thoth.

The Wave of Darkening

The wave of darkening takes second place only to Martian canals in historical development of the "life on Mars" hypothesis. First noted by Schiaparelli in the 1880s, and described in more detail by Lowell a few years later, an apparently dark zone advances equatorward from the pole as the polar ice cap retreats. Not surprisingly, Lowell thought that this was a spring-time growth of vegetation, nourished by water that was associated with the retreating cap. Other investigators, slightly more cautious, attributed the darkening either to percolation of water through otherwise dry soil, to formation of optically dark freeze-thaw soil structures (Otterman and Bronner, 1966, critique by Anderson et al., 1967), to chemical interaction between water and soil particles, or to seasonal movement of highly reflecting sediment (Sagan and Pollack, 1967; Pollack et al., 1967; Sagan and Pollack, 1969).

In recent years the very existence of a wave of darkening has been questioned. Even its advocates admit that the darkening surges ahead as a function of longitude. The generalization of that data to determine a planet-wide relationship between latitude and season is hazardous. Perhaps the most convincing data presented in support of the wave are by Focas (1961). His photometric measurements for more than 20 regions throughout the Martian year show a modest tendency toward an equatorward-drifting dark zone.

A dark collar that indisputably rims each polar cap probably is caused by seasonal stripping and accumulation of surficial dust to form, respectively, darker and brighter landscapes (Sagan and Pollack, 1967). This model is supported by Mariner 9 data that indicate that atmospheric thermal contrast between polar cap regions and frost-free terrain may induce strong cyclonic storms (Briggs and Leovy, 1974).

Polar Caps

The retreat and advance of the caps has been carefully monitored up to the present, ever since Herschel's 1784 interpretation of the "white polar spots" as seasonal accumulations of ice and snow. This is particularly true for the south cap, which is tilted toward Earth during those perihelic oppositions when the Earth-Mars distance is at a minimum and consequently observations are best (fig. 1.19).

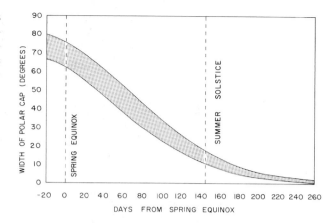

Figure 1.19. Width of south polar cap with the change in season for various oppositions from 1798 to 1924. Simplified from Slipher (1962).

Because Mars has an orbit with significant eccentricity, the seasons in the two hemispheres are of unequal length and intensity. When Mars is at aphelion its angular velocity is relatively low; this period corresponds with fall and winter in the southern hemisphere, a total of 382 days. When Mars has revolved 180° to its perihelion position, its orbital velocity has increased so that the equivalent fall and winter in the northern hemisphere is only 305 days. Because of this difference the south cap grows much larger than the north cap. At its maximum, the former has a width of 65°. The northern cap usually measures no more than 50° across (fig. 1.20). The north cap is centered almost precisely over the rotational pole (Soderblom et al., 1973a), whereas the south cap is offset by a distance of approximately 5° (Murray et al., 1972).

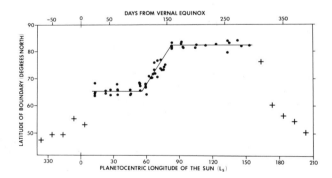

Figure 1.20. Position of the north polar hood (crosses) and cap (dots) with the changing season. For definition of L_s refer to figure 1.8. Note that the hood dissipates in early spring and that the cap recedes rapidly during the late spring. By early summer the cap has shrunk close to its minimum size. The hood reappears in early autumn. Taken from Fischbacher et al. (1969).

From one Martian year to the next the large-scale retreat patterns of the caps are remarkably consistent. The same is presumably true for buildup of the caps, but this has not been documented because of the presence of the polar hoods. The south cap shrinks regularly at a rate of about 10 or 20 km per day. At its smallest extent, just before it disappears under the developing cloud bank of the polar hood, it is less than 5° in width. The northern cap is stationary in early spring, recedes rapidly in late spring, and then retreats gradually during the summer. By late summer the cap is about 10° in width. The fact that both caps shrink so rapidly indicates that the ephemeral portions are no thicker than a few centimeters. The southern cap is seasonally reduced to a size smaller than the northern cap because Mars is relatively close to the sun during summer in the southern hemisphere.

The detailed boundaries of the polar caps are extremely ragged. When the ice recedes, some prominences persist for several days as completely isolated bright patches. The most famous of these are the Mountains of Mitchel, named for the American astronomer who was the first to observe them. Confusion has attended the precise location of Mitchel's Mountains, but they are most probably correlated with rugged terrain photographed by Mariner 7 in the vicinity of 320°W, 75°S (Cutts et al., 1972).

At least three mechanisms may account for the irregular retreat of the polar caps: (1) changes in frost stability as atmospheric pressure and temperature change with elevation, (2) variations in solar heating of the surface caused by irregular topography, and (3) wind-blown removal of frost in some areas and deposition in others.

In regions of irregular topography, sun-facing slopes receive a greater amount of solar energy than those in shadow and, hence, will be the first to lose their snow cover (Balsamo and Salisbury, 1973). The effect is probably responsible for the formation of alternating light and dark sinuous bands along the retreating edge of both the south (Murray et al., 1972) and north polar caps (Soderblom et al., 1973a). Frost deposits can also persist in crater interiors, which, in addition to being topographic sinks, are partially shadowed (Sharp et al., 1971a). Yet to be satisfactorily explained are the dark interiors of some craters situated in the interior of the south polar cap. This is in apparent contradiction to relationships observed along the cap edge where some crater interiors are bright and surrounding regions are dark.

In regions where surface winds are sufficiently strong to move frost, slope-related insolation effects will be modified by wind-blown frost redistribution. Flat surfaces will tend to be swept free of frost, which will preferentially pile up by trapping in more

rugged terrains. It is this process that Cutts et al. (1972) favor for retention of frost in the Mountains of Mitchel.

On Earth, we associate year-round snow with high mountains, where the air is cool enough to permit water to remain as snow and ice. However, decrease in temperature with increase in elevation may not be as marked on Mars (Sagan and Pollack, 1968). If the Martian frost is CO_2, the most likely candidate, then the terrestrial analogy is particularly inappropriate. Stability of CO_2 as frost is a function of both pressure and temperature. Increased partial pressure of atmospheric CO_2 causes condensation at *increased* temperatures. The result is that frost may persist at lower elevations where the atmospheric pressure is highest.

Before the Mariner missions, telescopic observers generally favored the idea that the polar caps are water ice, an exact duplicate of the terrestrial situation. A minority view, namely that solid carbon dioxide—or "dry ice"—was a likely candidate, was advanced almost a century ago. This idea received little support, principally because the requisite temperature, $-125°C$, seemed incompatible with the expected temperatures at the poles. Then, in 1966, Leighton and Murray showed that these low temperatures do, in fact, persist to the same latitudes as the polar caps extend. Not only that, but the thermodynamically calculated rate of CO_2 sublimation is in good agreement with the observed recession rates of the caps. Their theoretical calculations were borne out by subsequent Mariner 6–7 and 9 observations which conclusively demonstrate that the *major* surface component of the caps is CO_2 ice.

Surface Materials

During the nineteenth and early twentieth centuries, students of Mars were preoccupied with exotica such as canals, waves of darkness, and clouds. The characteristics of the material covering the Martian surface received little attention. Even those who believed that the planet was inhabited assumed that the bright regions were deserts. As Percival Lowell looked out across the painted desert of Arizona with its splashes of red, violet, and pink he imagined that he might just as easily be gazing across the Martian landscape. Similar panoramas were envisioned by S. A. Arrhenius, who was awarded a Nobel Prize in 1903 for demonstrating that ions carry electric charges. He speculated that the Martian surface is covered with meteoritic iron. According to Arrhenius, subsequent oxidation of the iron and sorting according to grain size led to generation of sediments with optical characteristics that vary from region to region.

The same ideas that provoked so little discussion

in the early part of the century have been argued at excessive length during the past decade. The modern approach is to determine the composition of Martian materials by reflectance spectrophotometry. This technique is based on the thesis that each mineral has a unique reflectance "signature" that can be measured in the laboratory and then compared to the data obtained remotely for Mars. Absorptions are caused, in part, by electron transitions. The wavelengths, or energies, of absorptions depend upon the types of ions present, and their crystallographic position and orientation. In the region of visible and near-infrared wavelengths there are several absorption features in silicates: a 0.9–1.1 micron band due to Fe^{++}, a 0.85–0.89 band due to Fe^{+++}, and $(OH)^-$ bands at 1.4 and 1.9 microns. Utilizing the principles just outlined, we should be able to determine the mineralogic properties of any planetary surface by measuring its reflectance properties.

That's the dream. What's the reality? First, there are problems in data collection. An accurate spectrophotometer is difficult to construct. This is especially true when the examined wavebands are small and consequently the total incoming energy is reduced. There are analytical stumbling blocks as well. Many minerals have spectra that are approximately the same and can be differentiated only at a very detailed level. Finally, there is a problem of composite measurement. The distinctive spectral responses just mentioned are for pure minerals of particular grain size. But natural surface materials, on Earth at least, are a complex mixture of minerals and rocks with varying grain size. Singling out the individual spectra is something like reading several typewritten passages, all overprinted on the same piece of paper.

Because of these difficulties, there has been ample room for disagreement in interpretation. Some scientists maintain that the spectral results have virtually no discriminatory value; others maintain that they are uniquely conclusive. Arguments have centered on both composition and grain size. Many telescopic observers have tacitly assumed that the Martian surface resembles terrestrial deserts. Because red, orange, and yellow colors characterize arid regions on Earth, analogous situations have been imagined for Mars. The argument was quantified by Dollfus (1961b), who showed that polarization measurements were consistent with the presence of limonite (a mixture of hydrated iron oxides).

Pollack and Sagan (1969) contend that photometric and polarimetric properties of Mars indicate that powder composed of the mineral goethite [FeO(OH)] is a significant constituent of the surface. Differences in surface brightness are attributed to differences in grain size. Smaller grains, present in the bright regions, scatter and reflect light more efficiently than do coarser grains in the darker regions.

A competitive model has been presented by Van Tassel and Salisbury (1964) and Binder and Cruikshank (1966). These authors point out that iron enrichment sufficient to form a substantial layer composed almost exclusively of goethite or limonite is geologically improbable. Instead, they argue that a thin limonite coating on silicate grains would produce the same colorimetric properties. Indeed, this latter situation is sufficient to explain the strikingly reddish hues of many terrestrial rocks and sediments. They contain only a few percent of iron, but it is present in a thin coating and cement of ferric oxides. According to plausible eolian mixing models it appears probable that goethite is stable under present Martian conditions (Pollack et al., 1970a, 1970b). Although oxidation is commonly related to the presence of water, it is also possible that contemporary photostimulated oxidation is occurring (Huguenin, 1974).

Salisbury and Hunt (1968, 1969; also Van Tassel and Salisbury, 1964) have pointed out that the change in reflectance properties with change in grain size is complicated. In many cases spectral structure will be removed with decreasing grain sizes, but, in some cases, the reverse is true. In any event, they contend that differences in grain sizes, by themselves, are inadequate to explain regional differences in Martian brightness.

Several exotic models have been presented based on telescopic reflectance spectra. Plummer and Carson (1969) hypothesize that carbon suboxide (C_3O_2) forms by photochemical reaction between CO_2 and CO. The compound polymerizes into heavier molecules that settle on the surface. It is an observational fact that the yellowish color of this polymer matches the reflection spectrum of Mars. However, above 1.0 microns the two spectra diverge (Cutts, 1971). Absorption features recorded by the Mariner 9 UV instrument during the occurrence of the 1971 dust storm indicate an absence of carbon suboxide (Perls, 1973). Even if it were formed, it seems certain that it would be unstable over geologically significant periods of time.

Sinton (1957) reported the detection of several absorption bands in the vicinity of 3 to 4 microns that are characteristic of carbon-hydrogen bands. Taken at face value, this would be a remarkable discovery that would strongly suggest life on Mars. However, Sinton's spectral evidence for Martian vegetation does not withstand close scrutiny. Most of the structure in the curves is due to HDO (hydrogen-deuterium-oxygen molecules) within Earth's atmosphere (Rea et al., 1965). The remaining data have no more than uncertain significance. More recently, Sinton (1967) has detected a broad absorption feature near 3.0 microns that may signal the presence of hydrated minerals.

Some of the more persuasive reflectance studies have been carried out by Adams and McCord (McCord, 1969; McCord and Adams, 1969; McCord et al., 1971). They find (Adams and McCord, 1969) that, between 0.4 and 1.1 microns, reflectance curves for the Martian dark regions are matched by laboratory data for basalt powder with grain sizes between 0.5 and 0.1 mm (fig. 1.21).

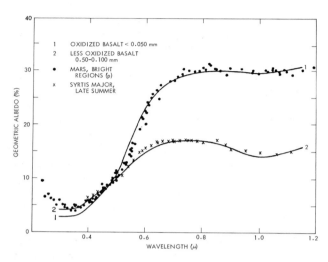

Figure 1.21. Visible and near-infrared spectra for bright and dark regions on Mars, compared with spectra for weathered and unweathered basalt. Reflectivity for both bright and dark areas increases toward longer wavelengths, accounting for the reddish color of Mars. A relatively greater increase for bright areas accounts for high contrast between surface features viewed in red light. Taken from Adams and McCord (1969).

Curves for the bright regions correspond to laboratory measurements for oxidized basalt powder with grain sizes less than 0.05 mm. A comprehensive survey of 150 regions in the spectral range 0.6 to 2.3 microns by Binder and Jones (1972) confirms that surface materials can be divided into two groups. Their general color can be attributed to mineral and lithic fragments coated by limonite. The desert units, more than twice as bright as the mare units, presumably have a more prominent limonite "mask." Although both bright and dark regions probably are underlain by basic rocks, the dark maria appear richer in olivine, pyroxene, or both.

The emerging picture of Mars includes a crust made up of volcanic basalts, the same rock commonly present in terrestrial volcanic provinces and throughout the lunar maria. Under oxidizing conditions, the iron-rich minerals of the basalt probably weather to form a mixture of yellow, limonite-coated grains and fine-grained clay minerals (Hunt et al., 1973). The movement and concentration of this fine-grained sediment is responsible for the establishment of the bright deserts, the variable appearance of darker bedrock lightly dusted with oxidized grains, and the formation of yellow clouds.

The Moons of Mars

Starting with Herschel in 1783, a number of astronomers searched for Martian satellites (moons). There was good empirical reason for thinking they existed since Galileo had already identified four moons revolving about Jupiter. Clearly Earth's moon was not a unique object. Kepler, ever interested in orderly geometric progressions, postulated Venus with no moons, Earth with one moon, Mars with two moons, and Jupiter with four moons.

However it was not until the particularly favorable opposition of 1877 that the two Martian moons were finally observed by Asaph Hall, an astronomer at the U.S. Naval Observatory. Hall searched for many nights in vain. Overtaken by frustration and depression, he quit the telescope in the middle of an observational period. Legend has it that his wife bolstered his spirits and urged him to return to the observatory. Within minutes he sighted a star-like object circling Mars.

The first moon to be discovered was the outer one; the inner one was detected a week later (fig. 1.22). Hall proposed that the two satellites be called Deimos (Terror) and Phobos (Fear) after the attendants of the god Mars mentioned by Homer in the Iliad. Both moons revolve in planes almost coincident with the equatorial plane of Mars. Phobos, the inner satellite, has a nearly circular orbit with radius of 9450 km. Since the equatorial radius of Mars is approximately 3400 km, Phobos is only

Figure 1.22. The two satellites of Mars, indicated by arrows. The disc of Mars is greatly overexposed in this telescopic image so that the faint satellites can be recorded. Photograph by Lowell Observatory, Flagstaff, Arizona.

6050 km above the surface. The diameter of Phobos is only 22 km but, because of its proximity to Mars, it would be easily observed from the planet's surface. Its apparent diameter would be about one-third that of our moon viewed from Earth.

Phobos makes one complete orbit in only 7 hours and 39 minutes. So rapidly does it move about Mars that Hall first entertained the idea that he was looking at two or three inner moons. The revolution of Phobos is in the same direction as the rotation of Mars. Consequently the time between two successive appearances at the same point in the Martian sky is 11 hours and 6 minutes, considerably longer than the sidereal period of 7 hours and 39 minutes. Since a Martian day is a little more than 24 hours, Phobos passes overhead two or three times daily, moving from west to east in contrast to the fixed stars and sun. At latitudes greater than 70°, the satellite would not be visible, since it is situated low in the sky and would be hidden below the local horizon.

Total eclipses of the sun by Phobos never occur on Mars since the apparent size of Phobos is only about half that of the sun. However, the intercession of Phobos between Mars and the sun is a common event. To an observer on Mars the rapid passage of the satellite's shadow would be hardly noticeable (fig. A.1). The situation for a resident of Phobos would be quite different. Because the satellite passes frequently through Mars' shadow, total eclipses would occur several times each day through much of the Martian year.

Deimos is less of a maverick satellite. Its orbital radius is 23,500 km, placing it about 21,000 km above the Martian surface. Because its diameter is only 12 km, its apparent size to a Martian observer would be about one-sixth that of Phobos, imparting the appearance of a bright star. Its sidereal period is 30 hours and 18 minutes, slightly in excess of Mars' rotational period. For this reason the satellite appears to move very slowly in the Martian sky, taking five and a half days to return to its former position.

Both Phobos and Deimos have synchronous rotations: that is, they rotate once during each revolution so that their same side always faces Mars (Pollack et al., 1973). The same phenomenon is displayed by Earth's moon.

The Canals

In the entire history of Martian telescopic observations, no item has been more widely proclaimed, more vehemently debated, and more abruptly forgotten than the canals. Schiaparelli was the first effective proponent of canals. True enough, other observers had noted linear markings, and Secchi had introduced the Italian word canale—channel—to

describe them. But Schiaparelli mapped a profusion of these features, gave each one a name, and tabulated their characteristics. For example:

> There are on this planet, traversing the continents, long dark lines which may be designated as *canali*. . . . Those lines run from one to another of the somber spots that are regarded as seas, and form, over the lighter, or continental, regions, a well-defined network. Their arrangement appears to be invariable and permanent. . . . Nevertheless, their aspect and their degree of visibility are not always the same. . . . They have a breadth of 2 degrees, or 120 kilometers, and several extend over a length of 80 degrees, or 4800 kilometers. . . . Every canal terminates at both its extremities in a sea, or in another canal; there is not a single example of one coming to an end in the midst of dry land. . . . In certain seasons these canals become double. . . . To the right or left of a pre-existing line, without any change in the course and position of that line, one sees another line produce itself, equal and parallel to the first, at a distance generally varying from 6 to 12 degrees.

Although Schiaparelli was adamant regarding the existence of canals—"I am absolutely certain of what I have observed"—he was, at first, equally specific in stating that their origin was unknown. He suspected that they were water-filled channels connecting larger seas and dividing continents. However, popularization of the more radical idea that they were literally canals, built by intelligent beings, was the special property of Percival Lowell (1855–1916), a prominent American astronomer.

Parenthetically it should be pointed out that the existence of a civilization on Mars was assumed by many pre-twentieth-century scientists. Huygens talked matter-of-factly about habitation of other planets. Herschel speculated that Martians enjoyed conditions roughly comparable to those on Earth. In the middle of the nineteenth century the mathematician Carl Friedrich Gauss proposed that we signal our presence to the Martians by creating a giant right-angled triangle with squares on each side. The figure would be lined by pine trees and "colored" by fields of grain. The supposition was that the Martians would recognize this Pythagorean symbol as the work of intelligent beings and would, in turn, signal us with sun-reflecting metal panels. Although Gauss' plan was never put to the test, many people were convinced that a successful exchange of messages was inevitable. In 1900 the French Academy of Science announced the Prix Pierre Guzman, a 100,000-franc prize donated by Madame Clara Goguet in memory of her son. The money was to be awarded to whoever first established communication with a world *other* than Mars. Apparently it was felt that Martian communication was more a certainty than a challenge.

Mars figured prominently not only in fact but also in fiction. At the turn of the century several popular novels appeared in Europe, having the common theme of interaction between Martians and Earthlings. A short time later Edgar Rice Burroughs, the creator of Tarzan, published the first of an eleven-volume series (1917–1964) recounting the Martian experiences of a John Carter. Captain Carter, dismissed from the defeated Confederate army, seeks his fortune as a gold prospector in the American southwest. On March 3, 1866, attacked by Apache Indians he undergoes a mysterious paralysis. An apparition separates itself from his rigid body. The terrified Indians retreat. With a final convulsive thrust, John Carter's unencumbered soul departs his body and, with disarming simplicity, undertakes an instantaneous journey to Mars, where Carter opens his eyes to sights "weird and strange." Indeed they are! Barsoom (Burroughs' name for Mars) is a decadent world where once-powerful civilizations have decayed under the impact of retreating oceans and the invasion of green barbarians, fifteen feet tall, with two sets of arms and tusks protruding from their lower jaws. Carter roams the Martian surface, triumphs over the green men, joins forces with the red men, courts a local princess, and—from time to time—returns to Earth to confound his great-nephew.

Perhaps the best known of all books about fictional Martians is *War of the Worlds,* by H. G. Wells. The Martians are a highly civilized race, but are living on a planet that is slowly dying as its air and water dissipate into space. This pressure "has brightened their intellects, enlarged their power, and hardened their hearts. And looking across space . . . they see . . . a morning star of hope, our own warmer planet, green with vegetation and grey with water." To invade Earth is their only hope for survival. They temporarily put humanity to rout with heat rays and poison gas, but eventually die, victims of terrestrial bacteria against which they have no immunity.

Unlikely, of course. But in 1938, forty years after the book was written, a radio adaptation was mistaken for an actual news account. Over a million people throughout the United States listened in frightened shock as the radio announcer reported:

Streets are all jammed. Noise in crowds like New Year's Eve in city. Wait a minute. . . . Enemy now in sight about the Palisades. Five great machines. First one is crossing river. I can see it from here, wading the Hudson like a man wading through a brook. . . . A bulletin's handed me . . . Martian

cylinders are falling all over the country. One outside Buffalo, one in Chicago, St. Louis . . . seem to be timed and spaced. . . . Now the first machine reaches the shore. He stands watching, looking over the city. His steel, cowlish head is even with the skyscrapers. He waits for the others. They rise like a line of new towers on the city's west side. . . . Now they're lifting their metal hands. This is the end now. Smoke comes out . . . black smoke, drifting over the city. People in the streets see it now. They're running towards the East River . . . thousands of them, dropping in like rats. Now the smoke's spreading faster. It's reached Times Square. People trying to run away from it, but it's no use. They're falling like flies. Now the smoke's crossing Sixth Avenue . . . Fifth Avenue . . . 100 yards away . . . it's 50 feet . . .

Thousands of listeners flooded the radio stations with calls, fled their homes, and clogged the highways. In a small southwestern college "the girls in the sorority houses and dormitories huddled around their radios trembling and weeping in each other's arms. They separated themselves from their friends only to take their turn at the telephones to make long distance calls to their parents, saying good-bye for what they thought might be the last time. This horror was shared by older and more experienced people—instructors and supervisors in the university. Terror-stricken girls, hoping to escape from the Mars invaders, rushed to the basement of the dormitory. A fraternity boy, frantic with fear, threw off dormitory regulations when he sought out his girl friend and started for home. Another boy rushed into the street to warn the town of the invasion" (Cantril, 1940).

But to return to Percival Lowell and his canals. Lowell was a member of a distinguished Boston family; one of his brothers became president of Harvard University and his sister, Amy, was a famous poet. A businessman by vocation, a Far Eastern scholar, classicist, and mathematician by avocation, he was fascinated by Schiaparelli's accounts, so much so that he established an observatory at Flagstaff, Arizona, for the special purpose of studying the planets. Lowell immediately confirmed Schiaparelli's observations and began charting hundreds of new canals (fig. 1.23) until Mars began to assume the appearance of a "spider's web" (Lowell's own words).

Lowell was not only literate but also impassioned, and his three volumes, *Mars, Mars and Its Canals*, and *Mars as the Abode of Life*, were best-sellers in the first decade of this century. Let him recreate his first recognition of the canals.

Introduction to the mystery came about in this wise, and will be repeated for him who is successful in his search. When a fairly acute eyed observer sets himself to scan the telescopic disk of the planet in steady air, he will, after noting the dazzling contour of the white polar cap and the sharp outlines of the blue-green seas, of a sudden be made aware of a vision as of a thread stretched somewhere from the blue-green across the orange areas of the disk. Gone as quickly as it came, he will instinctively doubt his own eyesight, and credit to illusion what can so unaccountably disappear. Gaze as hard as he will, no power of his can recall it, when, with the same startling abruptness, the thing stands before his eyes again. Convinced after three or four such showings, that the vision is real, he will still be left wondering what and where it was. For so short and sudden are its apparitions that the locating of it is dubiously hard. It is gone each time before he has got its bearings.

By persistent watch, however, for the best instants of definition, backed by the knowledge of what he is to see, he will find its coming more frequent, more certain and more detailed. At last some particularly propitious moment will disclose its relation to well-known points and its position be assured. First one such thread and then another will make its presence evident; and then he will note that each always appears in place. Repetition *in situ* will convince him that these strange visitants are as real as the main markings, and are as permanent as they.

Such is the experience every observer of them

has had; and success depends upon the acuteness of the observer's eye and upon the persistence with which he watches for the best moments in the steadier air. Certain as persistence is to be rewarded at last, the difficulty inherent in the observations is ordinarily great. Not everybody can see these delicate features at first sight, even when pointed out to them; and to perceive their more minute details takes a trained as well as an acute eye, observing under the best conditions. When so viewed, however, the disk of the planet takes on a most singular appearance. It looks as if it had been cobwebbed all over. Suggestive of a spider's web seen against the grass of a spring morning, a mesh of fine reticulated lines overspreads it, which with attention proves to compass the globe from one pole to the other. The chief difference between it and a spider's work is one of size, supplemented by greater complexity, but both are joys of geometric beauty. For the lines are of individually uniform width, of exceeding tenuity, and of great length. These· are the Martian canals.

Lowell noted that the canals spread in regular pattern from polar caps to equator, that they frequently appeared as doublets (gemini), that they maintained constant width throughout their length, and that dark spots or "oases" frequently occurred at their points of intersection. He was convinced that features with these characteristics were necessarily the work of "intelligent creatures." The Martians were living on an aging, desert-like planet. Most of the water was trapped in the polar caps; the canals, therefore, were an artificial device to transport water into the equatorial zones.

Piling inference on inference, Lowell indulged in some sociological speculations.

> Girdling their globe and stretching from pole to pole, the Martian canal system not only embraces their whole world, but is an organized entity. Each canal joins another, which in turn connects with a third, and so on over the entire surface of the planet. . . . The first thing that is forced on us in conclusion is the necessarily intelligent and non-bellicose character of the community which could thus act as a unit throughout its globe. War is a survival among us from savage times and affects now chiefly the boyish and unthinking element of the nation. The wisest realize that there are better ways for practicing heroism and other and more certain ends of insuring the survival of the fittest. It is something a people outgrow. But whether they consciously practice peace or not, nature in its evolution eventually practices it for them, and after enough of the inhabitants of a globe have killed each other off, the remainder must find it

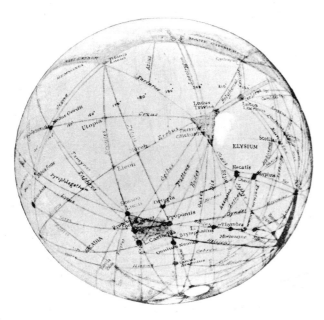

Figure 1.23. Map of Mars by Percival Lowell (1909) showing numerous canals and oases.

more advantageous to work together for the common good. Whether increasing common sense or increasing necessity was the spur that drove the Martians to this eminently sagacious state we cannot say, but it is certain that reached it they have, and equally certain that if they had not they must all die. When a planet has attained to the age of advancing decrepitude, and the remnant of its water supply resides simply in its polar caps, these can only be effectively tapped for the benefit of the inhabitants when arctic and equatorial peoples are at one. Difference of policy on the question of the all-important water supply means nothing short of death. Isolated communities cannot there be sufficient unto themselves; they must combine to solidarity or perish.

Lowell's visions elevated the Martian canals to a zenith of popularity. There were many believers, particularly in this country. Even Schiaparelli conceded that the arrangement of canals presented a scene of simplicity and symmetry "which cannot be the work of chance." However, many observers refused to succumb to canal fever. Slowly, inevitably, they began to assemble their arguments. Two English scientists, Evans and Maunder, showed pictures with discontinuous blotches and streaks to English children, untainted by any awareness of the problem they were addressing. Sure enough, under conditions of poor resolution the children connected the markings into regular straight lines. Lowell was indignant about what he contemptuously called the "small boy theory" and went on to question the relative intellectual capabilities of English and French schoolboys.

This theory attacks the reality of the doubles by questioning that of the canals *en bloc*. Because some boys from the Greenwich (Reform or) Charity School, set to copy a canal-expurgated picture of the planet, themselves supplied the lines which had preceptorily been left out, the Martian canals have been denied existence; which is like saying that because a man may see stars without scanning the heavens, therefore those in the sky do not exist. As to the instructions the boys received we are left in the dark. It looks as if some leading questions had unconsciously been put to them. At all events, English charity boys would seem to be particularly pliant to such imagination, for when Flammarion retried the experiment with French schoolboys, and even inserted spaced dots for the canals in the copy, not a boy of them drew an illusory line.

One obvious way to settle the argument was by taking a picture through a telescope and thereby preserving the debatable features for universal examination. It was a difficult task, but Lowell and his assistant persevered.

Many pictures were taken on each plate one after the other, both to vary the exposure and to catch such good moments as might chance. Seven hundred images were thus got in all; the days of best definition alone being utilized. The eagerness with which the first plate was scanned as it emerged from its last bath may be imagined, and the joy when on it some of the canals could certainly be seen. There were the old configurations of patches, the light areas and the dark, just as they looked through the telescope, and never till then otherwise seen of human eye, and there more marvelous yet were the grosser of those lines that had so piqued human curiosity, the canals of Mars. By Mr. Lampland's thought, assiduity, and skill, the seemingly impossible had been done. Upon the many images thirty-eight canals were counted in all, and one of them, the Nikokeras, double. Thus did the canals at last speak for their own reality themselves.

Half a century later similar evidence was acquired by Earl Slipher, astronomer at Lowell Observatory. He published a photographic atlas (1962) providing "each reader with the best possible opportunity to distinguish the fine lines and to judge their reality for himself." Anticipating the reader's verdict, he asserted that "while there is room for difference of opinion as to the interpretation of the canals, their existence as true markings on the planet has been clearly established."

Unfortunately the claim of victory was premature. Skeptics, squinting at the pictures just as they had previously squinted through the telescope were reminded of the fable of the Emperor's clothes—beautiful to some but distressingly nonexistent to others. Increasingly the weight of evidence suggested that, although the dark markings on Mars were real enough, there is no basis for uniquely defining them as narrow continuous lines. To reach this conclusion one must stop for a moment to consider that the effective resolution of the best telescopes is about 100 km on the Martian surface. This does not mean that objects of smaller size with great brightness contrast will be undetectable, only that they will be indistinguishable from larger objects of intermediate contrast. Because of this phenomenon a string of small discontinuous patches will appear as a continuous line. The visual effect is undeniably real, but not necessarily an accurate indication of Martian surface markings. Lowell's claim for identification of canals 50 km and less in width is simply not compatible with optical principles.

Although some of the canal enthusiasts were overly zealous, there remained a collection of telescopic observations that, at the very least, were puzzling. Several alternate explanations were advanced. Perhaps the canals were giant fracture systems in the Martian crust (Fielder, 1963a) made more prominent by upward seepage of water and attendant growth of vegetation (Jamison, 1965). In the same vein, it was hypothesized that some canals were actually giant rift zones equivalent to terrestrial midoceanic ridges (Sagan and Pollack, 1966). Maybe they were igneous dikes—dark spines of rock which resisted erosion and formed giant cliffs (Fairbridge, 1972). Might they not be chains of sand dunes, similar to the linear seif dunes of Saudi Arabia (Gifford, 1964)?

Since the critical problem in all interpretations was one of adequate spatial resolution, pictures taken from spacecraft showed promise of settling the issue. Mariner pictures have fulfilled this promise, if only in a negative sense. No regular, linear elements of the appropriate size are observable, either as low-albedo features or as topographic structures.

The planet does have a pattern of irregular and splotchy bright and dark markings at a scale much more detailed than 100 km, so previous arguments that the canals are a visual synthesis of discontinuous elements becomes even more persuasive.

During publication of Mariner results there has been virtually no mention of this canal controversy which historically has been so prominent. Two reasons come to mind. First, since the canals do not exist—at least in the form Lowell imagined them—there are no new data to review. Second, many modern scientists want to steer clear of the whole subject. It is a controversy without attraction for "objective" people unwilling to indulge in flights of fancy. Although this point of view is scientifically defensible, it must cause puzzlement among those older people who can recall the heyday of Lowell's Martian civilization and now find it purged from public discussion.

2.

Missions to Mars

THE ORBIT of Earth is approximately a circle. Its major axis is 299.20 $\times$ 10⁶ km and its minor axis 299.16 $\times$ 10⁶ km. Mars, however, has a notably elliptical orbit. The two axes are 455.9 $\times$ 10⁶ km and 453.9 $\times$ 10⁶ km, a difference of 2 $\times$ 10⁶ km. The orbital traces for the two planets thus tend toward convergence and separation. The point at which the orbits approach most closely is approximately at perihelion or closest approach to the sun. The perihelion distance to Mars is 207 $\times$ 10⁶ km and to Earth 152 $\times$ 10⁶ km, so the closest approach between the two planets is 55 $\times$ 10⁶ km—actually 55.5 $\times$ 10⁶ km since the major axes of the two orbits are not precisely coincident. The aphelion distances are, respectively, 250 $\times$ 10⁶ km and 147 $\times$ 10⁶ km, yielding a separation of 103 $\times$ 10⁶ km.

What happens if we chart the positions of the two planets as a function of time? Earth completes one sidereal revolution every 365 days. The equivalent figure for Mars is 687 days. To a first approximation then, two complete revolutions of Earth equal one revolution of Mars. Calculated with more precision, Mars, Earth, and the sun are positioned along a straight line projected on the ecliptic plane every 780 days. This is a mean value. Individual measurements range between 763 and 810 days, depending on the positions of the two planets when the alignment occurs. When Earth is between Mars and the sun, Mars is said to be in opposition; when the sun is between the two planets, they are said to be in conjunction.

Oppositions are important because these are the times when Mars can be best viewed from Earth, both because Mars is closest and because the side facing Earth is almost completely illuminated. But not all oppositions are of equal usefulness. Remember that the periods of revolution for Earth and Mars are not identical, nor simple multiples of one another. This means that, over a long period of time, oppositions occur at a sequence of points along Mars' orbit. Those oppositions when Mars is close to perihelion are termed favorable; those when it is close to aphelion are termed unfavorable (fig. 2.1). The differences in the size of the apparent disc are extremely important in distinguishing surface markings.

We can go further and integrate the information on seasons with data on oppositions to produce a more refined characterization of each viewing period. Because the axes of rotation and orbital paths of both planets are fixed in sidereal space, their orientation for particular oppositions is predictable. For example, for the most favorable perihelion opposition it will be late summer in the northern hemisphere of Earth and late autumn on the northern hemisphere of Mars. For the least favorable it will be late winter on Earth and late spring on the northern hemisphere of Mars (fig. 1.8).

The orderly progression of oppositions on an approximate 26-month cycle has important consequences both for telescopic observers and for space travelers. At the present time we are extremely limited in the power of the engines that launch spacecraft into their planetary voyages. One way of circumventing this problem is to miniaturize the spacecraft. Anyone who has examined firsthand the Gemini, Mercury, or Apollo spacecraft need not be reminded of our success in this area. Even after making everything as small and as light as possible, problems persist. One does not simply point the rockets at a distant planet and let go. That's a very inelegant and costly route that requires excess energy to escape Earth's gravitational field and then to catch up with the potential landing field, which may well be moving away from Earth at high speed—almost as fast as the rocket is moving.

Surprisingly enough, a solution to this problem of interplanetary travel was worked out almost fifty years ago by a German engineer and school teacher, W. Hohmann. He calculated that the path between two planets requiring the least energy is one in which the spacecraft leaves the first planet in a direction tangential to its orbit and arrives at the second planet along a similarly tangential course. These specialized routes became known as Hohmann transfer ellipses. An ideal, least-energy ellipse for the hypothetical situation in which planetary orbits for Earth and Mars are considered circular and coplanar involves a transit time of 260 days and a transit angle of 180°. The journey has to be carefully planned so that when the spacecraft spirals out to Mars' orbit, the planet is located at the proper position for an encounter. We can easily calculate the required starting position at launch. In 260 days Mars will move through an angle of 136°. If it leads Earth by 44° at launch, then the sum of these two figures equals the 180° angle traversed by the spacecraft. This is tantamount to saying that the launch should precede a Martian opposition by two or three months.

Hohmann trajectory constraints restrict our exploration of Mars, and indeed all other planets, as was not the case for the moon, where each mission could be scheduled more or less when the equipment was ready. The periodicity of launch opportunities to Mars is similar to that for oppositions and is based on the same criteria of orbital position and angular velocity. However, it is misleading to say that one can travel to Mars only when it is close to Earth. In fact, launches precede opposition, and encounters may occur substantial periods after opposition, at which time the planets are widely separated.

Because planetary orbits are neither circular nor coplanar the interplanetary trajectories actually used deviate somewhat from the ideal ellipse just described. But the general constraint remains. Considering our present launch capabilities, journeys to Mars are possible only on a 26-month cycle. The launch dates of the Mariner missions were: November 1964, March 1969, and May 1971. The January 1967 opportunity was not used. When that is inserted, the 26-month cycle stands out clearly. Looking into the future, the same pattern continues. The Viking mission was originally scheduled for 1973 but was delayed for budgetary reasons. The launches occurred on August 20 and September 9, 1975.

This launch restriction, imposed by the relentless progression of the planets in their orbits, imparts peculiar excitement to planning and development of planetary missions. A "best effort" is not good enough. Schedules have to be prepared and rigorously followed years in advance so that, when the brief interval of time known as the launch window arrives, the spacecraft is delivered to the launch pad in working order. A few days too late and the mission is delayed for 26 months. Under these demanding conditions, it is no wonder that a schedule slip of only a week several years before launch is a major concern. Any delay can be fatal.

Mariner 4

U.S. exploration of Mars began inauspiciously on November 5, 1964, when Mariner 3 was launched from Cape Kennedy. Although the launch was successful, shortly after lift-off a plastic shroud that pro-

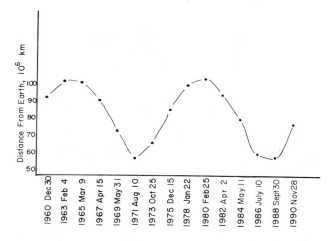

Figure 2.1. Opposition distances and times. Note the period of approximately 16 years between the most favorable oppositions. Times of closest approaches do not correspond precisely to opposition dates. Differences may be as much as 10 days.

tected the spacecraft failed to separate. Disabled and unrepairable, the spacecraft went into orbit about the sun. Following an intensive around-the-clock effort to design and build a stronger shroud, the companion spacecraft, Mariner 4, was launched on November 28.

With an initial thrust by an Atlas rocket and an additional nudge by an Agena rocket, the Agena-spacecraft combination was placed in parking orbit around Earth. Then, at the optimum point for achieving the required trajectory and escape velocity, the Agena engine was restarted, kicking the 260 kg Mariner spacecraft out of Earth orbit and into a transfer ellipse, headed for Mars (fig. 2.2). Its umbilical cord to Earth cut, the spacecraft deployed solar panels to provide power. With the sun as one

navigational beacon, Mariner 4 began a systematic search for the star Canopus. After several false identifications and with some help from ground commands, the onboard photosensor finally located Canopus. These two stars provided the orientation for all subsequent spacecraft operations.

On December 5 a midcourse maneuver was executed. A brief motor burn changed the angle of flight by ¼°. Depending upon one's point of view, the targeting change was radical or insignificant. The flyby distance at Mars was reduced from a 250,000-km miss distance on one side of the planet to a 10,000-km miss distance on the opposite side. This correction is more than half the distance from the Earth to the moon. However, viewed in the context of the total voyage the change could be compared

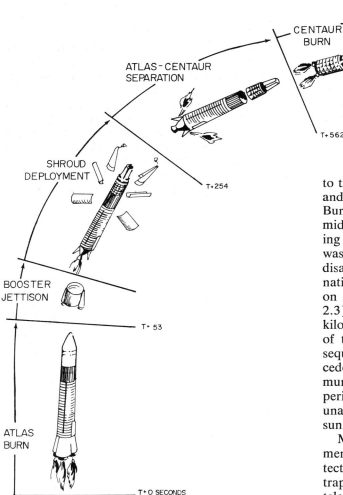

to traveling from Washington, D.C., to Los Angeles, and shortly after takeoff, altering course to land at Burbank rather than Los Angeles. Incidentally, the midcourse maneuver was not to correct launch aiming errors. The wide miss on the initial trajectory was consciously planned to avoid the possibility of a disabled spacecraft hitting Mars, possibly contaminating the planet with terrestrial organisms. Finally, on July 16, 1965, Mariner 4 flashed by Mars (fig. 2.3). After a journey of 228 days and 523 million kilometers, the spacecraft passed within 9780 km of the Martian surface. Following a picture-taking sequence that lasted barely half an hour Mars receded into the distance. On October 1 radio communication was lost (although contact was regained periodically for several years), and Mariner 4, left unattended, continued on its long orbit about the sun.

Mariner 4 carried numerous scientific experiments: a helium magnetometer, cosmic dust detector, ionization chamber, cosmic ray telescope, trapped radiation detector, solar plasma probe, and television camera. All experiments save the television were directed more at monitoring the distribution of dust, ions, high-energy particles, and magnetic fields in the inner part of the solar system than in making measurements directly applicable to an

Figure 2.2. Launch maneuvers for the Mariner 9 spacecraft. Equivalent maneuvers were carried out for Mariner 4, except that an Agena rocket was used in place of the Centaur rocket.

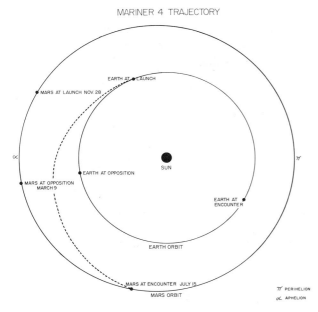

MARINER 4 TRAJECTORY

Figure 2.3. Flight trajectory for Mariner 4.

serve as an electro-magnetic dynamo. The cosmic dust detector recorded approximately the same density of dust throughout the belt between Earth and Mars. A small decrease of impacts in the vicinity of Mars suggested that the planet may have swept the adjacent area clear of debris.

The gross structure of the Martian atmosphere was revealed by an occultation experiment utilizing measured changes in the radio signal. Shortly after encounter, the spacecraft passed behind Mars and was hidden—or occulted—from Earth. But, just before entry and again just following exit from occultation, the radio signals from the spacecraft grazed the solid planet. In so doing they necessarily traveled through the Martian atmosphere and were refracted. This perturbation could be measured as a phase and amplitude change at Earth-based receivers. With these data in hand it was possible to construct a model of atmospheric density and temperature as a function of height. Surface pressure was determined to be in the vicinity of 5 millibars, lower than many estimates based on telescopic observations. Scale height was determined to be 8–10 km, indicating that the pressure decreased more rapidly with height than previously suspected.

understanding of Mars. Nonetheless, a few important observations were made. Magnetometer readings indicated that, if Mars has a magnetic dipole moment, it is no more than 0.03 percent the strength of that for Earth. Determination of this upper limit for Mars' magnetic field argued against the possibility of extensive differentiation within the planet and formation of an iron-nickel core that would

The Mariner 4 mission is best remembered for the pictures acquired. A single television camera took a total of 22 pictures, the best with a nominal ground resolution of about 3 km. The sequence was acquired along the track of spacecraft flight, sweeping across the equator, from 187°W, 37°N to 89°W, 35°S (fig. 2.4). To everyone's disappoint-

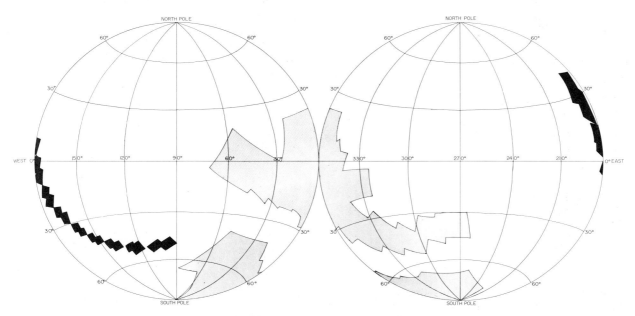

Figure 2.4. Positions of Mariner 4 (black) and Mariner 6-7 (shaded) near-encounter pictures.

ment the pictures showed remarkably little contrast. Even after the data were enhanced by computer processing the scenes were murky and ambiguous. This prompted a continuing controversy as to whether some stray light had entered the camera optics or whether the Martian surface actually had little contrast, possibly due to scattering of light in a dust-laden atmosphere.

Indistinct though the pictures were, a large number of craters—more than 300—was revealed. Crater densities were calculated to be intermediate between lunar maria and highlands. The craters were announced to be a feature "not expected by most scientists," although, in retrospect, it is difficult to imagine what they were expecting. Presumably they were looking forward to a scene more Earth-like, with mountains, plateaus, basins, and deserts. Be that as it may, the discovery of craters swung the pendulum in the opposite direction. It was concluded that the Martian surface closely resembles that of the moon. So began a misleading generalization that persisted through the Mariner 6–7 missions, not to be dissipated until the comprehensive photographic survey of Mariner 9.

Looking back, we realize that the few pictures acquired by Mariner 4 were clustered in a relatively uncratered and bland region (fig. 2.5). Because the small sample of pictures included none of the exotic features now known to us—volcanoes, canyons, and the like—geological and biological interest in Mars fell off sharply. After all, we were already involved in a comprehensive investigation of the moon. What was the sense in repeating the exercise?

Mariner 6–7

No missions were scheduled for the next launch opportunity in 1967, but in 1969 two more flyby missions were undertaken. On February 24 Mariner 6 was launched from Cape Kennedy, perched atop an Atlas/Centaur rocket. A month later Mariner 7 was launched, and both spacecraft were on their way to Mars. Although Mariners 6 and 7 were generically related to Mariner 4, they were substantially richer and better dressed than their humble ancestor. Spacecraft weight was increased from 260 to 380 kg, but, more importantly, science instrument weight was almost tripled. The instruments themselves were completely new. Infrared and ultraviolet spectrometers were included to determine atmospheric composition. An infrared radiometer was designed to measure ground temperatures. Two television cameras were included, with different fields of view and resolutions. The wide-angle camera covered an area about 12 times larger than the Mariner 4 camera but with the same ground resolution, ap-

proximately 2 km. The narrow-angle camera had a field of view one-tenth that of the wide-angle camera but with a tenfold increase in resolution. The cameras were designed to shutter alternately so that the wide-angle frames overlapped and the narrow-angle frames fell within the overlapping strips. All of the science instruments were mounted on a movable platform that could be commanded to point toward different parts of the planet. In addition to their finer scientific trappings, Mariners 6 and 7 were able to take advantage of engineering advances not available to earlier missions. In the older spacecraft most of the instructions and mission sequences were preset before launch. Mariners 6 and 7, however, contained onboard computers that could be reprogrammed during the mission. This flexibility almost certainly saved the scientific value of Mariner 7. The same concept of mission flexibility allowed Mariner 9 to wait out a dust storm and complete a successful mission while a Russian orbiter blindly followed its preprogrammed instructions and returned no useful pictures.

On August 1, five months after launch, Mariner 6 passed by Mars. Fifty hours before closest approach the narrow-angle camera was pointed toward Mars, and a sequence of 33 pictures was taken over the next 20 hours at distances from 1,240,800 to 724,990 km. After these were transmitted to Earth, a second sequence of pictures was taken at distances ranging from 561,000 to 175,000 km. With a best resolution of only 24 km, the sequence nonetheless greatly exceeded in detail pictures obtainable through Earth-based telescopes. The near-encounter sequence was initiated on July 31. For 30 minutes the two cameras took 25 pictures along an equatorial girdle from 10°W to 265°W (fig. 2.4). At its closest approach, the spacecraft passed within 3431 km of Mars. Eleven minutes later it disappeared behind the planet. When, after 25 minutes, it reappeared, all that remained was to play back data previously recorded on the onboard tape recorder. The Mariner 6 mission was over.

Just as Mariner 6 was passing Mars, Mariner 7, still several million kilometers from the planet, suddenly and inexplicably ceased to communicate. After hours of anxious conjecture and experimentation a signal was regained, but in a partly garbled form. Apparently the spacecraft was spinning out of control so that the radio signal was not properly beamed to Earth. After several unsuccessful efforts, the navigational sensors locked on Canopus, and the spacecraft stopped spinning. The reason for this misadventure was never definitely established, but it may have been caused by a bursting battery. Contents spewing over the inside of the spacecraft disrupted some of the electrical components causing the spacecraft to spin. Although radio contact was reestablished after several hours, the information giving the pointing direction of the instruments was not being received. In circumventing this problem the ability to reprogram the onboard computers, and thus alter the sequence of spacecraft operations, became the saving factor. A special sequence of pictures was initiated. The first contained nothing but black space; the platform was moved around until Mars finally came into the field of view. Mars was then centered in the frame to provide the pointing information needed to continue the flyby.

On August 2, several days after Mariner 6 encounter, Mariner 7 began its own close inspection of Mars. Before closest approach the Mariner 7 narrow-frame camera completed three far-encounter sequences including a total of 91 pictures. Study of Mariner 6 far-encounter pictures had shown indistinct but provocative details around the south polar cap area. For that reason the picture sequence on Mariner 7 was reprogrammed to include some near-encounter pictures of this region. Thirty-three more pictures were obtained, one group in a strip from 10°W, 14°N to 265°W, 32°S and a second group from 32°W, 53°S to 271°W, 75°S (fig. 2.4).

Data returned by Mariners 6 and 7 modified our previous understanding of the planet but did not radically change it. The majority of pictures revealed cratered terrain and reinforced the impression gained from Mariner 4 pictures that Mars is essentially a moon-like planet. True enough, some new landforms were recognized. A belt of chaotically jumbled ridges and irregular depressions in the vicinity of 40°W, 15°S was appropriately dubbed chaotic terrain. Apparently it formed by withdrawal of subsurface material and collapse of overlying sediments and rocks. Even more speculatively, it was suggested that the removed material may have been permafrost. Elsewhere, notably in the interior of Hellas, terrain with no observable topography, the so-called featureless terrain, was interpreted to be the result of sedimentary infilling and obliteration of pre-existing topography. In the vicinity of the south pole distinctive features were observed: sinuous ridges, polygonal pits, and sets of parallel grooves. Most of these defied persuasive interpretation except for the fact that they were somehow related to polar processes.

The inferences made from the photographic data were: (1) that Mars had few if any Earth-like forms and therefore had never experienced any of the internal processes that affect Earth, (2) that the dominant terrain type was a primordial cratered surface formed in approximately the same way and at the same time as the lunar highland surface, (3) that Mars had never had an Earth-like atmosphere, and (4) that water had never played an important role

in modifying the surface. The last inference was particularly important because it shifted the odds against finding life on the planet. Biological exploration of Mars was then, as now, a controversial subject. The central question is: "Should most of the money in NASA's planetary program be directed toward extensive exploration of Mars with experiments designed to test the (blank) proposition that life exists on Mars, or should emphasis be switched to smaller, cheaper missions directed to all the planets?" Depending upon one's prejudices and the data in hand, he might insert, in the blank, any number of qualifying adjectives: reasonable, unreasonable, exciting, stupid, probable, unlikely, incredible, etc.

Even before the Mariner flights, an ambitious program to explore Mars with unmanned landers—Voyager—had been seriously discussed. The design of the spacecraft proved too advanced for the resources of NASA, and the program was abruptly terminated. In its stead a more modest but still very elaborate and expensive venture, Viking, was planned for the 1970s. But if one concedes that development of life depends on water, and if it is near certain that there is no water on Mars, then why waste all that money on Viking? It is a treacherous chain of logic, and the Mariner investigators were most conscious of it. Though one gains the impression that their personal convictions weighed against life on Mars, they nonetheless struggled to preserve some semblance of neutrality. Witness this sort of fence-walking exercise: "While one cannot exclude, on the basis of television data, the possibility that a comparably brief, aqueous epoch occurred during the early history of the planet, the effect of the television results so far is to diminish the a priori likelihood of finding life on Mars. However, if Mars is to be a testing ground for ideas regarding the origin of life, we must avoid using these same ideas to disprove in advance the possibility of life in that planet" (Leighton et al., 1969b, p. 76).

The other Mariner investigations provided data that added more fuel to the biological fire. Infrared spectroscopic results confirmed, as previously determined, that CO_2 was the dominant atmospheric constituent. But several absorption peaks were not compatible with the CO_2 spectrum. Laboratory comparisons indicated the possibility that they might be formed by CH_4 (methane) and NH_3 (ammonia). On Earth these gases are largely formed through biological processes. If present on Mars, they would be the most persuasive indicators of life yet discovered. The news was too exciting to keep quiet, and the inevitable briefings and news conferences were held. Most scientists, however, remained skeptical. The presence of these gases was inconsistent with all other Mariner measurements; the interpretation of

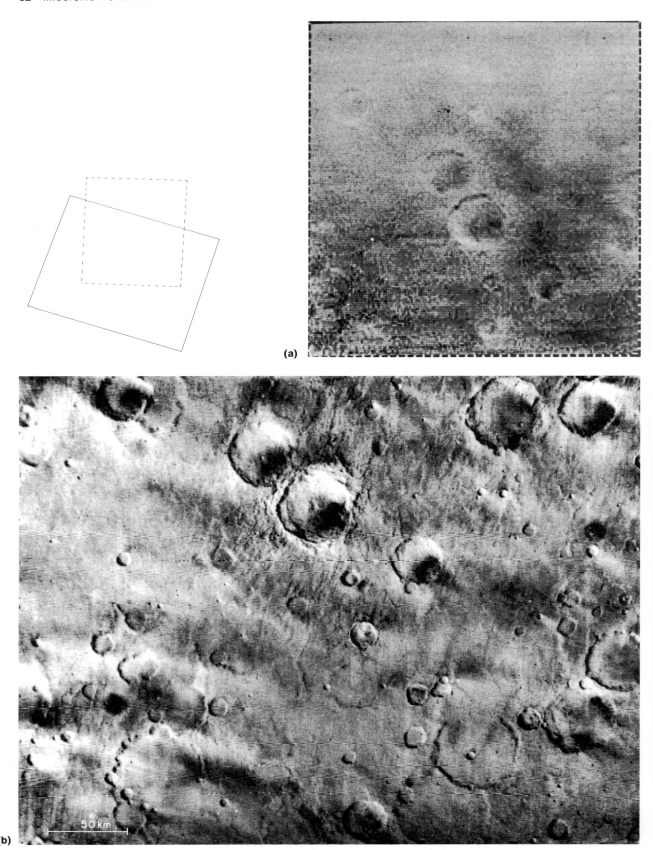

(a)

(b)

50 km

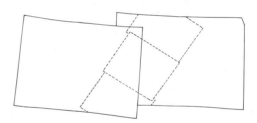

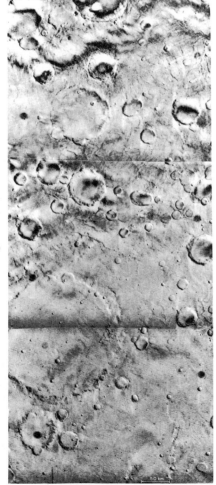

(c)

(d)

Figure 2.5 (a). Mariner 4 picture 8, centered at 174°W, 19°S. About a half-dozen craters are dimly visible.

(b). Mariner 9 picture of the same region shown in (*a*). Insert shows the overlap relationship between Mariner 4 and 9 pictures. A large number of additional craters are visible in the Mariner 9 picture. Hummocky rim deposits and furrowed terrain is identifiable. (DAS No. 6678668, rev. 142, A camera, center at 174°W, 20°S.)

(c). Mariner 6 pictures (6N21 and 6N23) of a region in the vicinity of 340°W, 15°S. Intercrater areas look bland and relatively featureless.

(d). Mariner 9 pictures of the same region shown in (*c*). Insert shows overlap between the two mosaics. Textured rim deposits and intercrater furrows are visible in Mariner 9 pictures but are difficult to discern in Mariner 6 pictures. This region is in the cratered terrain, just south of the contact with northern plains. (DAS No. 06786748, rev. 145, A camera, center at 336°W, 10°S. DAS No. 06786678, rev. 145, A camera, center at 338°W, 15°S. DAS No. 06786608, rev. 145, A camera, center at 340°W, 20°S.)

the infrared results, at best, argumentative. The skepticism was justified. Additional analysis demonstrated that the mysterious peaks could be more reasonably correlated with solid CO_2, a common material in the Martian polar caps (the locality for the disputed spectra) but material whose infrared spectrum had previously received little attention. The ultraviolet spectrometer confirmed the dominance of atmospheric CO_2, and established that nitrogen—an important biological indicator in the terrestrial atmosphere—could not be present in amounts greater than 1 percent. So, although the question of life on Mars remained unresolved, the odds against life were getting discouragingly high.

With Mariner 9 information in hand, it is easy to second-guess the Mariner 6–7 view of the planet. One might also accuse the Martian gods of a cruel hoax in failing to reveal many of their more exotic landscapes, and instead showing us more and more craters. The Mariner 6–7 investigators responded to these data in the only way they could; scientists don't believe in cruel hoaxes. Even so, and to the credit of the investigators, a note of disbelief threads through all their conclusions that Mars is only a moon-like impact counter. The chaotic terrain is strange, hinting of processes unknown, and so they state. The polar terrains are distinctive, somehow related to glacial processes, and so they speculate. The bright and dark markings hint of shifting dust and sand sheets. None of these processes was clearly perceived. They seemed just beyond the limit of our resolution, visual and intellectual.

Mariner 9

Barely six months before launch of Mariners 6 and 7, the National Aeronautics and Space Administration gave final approval to a much more ambitious project to map and observe the planet Mars. Two identical spacecraft would be placed in orbit in November 1971, returning data from six experiments.

Although the development of this orbital mission was little affected by Mariner 6–7 scientific results, the engineering legacy was profound. Indeed, the Mariner 1971 spacecraft looked almost identical to their 1969 predecessors (fig. 2.6). The greatest change was in the propulsion system. Mariner 1971 spacecraft were fitted with engines capable of changing the spacecraft velocity by 1700 m/sec, enough for orbital insertion with fuel remaining for later orbital changes and trim maneuvers. Other changes of the Mariner 6–7 design were required to accommodate the heavier and more versatile instrument payload and to provide reliable performance for the nominal 90-day mission. Mariner 7 experience was valuable in perfecting an onboard computer that

could be reprogrammed daily to meet changing requirements for operation of scientific instruments. In operating the flyby mission, engineers learned how to transmit and receive massive amounts of information. Tracking stations on four continents were used to monitor continually spacecraft functions, receive data, and reprogram flight sequences. Computer techniques were developed to squeeze more information from television pictures by automated filtering and contrast enhancement. Were it not for the requirement to gain some firsthand experience with this class of engineering problem before the Mariner 1971 flights, one could argue convincingly that the previous missions were not necessary. The scientific data were fragmentary, misleading, and totally eclipsed by Mariner 1971 results.

The stated objectives of the Mariner 1971 mission were to map about 70 percent of the planet, to gather information on the composition of the surface and atmosphere, and to study the topography and temperature of the surface. To accomplish these science objectives, six experiments were chosen: television, ultraviolet spectroscopy, infrared spectroscopy, infrared radiometry, S-band occultation, and celestial mechanics. A mapping, or wide-angle, camera was designed to provide pictures with approximately 1 km ground resolution from periapsis altitude. A narrow-angle camera acquired pictures with ten times better resolution—approximately 100 m. Pictures from the two cameras are commonly referred to as A and B frames, respectively. A detailed description of the camera characteristics is contained in appendix A. The ultraviolet and infrared spectrometers were designed primarily to provide atmospheric information: chemical composition, density, and temperature. Local variations in atmospheric density could be used to chart surface topography, assuming that pressures are highest over depressions and lowest above elevated regions. The

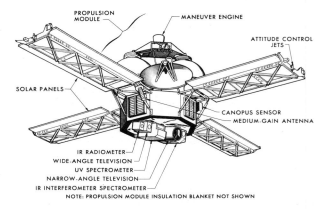

Figure 2.6. Mariner 9 spacecraft. Distance between solar panel tips is 6.9 m.

infrared radiometer was designed to measure ground temperatures and, by inference, to determine the physical properties of surface materials.

Much thought was given to the best strategy for using the two identical spacecraft. It was finally decided to use the first as a reconnaissance vehicle, mapping the entire planet, and to use the second in a more specialized way, studying the possible variation in selected regions with time. The first spacecraft was to be directed into a high inclination orbit, approximately passing over both poles. In this way the planet would rotate beneath the spacecraft so that, over a period of weeks, a complete longitudinal girdle would be visible from the periapsis position 1250 km above the surface. Since the ground trace of the orbit approximately paralleled the terminator, all pictures could be taken with constant sun angle. To assure best delineation of surface details, the orbit was selected so the sun was low in the Martian sky, about 30° above the horizon. In addition to providing a good platform for acquiring a mosaic of pictures that would cover the entire planet, the orbit was ideally suited for repetitive photography of the polar regions.

The planned orbit for the second spacecraft was at a low inclination to the equator. With an orbital period of 20.5 hours, the spacecraft would return to the same position at the planet every five days. The situation was ideal for repeatedly monitoring a small part of the surface to search for any variations possibly related to wind activity or seasonal processes. Other specialized tasks assigned to this vehicle included limb measurements of the atmosphere and photography of the two Martian satellites, Phobos and Deimos.

The original plan called for an extensive sequence of pictures to be taken prior to orbit insertion. These far-encounter pictures were designed to provide global albedo coverage and to reveal any large-scale changes in the light and dark markings since the 1969 missions.

In the series of Mariners 4 through 9 a natural evolution and improvement in space technology is documented. The earlier missions were well-designed to carry out their limited goals. A flyby mission has about 30 minutes to collect planetary data. The round-trip light time required to transmit a command to the spacecraft and receive verification is about 30 minutes. The time required to react to an emergency would have been about one hour for Mariner 4. Thus a flyby mission cannot make use of complicated adaptive capabilities. Another factor is telecommunications—the ability to send and receive data. With the completion of a 64-m antenna at Goldstone in California's Mojave desert, Mariner 1971 was able to take advantage of a vastly improved communications network. This improvement

is best shown by a comparison of the times required for Mariners 4 and 9 to transmit a picture, eight hours versus one hour—and a Mariner 9 picture contains more than 20 times as many bits as a Mariner 4 picture. At Mariner 4 transmission rates it would have taken more than 150 years to send back all the Mariner 9 data!

The first of the 1971 missions to Mars, Mariner 8, was launched from Cape Kennedy on May 9, 1971. Catastrophe! The second stage of the Atlas/Centaur launch vehicle failed, and Mariner 8 fell into the Atlantic Ocean 1500 km from its launch pad. This first orbiter was the one earmarked to map the planet, study the polar regions, and take global pictures for geodetic purposes.

The loss of Mariner 8 required quick action to salvage most of the science objectives with a single spacecraft. Three weeks remained until the launch of Mariner 9. At the Cape and elsewhere security was tightened to protect the remaining launch vehicle and its now invaluable cargo. The experimenters and project personnel went to work on a new plan to accomplish most of the goals of the mission with the remaining spacecraft. How could the major objectives be retained without severely curtailing the scientific return? After numerous difficult sessions, the team of experimenters developed a substitute plan. The remaining spacecraft would be placed in a 12-hour orbit inclined at 65° to the Martian equator. Periapsis altitude would be 1350 km. As in all compromises, this one could not be perfectly acceptable to all the experimenters. The mapping coverage would be at higher altitude and higher sun elevation than was previously acceptable. For variable features and albedo determinations the sun elevation was not high enough; for best topographic definition it was too high. But this new orbit remained the best compromise.

Mariner 9 was successfully launched on May 30, 1971 (fig. 2.2). An adjustment was made in the trajectory six days later, and Mariner 9 was following an orbit that would lead to intersection with Mars 161 days and nearly 400 million km later.

Not all the drama centered on the uncertainties of Mariner 9 launch. Two Soviet spacecraft, Mars 2 and Mars 3, had been launched May 19th and May 29th to take advantage of the 1971 launch window. These spacecraft were massive by comparison, each weighing nearly five times as much as Mariner 9. This prompted speculation that the Russians would attempt to land on Mars. Because of their slightly different trajectories, the U.S. spacecraft would arrive at Mars nearly two weeks before the first Soviet spacecraft.

There was a five-month respite from the long hours of prelaunch meetings; Mariner 9 was transmitting data to be used for celestial mechanics and some housekeeping data, but it was essentially coasting to Mars. In late September and again in early October the instruments were activated to check the scan platform pointing accuracy and to provide inflight calibration and testing of the TV cameras and other instruments. Pictures of Saturn with its rings vaguely showing, and the tiny disc of the approaching planet Mars were transmitted back to Earth.

As the time of encounter approached, Earth-based telescopic observations by the International Planetary Patrol began to worry some of the Mariner investigators. Late in September what appeared to be one of the typical perihelion dust storms was observed in Mars' southern hemisphere. Within a few weeks the storm spread over the planet, completely obscuring the classical markings. Mariner 9 might arrive during a massive global dust storm. The 90-day mission was in jeopardy.

On November 10, as the spacecraft approached 800,000 km from Mars, the first of the preorbital science sequences of pictures was shuttered. Mars was imaged at one-hour intervals to provide full global coverage of the disc. As feared, the dust storm completely obscured the planet. The next day a similar set of pictures was equally disappointing. These pictures were to have provided information on the global distribution of albedo features. A third sequence was taken just before orbit insertion. Some of the frames showed the south polar cap, its CO_2 frost in sharp contrast with surrounding terrain. Other of the frames showed four dark spots, three aligned in a northeasterly direction and a fourth to the west of these located at a classical feature, Nix Olympica. Little else could be seen. But Mariner 9 was at Mars.

At 2740 km from Mars, the spacecraft was properly oriented and the engine was fired to slow the craft for orbit. The retrorocket fired for 915.6 seconds, 4.4 seconds less than the 920 planned. Another instrument, an accelerometer, had determined that the correct velocity had been reached, and the engine was automatically shut off. The spacecraft swung around Mars, communication cut off by the occultation. Thirty-six anxious minutes later, the radio signal was acquired on schedule. It was quickly confirmed that Mariner 9 was in orbit. Four revolutions later, with the initial orbit parameters determined, the engine was restarted and fired for six seconds. This slight adjustment brought the mean period to within a minute of the planned twelve-hour period. Periapsis was 1387 km; the inclination was 64.4°. A slight error remaining after the first orbital trim maneuver was caused by unpredictable irregularities in the gravitational field of Mars. Although small, the accumulated error in the time of periapsis would amount to 30 minutes within a month. The effect of this was to reduce the time between daily appearance

of Mars on the horizon at Goldstone and the time of periapsis, in turn reducing the number of pictures that could be received at Goldstone. The solution to this problem would come later; more immediate was the problem of adjusting the mission plan because of the dust storm. The systematic mapping sequences had to be postponed and the mission plan adapted to monitor the storm, searching for breaks or signs of clearing.

On each orbit several pictures were taken from high altitudes to track the global course of the storm. Features observed in this way were then targeted for later high-resolution pictures. During this initial period the brightly reflecting south polar cap was one of the few surface features consistently visible. Daily pictures revealed details of the boundaries of the bright cap as it slowly shrank and finally stabilized.

On November 27, two weeks after the arrival of Mariner 9, the Soviet Mars 2 braked for Mars orbit. Later it was learned that, just before orbiting, Mars 2 ejected a capsule that landed somewhere in Hellas. If the Mars 2 capsule had instruments, it was not reported. The second Soviet probe, Mars 3, ejected a lander capsule and went into orbit on December 2. The Mars lander used a combination of parachute and descent rockets to soft-land in the vicinity of 158°W, 45°S (fig. 2.7). The capsule transmitted back a brief 20-second signal, then ceased. Soviet scientists speculated that unknown surface conditions were responsible for the failure of the lander. Perhaps the capsule sank into the dusty soil or was slammed into rocky outcrops by

Figure 2.7. Landing of Russian "Mars-3" capsule. After separation from a larger vehicle (1) and passage through the upper atmosphere (2, 3) a parachute is deployed (4). Rockets provide final braking (5). Following impact, the shell of the round capsule unfolds to provide a stable base (6).

winds accompanying the raging dust storm. The abrupt termination of the signal and the fact that the Soviets had designed the lander to withstand an impact velocity of 200 mph led U.S. experts to suspect a component problem, possibly the relay system aboard the orbiting Mars 3. Otherwise the Soviet orbiters apparently functioned as planned (Marov and Petrov, 1973). Unfortunately, their science sequences could not be altered to wait out the dust storm. The few orbital pictures that were released show practically no detail. The 1971 Soviet missions to Mars were apparently successful from an engineering point of view, even if they were scientifically disappointing.

Shortly before encounter, the Soviet Union and the United States began to implement an agreement to share data returned by their respective missions. A "hot line" was established between the Jet Propulsion Laboratory and a Soviet laboratory to facilitate this data sharing. In fact, very little information was exchanged—mainly because few requests were made. The Mariner experimenters sent pictures of the Mars 3 landing site and some other data. The Soviets were later to receive a complete set of the Mariner 9 pictures.

As Mariner 9 continued to orbit Mars there was little evidence of clearing. Except for the outlines of the bright polar cap, the first few orbits revealed nothing but "computer-enhanced dust." (A full description of enhancement techniques, as well as techniques used for picture identification and indexing is contained in appendix A.) The first signs of a break came on the eighth revolution. Near the south polar cap, vague circular forms could be distinguished. This clearing continued, providing hope that the mapping mission might be resumed in a few weeks. One of the fears developing now was that Mars would move so far from the Earth before the storm cleared that it would be necessary to transmit at reduced data rates. This would severely limit the number of pictures received at each Goldstone pass.

Some relief from the monotony of vague surface pictures was obtained from the occasional pictures of Mars' two natural satellites, Phobos and Deimos. B-camera frames of these rugged potato-shaped bodies proved the resolution capabilities of the camera system as well as its pointing ability.

The 13th day revealed details of Nix Olympica. The central region was shown to be a complex crater. On the 14th day in orbit, images were returned of the southernmost dark spot. Previous frames of these spots had revealed little, but these new pictures showed a wide crater ringed by concentric scarps—suggestions of volcano-tectonic origin. The entire region, called Tharsis on telescopic maps, was known from Earth-based radar data to be elevated relative to the rest of Mars. This accounted for its relatively clear appearance. The combination of facts—elevated region, alignment of three craters, morphology of the central craters—led to the conclusion that the mountains of Tharsis must be enormous volcanic structures.

The first phase of monitoring the planet, called RECON I (fig. 2.8), consisted of sequences of A and B frames in overlapping groups of 4, 5, and 6 pictures. The atmosphere became progressively clearer with each day. Certain features became familiar landmarks in repeated pictures of the same regions. The Greek- and Latin-based names yielded to more pronounceable tags—"salt and pepper," "south spot," "middle spot," "north spot," "Inca City," "elephant hide," "the chandelier." By revolution 64 the region around the south pole had cleared sufficiently to begin some mapping. After one month in orbit the first systematic mapping coverage was begun. Shortly before the new year we were briefly reminded that we are subject to the vagaries of weather on Earth as well as Mars—snow and ice on the antenna at Goldstone nearly prevented playback of data for a day.

As mentioned earlier, unexpected irregularities in Mars' gravitational field were progressively chang-

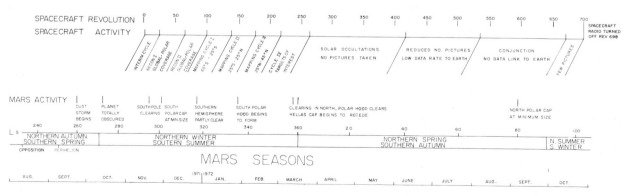

Figure 2.8. Sequence of spacecraft activity for Mariner 9.

ing the time of periapsis and reducing the time available to transmit pictures to the 64-m Goldstone dish. Another potential problem was the increasing distance from Mars to Earth. It was decided that an orbit trim to change the time of periapsis and to optimize the Goldstone viewing window could also be used to increase slightly the orbit period and to raise the periapsis altitude to 1650 km. This would increase the field of view of the cameras to provide sufficient overlap between successive mapping passes. The complete band around the planet would be mapped in 38 revolutions. This first sequence, called Cycle I, would map the planet with 10 mapping frames per pass between 25° and 65°S.

The orbit trim maneuver to put the spacecraft into an orbit that would meet the requirements was the most difficult yet attempted. The pointing accuracy was extremely critical. The spacecraft motor must be fired for 17 seconds about halfway between periapsis and apoapsis with the thrust direction nearly perpendicular to the flight path. On revolution 94 the trim maneuver was executed perfectly with the new orbit only 4 km higher and 4 seconds shorter than planned.

With revolution 100, the mapping officially began. The atmosphere was not as clear as desired, but spacing of pictures was good. Every day brought two strips of frames 180° apart and adjacent to the previous day's coverage (fig. 2.9). To a geologist,

each day's pictures were like a new traverse across unfamiliar terrain. Progressively revealed was a new planet, detail emerging from obscurity with overwhelming rapidity. Immense fracture systems spread across the planet as a new geologic map was constructed. Near the telescopic feature, Argyre, a basin the size of the lunar Imbrium Basin, was identified. Hellas, whose western margin was seen in pictures taken by the Mariner '69 mission was revealed in its entirety. Strange tributaries and fluvial-like forms were seen scattered across the equatorial region. Sinuous ridges resembling lunar mare ridges were seen. Lobate flow fronts suggesting basaltic outpourings were observed. The new pictures quickly erased from memory the "old" Mars as though the countless hours of previous speculation had been little more than science fiction.

The only problem with the television cameras came on January 12, revolution 118. The color filter wheel became stuck with the 60° polarizing filter in place before the lens. This meant that no color pictures could be taken. Certain atmospheric sequences designed to measure polarization were curtailed.

Cycle II was begun on revolution 139, January 22. This extended the continuous mapping coverage to 25° north latitude. Now the picture sequence swept across Nix Olympica and the three spots, revealing a number of other volcanic and structural features. The biggest surprise came when, just to the

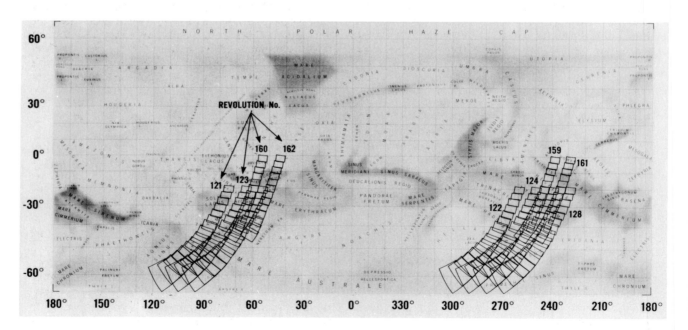

Figure 2.9. Footprints of Mariner 9 pictures during several revolutions of the mapping cycle. The orbital period of the Mariner spacecraft is approximately 12 hours, but the rotational period for Mars is approximately 24 hours. As a result, the planet rotates through approximately 180° between successive periapsis passes. This explains why strips of photographs on successive revs. are also separated by approximately 180° in longitude.

east of the spots, a vast canyon was seen to extend some 5000 km to the east, merging with chaotic terrain identified in 1969. This canyon had been glimpsed earlier in the mission, but its extent was not then appreciated. After Cycle II was completed, Cycle III was initiated to fill some gaps that developed around the equator and to extend the coverage above 25° north. By March 9 all the planned mapping and some additional northern coverage had been completed. The north polar haze had prevented much mapping above 40° north latitude.

On revolution 218, Cycle IV began. During this sequence a reduced number of pictures was taken of selected features. From mid-March to June 8, no data were obtained. This was due to other assignments for the Goldstone antenna in late March and also because the spacecraft was in shadow for a significant period of time during each orbit between April 1 and June 4. There was enough power to maintain the spacecraft through this period but not enough to operate the instruments and transmit back the data.

After the solar occultation period, the spacecraft was re-activated June 8 on revolution 416. It was decided to complete mapping of the north polar regions and to obtain better pictures of candidate Viking landing sites. About 60 pictures a week were now being returned.

The Mariner 9 mission finally ended on October 27, 1972. After 698 orbits of Mars, 7,329 pictures had been returned. In the middle of a maneuver to point the high gain antenna toward Earth the attitude control gas was exhausted. Mariner 9 went into a slow roll, breaking radio contact. The final commands were to turn off the spacecraft transmitter to eliminate the possibility of interference with other space projects. This was done by repeatedly transmitting the same command. One of these silenced the spacecraft.

Soviet 1973 Missions

The Soviets took advantage of the 1973 opportunity to attempt, again, a soft landing on Mars. Because more rocket thrust was required to reach Mars during this relatively unfavorable opposition, the Soviets were forced to split their payload. Four launch vehicles were used, two for the orbiters and two for the landers. Mars 4 and 5, the two orbiting spacecraft, reached the planet Mars in March 1974. A retrorocket on Mars 4 failed to fire and the spacecraft flew by the planet. Mars 5 was successfully placed in an orbit with a period of about 25 hours and a periapsis of 1500 km, very similar to the proposed Viking orbits. It remained functional for 20 orbits. On 10 of these orbits scientific data were collected.

As in the preceding 1971 missions, direct-entry landings were attempted with the remaining two spacecraft. Mars 7 failed to intersect the planet. Mars 6 completed a successful descent to its planned landing site at 20°W, 24°S, northeast of the Argyre Basin. Then inexplicably, only three-tenths of a second before touchdown, contact between the lander and orbiter was lost.

The Soviets are less willing than the Americans to describe their missions beforehand. They tend to be equally vague in describing the detailed operational and scientific results. Nevertheless, it is apparent that the scientific instruments on board the 1973 spacecraft were more numerous and sophisticated than in 1971. Equally important, a more flexible plan for their use had apparently been incorporated.

A total of 14 instruments were contained on each orbiter. Two television cameras provided image resolutions of 1 km and 100 m, very similar to the Mariner specifications. The 70 pictures that were acquired compare favorably with Mariner 9 images (fig. 2.10). Measurement of atmospheric water content by spectroscopic techniques indicated amounts as high as 60 precipitable microns. Equally high amounts have been measured with Earth-based instruments although the mean values are appreciably lower.

Magnetometers carried on Mars 2 and 3 previously had measured magnetic fields on the order of 30 gamma. The measurement was confirmed by Mars 5. This field strength is several times that which can be accounted for by interferences with the solar wind. This discrepancy suggests the presence of an internal field. However, the measured values are on the edge of significance. For example, Earth's magnetic field, measured at the surface, varies from 25,000 to 75,000 gamma. At an altitude of 1500 km, the field strength is in the range of 13,000–40,000 gamma, greatly reduced but still much higher than Martian measurements.

The most provocative of all the Soviet measurements were made on the entry vehicle. Although a mass spectrometer designed to measure atmospheric composition did not function as planned, returned engineering data that characterized the behavior of an associated ion pump indicated the presence of approximately 20 percent of an inert gas. The only likely candidate is argon. Although this result was initially greeted with skepticism, a review of previous atmospheric measurements demonstrates that, although they do not directly indicate the presence of large amounts of argon, neither do they preclude that possibility. If, in fact, argon is a major constituent, then the present atmosphere probably is a residuum from a much denser primary atmosphere. Most other components either have been trapped in surface materials or have been lost by achieving

Figure 2.10 (a). Photograph of Mars acquired by the Soviet spacecraft, Mars 5. The center of the picture is at approximately 38°W, 35°S. The sinuous channel at the top is Nirgal Vallis. Arrow points to a large crater also visible in (b).

(b). High-resolution Mars 5 picture of part of the same area photographed in (a).

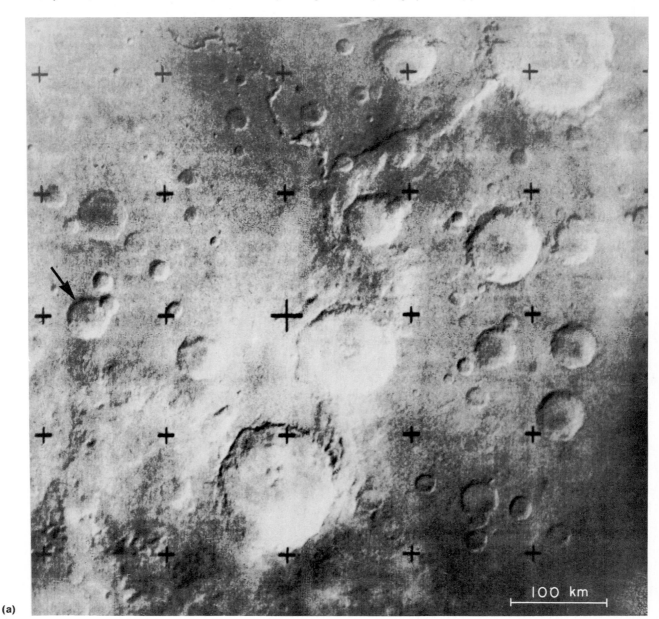

(a)

escape velocities. The implications are further discussed in chapter 8.

Viking

The Mission Plan

So many years are required to design and build spacecraft that the mission to follow Mariner 8–9 was conceived in 1968, almost three years before the Mariner 9 launch. The follow-on venture, Viking, includes a total of four spacecraft, two landers and two orbiters. Each lander-orbiter combination was launched by a Titan III/Centaur vehicle, intermediate in capability between the Atlas/Centaur used for Mariner and the familiar Saturn V, which launched Apollo spacecraft (fig. 2.11). The Viking mission was first scheduled for a 1973 launch, but

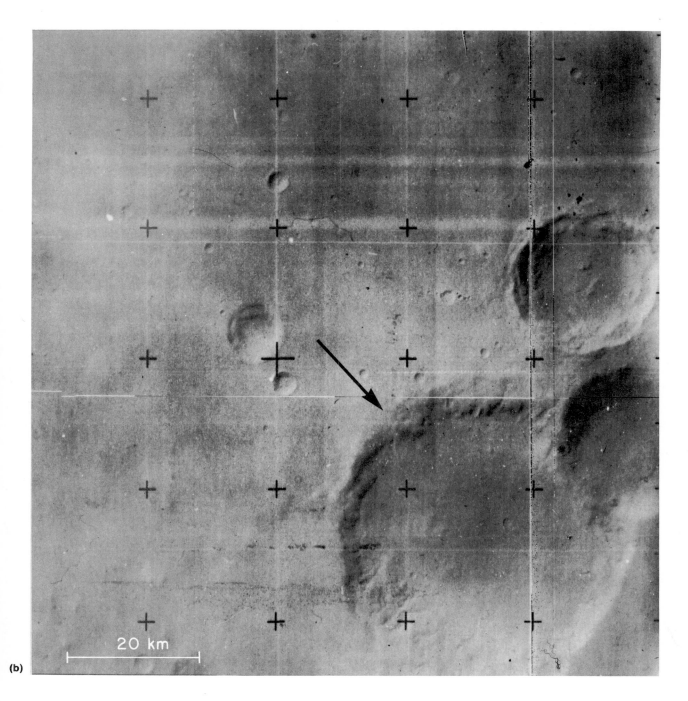

(b)

it was decided to delay two years for budgetary reasons. To be less euphemistic, the only way to absorb the costs of the mission was to schedule the payments over a longer period of time. The delay was initially a bitter disappointment for involved engineers and scientists, but it subsequently became clear that the elaborate scientific instruments that are an integral part of Viking simply never could have been completed by 1973.

The launch date for the first vehicle was August 20, 1975; the second followed several weeks later, on September 9. The trip to Mars will take almost a year, approximately twice the period of time required for Mariner 9. This is caused by a relatively unfavorable orientation of Mars and Earth during the 1975 launch opportunity. In selecting a minimum energy trajectory it will be necessary to follow a course that travels around the sun more than

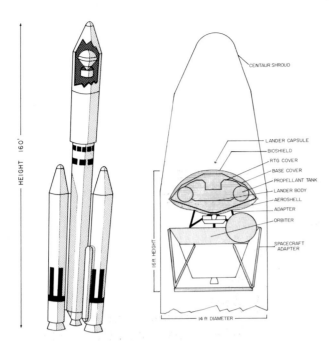

Figure 2.11 (a). The rockets that launched the Viking spacecraft. The lower vehicle is a Titan with two booster rockets strapped to its side. The upper vehicle is a Centaur rocket. Viking spacecraft is shown schematically atop the Centaur vehicle.

(b). The configuration of Viking lander and orbiter within the Centaur shroud.

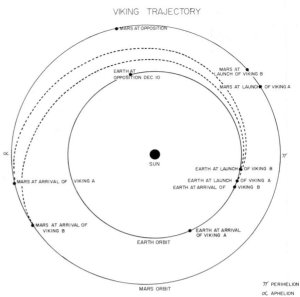

Figure 2.12. Flight trajectories for the two Viking spacecraft.

180°, a so-called Type II trajectory (fig. 2.12). Arriving at Mars, the two lander-orbiter vehicles will go into orbit just as did Mariner 9. But from then on, the sequence will be completely different. After several orbits the lander will be detached from the orbiter. Kicked out of orbit by the firing of small rockets, it will follow an elliptic trajectory to the Martian surface (fig. 2.13). As the spacecraft begins to encounter the effects of the Martian atmosphere, a deceleration from about 5 km/sec to 500 m/sec will result from atmospheric drag. At an altitude of about 5 km, a parachute will be deployed and will further slow the descent (fig. 2.14). At about 1.6 km above the surface, three rocket engines will be ignited to provide final deceleration. The rockets will continue to fire until contact is made with the surface. The landing is termed "soft," even though the vertical velocity at touchdown will be about 2.4 m/sec. The complete landing operation is an amalgam reminiscent of Apollo landings on the moon and returns through Earth's atmosphere. The chief difference is that, in the case of Viking, the sequence of events has to be carried out automatically, with no opportunity for human intervention. All of the hazards have to be anticipated in advance.

Once on the surface, the landers are designed to operate for a minimum of 60 days. Information acquired from the scientific experiments either will be transmitted directly to Earth or will be relayed through the orbiter as it passes overhead.

The Viking Experiments

It is probably no exaggeration to say that the Viking scientific instruments, in their diversity and complexity, exceed the payloads included on any previous space mission, manned or unmanned. The

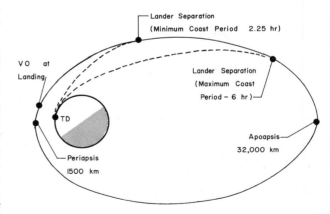

Figure 2.13. The separation of the Viking lander from orbiter following rendezvous with Mars. Note that the lander is released at a substantial distance from the planet and takes between 2.25 and 6.0 hours to reach the surface.

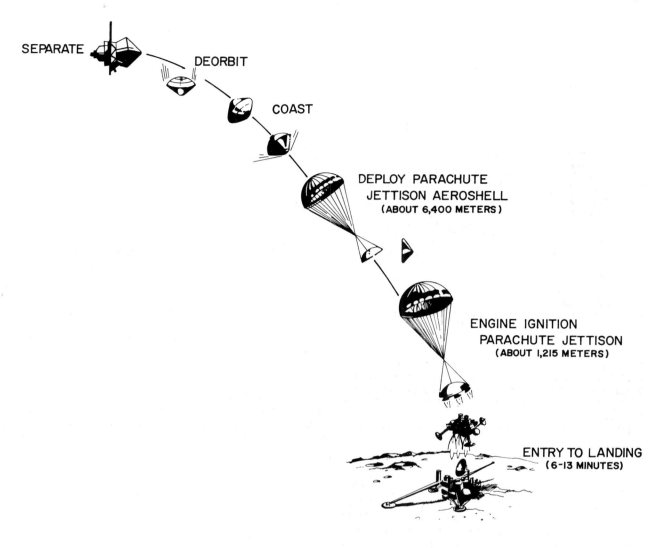

SEPARATE

DEORBIT

COAST

DEPLOY PARACHUTE
JETTISON AEROSHELL
(ABOUT 6,400 METERS)

ENGINE IGNITION
PARACHUTE JETTISON
(ABOUT 1,215 METERS)

ENTRY TO LANDING
(6-13 MINUTES)

Figure 2.14. Landing maneuvers for the Viking spacecraft.

instruments can be divided into three groups: those on the orbiter, on the entry capsule, and on the lander.

ORBITAL EXPERIMENTS

The orbiters will contain television cameras and two infrared instruments specially designed to detect atmospheric water and to measure surface temperature. The strategy for the television experiment will be different from that for Mariner 9, where the goal was to obtain a complete low-resolution survey of the planet and isolated high-resolution pictures covering only a very small part of the surface. The two Viking cameras are identical instruments with an object resolution of about 100 m at periapsis altitude of 1500 km. This is similar to the resolving power of the Mariner 9 high-resolution camera. During a single periapsis pass the two cameras, operated in rapid succession, will be able to take about 25 pictures, each 40 km on a side. The pictures, if arranged in two contiguous strips, will cover an area of about 80 km by 500 km. Alternate patterns can be selected to cover areas of different shapes or to provide overlapping stereoscopic images. The primary goal will be to gain detailed information for landing sites. Large regions must be photographed because there is a comparably large uncertainty in pinpointing the exact latitude and longitude of the landing point. The actual touchdown site may differ from the preselected location by as much as 300 km in a downtrack direction and 100 km in a crosstrack direction. There are additional scientific reasons for obtaining contiguous mosaics of high-resolution pictures. Imagine, for example, the difficulty a resident of

Mars might have in interpreting our interstate highway system from a single picture showing one white streak. (It should be acknowledged that our argument could be turned around by some gadfly. A single line, if perfectly straight, would bespeak the presence of intelligent persons with advanced surveying techniques. An extensive mosaic showing the spaghetti-like road system that connects our cities might give our Martian observer second thoughts about our intelligence.)

An orbital spectrometer operates in the vicinity of 1.4 microns, a region of the spectrum corresponding to a prominent water absorption band. If water is present in the atmosphere there will be diagnostic irregularities in the amount of infrared radiation transmitted through the atmosphere. Amounts of water as small as 1.0 precipitable microns can be detected. The instrument has a small field of view, 3×24 km at periapsis, so local concentrations of water vapor, if present, can be observed.

The second infrared instrument is designed to obtain different information. It detects radiation with wavelength from 6 to 35 microns. Within this spectral range, measurements are made in six fairly broad channels. The purpose is to measure surface temperature. Analogous experiments were conducted on Mariner 6–7 and 9, but those instruments had relatively poor spatial resolution. The Viking instrument can look at an area 5 km across from a periapsis altitude of 1500 km.

Thermal properties of potential surface materials vary widely. For example, CO_2 frosts cannot have temperatures greater than 150°K whereas H_2O frost temperatures will be in the vicinity of 200°K. Another example: thermal inertia, a property related to cooling rate, differs by a factor of ten between solid rock and the unconsolidated debris that is thought to cover much of Mars. Measured late in the evening, the rock will be relatively warm, radiating energy from sunlight of the preceding day.

ENTRY EXPERIMENTS

In many respects an understanding of the composition and dynamics of the Martian atmosphere supplies the key for unlocking larger questions of planetary evolution. On the one hand, certain atmospheric compositions will preemptorily rule out the likelihood of life. On the other hand, the existence of an extensive biota will alter the atmospheric chemistry. Even if we discount the biologic arguments, there is the question of whether the atmosphere formed recently or continuously by volcanic degassing, or whether it is a relic from some earlier differentiation of the planet. Mariner 4 occultation experiments demonstrated that a substantial ionosphere, with O^+ the dominant ion, overlies a lower atmosphere composed principally of CO_2. Bom-barded by ultraviolet radiation from the sun, the CO_2 of the upper atmosphere dissociates to form CO and O_2. The O_2 is further dissociated to form the abundant O^+ of the ionosphere. These reactions are straightforward and well known from terrestrial experience. Indeed, the problem is to figure out a way to retain the measured CO_2-rich atmosphere, which, according to the previous arguments, should long ago have been reduced to CO and O_2. A number of recombination schemes have been proposed, and Viking entry experiments are designed to test the several competitive models.

The entry experiments are necessarily "one shot" affairs, operating during descent to the surface. As the aeroshell falls through the upper ionosphere, the ion/electron plasma will flow past a sandwich of grids with different electrical potentials. Particles with different voltages are separated, and their aggregate charge measured by an electrometer. This instrument, termed a retarding potential analyzer, will be useful in determining the vertical structure of the ionosphere. Complementary measurements will be made repeatedly every few seconds by a mass spectrometer that will identify neutral species: notably CO_2, CO, O_2, N_2, and Ar. These two instruments will operate only during the aeroshell phase, before the parachute is deployed, but measurements of atmospheric temperature and pressure will be made all the way to the surface.

LANDED EXPERIMENTS

The most provocative instruments on the lander are three that attempt to record the metabolism of a Martian biota under controlled conditions. All three instruments are contained within the body of the lander. A scoop on an extendable arm reaches out, collects some soil, and then dumps it into a hopper leading to the biology instruments (fig. 2.15). From that point on the subdivided sample will be handled in three different ways. The first experiment measures carbon assimilation, the primary mechanism being photosynthesis of organic compounds from CO_2 and H_2O by green plants, using the sun as a source of energy. The sample is placed in a chamber with ambient Martian atmosphere to which a small amount of radioactive $^{14}CO_2$ and ^{14}CO has been added. After an incubation period of several hours to several days, during which the soil is exposed to simulated sunlight, all gases are flushed out. The solid sample is then heated to 625°C, and the liberated gases are measured with a radiation detector. If any of the labeled ^{14}C has been incorporated into organic compounds, that fact will be recorded (fig. 2.16). Transfer of large amounts of ^{14}C from gas to soil would, of course, be strong evidence in support of Martian life.

The attractiveness of this experiment is that it

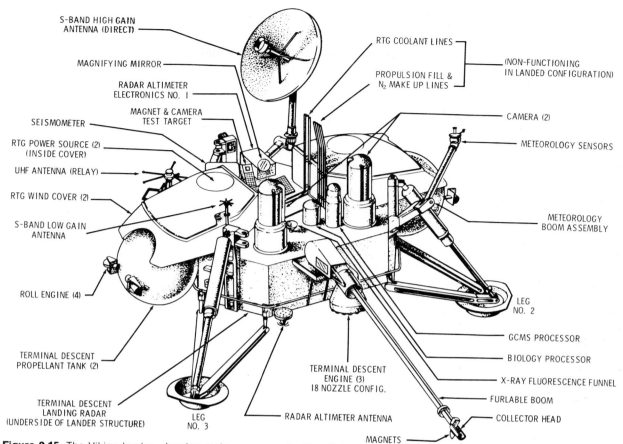

Figure 2.15. The Viking lander, showing major components. The distance between footpads is approximately 2 m.

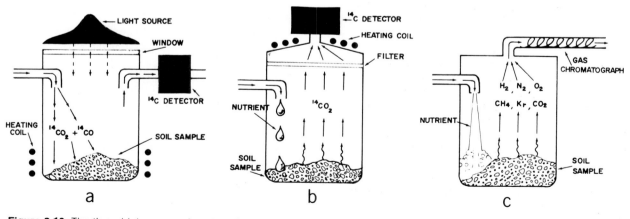

Figure 2.16. The three biology experiments to be carried out by Viking.
(a). carbon assimilation
(b). labeled release
(c). gas exchange. Experimental strategies are described in the text.

makes very few assumptions regarding the character of Martian life. The only built-in prejudices are, first, that life is carbon-based and, second, that it exchanges carbon with the atmosphere. No special nutrients or solvents are introduced. The soil sample is exposed to normal Martian atmospheric, thermal, and sunlight conditions. The investigator just sits back, waits, and watches.

The second experiment is, in a procedural sense, a reverse image of the first. Soil is wetted with a dilute medium of simple salts or more complex organic compounds to which radioactive organic matter has been added. If organisms are present, it is reasoned that they will consume the labeled organic matter, which will ultimately be converted into gas. By monitoring the liberated gas it will be possible to document the metabolic processes. The presumed biologic activity is respiration and evolution of CO_2 waste, a common phenomenon in terrestrial animals. The biologic assumptions for this experiment are more numerous than for the first. Living forms capable of degrading organic matter and producing gas are presumed. It is further presumed that the biochemical reactions occur in the presence of water and that the organisms thrive on the selected media and nutrients. However, the media are purposely kept dilute so that any compound that is potentially injurious will not be present in overwhelming amounts.

The third experiment incorporates a "smorgasbord" approach. The soil is thoroughly wetted, almost to the point of submersion, with a medium enriched in a wide variety of amino acids, salts, and vitamins. In recognition of its hopefully wholesome characteristics, the medium is sometimes referred to as "chicken soup." At periodic intervals the enclosed gas over the soil is analyzed for H_2, N_2, O_2, CH_4, Kr, and CO_2. If any metabolism is taking place, the composition of the gases will change as the nutrient material is biologically consumed and related gases are released. Of the three experiments, this has the strongest Earth orientation. Terrestrial organisms flourish as they splash through the chicken soup, but it is by no means clear that Martian organisms have a similar fondness for swimming pools and for our particular foods.

What are the chances that, in the summer of 1976, newspaper headlines will read: "Viking biology experiment proves life on Mars"? Many scientists are skeptical, so much so that they criticize NASA for spending money on an experiment that, in their opinion, is guaranteed to give negative or ambiguous results. Rather surprisingly most biologists, even those who support the Viking mission, are doing little to dispel this skepticism. They minimize the goal of "life detection" and, instead, talk vaguely about determining the "life-supportive" aspects of the Martian environment. Their apologia are needlessly circumspect. First, nothing that we know about Mars rules it out as a habitat for life. Second, the three biology experiments are designed to give unambiguously positive results. True, one can imagine situations in which a borderline positive result will be difficult to interpret; a certain amount of background radiation from the radioisotope power generators on the lander will mask a very weak signal from the two ^{14}C-measuring instruments. But a strong positive response from any of the three experiments will be difficult to discredit. The possibility of detecting organisms inadvertently carried from Earth to Mars is discussed later in this chapter. The chance that the measured gases form from inorganic chemical reactions can be evaluated by analyzing two identical soil samples, one of which has been heat-sterilized. If the signal is truly biologic it will not be observed in the sterilized sample. In sum, then, the Viking strategy is sound. An important question is being asked and the way made clear for a persuasively affirmative response.

Almost as important as actually detecting living systems is determination of organic compounds indicative either of biologic or protobiologic materials. To make these analyses the lander contains a combined gas chromatograph-mass spectrometer— or GCMS. Soil samples are heated by stages to 150°, 200°, and 500°C. At each stage any vaporized material is first separated by the gas chromatograph and then analyzed by the mass spectrometer. Organic compounds in the mass range 12 to 200 can be identified. The mass spectrometer also can be used to make atmospheric analyses. Minor constituents, present in concentrations no more than 1 part in 1 million, can be identified. Although this brief description does little to suggest it, the mechanical and electronic complexity of the GCMS is staggering. Consistent with this complexity the developmental history of the experiment reads something like "The Perils of Pauline." One can only hope that, when the climactic scene takes place, the GCMS is as resilient as the fictional heroine.

The trio of major investigations on the Viking lander is completed by the cameras. Almost all pictures previously taken from unmanned spacecraft make use of television systems. The two cameras on the lander are conspicuously different from previous devices. The imaged scene is reflected from a rocking mirror through lenses and onto a very small photodetector in the focal plane (fig. 2.17). The detector is so small that it senses only a small part of the total scene, a unit known as a picture element. With a single oscillation of the mirror, the picture elements within one vertical line are sequentially recorded. Then the entire camera is rotated in azimuth so that the mirror projects the adjacent vertical

VIKING LANDER IMAGING CONCEPT

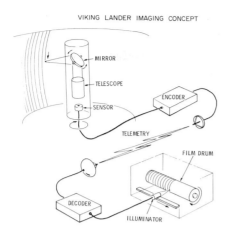

Figure 2.17. A highly schematic view of the Viking lander camera. The scene is reflected from a rocking mirror through focusing optics onto a small sensor. Different points in the scene are successively recorded and encoded on board the lander. Data are then telemetered back to Earth, where they are decoded and recorded on a film drum.

Figure 2.18 (a). Panorama taken with the Viking lander camera at Great Sand Dunes National Monument, Colorado. Dunes in the background are approximately 200 m high.

(b). High-resolution picture taken with the Viking lander camera and positioned at the right of the panorama shown in (a). All reflected energy with wavelength between 0.4 and 1.1 microns is recorded. Foliage of plants appears bright because chlorophyll is highly reflective in infrared wavelengths (0.7–1.1 microns).

(a)

(b)

line on the detector. Successive signals from the detector are stored and radioed back to Earth, where a picture is produced by essentially reversing the process. Electrical signals are converted to pulses of light, each of which exposes a small part of a photographic negative (fig. 2.18).

Each camera has twelve photodetectors, any one of which can be selected for use. Three image in the blue, green, and red part of the visible spectrum. The composite image will duplicate "natural" color. Three other detectors image in the near infrared. The remaining six are used for black-and-white imaging at different focal positions. The size of the picture element, that part of the field of view that is instantaneously recorded on the detector, determines the spatial resolution of the complete picture. This solid angle is 0.04° for black-and-white and 0.12° for color. At the closest focus position of several meters the 0.04° picture element corresponds to a linear resolution of 2–4 mm.

By contrast with a television camera, the Viking camera requires a long time to take a single picture. Eleven minutes is necessary to acquire a complete panorama of 360°. This slow operation leads to some interesting anomalies. A rapidly moving object will be recorded on only one to two vertical lines. Obviously, its true shape will not be represented. At first glance this appears to be a liability, but it can be converted into an asset. The azimuthal drive of the camera can be inhibited and the same vertical line scanned repeatedly. Assume that the stationary background recorded in this single line is brightly reflecting sand. Each successive vertical scan will display identical information, discounting the slight change in the position of the sun. But, if any dark object crosses this line, the reflectance values will be suddenly and dramatically reduced. In this way the single-line scan mode of camera operation becomes an interesting form of life detection.

Two instruments on the lander will probe the inorganic or geologic characteristics of Mars. An X-ray fluorescence spectrometer will examine soil samples that are scooped up and conveyed, through vibrating screens, into the spacecraft. When the soil is bombarded with X-rays from a radioisotope source, the soil particles will emit energy characteristic of their elemental composition. This energy, or fluorescence, will be measured by proportional counters that can differentiate most of the major rock-forming elements. Additional information on the composition of the soil will be obtained by taking pictures of an array of permanent magnets attached to the scoop. The amount of adhering soil particles will reflect the iron content of surficial materials. A passive seismometer, comparable to those previously flown to the moon, will monitor the planet both for meteoroid impacts and for internal marsquakes.

The complement of lander experiments is completed by a miniaturized meteorological station that will record atmospheric temperature, pressure, wind speed, and wind direction.

The Viking Landing Sites

The sites for the two Viking landings have already been chosen (figs. 2.19–2.21). True enough, these are only "nominal" sites, subject to change at any moment up to the actual descent from orbit, but the original sites will probably remain the final ones. So much time and effort are being spent in planning the mission that is uniquely associated with these sites that future candidates will face an almost insurmountable handicap, especially if they are proposed at the last minute.

The landing spot for the first spacecraft is in a region known as Chryse (34.0°W, 19.5°N), where several large canyon systems disgorge into a low basin (fig. 2.20). The backup site, to be used if the primary one proves unacceptable, is located at 252.0°W, 20.5°N, a region of level, windswept plains. The second spacecraft will be targeted toward 10.0°W, 44.3°N, a region with relatively large amounts of atmospheric water anticipated during the summer season of the Viking landing (fig. 2.21). The backup site is at 110.0°W, 44.2°N.

How were these sites chosen? Most importantly they satisfy certain engineering criteria. They are in a latitudinal belt where it is easiest to effect a landing from orbit, to maintain communications with Earth, and to keep the lander within the surface temperature extremes for which it has been designed. The sites are relatively low, with attendant high surface pressures. Accordingly, parachutes will be effective in slowing the descent of the entry vehicle. As nearly as one can judge from Mariner 9 pictures, the terrain at the two sites is smooth. However, this is a questionable conclusion. The best resolution of Mariner pictures is several hundred meters, but the Viking lander is endangered by irregularities on the order of a meter. The surface conditions at the primary site for the second spacecraft are further obscured because this locality was close to the edge of the northern polar hood at the time of Mariner photography. The pictures reveal irregular blotches and patches that might be more indicative of atmospheric conditions than surface topography.

There are scientific as well as engineering reasons for the selection of these particular localities. The Chryse site is located in what may be a depression filled with sediment eroded from the canyons that empty into the depression. For this reason the soil may be a fortuitously mixed and homogenized sam-

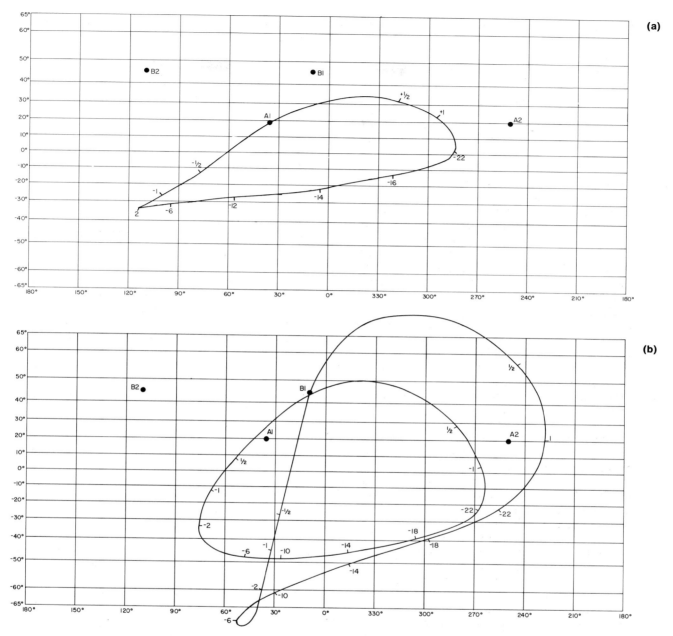

Figure 2.19 (a). The groundtrack of the Viking orbiter A. Times from periapsis are shown in hours. Because the orbital plane is inclined only 34° to the equatorial plane, the sub-spacecraft point stays between 37°N and 35°S. Primary landing sites are A1 and B1. Backup sites are A2 and B2.

(b). The groundtrack of the Viking B orbiter. Following a successful landing maneuver, the orbiter will be injected into a new orbital plane that has an inclination of 70°, permitting good observations of the north polar region.

ple representative of a large part of the Martian crust. As previously mentioned, the northern site is located in a latitudinal belt where some of the highest amounts of atmospheric water have been detected by ground-based measurements during the Martian summer, the same season at which Viking will land.

We have just mentioned some arguments in support of the chosen sites but, in truth, we have sidestepped the question: "How were the sites chosen?" They were selected by the traditional democratic processes, with all the same uncertainties that surround any political choice. One of the primary

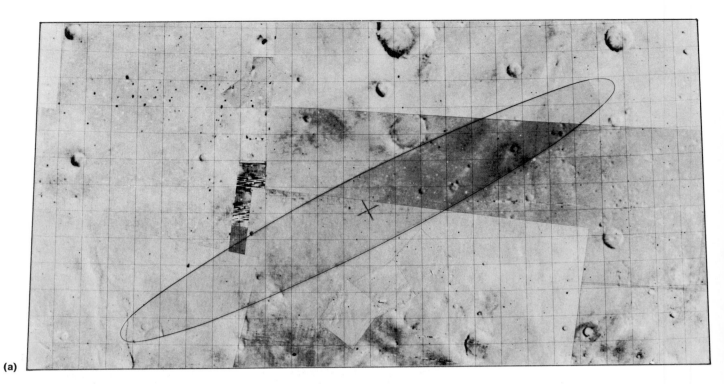

(a)

Figure 2.20 (a). Photomosaic of the Chryse site, proposed for the first Viking landing. The ellipse, 640 by 100 km in size, indicates the aiming uncertainty. If the lander is directed to the central cross, there is a 98 percent probability that it will touch down within the ellipse. **(b).** Geologic map of the Chryse site. Adapted from map by N. J. Trask (1972).

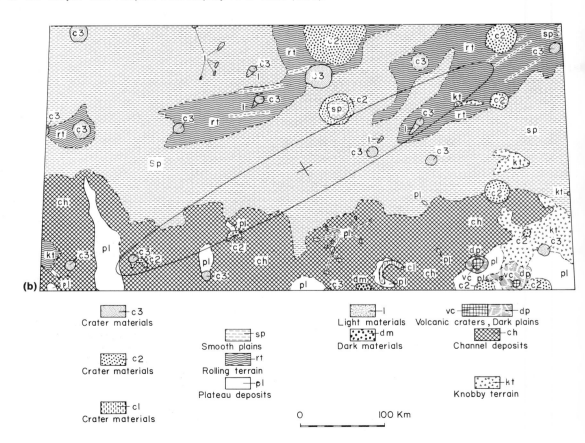

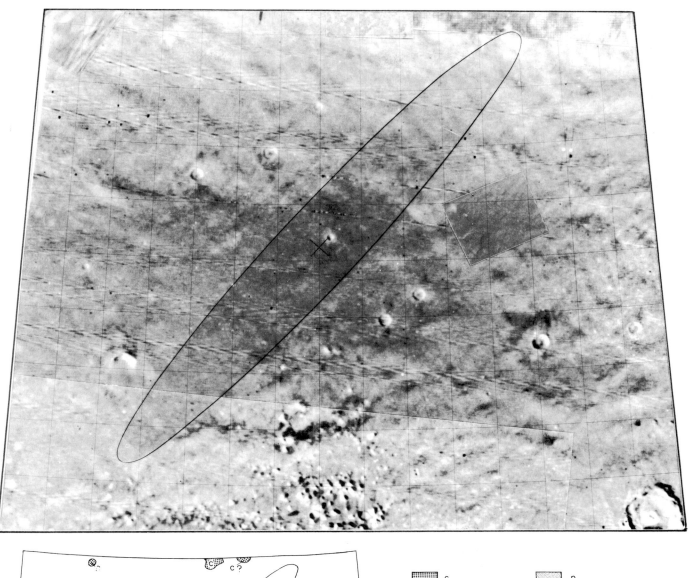

(a)

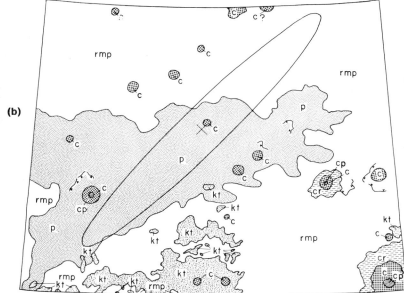

(b)

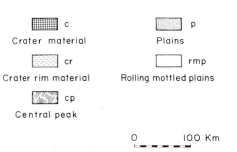

	c		p
	Crater material		Plains
	cr		rmp
	Crater rim material		Rolling mottled plains
	cp		
	Central peak		

kt

Knobby terrain

0   100 Km

Figure 2.21 (a). Photomosaic of the Cydonia site, proposed for the second Viking landing.
 (b). Geologic map of the Cydonia site. Adapted from map by D. H. Scott (1972).

dilemmas was to effect a balance between engineering safety and scientific interest. Mars has a hostile environment with surface materials that have a treacherously low density indicated by radar measurements. But the interpretation of the radar measurements is uncertain. How much weight should they be given? Will the spacecraft really sink out of sight upon landing? And we should consider Martian winds. Are some regions more storm-crossed than others? The list of engineering hazards goes on and on. So overpowering do these constraints become that some scientists forsook their special-interest arguments and contended that we must land at whatever spot looks safest. In something like the atmosphere of a New England town meeting, heated discussions were followed by a series of votes. Particularly in the area of biology, dissonant opinions were heard. Where on the planet does one have the best opportunity to detect life? The polar caps are regions where substantial amounts of H_2O ice may be present; elsewhere ice may exist as diurnal frost. Can ice perform the same life-supporting functions as water? Does it locally melt to form transient pockets of water in the soil? In the vicinity of the 45°N latitudinal belt, there are the largest amounts of atmospheric water. But what necessary connection does atmospheric water have with surface water available for use by organisms? Perhaps the ancient river beds in the equatorial region provide the best site for water, or maybe it occurs close to the fresh-looking volcanoes. But in both cases, how recent was the activity? If it occurred millions of years ago, as well may be the case, then there is little reason to expect water to be present now.

Very few of the issues are resolvable at the present time. There simply are not enough data. The best one can do is to seek a consensus—that is, vote and abide by the majority decision. But the arguments of the past few years are certain to be rekindled when Mars is reached in 1976. Before the lander is released from orbit, the landing site must be certified. Swaths of contiguous pictures will be taken that completely cover the landing ellipse with stereoscopic perspective. Topographic maps compiled from pictures will confirm the smoothness of the terrain. Comparison of pictures taken on successive days will establish any variable features caused by shifting sediments, driven by high surface winds. Simultaneously the IR spectrometer will measure the atmospheric pressure and amount of atmospheric water.

If the primary site looks unacceptable, then the lander will be directed to the predetermined backup site. What if that, too, looks unacceptable? What, indeed? At the very least one can predict frenzied activity and confusion.

Viking Mission Operations

Once the landers are on the surface of Mars, the scientists and engineers directing the mission—a group in excess of 700 people—will be preoccupied with the task of directing the four spacecraft (landers and orbiters) to carry out their scientific tasks for the next several months. As in the case of site selection and certification, a nominal plan is already in existence. Specific days within the mission have been set aside for acquisition and analysis of three biology soil samples, three organic analysis samples, and six inorganic analysis samples (fig. 2.22). Before the acquisition of each sample, time has been set aside for taking pictures that will reveal the characteristics of the sample area, leading to a decision as to exactly where the sample should be collected.

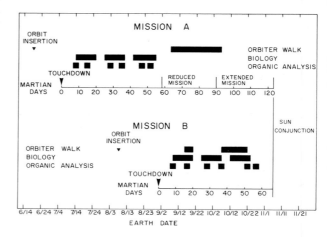

Figure 2.22. A proposed sequence for scientific analyses during the Viking mission.

After the landing, the nominal sequence will have to be updated, possibly on a daily basis. This procedure is distinctly more difficult than for lunar missions, where the commands, transmitted over a distance of 384,000 km, are essentially real time (actually less than 2 seconds). For a radio signal to travel from Earth to Mars takes about 20 minutes during the period of Viking mission operations. If the Viking surface sampler is directed to go to a particular location, scoop up a sample, and return it to the biology hopper for analysis, there is no way of observing the sequence of events in time to modify them. The problem is further compounded by the fact that the up-link or command-link from Earth to Mars can be used for only a few hours each day, and most commands must be verified by a return read-out to Earth before they are executed. So, even though Viking is an adaptive mission, the adaptive plans must be thought out days in advance.

Clearly, the first few days on the surface will be the most critical. The same high winds that apparently silenced the 1971 Soviet lander might plague the Viking spacecraft, even though telescope observations indicate that summer in the northern hemisphere is not a period of dust-storm activity. Damage caused by wind-blown dust is a related danger. Wind-tunnel tests under Martian conditions demonstrate that the protective windows on the cameras are frosted by sandblasting, that the bearings that allow the camera to turn in azimuth are frozen by sifting dust, and that the outer housing of the camera can be badly abraded. If the surface winds blow continuously at 100 m/sec, there is little likelihood that the spacecraft will survive one or two days, let alone several months. However, if meteorological measurements indicate some sort of diurnal or seasonal cycle, it may be possible to devise an experimental strategy that reduces interference from periodic storms.

Advanced Missions to Mars

After the Viking '75 mission is completed, there are, at present, no plans for additional exploration. There are a number of choices but no clear favorite. In order of increasing ambitiousness—and cost—we could send smaller spacecraft to perform limited experiments, repeat the Viking mission, return a sample of Martian soil to Earth, or explore Mars with a manned spacecraft.

Pioneer Missions

Small spacecraft of the Pioneer class have proved useful in exploration of Venus, and some scientists argue that they could be modified easily for a Martian mission. The spacecraft are small, weighing only one-third as much as Mariner 9. Accordingly they can be launched with smaller rockets at less cost. Once at Mars, a number of options exist. The capsules could be used as orbiters, entry probes, or landers. Within the weight limitations one or more landers or entry probes might be released from a single orbiting vehicle. An orbiter would be particularly attractive for monitoring atmospheric or surficial conditions throughout an entire Martian year. The usefulness of an entry probe is more problematical, if only because it makes a single suite of measurements over a brief period of time before crashing on the surface. The aeroshell phase of the Viking '75 mission will provide two atmospheric profiles, and a repetition, even with upgraded instruments, might be difficult to justify. Most controversial are the hard landers. One variety, encased in a protective shell, is designed to survive a parachute descent.

A second, shaped like a bullet and termed a penetrator, actually knifes through the Martian soil. The small payloads of penetrators and the rapid deceleration during "landing" present severe restraints for development of useful scientific investigations.

The chief advantage of a Pioneer mission is that it is inexpensive. But for that financial benefit one pays a handsome scientific price. The weight available for scientific instruments is small, and the power to operate them limited. In the case of the lander, instruments have to be extremely simple, able to survive high shock levels, and capable of returning information within a few days. Of course, none of these constraints is inflexible. The scientific payload can be increased, the power generator enlarged, and the communication system made more sophisticated. But, as soon as that is done, the mission loses its chief attractiveness. It is no longer inexpensive.

Pioneer missions would be compellingly attractive if we could think of particular measurements, limited in character, that hold the key for an understanding of Martian history. For example, we might assume that adoption of an all-encompassing model for geologic evolution hinges only on two determinations of chemical composition, one in the ancient cratered terrain of the southern hemisphere and one in the volcanic plains of the northern hemisphere. Simple X-ray fluorescence or alpha back-scatter instruments could be put aboard two penetrators, accompanied perhaps by seismometers and temperature-measuring devices. One penetrator might be directed to the uplifted crust northwest of the Hellas basin, the second to the central caldera of Olympus Mons. For about 100 million dollars the experimenters would get two chemical analyses and some information on thermal and seismic properties of the interior. Then they'd have to set about constructing that persuasive model promised some months earlier. Our experience on the moon suggests that they'd have trouble. Before the Apollo 11 mission there was widespread anticipation of simple answers to complicated questions. The moon was frequently referred to as the Rosetta stone of the solar system, the implication being that words of genesis would be writ large and clear on the returned rocks. It was not to be. The early sweeping syntheses, based on a few measurements, were also the last syntheses. With more and more complexity revealed by each mission, most investigators began retreating into the familiar confines of their own disciplines. To the dismay of scientist and layman alike, the moon gives its secrets grudgingly. There is no reason to think that Mars will be any different.

Advanced Viking Missions

Now that Viking spacecraft have already been designed and built for the 1975 mission, it would seem logical to use the same model for additional missions. Indeed, enough spare parts will be left over after the 1975 launches to assemble at least one additional vehicle. Logical though it may be, a second Viking mission has its weak points. The first involves money. To fly another Viking mission would cost anywhere from one-fourth to one-third of the original effort. Even though that is a substantial reduction, it still represents a cost of 200 million dollars or more. So large is the cost that almost all other planetary missions would have to take a back seat. Under those conditions a repeat Viking must have strong scientific justification.

If exploration of Mars is to be an orderly, progressive affair, each mission must stand on the shoulders of the preceding mission. Accordingly, it would be

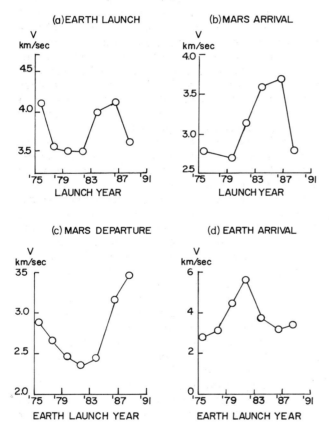

Figure 2.23. The energy requirements for traveling to Mars and returning to Earth during the next 15 years. Velocity requirements dictate the size of rocket and amount of fuel. High velocities require a large amount of thrust, either for acceleration during launch or for braking during rendezvous. Adapted from Pritchard and Harrison (1973).

desirable to have Viking '75 results in hand before specifying the instruments to be carried on the next Viking mission. This argument is particularly persuasive for the biologic investigations. A launch in 1977 would be out of the question since little more than a year would be available for the time-consuming job of instrument modification. Looking further into the future, the relative positions of Earth and Mars in the mid-1980s are so unfavorable that it would be difficult for a Titan/Centaur rocket to deliver the Viking spacecraft to Mars (fig. 2.23). By default, then, the best time for a repeat performance is 1979 or 1981.

In mentioning the desirability of responding to Viking '75 results, we have opened Pandora's box. Even without those results available, one can think of many improvements: new instruments to measure mineral composition, ground water, magnetic fields and gravity fields; spacecraft modifications to permit a polar landing; a roving vehicle to visit selected localities around the landing point. Each improvement has vocal scientific support. But, if they were all implemented, they would more than double the cost of the original "carbon-copy" mission. The financial problems would become overwhelming.

It is probable that very limited plans for a Viking '79 mission will proceed in the next several years, but that a final decision will not be made until after the Viking '75 landings. If the results of that mission are unusual and exciting, then the next Viking will move forward on a groundswell of enthusiasm. If the results are bland and unimpressive, then no scientist, engineer, or NASA administrator will be able to make a case for repeating the exercise, with or without variations.

Returned Sample Missions

There are excellent scientific reasons for bringing a sample back to Earth on an unmanned spacecraft (Young and DeVincenzi, 1974). Combined visual, chemical, and mineralogical analyses will yield results greatly exceeding anything possible through remote examination. Especially valuable is the opportunity to replace a single bulk analysis with thousands of grain-by-grain analyses. Identification of a single exotic grain may document a process that would be completely hidden by the masking effects of bulk analysis. Our lunar experience provides ample supporting evidence. Hundred-gram samples acquired by unmanned Soviet spacecraft under conditions similar to those anticipated for a Mars mission provided, in miniature, much of the same information acquired during the much more extensive Apollo sampling program.

A returned sample mission presents intriguing operational possibilities. At present the spacecraft would have to be launched from Earth, but in the future we may have an Earth-orbiting space station that will act as a launching platform. From this rapidly moving launch station a relatively large spacecraft could be placed on a trajectory headed to Mars. Once in the vicinity of the planet, the spacecraft could either descend directly to the Martian surface or go into orbit and, from that intermediate position, deploy a lander to the Martian surface.

The return to Earth is considerably more complicated. The lander could be fired into Martian orbit or could rendezvous—Apollo-like—with a vehicle already in orbit. In either case the spaceship would have to circle Mars for 300 or more days before Earth would be in a position to be reached with the limited fuel available for the return journey. The length of the total mission, then, is almost three years: one year going out, one year in residence, and one year coming home.

The most important choices occur once the spacecraft approaches Earth, carrying its cargo of Martian soil. The same two choices exist as before: direct descent or rendezvous in orbit. A direct descent is simple but potentially disastrous. If the spacecraft crashes instead of landing gently, the soil cannister could break open, spreading any Martian biota far and wide with potentially disastrous results. A rendezvous in orbit is less objectionable but the finite, if small, possibility that the maneuver will go awry and the spacecraft will spin in toward Earth alarms many critics.

If successful orbital insertion is achieved, the next step will be to transfer the sample from the Mars craft to a permanent space station. The operation is made difficult by the fact that the vehicles will be in different orbits. A third auxiliary vehicle, sometimes referred to as a tug, will have to be launched from the shuttle, rendezvous with the Mars craft, and then bring the soil sample back to the shuttle. From this point on the sample could be handled in a variety of ways. It could be heat sterilized; indeed, this could be automatically done much earlier, either while the return vehicle was in orbit about Mars or while it was on the journey back to Earth. The advantage of delaying the sterilization is in preserving opportunity for preliminary study of any live organisms in the space laboratory, a station sufficiently far from Earth so that it might be considered safe. The sterilization temperature controls the amount of potential scientific data destroyed. At 200°C organic compounds will be partially volatilized but not completely destroyed. At 400°C almost all organic compounds will be decomposed but major chemistry of

inorganic minerals will be largely unaffected. Some volatilization of loosely bound water will occur, a certain amount of diffusion will take place, and certain mineral structures such as carbonate and sulfate will be broken down. Sample thermoluminescence will be severely degraded, and fossil cosmic ray and fission tracks will be annealed.

A palatable compromise might be to divide the sample into two parts and to sterilize only that fraction that will be transferred to Earth. The biologically active remainder of the sample could remain in orbit until it was ascertained that no risk was involved in bringing it down to Earth.

Manned Missions to Mars

To the surprise of many laymen who ask the question, NASA has no plans for any manned mission to Mars. To launch a capsule toward Mars capable of serving as a temporary home for two years or more would require rockets more powerful than any existing today. Living in space for such a long period of time is possible, but not a demonstrated reality. If astronauts ever do set out for Mars, they probably will use an Earth-orbiting space station as a staging platform. In the 1980s NASA looks forward to orbiting such a satellite that will be outfitted to assemble and launch planetary probes. The obvious advantage is that the orbital velocity of the launch platform will supply a major factor of the escape velocity, thereby reducing the need for large rockets.

The journey to Mars will closely resemble an Apollo mission, except for its length. Following departure from Earth, the spacecraft will require about half a year to reach Mars. After going into orbit around the planet, a landing module will be deployed. Because of unpredictable hazards on the surface, the landing party will probably stay no more than a few days—at most several weeks. Then they will rendezvous with the orbiter and wait for a year or more until Earth is favorably positioned for a "short" half-year return trip.

That we have the technological ability to assemble, launch, and recover a Mars ship is certain. That we will eventually do it is almost equally certain. That we will do it within the next 25 years is extremely uncertain. The expense will be so great that scientific justification will never be more than incidental. The journey will be undertaken only if there is a national commitment similar to that sparked by President Kennedy's pledge to put a man on the moon by 1970.

3.

Physiographic Provinces

Introduction

Now THAT Mariner 9 has provided virtually complete photographic coverage of Mars with ground resolution of 1–3 km and about one percent coverage with 100-m resolution, it is apparent that the planet is not the moon-like body suggested by Mariner 6–7 pictures. In this chapter we will briefly catalogue the various features seen on Mars and the physiographic provinces that can be delineated. Our intent is to provide an overview and frame of reference for more detailed discussions to follow. Many of the larger craters, canyons, volcanoes, and other distinctive features have been assigned names by the International Astronomical Union (fig. 3.1). To facilitate discussion, these names will be used in addition to the nomenclature established by telescopic observation.

The physiographic map presented here (fig. 3.2) has been compiled exclusively from photographic data. It does not benefit from "ground truth" or other corroborating information. It is physiographic in the sense that none of the provinces is distinguished on any basis other than surface morphology and topography. The names, however, are drawn from geologic interpretation where such interpretation is reasonably secure. For example, certain conical mountains are termed volcanoes. Deposits at the south pole of Mars are almost certainly layered, so the name of these landforms reflects that geologic fact.

In general, the planet is divided into two morphologically distinct hemispheres, plains in the north and cratered terrain in the south (figs. 3.2, 3.3). The approximate boundary between the two hemispheres is a great circle inclined 35° to the equator. Scattered throughout these broad regions are particular features produced both by endogenic and exogenic processes. Most of the volcanic structures are in the plains. Much of the chaotic terrain has been formed by modification of cratered terrain.

There is no detailed correlation between telescopically determined albedo markings and physiographic provinces (Cutts et al., 1971; Frey, 1974) although, in general, plains regions appear bright. Similarly, there is no detailed correlation between albedo and topography (Inge and Baum, 1973), although the cratered terrain of the southern hemisphere generally stands higher than the plains of the northern hemisphere.

Ages

It is difficult to identify from orbital photographs the processes that are modifying a planetary surface. It is doubly difficult to place those processes—and

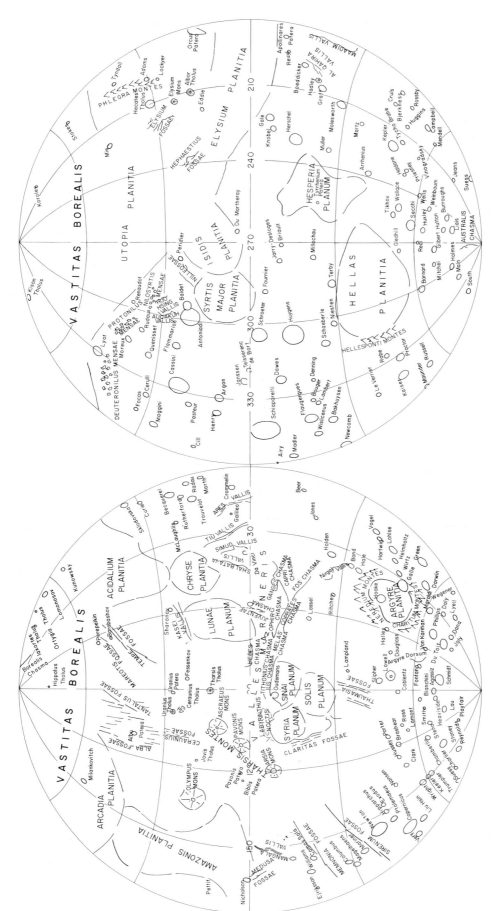

Figure 3.1. Major landforms on Mars. The reference point for the coordinate system is a small crater, Airy-zero, within the crater Airy. Names for features have been established by the International Astronomical Union. *Catena* is a crater chain (most of volcano-tectonic origin). *Chasma* is a large canyon (most of structural origin). *Dorsum* is an elongate prominence. *Fossa* is a linear depression (most similar to terrestrial graben). *Labyrinthus* is a set of intersecting linear depressions. *Mensa* is a flat-topped mesa. *Mons* is a large isolated mountain (most are of volcanic origin). *Patera* is an irregular or complex crater with scalloped edges (most are volcanic calderas). *Planitia* is a plain that is low relative surrounding terrain. *Planum* is a relatively smooth, uncratered plateau. *Tholus* is an isolated domical small mountain or hill (most are steep-sided volcanoes). *Vallis* is a sinuous valley (most are of probable fluvial origin). *Vastitas* is an extensive plain.

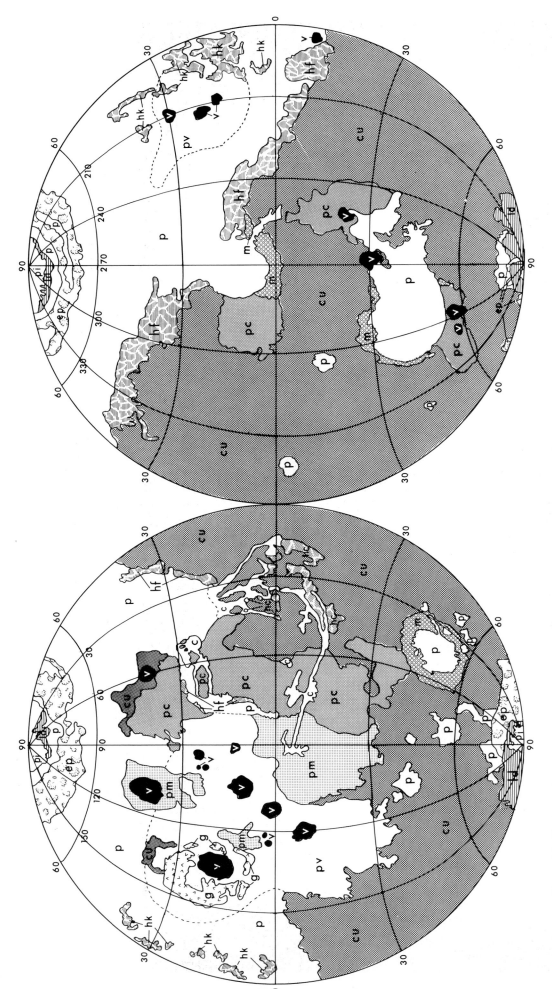

Figure 3.2. Physiographic provinces of Mars. Map units are adapted from Carr et al. (1973); Murray et al. (1972); and Soderblom et al. (1973a); and are plotted on a Lambert equal-area base. *Polar units* include *pi* (permanent ice), *ld* (layered deposits), and *ep* (etched plains). *Volcanic units* include *v* (volcanic constructs), *pv* (volcanic plains), *pm* (moderately cratered plains), and *pc* (cratered plains). *Modified units* include *hc* (hummocky terrain, chaotic), *hf* (hummocky terrain, fretted), *hk* (hummocky terrain, knobby), *c* (channel deposits), and *g* (grooved terrain). *Ancient units* include *cu* (cratered terrain, undivided) and *m* (mountainous terrain). See text for discussion of individual units. See p. 341 for extended legend.

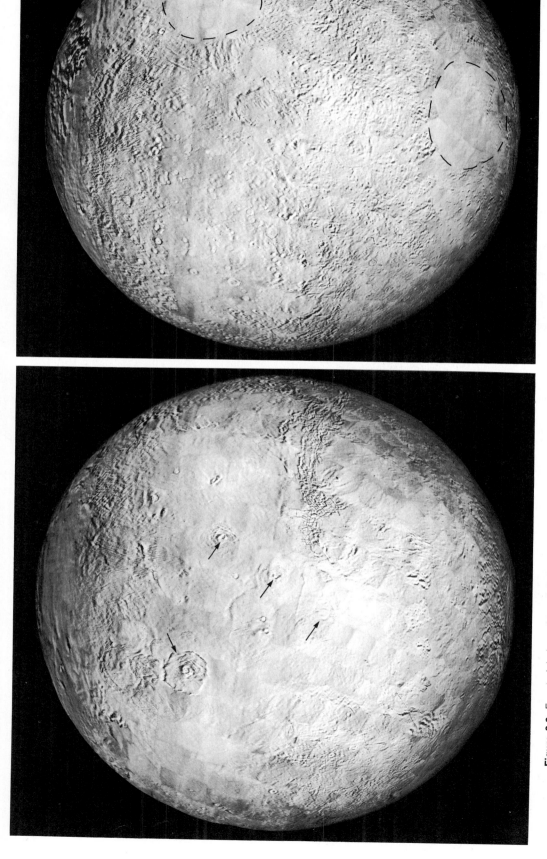

Figure 3.3. Equatorial view of two hemispheres of Mars photomosaic globe. The left view is centered at 120°W longitude. A variety of massive volcanic features occur in this hemisphere. Olympus Mons is seen in the upper left quadrant (arrow); a line of three volcanic shields constituting the Montes Tharsis is in the center (arrows). At center right is the valley network of Labyrinthus Noctis, leading into Valles Marineris. The right view is centered at 300°W longitude. This hemisphere is dominated by ancient craters and impact basins. In the upper right is the basin occupied by the Isidis Planitia (dashed lines). The classical bright feature Hellas is seen at the bottom (dashed lines). Hellas is the largest impact basin on Mars. Both views of the planet are from a distance equivalent to several planetary radii. Photograph of JPL photomosaic globe.

their attendant material products—into a temporal sequence.

At the outset we should make some semantic distinctions. When we speak of the age of a terrain unit, or physiographic province, we refer to the age of that process that imparted a distinctive topographic aspect to the unit. This is not necessarily the same age as that of the materials that appear at the surface. A terrestrial example will illustrate the point. The Appalachian Mountains of the eastern United States are underlain by sedimentary rocks that were deposited in linear troughs during the Paleozoic era, some 200–600 million years ago. Volcanic activity occurred intermittently during the same period. The rocks were deeply buried, metamorphosed, folded and faulted by compressional forces. Subsequently, the region has been uplifted, dissected by streams, and scalloped by glacial activity. Given this history, how does one specify the "age" of the Appalachian Mountains? In a topographic sense they are relatively young. However, the rocks that underlie the present mountains are much older. Sedimentary rocks are dated according to the time of their deposition, igneous rocks according to the time of their crystallization from a melt, or magma. Structural events are intermediate in age, younger than the depositional events but more ancient than the topographic events. The extent of this complicated Appalachian history that can be determined from examination of aerial photographs depends on the fortuitous presence of certain diagnostic relationships revealed in landforms.

A few general stratigraphic-structural relationships observed on Earth have application to remote analyses of other planets. In some instances one unit can be seen to overlie an older unit. This relationship of superposition provides a foundation for the discipline of stratigraphy, the interpretation of sediments and their contained fossils by evaluating sequences of rocks that become progressively older with depth. The simplest situation occurs when horizontally stratified rocks are exposed along steep cliffs. The Grand Canyon, cutting through Paleozoic rocks of the Colorado Plateau, is a frequently mentioned example. Some sedimentary layers are less resistant to erosion than others. As a result, the canyon walls consist of alternating cliffs and benches, which can be detected in aerial photographs.

Where younger deposits blanket an older irregular terrain, preferentially accumulating in depressions, the depositional age can be correlated with topography. The younger deposits occur at lower elevations and form smooth surfaces. In some instances the younger deposits partly cover older structures.

Structural elements—faults and folds—obviously are younger than the rocks that are deformed. In some instances it is possible to unravel cross-cutting relationships and thus determine a sequence of structural events. For example, the trace of an ancient fault will be offset where it is crossed by a younger fault.

In almost all instances a topographic form is younger than the material units that appear at the surface. However, there are exceptions. In some cases reworking of surficial materials to produce a topographic form is so pervasive that it can also be called a depositional event. Obvious examples are large-scale landslides. In other instances, the surficial materials can actually be younger than the topographic forms. Clearly this is the case if one counts soil as a sedimentary unit. A less controversial example would be a volcanic ash that mantles a rough terrain to a depth of a few centimeters without affecting the subjacent topography.

Relationships just described serve to establish *relative ages*, a rank ordering of a series of events. An alternate chronology involves *absolute ages*, generally measured as years before the present and, in the case of Earth and the moon, often based on radiometric determinations. Lunar and Martian surfaces are sometimes assigned absolute ages based on the measure of time they have been exposed to impacting meteoroids. If the impact rate has been constant, then the crater density is directly proportional to age, discounting problems of saturation, simultaneous erosion, and crater formation by processes other than impact. This technique, even though beset by numerous assumptions and observational uncertainties, has proved extremely useful in dating planetary surfaces. Relative ages can be established by comparing crater densities. Since it makes no difference if the regions under study are adjacent or widely separated, planet-wide correlations are possible. As just noted, the technique offers the additional bonus of providing absolute ages. In contrast, superposition, transection, and cross-cutting relationships indicate only relative ages—and this only for contiguous units.

Intertwined depositional, structural, and erosional events are obviously difficult to disentangle. On Earth the problem is made easier because we can roam the surface, drill to considerable depths, and analyze rocks in the laboratory. For other planets, where most of our information is photographic and where we have never directly sampled the surficial materials, the task is exceedingly difficult. The most useful data are topographic, but only rarely do they yield separate sets of sedimentary, structural, and erosional data. For this reason construction of geologic or stratigraphic maps (as distinct from topographic or terrain maps) almost always involves subjective assumptions.

Global Topography

A topographic map of Mars has been constructed using, primarily, Earth-based radar observations and Mariner 9 radio occultation measurements (fig. 3.4). These have been supplemented by infrared and ultraviolet measurements. Details of the map compilation are further discussed in chapter 6.

It is apparent that the topographic figure of the planet departs significantly from a spheroid. Differences in elevation of more than 30 km exist. Prior to the radar and spacecraft measurements it had been variously postulated that the broad albedo markings were correlated with elevation. The earliest observers assumed that the dark areas were analogous to those on the moon and thus low (Tombaugh, 1966). Later it was argued, with some force, that the bright regions were low relative to the dark areas (Sagan and Pollack, 1968). With the availability of radar measurements the arguments became complicated, since no apparent correlation could be discerned between elevation and albedo. In fact, there was some slight evidence that the dark regions occurred on slopes (Pettengill et al., 1969). This conclusion was based on a small and unrepresentative sample. With present data in hand, it is seen that there is no simple correlation between telescopic image and topographic map (Inge and Baum, 1973). Some of the highest as well as the lowest regions appear bright, and probably for different reasons.

The major region of high elevation on Mars is approximately coincident with a bright area classically named Tharsis. The major features are three volcanoes, the Montes Tharsis. We will refer to this broad region as the Tharsis plateau. It is a crustal bulge rising about 6 km in elevation. Straddling the equator at 110°W, Tharsis extends for 4000 km from north to south and 3000 km from east to west. By comparison with Earth, its proportions are continental. Another distinctively elevated region is a 1500-km diameter dome at 210°W, 25°N, in Elysium Planitia. This broad plateau rises 4–5 km above the plains at its base. Both of these regions, Tharsis and Elysium, are characterized by major volcanoes and lava plains.

An elevated region about 2000 km wide and 4 km high is located at 90°W, 40°S. It is here called the Thaumasia plateau. In contrast to the other plateau regions, Thaumasia is characterized by intense fracturing rather than by volcanic features. There is also geophysical evidence that the Thaumasia plateau is not simply an extension of the Tharsis plateau. The Thaumasia region is largely in isostatic balance, while Tharsis is not (Phillips and Saunders, 1975). The significance of this observation is discussed in chapter 6.

There are several topographic basins on Mars. The low regions at Hellas, Argyre, and Isidis have rim structures and other features suggestive of impact origin. Other basin-like features occur in Chryse and Amazonis. These are adjacent to the Tharsis ridge and may in some way be dynamically related to Tharsis. The low region at Chryse lies along a major topographic low, the Chryse trough, that extends for more than 5000 km from the north rim of the Argyre basin to the plains of Mare Acidalium. This trough, in common with the Tharsis plateau, appears to be isostatically uncompensated (Phillips and Saunders, 1975).

Volcanic Units

Volcanic Plains

The volcanic plains (pv on the physiographic map) are widespread, sparsely cratered surfaces in the Tharsis and Elysium regions. They occur in the same regions where volcanic shields and domes are present. The plains are almost certainly surfaces produced by volcanic flooding. The chief evidence of volcanic origin are lobate scarps inferred to be lava flow fronts (fig. 3.5). The unit has been delineated on the physiographic map solely on the basis of B frames. Plains beyond the margin of the volcanic plains do not contain lobate scarps, although at A-frame resolution they are commonly indistinguishable from volcanic plains. Since B-frame coverage is not continuous or of uniform resolution, there is uncertainty in the exact boundary.

Almost all craters on volcanic plains, and on other plains units, have a crisp bowl shape. At A-frame resolution, craters exhibit little or no degradation, but at B-frame resolution, a few craters show infilling and erosional modification.

The volcanic plains of Mars resemble in some aspects the mare regions of the moon. The lunar maria contain lobate scarps and generally have been interpreted to be extensive volcanic plains formed from low-viscosity basaltic lava. This photographic interpretation has been verified by results of the Apollo program. Such low-order questions as volcanic versus something else for the lunar maria have been replaced by much more sophisticated questions concerning the geochemical evolution of the mare basalts.

There are, however, some important differences between the Martian volcanic plains and the lunar maria. The maria occupy topographic basins whereas, on Mars, volcanic materials are relatively elevated. Volcanoes on the moon are rare and of ambiguous morphology; volcanic plains on Mars are associated with constructs so distinctive that their interpretation as volcanic is virtually unopposed.

Figure 3.4. Global topographic map of Mars, based on spacecraft occultation data, Earth-based radar observations, and Mariner 9 ultraviolet and infrared spectral measurements. Contours are in kilometers from approximate 6.1 millibar pressure surface. The vertical error may exceed 1 km in regions north of 20°N latitude. Adapted from Christensen (1975).

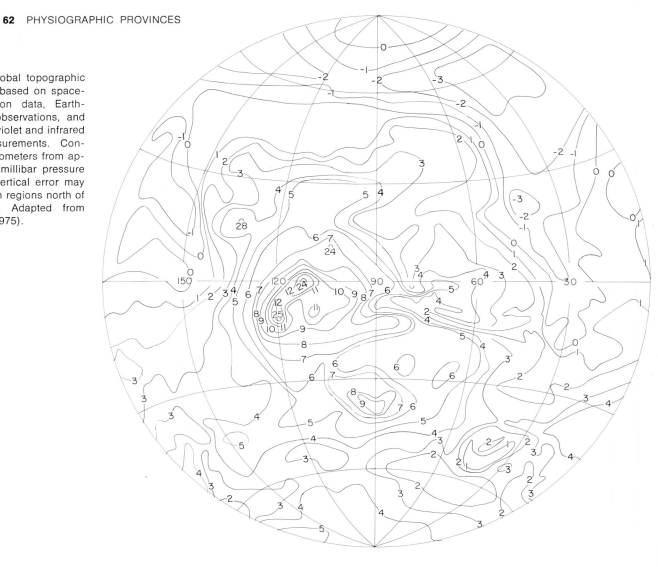

Figure 3.5. Volcanic flow fronts in Tharsis volcanic province. Regions of volcanic plains on physiographic province map (fig. 3.2) typically display these lobate flow fronts resembling those of the lunar maria and certain volcanic regions of Earth. As on Earth and the moon, these regions of extensive flows are inferred to be of basaltic composition. Some areas of plains are among the least cratered, and therefore youngest, regions of Mars. (DAS No. 6966613, rev. 150, B camera, center at 136°W, 17°S.)

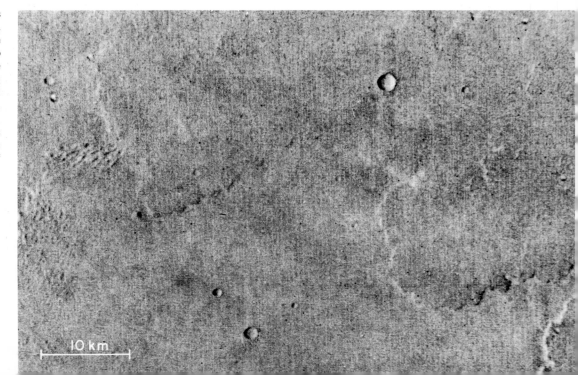

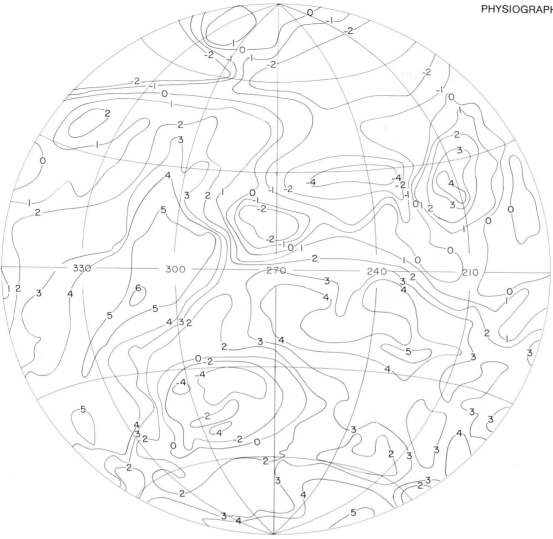

Lunar maria apparently were formed by volcanic infilling of ancient basins. These basins may have been nearly in isostatic equilibrium when the volcanism occurred. The result was a super-isostatic load—the additional mass of the lavas is supported by the strength of the crust (Phillips et al., 1972). This combination of processes results in slightly higher gravitational attraction, termed positive free-air gravity anomalies (mascons) over lunar basins. The Martian volcanic plains may also provide a super-isostatic load on the crust, but they are more analogous to great continental flood basalts on Earth, which occur at relatively high elevations. This will be discussed in more detail in chapter 6.

Lunar maria are dark relative to the surrounding heavily cratered highlands. The Martian situation is approximately reversed. Plains, both volcanic and undivided, generally correspond to telescopically bright regions. The low albedo of lunar maria is due primarily to the fact that they are composed of basalt fragments, small iron- and titanium-rich glass droplets, and agglutinates (Conel and Nash, 1970). These form by impact, brecciation and melting of correspondingly iron- and titanium-enriched basalts. Apparently, on Mars, weathering and depositional processes account for a veneer of highly reflecting sediment that overlies darker basalt. Many undivided plains areas (p) may also be underlain by volcanic rocks, the only distinction between these plains and the volcanic plains being in the thickness of the sediment blanket or in the extent of erosional modification.

Volcanic Mountains

Perhaps the most provocative and instructive class of features revealed in Mariner 9 pictures are the volcanic constructional forms (v), including shields, domes, and craters (figs. 3.6, 3.7, 3.8). The shields

Figure 3.6. Tharsis volcanic province. In the upper left is Olympus Mons, the largest known volcano in the solar system. The summit towers at least 21 km above its base (Davies, 1974) and is some 30 km higher than the floor of the Hellas basin. Three other major shield volcanoes, indicated by arrows, lie along a line from the lower center of the picture to the upper right. Photograph of a part of the JPL photomosaic globe centered at 120°W, 10°N.

range in diameter up to 600 km, massive in comparison with similar forms on Earth. These structures are extensively discussed in chapter 5.

The largest Martian volcanoes have been called shields (Carr, 1973). The best known shield volcano on Earth is Mauna Loa on the island of Hawaii. By observing eruptions over the past 200 years and by making field observations in areas where the shield has been cut by erosion, geologists have concluded that Mauna Loa and similar shield volcanoes have been built by numerous eruptions of highly fluid basaltic lavas. Eruptions occur from vents near the summit or on the flanks, and lavas spread out in long tongues or in thin sheets that flow for great distances. Many hundreds of eruptions over the past

half-million years have combined to produce the present topography of Mauna Loa. Its summit caldera, like that of other terrestrial volcanic shields, forms by collapse following removal of magma from below.

The striking feature of Martian shield volcanoes and their associated calderas is their enormous size. Olympus Mons is 600 km across and rises 21 km above its base (Davies, 1974). The central caldera is 70 km in diameter. By comparison, the island of Hawaii, which is made up of several shield volcanoes, has a diameter of 200 km and a height of 9 km above the sea floor. Its total volume represents less than 10 percent of the volume of Olympus Mons.

Figure 3.7. Olympus Mons, a massive scarp-bounded volcanic shield nearly 600 km across and having an average height of at least 21 km above its base. Note the region of plains adjacent to the scarp and the annulus of fractured ridges that make up the grooved terrain (fig. 3.2). Oblique photograph of JPL photomosaic globe.

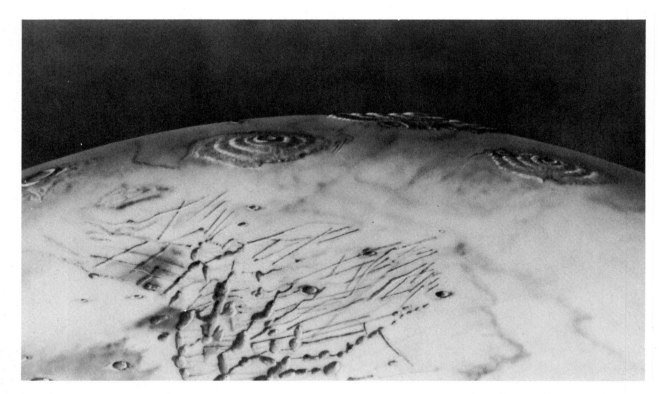

Figure 3.8. Artistic reconstrucion of Mars as viewed from several hundred kilometers above the surface, looking toward the northwest. Olympus Mons is on the horizon. In the middleground, Arsia Mons, Pavonis Mons, and Ascraeus Mons are visible. Labyrinthus Noctis is in the foreground. Artwork by R. Jennings.

Martian volcanoes are generally confined to three broad regions: Tharsis, Elysium, and Hellas. The Tharsis volcanoes appear the freshest, the Hellas structures are the most degraded. In many of the presumably older volcanoes, the conical form that characterizes Olympus Mons is replaced by a radial pattern of channels and ridges (fig. 3.9). These features have no known analog on Earth or the moon, and may represent an important and previously unknown volcanic process.

Cratered Plains

In addition to the volcanic plains previously discussed, two additional plains surfaces can be objectively distinguished by their crater populations. These units are presumed to be of volcanic origin because they are planar deposits that blanket ancient cratered terrain in regions adjacent to obviously volcanic terrains. However, similar relationships are seen on the moon for dark plains (maria) and light plains of very different origins, so caution must be exercised in this interpretation.

The moderately cratered plains (pm) are more cratered than the volcanic plains (fig. 3.10). They occur in the vicinity of the telescopic feature Syria,

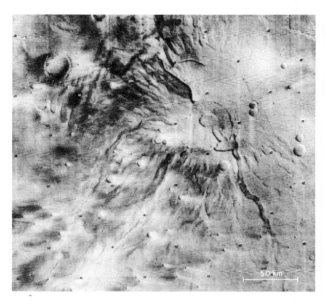

Figure 3.9. Tyrrhenum Patera is a radial volcano typical of several such ancient features. A broad channel, 5 km wide, can be followed from the central caldera to the south through a second caldera and continuing for nearly 200 km. (DAS No. 8909224, rev. 204, A camera, center at 254°W, 21°S.)

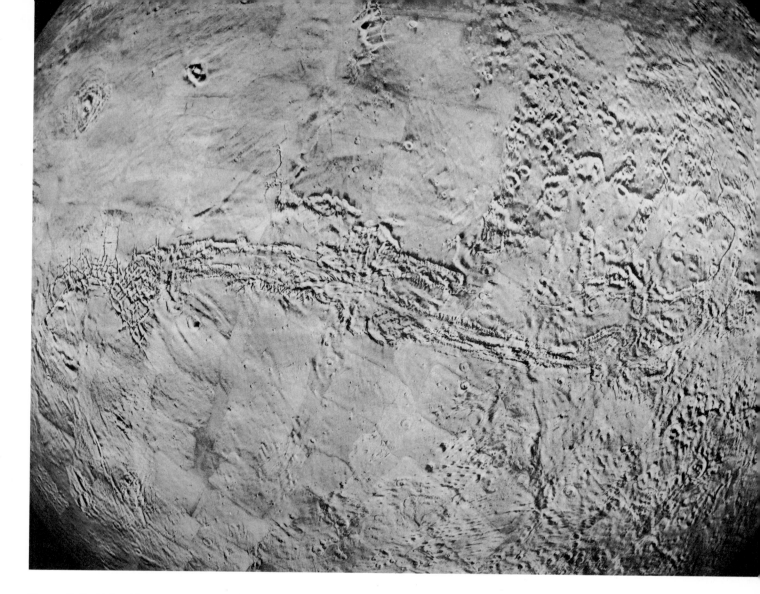

Figure 3.10. Valles Marineris. This great Martian chasm extends, in various forms, more than a quarter of the way around the globe, nearly 6000 km. In the west it terminates in intricately patterned fractures of Labyrinthus Noctis. The central part is a more regular system of deep, wide chasms. In the east, the chasms give way to broad riverlike valleys that meander northward to the Chryse basin. Volcanic plains occur in the upper left, moderately cratered plains in the lower left, densely cratered plains in the center top and bottom, and cratered terrain to the right (fig. 3.2). Photograph of JPL photomosaic globe.

between the volcanic plains and more densely cratered plains (pc). The moderately cratered plains do not appear to have lobate flow fronts as do the volcanic plains. In the region around 105°W, 15°S, irregular dark markings occur at the surface (fig. 3.11). Many form closed contours. B frames show irregular scarps. It is possible that the markings are produced by eolian deposits concentrated along irregularities in the surface. The irregularities may be caused by fracturing that has not broken up the surface as much as the tectonic activity to the north that formed Labyrinthus Noctis.

Densely cratered plains (pc) occur in four regions. The most extensive exposure is along the 60°W meridian, including Lunae Planum, Sinai Planum, and Solis Planum. A second region, termed Syrtis Major Planitia, is generally coincident with the prominently dark telescopic marking, Syrtis Major. The last two regions are to the northeast and southwest of the Hellas basin.

In addition to having crater densities higher than all other plains units, densely cratered plains are characterized by relatively smooth intercrater areas. A few degraded craters may be modified remnants

Figure 3.11. Irregular polygonal markings in cratered plains in Syria Planum. The proximity of fractured ancient terrain suggests that the markings may be irregular fractures along which eolian sediment has collected. (DAS No. 7254628, rev. 158, A camera, center at 101°W, 18°S.)

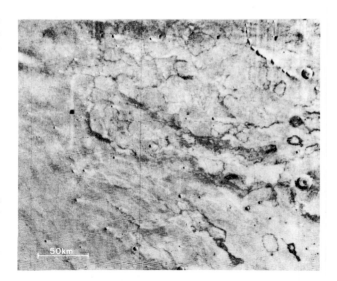

Figure 3.12. Cratered plains in the west part of Sinai Planum. This region contains clear evidence that intense fracturing of the crust preceded the emplacement of the cratered plains. In the lower right is an elevated region of fractured crust not covered by the plains. Observation of this relationship increases the likelihood that some large craters in cratered plains belong to an earlier pre-plains population. (DAS No. 7398618, rev. 162, A camera, center at 81°W, 14°S.)

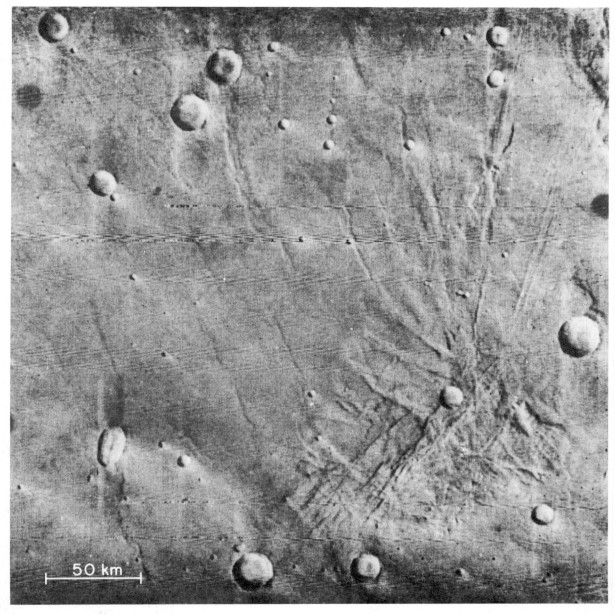

of an older surface that has been blanketed by smooth material (fig. 3.12). The densely cratered plains are also characterized by long sinuous ridges, morphologically similar to those on the lunar maria (fig. 3.13). In both cases the ridges may indicate emplacement of volcanic rocks or deformation uniquely related to volcanic activity. The arguments are further developed in chapter 5.

Many Martian craters have associated bright and dark streaks extending many kilometers, presumably formed by erosion and deposition in the downwind direction. The best-developed streak clusters occur on cratered plains (fig. 3.14). There are a number of contributing reasons. A regionally flat surface broken by circular ramparts of bowl craters, dark basaltic bedrock, and a thin veneer of fine-grained sediment may create an environment favorable for formation and identification of streaks.

Grooved Terrain

Grooved terrain (g), although classified as a modified unit, is discussed with the plains and volcanic features because of its association with Olympus Mons. It occurs in large arcuate patches of rugged terrain around the volcanic shield (fig. 3.7). The inner zone, nearest the shield, is characterized by linear mountains 1–5 km wide and typically 100 km long (fig. 3.15). An outer zone has finer surface texture with approximately equidimensional blocks whose dimensions decrease until they are indistinguishable from similar protuberances in the surrounding plains. Geometric relationships suggest

that plains, in part, cover the grooved terrain at the margins. Additional evidence is provided by graben that cross the grooved terrain but not the smooth plains.

The age relations are demonstrated in figure 3.16. The oldest surface is a plains unit crossed by several sets of fractures. The grooved terrain is inferred to be younger than the older fractured plains. Superposed on both units are the sparsely cratered volcanic plains.

Ancient Units

Cratered Terrain

The most widespread physiographic unit is undivided cratered terrain (cu). The surface of more than half of the planet is nearly covered with large craters and multi-ring basins (fig. 3.17). Crater morphologies and distributions are discussed at length in chapter 4.

It is likely that the cratered terrain is very ancient, much like the lunar highlands, the craters being a remnant of the final stage of planetary accretion. If true, then most were formed at a time when the rate of bombardment was much higher than at present. Implicit in this interpretation is the assumption that the majority of craters of the cratered terrain were formed by impact and that the impact rate has been approximately the same as the lunar rate.

As first noted during the Mariner 4 mission and confirmed by examination of Mariner 6–7 pictures,

Figure 3.13 (a). Ridges in cratered plains. Individual ridges are 4 to 5 km wide and hundreds of kilometers long. The morphology is similar to that observed on the lunar maria, suggesting the possibility that these features on the two planets may have a similar origin. Note the *en echelon* frac-

tures, indicated by arrows, near the right center of the picture. (DAS No. 7470508, rev. 164, A camera, center at 74°W, 18°S.)

(b). Part of ridge in cratered plains. (DAS No. 7542363, rev. 166, B camera, center at 66°W, 22°S.)

(b)

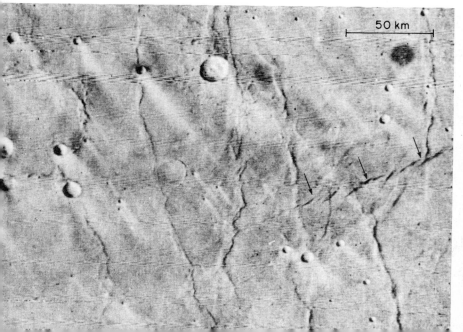

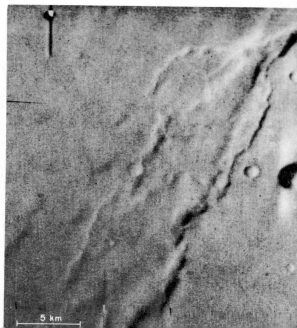

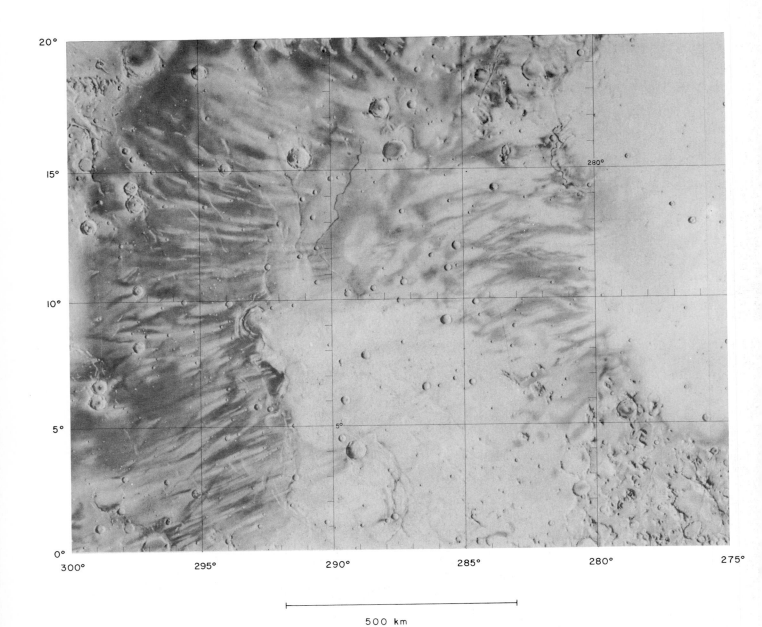

Figure 3.14. Bright and dark streaks on Syrtis Major Planitia. This features corresponds almost exactly with the telescopically observable dark, triangular region, Syrtis Major. Most streaks are associated with craters. Dark and bright streaks trend in different directions, indicating changing wind patterns. This is an airbrush map based on photomosaics. It was prepared by the U.S. Geological Survey as part of a NASA-supported mapping program.

(a)

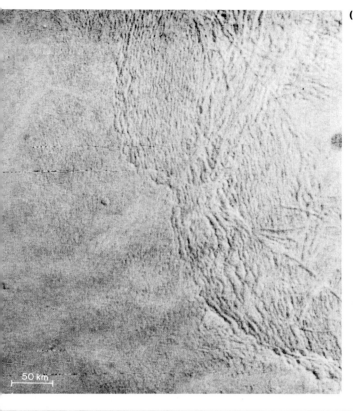

50 km

Figure 3.15 (a). Grooved terrain west of Olympus Mons. This is a rugged mountainous region of linear ridges 5 km across and 50 km or more in length. The region of prominent ridges is fringed by a zone of smaller ridges. (DAS No. 6751818, rev. 144, A camera, center at 148°W, 23°N.)

(b). Wind-sculpted region of grooved terrain. Ridges are about 1 km wide and 5 to 20 km long. (DAS No. 6823393, rev. 146, B camera, center at 142°W, 14°N.)

(b)

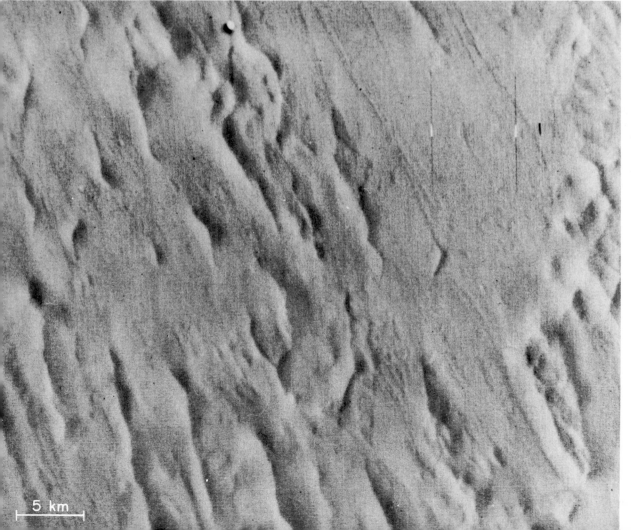

5 km

Figure 3.16. Region east of Olympus Mons illustrating age relations among surface materials. Oldest are the fractured plains near the center (arrow). Youngest are the smooth plains in the central and northern regions. Grooved terrain is present in the upper left. (DAS No. 6967418, rev. 150, A camera, center at 124°W, 17°N.)

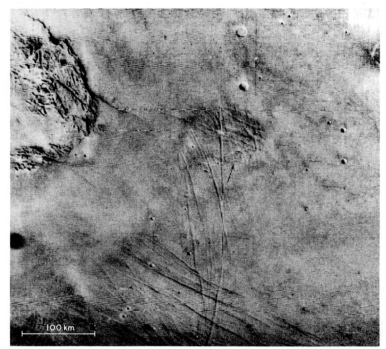

Figure 3.17. Cratered terrain. Ancient terrain between the equator and about 30° south latitude and between 310° and 345° west longitude. Cluster of craters in lower left resembles famous Disneyland character. Photograph of JPL photomosaic globe.

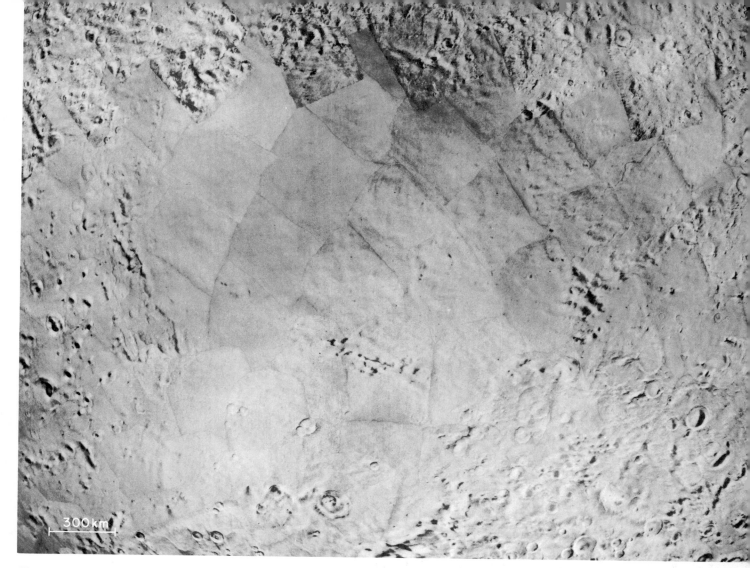

Figure 3.18. The Hellas basin. One of the earliest recognized and best known albedo features, this bright-floored basin records a colossal impact event. The ancient basin, now highly degraded, is approximately 1600 km in diameter and 4 to 6 km deep. Photograph of JPL photomosaic globe.

almost all large Martian craters are shallow, with flat floors. Ejecta deposits are not as prominent as for fresh lunar craters. Notably absent are the bright rays that characterize lunar craters such as Tycho, Copernicus, and Aristarchus. These several observations indicate that Martian craters have been much more modified than their lunar equivalents. Ejecta deposits have been either blanketed or stripped, and crater interiors have acted as traps for eolian sediment.

Flat intercrater surfaces are more prominent on Mars than in the most densely cratered lunar terrains. Wilhelms (1974) attributes the formation of these flat plains to a depositional event, possibly volcanic, that occurred in the first billion years of Martian history when an initially high meteoroid flux was declining.

Basins and Mountainous Terrain

Three large multi-ring basins and approximately twenty smaller two-ring basins have been recognized on Mariner 9 photographs. In some respects these features resemble the large lunar basins such as the Imbrium basin. They have broad, relatively smooth floors and rugged raised rims. As suggested by the adjective "multi-ring," surrounding mountains form a series of concentric ranges. The three major basins contain named plains—for example, the Argyre Planitia. We will refer to the entire structure, including the mountainous rim, as a basin. The interior plains are a more recent deposit. On the moon a similar distinction exists between basins and mare deposits.

The Argyre basin is centered at 43°W, 50°S. The

interior plains are approximately 900 km in diameter. The mountainous rim deposits (m) have a diameter of about 1400 km. Radial graben-like structures cut the rim of this basin in several places. The interior plains appear to be little cratered, but, as with Hellas, this may be because the pictures of this basin were taken early in the mission before low-lying regions were clear of the obscuring dust in the atmosphere.

The Hellas basin, centered at 293°W, 42°S, has an incomplete rim (fig. 3.18). The floor of this elongate basin extends some 2000 km in an east-west direction and 1600 km north-south. The rim is irregular in width, ranging from 50 to 400 km.

The third large basin, Isidis, lies on the east flank of Syrtis Major (fig. 3.19). The plains-filled floor is approximately 1100 km in diameter, open to the northeast. It is flanked by a 300 km wide section of rugged rim preserved along the southern margin (fig. 3.20). On the northwest and southwest there are concentric graben.

These basins, by analogy with those on the moon, are interpreted to be the degraded remnants of ancient impact basins. The mountainous terrain unit is interpreted to be the remnants of uplifted rim material and basin ejecta. The proximity of the mountains to the basins suggests that most of the primary morphologic characteristics have been destroyed, leaving only the more resistant blocks of uplifted crust.

Figure 3.19. Isidis basin. An 1100-km-diameter basin at 270°W, 10°N, this is the third largest basin on Mars. Except for the subdued remnants on the southern margin, evidence of basin rim deposits has been destroyed by erosion. Arcuate troughs northwest and southeast of basin are controlled by basin structure, although they formed relatively recently. Clusters of both bright and dark streaks occur on Syrtis Major Planitia, a region of cratered plains to the west of Isidis Planitia. Photograph of JPL photomosaic globe.

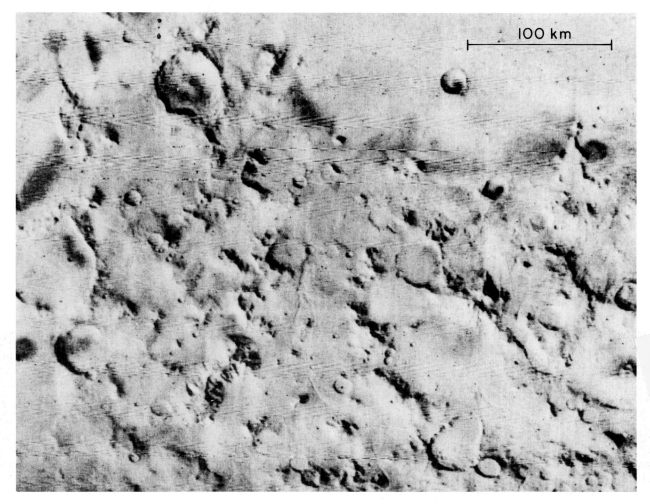

Figure 3.20. Mountainous terrain along the southern margin of the Isidis basin. Distinctive, rugged topography such as this is indicative of impact origin of the basin. (DAS No. 7219208, rev. 157, A camera, center at 278°W, 3°N.)

Modified Units

Canyons

A major rift system stretches for some 5000 km across Mars from approximately 20°W to 100°W and between the equator and 15°S. The extent of this system is comparable to that of the great rift valleys of East Africa, although the rifts of Mars are wider and deeper than those on Earth. Graben on Earth are generally on the order of 50 km wide and perhaps as much as 2 km deep. Those on Mars surpass these dimensions severalfold (figs. 3.10, 3.21, 3.22).

The character of the canyons of Mars varies considerably with longitude. The central part includes huge straight rifts that almost certainly result from faults. To the west, the rifts grade into a broad zone of branching troughs in the vicinity of 90°W (fig. 3.23). Early in the Mariner 9 mission this maze of interconnecting valleys was named the Chandelier; later, it acquired the formal name Labyrinthus Noctis. The troughs appear to be graben that form a pattern similar to fractures in the central part of a domical uplift (Beloussov, 1962).

At their eastern end the large, centrally located rifts merge with an extensive region of chaotic terrain (fig. 3.24) as they branch and curve to the north. Several meandering valleys thread their way through this region of irregular depressions, emptying onto the plains of Chryse. About a dozen similarly sinuous channels—termed valles—are located elsewhere in the cratered terrain close to the boundary with the northern plains (figs. 3.25, 3.26). A few stubby tributary channels can be seen entering the main channels. Some of the valleys have superposed channels that impart a braided pattern. Teardrop-

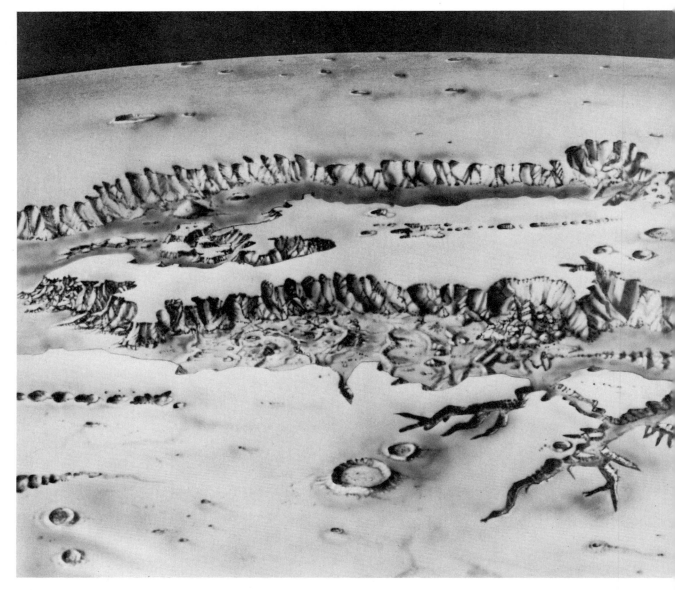

Figure 3.21. Artistic reconstruction of Mars as viewed from several hundred kilometers above the surface, looking toward the north. The two prominent canyons are Melas Chasma and Ophir Chasma. The Sinai Plains are in the foreground, Lunae Plains in the background. Artwork by K. A. Travis.

shaped islands appear within channels. These features, along with a positional correlation between valleys and hummocky terrain, which possibly formed by permafrost melting, suggest that the valleys were formed by flowing water. The arguments are further developed in chapter 8.

The floors of the major canyons (c) are generally smooth at A- and B-frame resolution, as are the channels that flow out from chaotic terrain. In many places it appears that the rough chaotic deposits are covered with smoother canyon floor materials. These may be alluvial deposits.

Hummocky Terrain

It has been previously noted that Mars is divided into a low northern hemisphere and a high southern hemisphere. Separating these hemispheres is a distinct topographic break (fig. 3.4). This slope can be traced for two-thirds of the way around the planet. It is obscured by more recent volcanic and tectonic structures in the region of the Tharsis ridge. Within the hummocky terrain there is abundant evidence that cratered terrain has been disrupted and destroyed. The zone of this disruption along most of the scarp is the hummocky terrain.

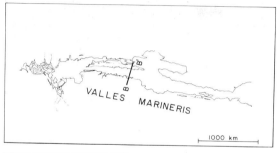

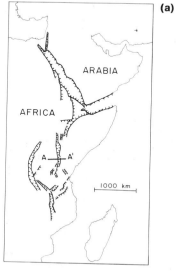

(a)

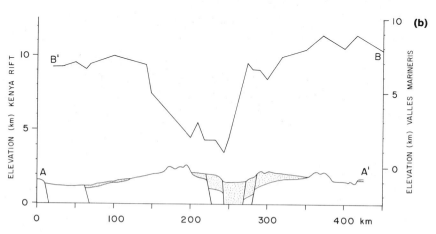

Figure 3.22 (a). Comparison of the rift system of East Africa (after Baker et al., 1972) and the Valles Marineris system. Positions of topographic cross sections shown in (b) are indicated.

(b). Topographic section across Kenya rift valley at about 36°E, 1.5°S (after Baker et al., 1972). The profile at the top shows the topography across Melas Chasma in the Valles Marineris (after Hord et al., 1972).

Figure 3.23 (a). West margin of the Valles Marineris. The complex of broad rifts in the right center is Labyrinthus Noctis. The volcanic plains of the Montes Tharsis region are in the upper left. The large calderas are Arsia Mons, the southernmost volcano, and Pavonis Mons, the central volcano of the three Montes Tharsis. The two small calderas in the extreme upper left are Biblis Patera and Pavonis Patera. The closely spaced fractures just left and below center are the Claritas Fossae, an elevated region of intensely disrupted cratered terrain. Below and to the right of the Claritas Fossae are the Thaumasia Fossae, a similar region of up-lifted and fractured ancient crust. Immediately below the Labyrinthus Noctis is a region of irregular dark markings shown in figure 3.11. The entire region pictured here is the most elevated region of the Tharsis plateau, pervasively fractured, and—for the most part—blanketed with volcanic deposits. Photograph of JPL photomosaic globe.

(b). Detail of Labyrinthus Noctis. A series of older north-trending graben are intersected by the broad reticulate forms of the Labyrinthus. The broad graben are approximately 10 km wide. (DAS No. 7182808, rev. 156, A camera, center at 105°W, 6°S.)

(a)

(b)

Figure 3.24 (a). Part of Simud Vallis. Breakup of the cratered terrain surface within the well-defined scarp-bounded regions has resulted in the formation of chaotic terrain in parts of this broad channel. The narrow smooth channel in the center of the picture connects two broader regions of chaotic terrain. The smooth-floored channel is nearly 20 km wide. (DAS No. 7686808, rev. 170, A camera, center at 37°W, 3°N.)

　(b). Chaotic terrain of Eos region. This region of chaotic terrain lies along the southern margin of one of the numerous chasms that form the eastern end of the Valles Marineris. Chaotic terrain here has formed by the breakup of cratered terrain. (DAS No. 7758208, rev. 172, A camera, center at 32°W, 11°S.)

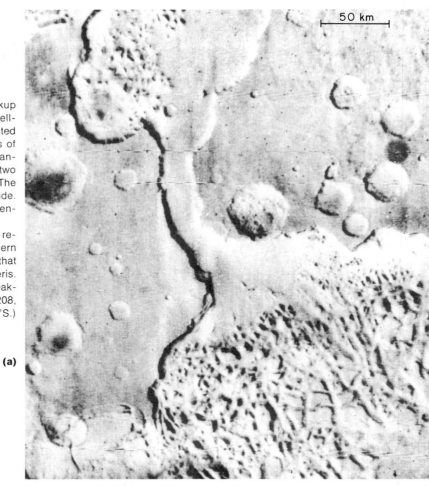

(a)

(b)

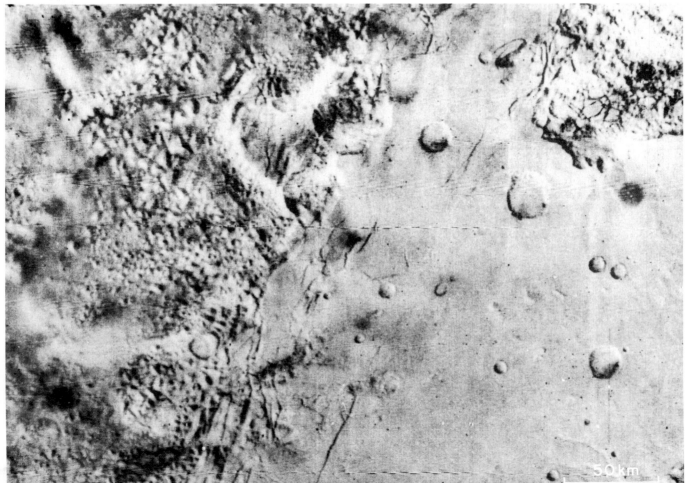

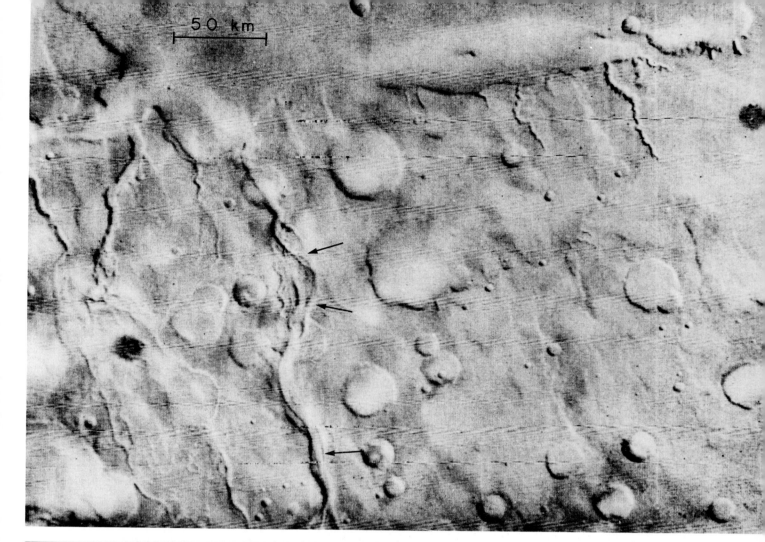

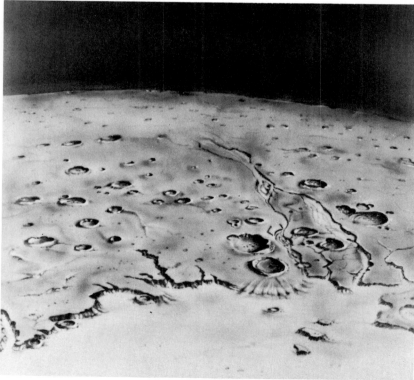

Figure 3.25. The sinuous valley near the center is Mangala Vallis. The valleys slope northward to the scarp and into lineated plains (a part of the undivided plains). The valleys terminate at the plains. It is possible that the valleys were once longer and that scarp recession has shortened them. Parts of the Mangala Vallis, shown by arrows, have a braided appearance. There is a suggestion of meander successions with progressive erosion of the outside of the channel bend and deposition on the inside. (DAS No. 6822798, rev. 146, A camera, center at 149°W, 6°S.)

Figure 3.26. Artistic reconstruction of Mars as viewed from several hundred kilometers above the surface, looking toward the south. The plains of Amazonis are in the foreground, cratered terrain in the background. The two regions are separated by a prominent escarpment. Mangala Vallis cuts through the cratered terrain. Artwork by K. A. Travis.

The hummocky deposits have been divided into chaotic terrain, fretted terrain (Sharp, 1973a), and knobby terrain (Carr et al., 1973). The chaotic terrain (hc) is associated with and adjacent to the major canyons. Fretted terrain (hf) occurs at the contact between cratered terrain and plains. The two units hf and hc are considered here to be gradational one into the other and to have similar origins. Knobby terrain (hk) may also be related, but the morphologic and geologic affinities are less clear.

The chaotic terrain (Sharp et al., 1971a) occurs as scarp-bounded areas of irregular, blocky topography resembling parts of the fretted terrain. In all instances the chaotic deposits are in locally low regions, occasionally in completely closed basins (fig. 3.24).

Fracture patterns on the margins of chaotic ter-

rain suggest that this unit is produced by mass wasting of material into low lying areas. The irregular outlines of the chaotic terrain may reflect the pervasively impact-fractured nature of the cratered terrain from which it has formed. It has been suggested that slumping occurs along a pre-existing set of fractures (Wilson et al., 1973). Sinuous channels drain some areas of chaotic terrain, suggesting fluid flow from these regions. As just mentioned, one possible source of fluid would be water released by melting of extensive subsurface ice deposits.

Fretted terrain occurs between the cratered terrain and the northern plains, along a great circle having a pole at approximately 145°W, 55°N. The zone is more than 500 km wide and extends halfway around the planet. The material of the cratered terrain has been eroded, creating scarp-bounded massifs with smooth regions between (figs. 3.27, 3.28, 3.29). There are numerous smooth-floored sinuous channels that originate in cratered terrain and flow onto the plains.

Fretted terrain is characterized by irregular blocks of all sizes, some as large as several hundred kilometers and others extending down to the limit of

Figure 3.27. Fretted terrain between Utopia Planitia and the cratered terrain. This complex region of channels and mesas has apparently formed by breakup and erosion of cratered terrain, as in the regions of chaotic terrain. Sinuous channels, as in other regions, suggest flow of liquid water at some time in the past. Regional slopes are northward. Photograph of JPL photomosaic globe.

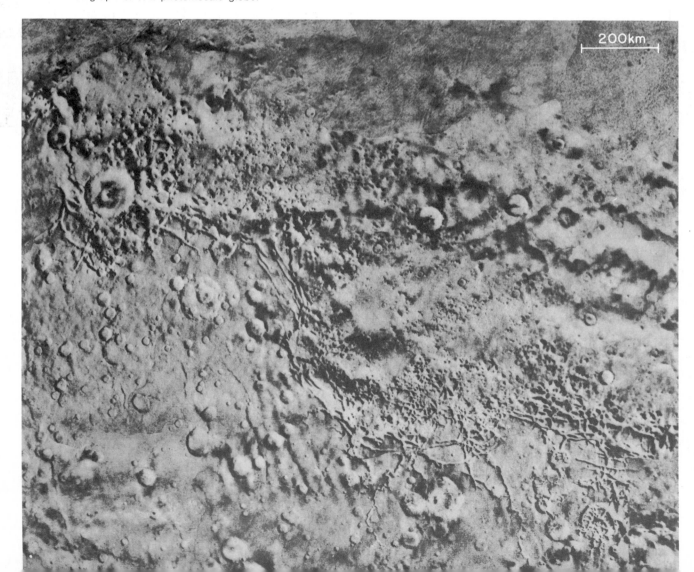

(a)

(b)

Figure 3.28 (a). Detail of fretted terrain. The large central-peak crater, Moreux, is approximately 130 km in diameter. (DAS No. 819-1164, rev. 184, A camera, center at 313°W, 43°N.)

(b). Domical hills in northern plains. Occurring further out on the plains than the mesa-like forms of the fretted terrain, these domical hills probably form by erosion of larger blocks. The larger hills are 6 to 8 km across. (DAS No. 8335049, rev. 188, B camera, center at 298°W, 38°N.)

Figure 3.29. Artistic reconstruction of Mars as viewed from several hundred kilometers above the surface, looking toward the south. Plains of Vastitas Borealis are in the foreground, cratered terrain in the background. Protonilus Mensae appear in the intermediate erosional zone. The large crater to the right of center is Moreux. Compare with figure 3.28. Artwork by R. Jennings.

A-frame resolution (1 km) at the outer fringes. In general the individual blocks are larger and have more linear boundaries and planar surfaces nearer the cratered terrain. Within the fretted terrain there are irregular depressions filled with jumbled blocks. The details of the scene are identical with those of chaotic terrain. Apparently the same processes of erosion are at work in both regions. The surface distribution of the blocks suggests that the unit is produced by breakup of crustal material on a massive scale and probably to great depth, as evidenced by the complete disappearance of the northern rim of the Isidis basin near 265°W, 20°N.

Knobby terrain occurs as isolated patches of small knobs on the undivided northern plains, between the Tharsis and Elysium volcanic provinces. The protuberances resemble distal knobs of the fretted terrain (fig. 3.30). This unit apparently results from the breakup of the ancient crust. Much of the northern undivided plains contain isolated blocks with debris aprons around them. Some erosion mecha-

nism, either mechanical or chemical, has reduced these blocks to finer debris. The hummocky-knobby terrains are regions where the destruction of crustal blocks and the complementary formation of plains have not been completed.

Furrowed Terrain

Superposed on the ancient cratered terrain in the equatorial region of Mars are numerous sinuous furrows (fig. 3.31). If the north pole were located at 110°W, 75°N, then the band would follow the 15°S parallel of the rotated coordinate system (fig. 3.32). Individual furrows typically are from 2 to 10 km wide and 50 to 100 km long, comparable in size to many terrestrial river valleys. They commonly occur in roughly parallel aggregates, imparting a corrugated texture to the terrain. Some furrows radiate from large craters. Others head along ridge lines and are traced down the facing slopes. Primarily because they appear to originate in topographically high regions and show no obvious association with

Figure 3.30. Knobby terrain. Numerous small clusters of hills occur in this region, classically known as Erebus. Individual hills are up to 10 km across. The knobby terrain is similar in appearance to the region of small hills between the fretted terrain and the plains, and probably forms by breakup of older crust. Bright streaks against a dark background are visible in the lower part of the picture. (DAS No. 7867128, rev. 175, A camera, center at 186°W, 24°N.)

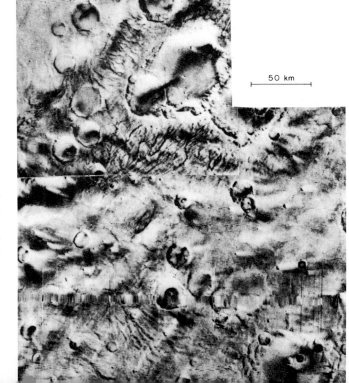

Figure 3.31. Furrowed terrain. Numerous small furrows occur on the cratered terrain in the equatorial regions. Furrows here are about 2-3 km wide and up to 30 km long. (DAS No. 8261584, rev. 186, A camera, center at 329°W, 6°S. DAS No. 8261514, rev. 186, A camera, center at 331°W, 11°S.)

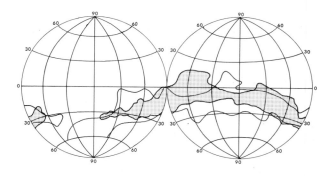

Figure 3.32. Shaded regions comprise furrowed terrain. Heavy lines indicate the edge of the telescopically dark equatorial belt with geometric albedo less than 0.18 (from de Vaucouleurs, 1967). Furrows are preferentially located in telescopically dark areas.

Figure 3.33. Surface irregularities, largely of erosional origin, in undivided plains (p), which are featureless at A-frame resolution.

(a). Eroded knobs and irregular grooves in the region west of the Phlegra Montes. Knobs are being modified by erosion or have been modified in the past. The largest knobs here are approximately 2.5 km across. A sinuous depression, shown by arrows, is approximately 2 km across and quite shallow. (DAS No. 7867303, rev. 175, B camera, center at 184°W, 43°N.)

(b). Eroded knobs with well-developed debris aprons, in the region north of Chryse Planitia. The largest knob is nearly 4 km long. (DAS No. 9018599, rev. 207, B camera, center at 35°W, 31°N.)

(c). Fractured surface in undivided plains. This region of plains appears to be irregularly fractured into blocks approximately 10 km across. (DAS No. 12188412, rev. 445, B camera, center at 23°W, 48°N.)

(d). Fractured surface in undivided plains. Bright ejecta around crater overlies the fractures. (DAS No. 11799626, rev. 430, B camera, center at 263°W, 51°N.)

(a)

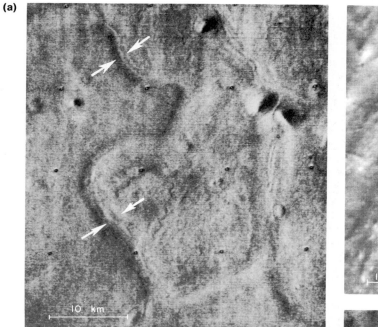

(c)

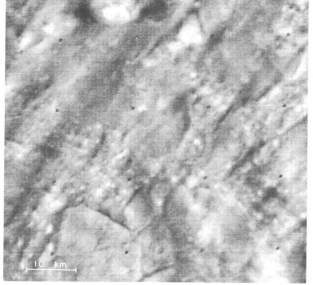

(b)

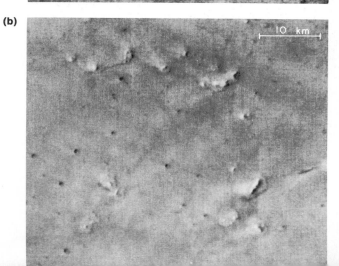

(d)

volcanic or permafrost features, some investigators feel they provide the strongest available evidence for erosion by rainfall at some former time in Martian history.

Undivided Plains

A large tract in the northern hemisphere along with the floors of the Hellas and Argyre basins are shown on the physiographic map as undivided plains (p). This province forms the bulk of the northern lowlands. B frames show numerous features suggesting modification by tectonism and erosion. These include small knobs, scarps, and furrows (fig. 3.33).

In many areas the surface has an etched and scoured appearance.

The floor of Hellas appears smooth in many pictures, so much so that it was termed "featureless terrain" during the Mariner 6–7 mission. However, late in the Mariner 9 mission, pictures taken in a region of the basin that had previously appeared featureless revealed rugged scarps (fig. 3.34). It now appears that much of the time the floor of Hellas is obscured by a haze of dust. By extension, the featureless character of any low-lying region of "smooth plains" is suspect.

Figure 3.34. Hellas basin floor. This region in the northwest part of the basin was photographed late in the Mariner 9 mission. The apparent lack of features in earlier pictures of the same region was due to atmospheric dust that remained from the great storm that obscured this region early in the mission. (DAS No. 11797001, rev. 430, A camera, center at 305°W, 37°S.)

Polar Units

The polar regions of Mars have revealed some of the most astonishing geological puzzles on the planet. The first inklings of the puzzle were suggested by Mariner 7. Strange frost-brightened "quasi-linear" features thought to be the upturned cuesta-like edges of plates were seen near the south pole (Sharp et al., 1971b). Elsewhere around the pole were irregular pits etched into the surface. Little could be learned from these pictures since much of the region was still covered with a thin veneer of CO_2 frost.

Mariner 9 coverage of the south pole was extravagantly good. Since the south polar region was traversed by every revolution of the spacecraft as it neared periapsis, and since the bright patch of permanent frost was about the only feature clearly seen through the dust for the first month of the mission, there is an almost embarrassing repetition of south polar views. The entire region has been imaged at 3–4 km resolution (fig. 3.35) with an extremely

Figure 3.35. View of the south polar region. Center at 90°W, 65°S. Photograph of JPL photomosaic globe.

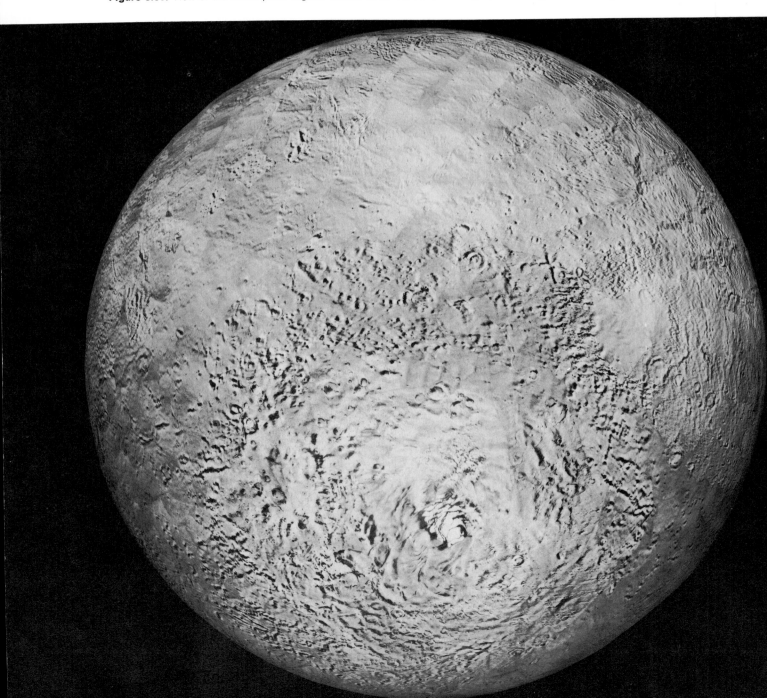

generous 20 percent covered by high-resolution B frames at 300–400-m resolution.

The north pole was photographed at a greater distance from the planet and much later in the mission, after the polar hood had cleared (fig. 3.36). Few pictures were shot since the extended mission was limited by lack of attitude control gas.

One of the most intriguing discoveries is that both poles have the quasi-linear, swirl-like features, compelling evidence that they are produced by conditions unique to the poles (fig. 3.37). Deposits of both polar regions display irregular etch pits, again suggesting some process unique to the polar regions.

Abundant pictures notwithstanding, controversy continues regarding the topographic character of polar terrains and the inferred geologic sequence of erosion and deposition. It has been particularly difficult to see through the dominant albedo patterns and distinguish the more subtle topography. For example, the dark quasi-linear markings suggest a swirling complex of valleys bounded by steep slopes. This is consistent with the presence of descending air over

Figure 3.36. North polar region of Mars, photographed late in the Mariner 9 mission after the clearing of the north pole hood. Picture center at 270°W, 65°N. Photograph of JPL photomosaic globe.

(a)

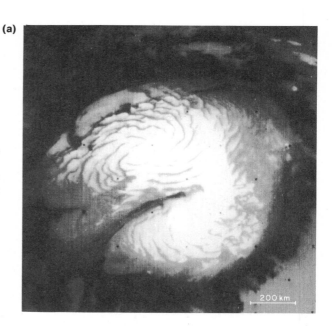

Figure 3.37 (a). Residual cap at the north pole of Mars. Shown near its minimum extent, the northern permanent cap is almost 1000 km in diameter, much larger than the permanent cap at the south pole. The spiraling dark lines are frost-free cuesta-like slopes similar to those observed at the south pole. (DAS No. 13317550, rev. 667, A camera, center at 90°N.)

(b). Stereographic projections of four A-frame pictures that show the retreat and stabilization of the south polar cap. (DAS No. 2568620, rev. 27, A camera, center at 28°W, 86°S. DAS No. 3071645, rev. 41, A camera, center at 40°W, 85°S. DAS No. 4508115, rev. 81, A camera, center at 17°W, 83°S. DAS NO. 8331864, rev. 188, A camera, center at 149°W, 89°S.)

(b)

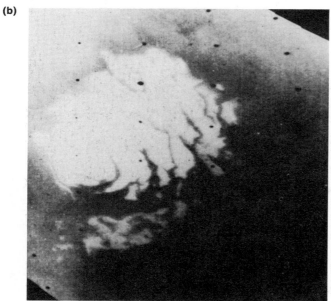
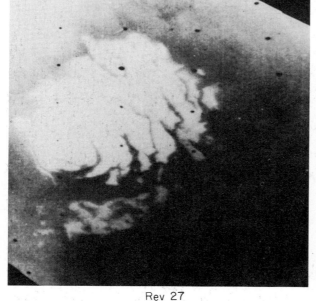

Rev 27

Rev 41

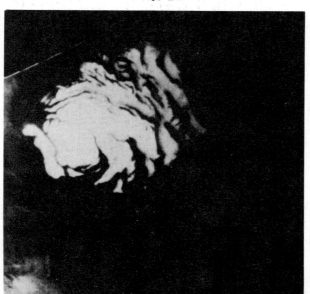

Rev 81

500 km

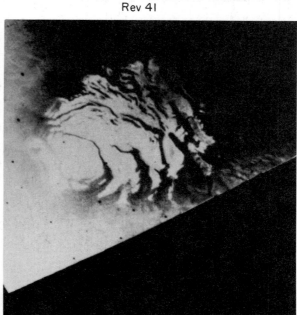

Rev 188

the poles, driven equatorward while under the influence of the Coriolis force. However, many of the dark linear markings, attributed to eolian scour, have extremely low slopes and do not resemble wind-scarred troughs. The model has yet another anomaly. The well-developed swirl at the north pole has a deflection opposite to that predicted from a Coriolis effect.

Layered Deposits

The nature of some quasi-linear features was revealed by the high-resolution B frames of the south

polar region. A cliff and bench topography has been produced by erosion of a large number of layered deposits (fig. 3.38). Individual layers are on the order of 100 m thick (fig. 3.39).

Two models for the depositional geometry of the layered deposits have been proposed. Murray et al. (1972) argue that each curvilinear marking is the edge of a separate plate made up of many individual layers. Cutts (1973a, also Nash, 1974) suggests that a single unit has been dissected by subparallel valleys that impart a pinwheel pattern on a regional scale.

Figure 3.38. A remnant of layered deposits in the upper right overlies irregularly dissected etched plains. The irregular sculpting of the etched plains may be a consequence of their homogeneous structure. (DAS No. 9231189, rev. 213, B camera, center at 85°W, 82°S.)

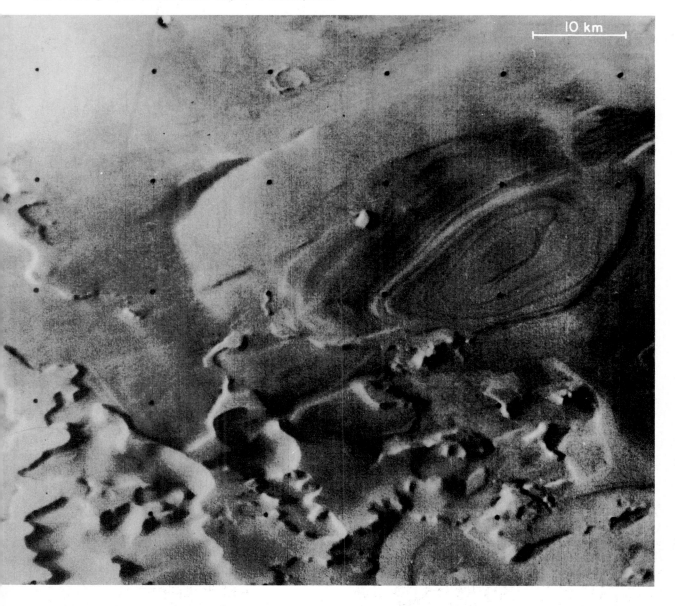

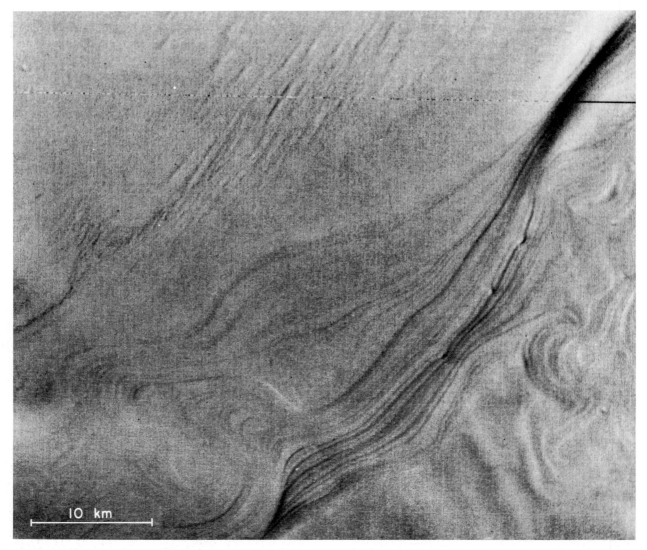

Figure 3.39. Layered deposits of south polar region. Lightly furrowed surface occupying the upper left part of the picture is topographically high. The surface with swirly texture in the lower right is topographically low. Horizontal layers are especially prominent along the regional slope joining these two surfaces. Layered deposits apparently have varying resistance to erosion, causing the more resistant layers and some individual blocks to stand out in relief. Layers are on the order of 100 m thick. (DAS No. 8080243, rev. 181, B camera, center at 229°W, 75°S.)

The origin of the layers is almost certainly sedimentary. Cutts (1973b) has argued that the layers are deposits of wind-blown dust carried in from the equatorial regions. The layered aspect suggests a cyclic variation in conditions of deposition. Models that yield 50,000-year and 95,000-year climatic cycles have been proposed to account for the layers (Murray and Malin, 1973a). Cutts (1973b) points out that a 2-million year periodicity in the orbital ellipticity of Mars may sufficiently modify perihelic insolation to modulate the intensity of global dust storms with a 2-m.y. period.

Estimates of the time required to deposit the layered polar deposits range from 50 million years (Murray and Malin, 1973a) to 500 million years (Cutts, 1973b). In view of the uncertainty in the time represented by each layer, and additional uncertainty as to whether the deposits form a single dissected group of many thin layers or a number of plates of such layers, any estimates of the age of the polar layered deposits are suspect.

The presence of a large crater with secondary craters superposed on layered deposits in the south polar region suggests that the entire depositional sequence may be older than first postulated (Cutts, 1975). The argument is based on the supposition that an impact of this size occurs infrequently.

Etched Plains

The layered deposits appear to overlie unconformably more massive material. These apparently unlayered blankets are extensively pitted and etched (Sharp, 1973b) and in many areas have a lineated surface (Murray et al., 1972) (fig. 3.40). Called pitted plains by Murray et al. and containing small regions specifically designated by Sharp as pitted terrain and etched terrain, we refer to this unit as etched plains (ep). The etched plains appear to be undergoing erosion, presumably by selective deflation, to reveal underlying cratered terrain.

The stratigraphic succession exhibited at the south pole is reasonably clear (Murray et al., 1972). The oldest surface is the ancient cratered terrain. Overlying this are one or more massive (i.e., unlayered) blankets probably composed of fine-grained, wind-transported dust. Cutts (1973a) cites evidence that the blankets may have been produced by local volcanism. The surface of these massive deposits was then irregularly eroded to form the etched plains.

Layered deposits were then laid down, some resting unconformably on eroded deposits of the etched plains. Finally, both layered and massive deposits were subjected to eolian erosion, an episode that apparently continues to the present.

Permanent Ice

Residual polar caps (pi) remain at both poles throughout the respective summers. These are distinct from the thin mantles of CO_2 ice that advance and retreat each year (Murray et al., 1972). The north polar permanent cap is presently larger than its south pole counterpart owing to the position of the axis of rotation with respect to perihelion. The recession of the south polar cap appears to proceed in a regular manner and then abruptly stops when it reaches a diameter that, from Earth observations, has remained fairly constant over the years. This, and the fact that summer surface temperatures do not favor the existence of CO_2 ice, has led to speculation that the permanent caps are largely H_2O ice, which would evaporate more slowly at polar temperatures than would CO_2 (Leighton and Murray, 1966).

Figure 3.40. Etched plains of south polar region. Irregular pits range from 1 km to 10 km across. The large pits apparently form by coalescence of smaller pits. They are presumed to be the product of wind erosion. (DAS No. 5453843, rev. 108, B camera, center at 68°W, 73°S.)

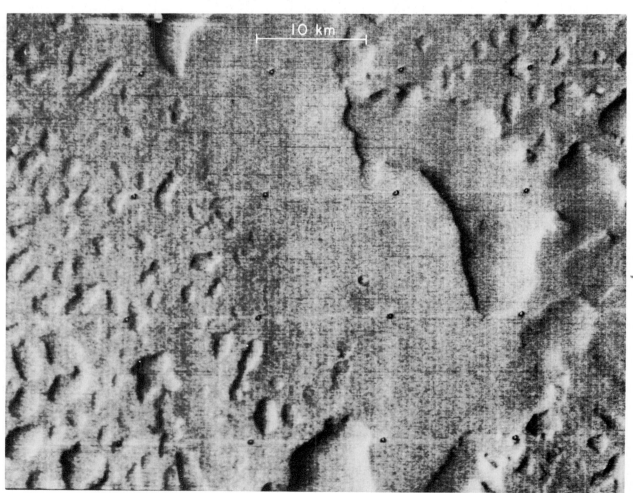

4.

Craters

Introduction

CRATERS are, at the same time, the least interesting and the most informative morphological features common to the Earth, moon, and Mars. They lack interest because each looks superficially the same as the next, a circular hole in the ground. Small wonder that, as our space exploration continued through the 1960s, many newspaper editors displayed current pictures of the moon first on the front page, then on the back page, and finally on no page. In their opinion, the succession of pictures, each showing a slightly different arrangement of circular pits, was hardly newsworthy.

Over the years a variety of explanations has been put forward to account for craters, but only two merit serious consideration: impact and volcanism. These two theories of origin drew investigators into two camps and fueled a raging debate. In this regard, the comments of W. M. Davis (1926, pp. 185–186) in his biography of G. K. Gilbert are interesting:

> It has been remarked that the majority of astronomers explain the craters of the moon by volcanic eruption: that is, by an essentially geological process—while a considerable number of geologists are inclined to explain them by the impact of bodies falling upon the moon—that is, by an essentially astronomical process. This suggests that each group of scientists finds the craters so difficult to explain by processes with which they are professionally familiar that they have recourse to a process belonging in another field than their own, with which they are probably imperfectly acquainted, and with which they therefore feel freer to take liberties.

Apollo exploration has shown that lunar impact craters are many times more numerous than volcanic craters, and, although the proportions may be slightly different on Mars, we will confine our attention in this chapter to that majority presumably formed by infalling meteoroids and comets.

A meteoroid falling from the skies is one of a class of phenomena that we readily accept in general but are inclined to reject in particular; phenomena that we routinely insert in some cosmic scheme but resist including in our own limited experience. For example, we accept without question or passion the observation that the universe contains countless solar systems, thousands of which may have Earth-like planets populated by plants and animals. But we are embarrassed by those who assert too loudly that, when we land on Mars in 1976, the biological experiments conducted within the spacecraft may provide clear evidence of life on that planet.

We know that matter exists throughout the universe but we find it convenient to assume that Earth is insulated from that matter. Meteoroids burning up in the atmosphere, leaving behind a fiery wake, are acceptable because they seem no closer than the nearest star. We tend to regard as academic calculations such as those showing that a meteor, streaking across the sky of western United States on August 10, 1972, came within 58 km of Earth's surface. Had it hit, the explosion would have rivaled the World War II atomic blasts that destroyed Hiroshima and Nagasaki (Rawcliffe et al., 1974). Small meteorites found in some farmer's field cause little wonder because they arrived at some other time. But the seeds of our incredulity are planted when we read that, on the afternoon of May 27, 1974, in Milford, Connecticut, nine-year-old Susan Leson, startled by a whistling sound, looked up from her game of kickball to see a small meteorite hit the pavement with enough force to make a hole one inch deep. The same seeds of doubt germinate when someone proposes that large cometary bodies struck Earth many years ago, causing disruption of continents and oceans. The seeds spring into full bloom when we read that a comet may, some day in the future, collide with Earth, causing chaotic upheaval.

In these several examples our disbelief is rooted in a misguided variety of uniformitarianism. We find it difficult to accept that any future day will be much different than the present one. The phenomena that are repeated daily or seasonally are familiar, and therefore real. When, in time and experience, we step outside our brief span of years, our scientific prejudices follow unpredictable patterns. Consider the controversy surrounding statements of biological evolution, a theory that—according to some— threatens the very foundations of our essential religious beliefs. However, discussions of expanding or steady-state universes, certainly a topic with theological overtones, occasion little fervor on the part of today's laymen.

It is interesting to note how fashions change. Five hundred years ago the cosmological models that are now regarded with such disinterest would have been heresy. How might our prejudices change in the future? Today we refuse to believe that catastrophic interplanetary collisions have warped Earth's history. A hundred years hence, when the large impact scars on other planets are familiar landscapes, will we feel the same?

Rates of Crater Production on Planetary Surfaces

Our knowledge of the density of debris in interplanetary space comes from several sources. The smaller particles, up to several millimeters in diameter, can be collected by high-flying airplanes, rockets, and balloons. They are, in addition, recorded as they strike specially designed targets mounted on spacecraft. The larger meteoroids are visible as they burn up in Earth's atmosphere. In the last several years a systematic and comprehensive search has been made for these "fireballs." When the results of all these experiments are integrated, a mass distribution is obtained as shown in figure 4.1 (Dohnanyi, 1972). Translating this into results observable on a planetary surface, we conclude that more than 300 bodies 1 kg or larger would strike each square km

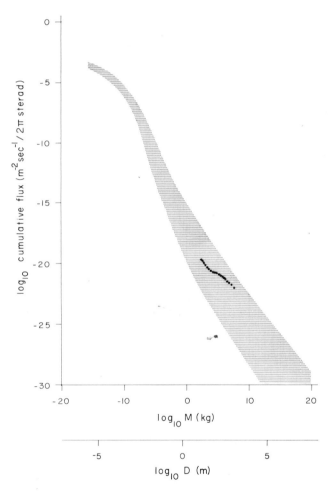

Figure 4.1. Cumulative flux (m⁻² sec⁻¹/2π sterad) of meteoroids and micrometeoroids into Earth's atmosphere. Simplified after Dohnanyi (1972). Projectile masses are converted into crater diameters by assuming velocity of 20 km/sec and using scaling data of Vortman (1968). For comparison, crater densities for Mare Tranquillitatis in the vicinity of the Apollo 11 site (Greeley and Gault, 1970) are plotted, using the radiometrically measured age of 3.7 b.y.

of that surface every million years. These bodies would form an equivalent number of craters with diameters larger than 2 m (Kaula, 1968). The mass ejected from a crater is much larger than that of the associated projectile, the exact ratio depending largely on the velocity of impact. For this calculation we made the simplifying assumption that all meteoroids strike the surface with a velocity of 20 km/sec. Although only an approximation, the results compare favorably with what is actually observed on the moon's surface (fig. 4.1).

Moving away from the Earth-moon system, determination of meteoroid fluxes is more speculative. It becomes necessary to consider the general problem of small-body sources, orbits, and lifetimes. The two contributing classes of objects are asteroids and comets. Asteroids comprise a swarm of mini-planets, the largest of which has a diameter of about 950 km. Although asteroids generally occur between Mars and Jupiter, some have elliptical orbits that carry them across Mars' orbit. A few drift even closer to the sun, intersecting Earth's orbit.

Comets follow orbits that are highly elliptical or hyperbolic. When a comet swings close to the sun condensed volatiles vaporize, producing a bright tail. Historically, the passage of Halley's Comet is most famous. It last appeared in 1910 and is scheduled for another performance in 1986. Many readers will recall that the Comet Kohoutek, which passed behind the sun on Christmas 1973, failed to live up to its advance billing. Advertised as an object of dazzling brilliance, it was, in fact, barely discernible close to the horizon in the night sky. The return of Halley's Comet may not be any more impressive to the casual observer (Roosen and Marsden, 1975).

Knowledge of asteroid and comet motions can be used to construct collision models for all of the terrestrial planets. Öpik (1951, 1963) developed the mathematical arguments. His theoretical conclusion that Mars should be heavily cratered was confirmed by the first Mariner 4 pictures of Mars. Leighton et al. (1965) first suggested that the heavily cratered Martian terrain resembled the lunar uplands and was, by analogy, very ancient, somewhere between two and five billion years old. Almost immediately this conclusion was challenged by Witting et al. (1965) and Anders and Arnold (1965), who derived collision models for Mars with impact fluxes 15 to 25 times as high as for the moon.

Recently, the pendulum has swung back again. Shoemaker (see Metz, 1974) has proposed that gravitational interactions between asteroids and the neighboring massive planet, Jupiter, perturb orbits and spray asteroids randomly across the inner part of the solar system. Accordingly, impact rates should be approximately the same for all of the terrestrial planets, a view shared by Wetherill (1974). The theoretical argument is bolstered by superficial similarities in lunar and Martian crater abundances (Soderblom et al., 1974) and by recent analysis of craters on Mercury (Murray et al., 1974; Gault et al., 1975; Murray et al., 1975).

It is generally agreed that cratering rates have not been constant with time. During the terminal stages of planetary accretion, sweeping up of interplanetary debris was marked by very high impact rates. The conventional view is that after about one billion years, or 3.5 billion years before the present, the flux had decreased to a value that has remained essentially constant to the present (fig. 4.2). However, the existence of one or more periods of high flux following an initial flux decrease cannot be ruled out. Tera et al. (1974) present isotopic evidence for a pulse of high-impact activity on the moon, occurring about 3.9 b.y. ago and extending for perhaps no more than 2×10^8 years. Those authors suggest that most of the relatively unmodified lunar impact basins, including Imbrium, Crisium and Orientale, formed at this time.

It is clear that our understanding of small-body distributions in the solar system is minimal. In large part this is because available data are few and the relevant gravitational interactions between solar system bodies are complicated. Models to date have emphasized the role of Mars-crossing asteroids and the "Jovian generator." Comets have received very little attention. Our ignorance is unfortunate because interpretation of the large volume of crater statistics produced for Mars would be immeasurably enhanced by confident normalization to an absolute time scale.

A Strategy for Interplanetary Comparison of Craters

In the immediately following four sections we will discuss the morphological characteristics of terrestrial, lunar, Mercurian, and Martian craters. Lunar craters are especially instructive for studying the effects of large impacts, generally undisturbed by other geologic processes. Terrestrial craters, on the other hand, illustrate the modification that occurs in an environment of very extensive geologic modification.

Appreciation of Martian craters is greatly enhanced when they are examined with lunar and terrestrial situations in mind. A morphologic classification applicable to all four bodies can be developed. Comparison of Martian craters with the "ideal" lunar examples facilitates detection of features originally associated with Martian craters that have been modified or destroyed. Some of the comparisons yield unexpected results. For example, it

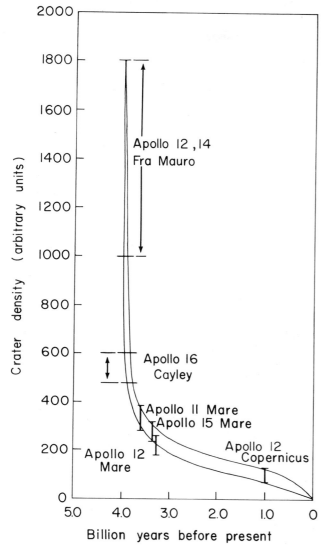

Figure 4.2. Plot of radiometric ages of lunar samples collected during the Apollo missions versus crater density for landing sites (Soderblom et al., 1974). Radiometrically determined ages are from Papanastassiou and Wasserburg (1971a, 1971b), Wasserburg and Papanastassiou (1972), Wasserburg et al. (1972), and Silver (1971). Crater density measurements are from Soderblom and Boyce (1972) and Soderblom and Lebofsky (1972). This graph indicates that impact rates were extremely high in early solar system history. A sharp dropoff occurred about 3.5 b.y. ago, and the rate has been approximately constant since then.

will be shown that Martian craters are not as shallow, relative to lunar craters, as commonly assumed by many investigators.

Review of lunar and terrestrial craters is also valuable because we can examine them at detailed scales far beyond the resolution of Mariner pictures.

In this way we can extrapolate from observable Martian landforms to much smaller scale landscapes. It is acknowledged that this analytical technique is speculative. The details of Martian craters certainly do not mimic perfectly those of lunar or terrestrial craters. Some deviations, for example those related to eolian erosion and deposition, can be anticipated. Other distinctions might prove to be unexpected by reference to terrestrial and lunar features.

In successive chapters, the same techniques of interplanetary comparative analysis will be used for volcanic landforms, structural features, eolian features, and channels. As in the case of craters, it will prove useful to consider Martian landforms against a background of potentially comparable features on Earth, Mercury, and the moon. In all four categories, there is a wide range of terrestrial examples. Relatively subtle volcanic features are observed on the moon. Lunar structural features are restricted to lineaments and possible faults. Volcanic rilles comprise almost all channels. Eolian features are, of course, absent.

Terrestrial Impact Craters

Terrestrial impact craters are relatively rare. Earth's atmosphere acts as an efficient shield for bodies smaller than 1000 kg; they are decelerated, broken up, and ablated before they reach the surface. Even more importantly, the evidence of an impact is destroyed quickly by erosion. The average rate of erosion for Earth is about 1 cm/1000 years (Gilluly et al., 1970); it follows that, to a first approximation, a crater 1 km wide and 100 m deep would be completely erased in 10 m.y. if subjected to continuous erosion. Empirically, Hartmann (1971b) derives shorter lifetimes, one million years or less.

Even when preserved, the faint circular outline of an ancient crater is difficult to recognize from a ground perspective. Many impact structures were first detected in aerial photographs. Finally, it should be remembered that Earth's surface is comparatively young. Over 70 percent is covered by water. Of the remaining 30 percent of land, approximately equal areas are covered by Cenozoic rocks (0–65 m.y.), Mesozoic rocks (65–230 m.y.), Paleozoic rocks (230–600 m.y.) and Precambrian rocks (> 600 m.y.). More precise measurements by Gilluly (1969) for North and South America indicate 36 percent Cenozoic rocks, 19 percent Mesozoic, 18 percent Paleozoic, and 27 percent Precambrian. As previously mentioned, meteoroid fluxes were highest in the first billion years of solar system history, the same period for which we have no rock record on Earth. (Minimum ages for the oldest radiometrically dated rocks, found in Greenland [Moorbath et al., 1973] and Minnesota [Goldich and Hedge, 1974] are approximately 3.8 b.y.)

In all, more than fifty terrestrial craters show strong evidence for impact origin (table 4.1). A very few, like Meteor Crater of Arizona, are readily recognized by their distinctive morphology. Geologic studies have revealed structurally deformed and overturned rocks on the rim, and scattered fragments of iron meteorite (fig. 4.3). Most craters, however, are quickly erased by erosion (fig. 4.4). Some are confirmed by the discovery of floor breccia beneath an eroded crater. Others are revealed by exposure of even deeper structural features, and exotic minerals that form under conditions of transient high pressures and temperatures that could exist only during an impact event.

In desert regions, impact craters are often mantled

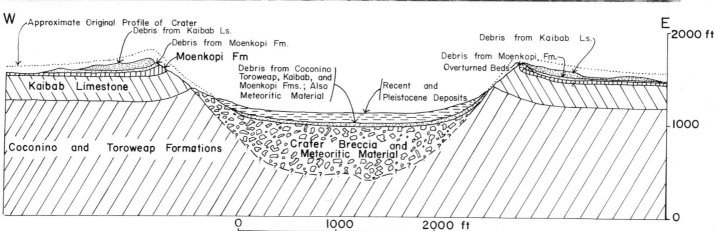

Figure 4.3. Meteor Crater, Arizona. This well-preserved impact structure was formed sometime during Pleistocene time. The crater is over 1 km in diameter and about 200 m deep. In this oblique aerial view, north is at right (courtesy of United States Air Force). The geologic cross-section is adapted from Shoemaker (1960).

Table 4.1. The worldwide distribution of impact structures.

Structure	Diameter (km)	Long.	Lat.	Structure	Diameter (km)	Long.	Lat.
United States				*South America*			
1. Crooked Creek (Mo.)	4.4	91°23′W	37°50′N	30. Campo del Cielo (Arg.)	0.1 (3)	61°40′N	27°40′S
2. Decaturville (Mo.)	4.4	92°43′W	37°54′N	31. Monturaqui (Chile)	4.5	68°17′W	23°56′S
3. Flynn Creek (Tenn.)	3.2	85°40′W	36°17′N				
4. Glasford (Ill.)	4.4	89°49′W	40°35′N	*Europe*			
5. Howell (Tenn.)	2.4	86°35′W	35°15′N	32. Chassenon (France)	15	0°45′E	45°54′N
6. Jeptha Knob (Ky.)	3.5	85°06′W	38°06′N	33. Kabbijärv (Est.)	0.1 (7)	22°40′E	58°24′N
7. Kentland (Ind.)	5.8	87°25′W	40°54′N	34. Köfels (Austria)	4.8	10°58′E	47°13′N
8. Manson (Iowa)	31.0	94°31′W	42°35′N	35. Lake Mien (Swed.)	6	14°55′E	56°25′N
9. Meteor Crater (Ariz.)	1.2	110°01′W	35°02′N	36. Lappajärvi (Fin.)	5	23°40′E	63°10′N
10. Middlesboro (Ky.)	5.8	83°44′W	36°37′N	37. Ries Kessel (W. Germ.)	23	10°37′E	48°53′N
11. Odessa (Texas)	0.15	102°30′W	31°48′N	38. Steinheim (W. Germ.)	2.8	17°25′E	58°46′N
12. Serpent Mound (Ohio)	6.4	83°25′W	39°02′N				
13. Sierra Madera (Texas)	12.0	102°55′W	30°36′N	*Africa–Asia*			
14. Wells Creek (Tenn.)	14.0	87°40′W	36°23′N	39. Ashanti (Ghana)	11	1°23′W	6°32′N
				40. Aouelloul (Maur.)	0.2	12°14′W	20°13′N
Canada				41. Lonar Lake (India)	1.8	76°51′E	19°59′N
15. Brent (Ont.)	4.0	78°40′W	46°04′N	42. Pretoria Salt Pan			
16. Carswell Lake (Sask.)	32	109°30′W	58°27′N	(S. Africa)	1.1	28°00′E	25°30′S
17. Charlevoix (Que.)	35	70°18′W	47°32′N	43. Richat (Maur.)	50	11°24′W	21°09′N
18. Clearwater Lakes (Que.)	26, 14	74°20′W	56°08′N	44. Roter Kamm (S.W. Af.)	2.4	16°18′E	27°46′S
19. Deep Bay (Sask.)	10	103°00′W	56°24′N	45. Sikhote-Alin (U.S.S.R.)	0.03 (122)	134°39′E	46°10′N
20. Holleford (Ont.)	1.9	76°30′W	44°47′N	46. Talemzane (Alg.)	17.5	4°00′E	33°20′N
21. Lac Couture (Que.)	10	75°20′W	60°08′N	47. Vredeford (S. Africa)	40	27°29′E	27°28′S
22. Lake Mistastin (Lab.)	19	63°18′W	55°53′N	48. Wabar (Arab.)	0.1 (2)	50°28′E	21°30′N
23. La Malbaie (Que.)	37	70°15′W	47°35′N				
24. Manicouagan (Que.)	64	68°37′W	51°28′N	*Australia*			
25. New Quebec (Que.)	3.5	73°40′W	61°17′N	49. Boxhole (North Terr.)	0.2	135°12′E	22°37′S
26. Nicholson Lake				50. Dalgaranga (West Aus.)	0.02	117°05′E	27°45′S
(N.W. Terr.)	12	102°41′W	62°40′N	51. Gosses Bluff		132°18′E	23°48′S
27. Pilot Lake (N.W. Terr.)	5.0	111°01′W	60°17′N	(North Terr.)	19		
28. Sudbury (Ont.)	60	81°01′W	46°30′N	52. Henbury (North Terr.)	0.2 (13)	113°10′E	24°34′S
29. West Hawk Lake (Man.)	2.4	95°11′W	49°46′N	53. Wolf Creek (West Aus.)	0.8	127°46′E	19°18′S

Data primarily from Short and Bunch (1968). Many more craters have been described. For example, Classen (1975) lists approximately 1000 craters at 216 localities. However, decisive evidence of impact is lacking for many of these structures.

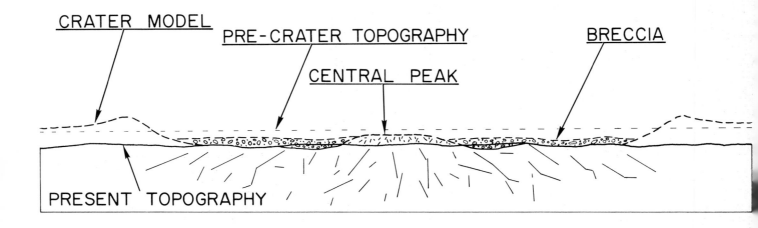

CRATER MODEL PRE-CRATER TOPOGRAPHY BRECCIA

CENTRAL PEAK

PRESENT TOPOGRAPHY

0   1   2   3 km

Figure 4.4. Diagrammatic profile of the Nicholson Lake impact structure in Canada, showing the present topography and a reconstruction of the original profile (after Dence et al., 1968). In addition to normal degradational processes, glaciation has played a major role in the exhumation and modification of this crater. A mid-to-late Paleozoic age has been assigned to this feature, based on degree of erosion.

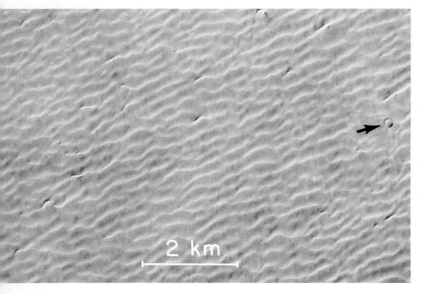

2 km

Figure 4.5. The Wabar Crater, nestled among transverse sand dunes in Saudi Arabia. This remote desert region, known as the Empty Quarter, was first traversed in 1933 by the British explorer, John Philby, and 18 Arab companions. There was, at the time, a widespread legend that the ruins of a vast city with giant statues of iron were buried in the desert sand. Philby discovered that the legendary city was actually a cluster of five impact craters, the largest 100 meters across. The reported statues were large fragments of an iron-nickel meteorite. Since their initial discovery, all but one of the Wabar craters have been engulfed by sand.

with sand encroaching from surrounding areas (fig. 4.5). Examination of deposits around recent nuclear explosions shows that eolian activity removes fine debris in ejecta deposits. The modification of the craters in figure 4.5 probably results from erosion of finer-grained crater deposits as well as deposition of sand on the coarser, more resistant crater deposits. Since eolian encroachment takes place rapidly on Earth, there may be some well-preserved craters buried beneath thick sand deposits in the world's larger deserts.

Rapid burial by sediments has helped to preserve many older terrestrial impact structures. For instance, craters on the Canadian Shield were covered by thick sedimentary deposits before they were noticeably degraded. Thus protected, they have only recently been exhumed by erosion. Perhaps the overlying sedimentary strata were stripped away by the Pleistocene ice sheets (Dent, 1973). This burial and subsequent exhumation has resulted in a large number of craters on the Canadian Shield (fig. 4.6). Craters as small as 2 km in diameter have been preserved for approximately a half billion years (Dence, 1972), much longer than they could have existed unprotected from erosion.

Examination of a worldwide map of impact craters shows that most are located in the ancient stable shield areas (fig. 4.7). Two reasons explain this distribution: first, surfaces formed relatively recent-

Figure 4.6. An Earth Resources Technology Satellite (ERTS) photograph of Clearwater Lake. West Clearwater Lake (upper left) is approximately 30 km in diameter and contains a central uplifted ring. The craters were formed approximately 285 million years ago (Dence, 1972) and were probably buried and subsequently exhumed. Note the light halo extending approximately a crater's radius into the surrounding bedrock. This probably represents a change in the physical properties of the material produced by the impact.

Figure 4.7. Worldwide distribution of impact craters, as listed in table 4.1. Shield areas, underlain by Precambrian rocks, and associated platforms are shaded.

Figure 4.8. ERTS photograph of the Sudbury basin, about 60 km long and 30 km wide. Three geologic provinces are visible. In the Superior Province, ancient Precambrian rocks display prominent northwest-southwest lineations. In the Southern Province more recent Precambrian rocks have been deformed into east- and northeast-trending folds. Rocks of the Grenville Province were intensely fractured during the most recent orogenic event. The Sudbury crater, presumably circular at the time of its formation, has been distorted by the same stresses that caused folding within the Southern Province.

Figure 4.9. The lunar farside showing overlapping craters of all sizes. Mare Crisium (at top), Mare Marginus, and Mare Smythii are in the upper left. As viewed from Earth, these same maria would appear on the extreme eastern edge of the lunar disc. Apollo 16 Mapping Camera photograph.

ly have been exposed to impacts for proportionally shorter times, and, second, many of the regions adjacent to shields are tectonically active. Deformation along with erosion may have obliterated or modified most of the craters. In that respect, the Sudbury structure in Ontario, Canada (fig. 4.8), is an anomaly since it was preserved despite tectonic activity. The present form does not fit our general concept of a circular crater, even allowing for the removal of much of the original deposits. Placed in its proper historical sequence, however, it becomes clear that the Sudbury impact event occurred prior to a major phase of crustal folding (Dietz, 1964; Brocoum and Dalziel, 1974). The Penokean orogeny affected the region, causing folding and overthrusting toward the northwest. As a consequence Sudbury deposits were deformed to their present form.

When we consider the effect of tectonic deformation on craters, it is intriguing to think about the opposite effect: that is, how cratering has influenced the tectonic development of planets. A glance at the early history of the moon shows that the majority of crustal structures—lineaments, mountain ranges, and regional basins—are related to large-scale impact cratering during the first billion years. What does this imply about the early history of Earth? How did the late-stage accretional cratering affect development of the primeval crust? Was the distribution of terrestrial protocontinents controlled by large impact basins? Might the breakup of protocontinents be related to fractures inherited during this early history of protocontinental bombardment? It is clear that, as paleocontinental reconstruction reaches earlier into Earth history, the role of large-body collisions should be considered. (For a discussion of the

Figure 4.10. An oblique view of Messier B, a 7-km-diameter fresh crater, displaying a gently upraised circular rim and a bowl shape. Downslope debris movement has formed streaks on the wall and created a pool of material at the crater center. (AS10-4254.)

beginning of continental evolution, based on insights gained from geochemical analysis of lunar rocks, see Wetherill [1972].)

Lunar Craters

Crater Morphology

At first glance, a photograph of the lunar surface appears to contain nothing more than craters of all sizes, crowding together shoulder to shoulder, often overlapping (fig. 4.9). There are so many craters that one is struck with their similarities, rather than their differences. However, continued study reveals morphologic differences, many associated with different sizes. In this vein, Apollo 15 Astronaut Dave Scott noted on his return from exploring the Hadley-Apennine region that his first impression of the lunar surface was a forest of craters. Before long, how-

Figure 4.11. Lunar Orbiter photograph covering half of the 4-km-diameter crater Mosting C and a part of its ejecta deposit. Note the concentration of blocks near the upturned crater rim and the dune-like morphology of the ejecta. (LO-III-112H.)

ever, he was able to identify various "trees" in the forest by their associated characteristics such as blockiness, presence of glass, and shape. After only a few hours on the lunar surface, he was able to use this knowledge of crater size and associated morphologic characteristics to anticipate and judge the best path for the lunar rover as he drove across the mare surface at 5–10 km/hour.

Crater morphologic characteristics are significant in several ways: 1) different morphologies for craters of similar size and age may indicate different origins (endogenic vs. exogenic) or a different type of impacting body (comet vs. meteoroid); 2) different morphologies for craters of similar size and origin but different age are important in deciphering crater degradation processes; 3) different morphologies for craters of similar origin and age but different size are useful in determining the behavior of lunar crustal material in response to different amounts of impacting energy and to different depths of penetration; and 4) lateral ejection of material from cratering events is a major process for building up the lunar stratigraphic column and for degrading surrounding terrain. Study of the surface in the vicinity of craters can help provide data on the mode of emplacement and extent of these stratigraphic units.

For terrestrial craters Dence (1968) observed a gradation in size from bowl-shaped craters to craters with central uplifts that often take the form of peaks. He called the smaller craters *simple craters* and the larger ones *complex craters*. Since there is a similar transition for lunar craters (e.g., Pike, 1968; Hartmann, 1972a) this convention is used here. Crater morphologic characteristics have been discussed by Moore (1971, 1972); El-Baz (1972); Pohn and Offield (1970); Howard (1974); Quaide et al. (1965); Pike (1971, 1972); Smith and Sanchez (1973). On the moon, simple craters usually are less than 15 km in diameter although there is a transition zone between 10 and 30 km where craters share the characteristics of both simple and complex craters. For purposes of simplification, a diameter of 20 km is here chosen as the boundary between lunar simple and complex craters (fig. 4.28). The most striking characteristic of simple craters is their bowl shape (fig. 4.10) and generally circular outline. The walls of less degraded simple craters are generally smooth, lack the terraces common in larger complex craters, and are often streaked by material moving downslope. The rims of young simple craters are characteristically strewn with blocks. Typical rim morphologies include hummocky texture near the rim and the development of concentric ridge-like structures or "dunes" (fig. 4.11).

In small lunar craters, 400–1000 m in diameter, abundant blocks in the 1–20 m size range are found on the rim (Moore, 1972). In addition, larger blocks, up to several hundred meters in diameter, are found around larger fresh craters. Figure 4.12 shows the characteristics of an extremely fresh crater about 5 km in diameter. The prominent ejecta and dense ray system spreads out one to two crater diameters, and discontinuous ray patches extend considerably further. Particularly interesting are the bright and dark rays extending radially outward from the rim and the pool of dark material on the crater floor. Association of the dark material with such a young highland crater strongly suggests that it is shock-melted material that has ponded on the crater floor, rather than material of volcanic origin. The Apollo 16 crew obtained photographs of South Ray crater, a small, fresh, rayed crater (fig. 4.13) that has been dated at 2 m.y. (Behrman et al., 1973).

Craters of the transitional size between 10 and 30 km share some characteristics of both simple and complex types. The 17-km-diameter crater Dawes has this sort of mixed appearance (fig. 4.14a). It has begun to lose some of its circularity and also has a rough crater floor. The loss of circularity appears to be related to the formation of slump terraces (fig. 4.14).

Complex craters generally fall between 20 and 200 km in diameter and include some of the most prominent craters on the moon—Copernicus, Aristarchus, Tycho. Beginning at about 20 km, interior terraces appear, and become well-developed with increasing size (figs. 4.15, 4.16). The 30-km-diameter crater Timocharis (fig. 4.15) shows several crater-wall terraces, a flat floor, and central peaks in the interior. Rim deposits are typically hummocky within a crater radius of the rim and are transitional to a radial facies that usually extends out an additional crater radius. A generally distinct change occurs at about a diameter from the rim, particularly evident in the low-sun photography of Timocharis. In this transitional zone the radial facies gives way to discontinuous clusters of secondary craters and crater chains. In contrast to the nearly circular form of smaller craters, many craters in the 20–200 km size range have rim crests made up of elongate segments fashioned by slump terraces. Many segments are concentrated along lines parallel both to the lunar grid and to directions radial to major basins. Pohn and Offield (1970) have noted that older craters show more polygonal crater rims than their younger counterparts.

Larger craters (200–300 km in diameter) are transitional in nature with still larger multi-ringed basins. The farside crater Tsiolkovskij (fig. 4.17) illustrates many typical characteristics. The crater wall, although still consisting of slumped terraces, appears steep and sharp. Although Tsiolkovskij has

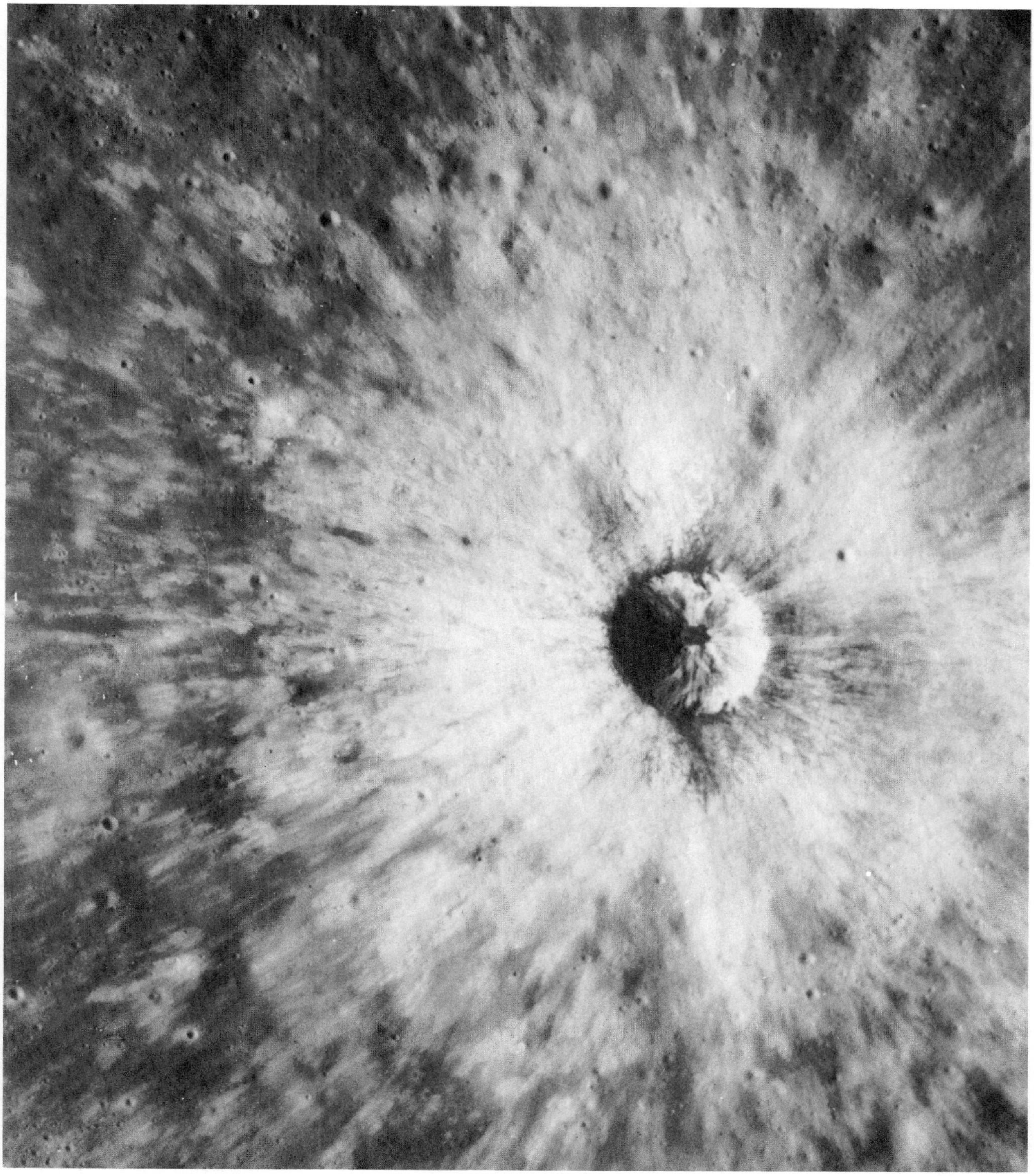

Figure 4.12. Fresh lunar crater, about 3 km in diameter, located on the rim of the large crater, Gagarin. The bowl shape, raised rim, and absence of terraces are characteristic of the simple crater class. The bright halo and ray system are typical for fresh craters in this size range. Note the streaks on the crater walls and the ponded material on the crater floor. (AS15-13156.)

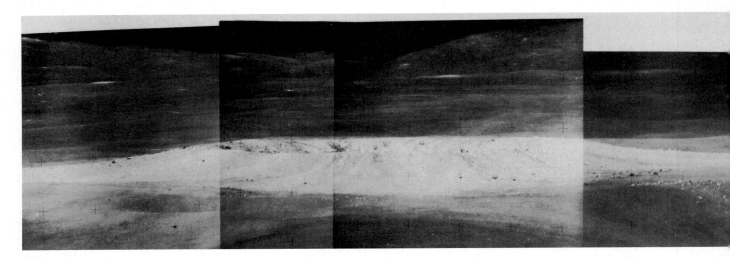

Figure 4.13. Mosaic of oblique photographs covering the 700-m-diameter South Ray crater. This sequence was taken by Astronaut Duke during one of the traverses at the Apollo 16 Descartes landing site. South Ray is young enough ($\sim$ 2 million years) to retain all primary impact features. (AS16-112-18246, 7; 18256, 7.)

(a)

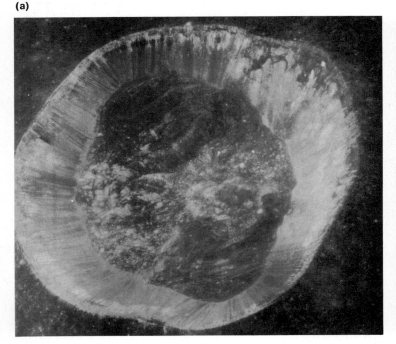

(b)

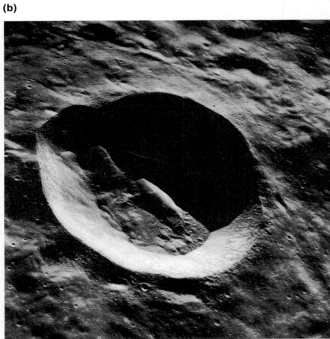

Figure 4.14 (a). High sun-angle view of Dawes Crater (17 km diameter) illustrates the transition from simple to complex crater morphologies. In contrast to the smaller craters in figures 4.10 and 4.11, Dawes has a subcircular rim outline. Post-impact slumping on northern and southern walls has exposed stratified bedrock and has contributed to the rough appearance of floor deposits. (AS15-9874.)

(b). This 10-km-diameter crater has been modified by collapse and slumping of wall materials, forming a hummocky floor and irregular rim outline. (AS15-13180.)

Figure 4.15. Near-vertical view of the 30-km-diameter crater Timocharis. Note the large slump terraces, flat floor, and central peaks. The ejecta deposit exhibits a dune-like texture near the crater and changes to a patchy ray system further out. (AS15-1005.)

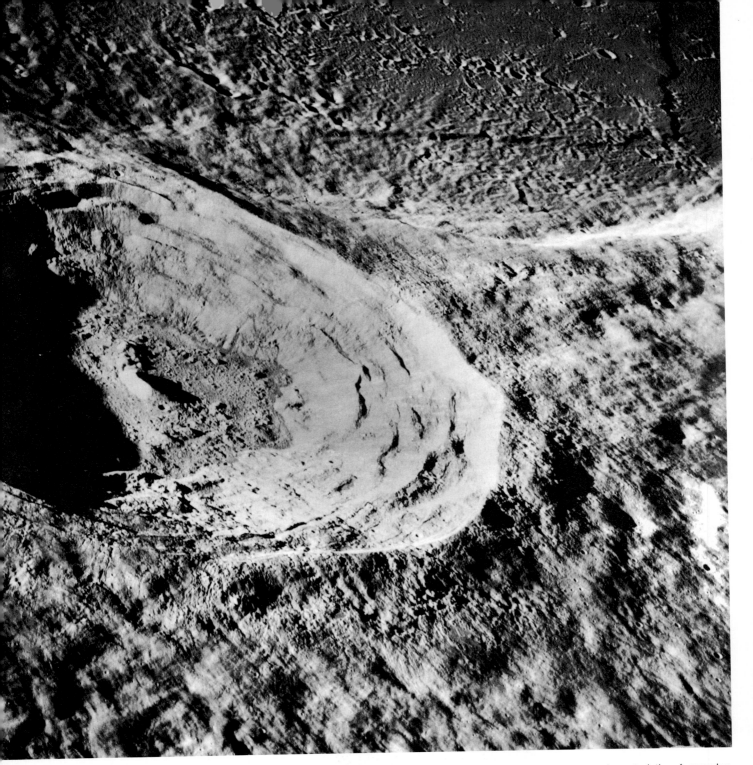

Figure 4.16. Oblique view of the 40-km-diameter crater Aristarchus, illustrating slump terraces characteristic of complex craters. Note the group of central peaks in the center of the crater floor. (AS15-13050.)

Figure 4.17. Located on the lunar farside, the 180-km-diameter crater Tsiolkovskij displays features typical of large, complex craters. Extensive slump terracing has modified the crater rim. Dark mare rocks have inundated much of the crater floor. (AS15-0758.)

a prominent central peak cluster, several other craters in this size range are characterized by rings of central peaks (Hartmann and Wood, 1971; Hartmann, 1972a) (fig. 4.18). The inner part of the rim typically has large planar surfaces and lobate flow fronts and flow channels that are interpreted to be produced by fluidized ejecta.

Craters with diameters larger than 300 km are termed *basins*. Most basins are fringed by several concentric mountain ranges and depressions; hence the name "multi-ringed basin" (Stuart-Alexander and Howard, 1970; Hartmann and Wood, 1971).

The freshest and most notable lunar example is the Orientale basin (fig. 4.19). This basin is a complex depression outlined by a prominent circular scarp, the Montes Cordillera, extending between 900 and 950 km in diameter. Outward from the Cordillera a broad blanket of radially patterned ejecta stretches for about 500 km (Moore et al., 1974; Howard et al., 1974). Inward from the Cordillera scarp are three other rings. Two form the inner and outer part of the Rook Mountains; the third forms a terrace immediately surrounding Mare Orientale. The Cordillera scarp is continuous, characterized by linear

Figure 4.18. An oblique view of Schrödinger, 320 km in diameter, illustrating rim and wall slump terraces similar to those of Tsiolkovskij. A circular set of central peaks occurs halfway toward the crater center, indicative of the transition between complex craters and multi-ringed basins. Amoeboid marking is a film defect introduced during automatic development of the film onboard the unmanned spacecraft. (LO-V-21H.)

fault scarps; the outer Rook Mountains are generally continuous, but are characterized by groups of massifs; the inner Rook Mountains are discontinuous and characterized by isolated peaks and groups of peaks (fig. 4.19). Comparison of the Orientale basin with slightly smaller central-ring craters suggests that the outer Rook Mountains represent the rim of the Orientale impact crater, 620 km in diameter. The inner Rook represents an expanded central peak ring, and the Cordillera represents a fault scarp formed by the partial collapse inward toward the newly formed crater after the outflow of ejecta (Head, 1974a). The region within the Cordillera ring is composed of three units. An outer domical facies, between Cordillera and outer Rook, is dominated by terrain composed of adjacent domes ranging from 2–8 km in diameter. A corrugated facies, between outer Rook and the basin interior, displays cracks and rilles developed on broad elongate domes. An interior smooth-plains facies is partially covered by mare material. The inner facies (corrugated and plains) appear to be produced by fallback and flow of material within the original crater (Head, 1974a), while the outer domical facies is related to readjustments of the inner ejecta blanket during formation of the Cordillera scarp. Thus, for multi-ringed basins, the actual crater is probably represented by one of the inner rings—the major scarp that defines the modified basin is often a fault-bounded ring and not the original basin (Head, 1974a; McGetchin et al., 1973).

A complex ejecta blanket lies outside the Cordillera scarp of the Orientale basin. Fractured, brecciated, and possibly shock-melted rock from the crater interior forms an intricately textured ejecta blanket, with radial and transverse constructs and furrows (Moore et al., 1974). In addition, secondary craters, formed by discrete masses of ejecta following ballistic trajectories, extend as far as 3000 km from the basin center in some areas. Great valley-chains of secondary craters emanate from Orientale. Both surficial flow and ballistic sedimentation apparently contribute to extensive mixing and modification of the lunar surface (Moore et al., 1974; Oberbeck, 1975). Evidence from the interior of Orientale suggests that the volumes of impact melt associated with the Orientale basin may be very great, between 75,000 and 255,000 km^3 (Head, 1974a).

(a)   **(b)**

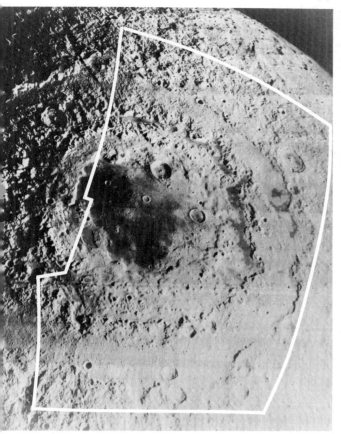

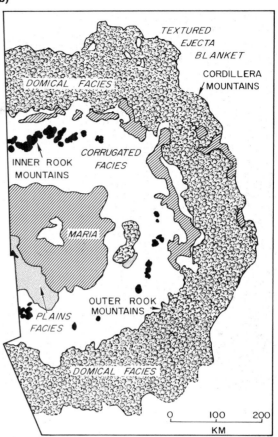

Figure 4.19 (a). Orientale multi-ringed basin. Location of (*b*) is outlined. The outermost scarp (Cordillera Mountains), which defines the basin, is about 900 km in diameter. (LO-IV-194M.)

(b). Morphologic units or facies found in the Orientale basin interior. Descriptions of the units are provided in the text.

Crater Degradation

The morphologies described in the previous section refer to young lunar craters. The vast majority of craters on the moon are, however, older and more degraded (fig. 4.20). Understanding the morphologic age sequence is important for several reasons: 1) morphologic characterization allows the determination of sequence among the many events in lunar history; 2) crater degradation analyses provide relative ages for specific craters and regions (Pohn and Offield, 1970; Offield and Pohn, 1970). Determination of rates for crater degradation will permit the assignment of absolute ages.

Crater degradation processes on the lunar maria have been studied by Marcus (1964), Ross (1968), Shoemaker (1966), Moore (1964, 1972), Soderblom (1970), and Soderblom and Lebofsky (1972). A scheme of mare crater morphology was devised by Trask (1970). Mare craters are thought to be removed by one of two processes: 1) catastrophic removal or modification by the formation of fresh craters superposed on the older one (Marcus, 1964); or 2) gradual erosion of craters by the impact of large numbers of small particles, any one of which would not have a marked effect on crater morphology (Ross, 1968; Soderblom, 1970, 1972). Mare near-terminator photography obtained on Apollo 14 (Head and Lloyd, 1971) provided evidence to assess this problem for the first time (Soderblom, 1972; Moore, 1972). In the near-terminator photographs, the mare crater populations appear to cover the entire surface as the sun elevation angle drops below about 1°. At the sun angle of about 2° the surface does not appear saturated. Therefore, as Soderblom (1972) concludes, the accumulated impacts necessary to saturate a surface with these 50–100-m diameter craters is approximately the same as that required to erode a 50–100-m diameter crater to a state where its interior slope is about 1°. Craters are eroded primarily by small impacts that, individually, produce small changes in crater form. When

Figure 4.20. The complex processes of crater degradation are illustrated by Guericke, 55 km in diameter (lower left). First, Imbrium ejecta filled part of the crater and sculptured the rim. Then, mare material modified the exterior. Crater impacts eroded Guericke throughout its history. An experiment boom of the Apollo spacecraft appears in the middle left. (AS16-2487.)

the interior slope is decreased to about 1°, modification is subsequently dominated by overlapping craters.

Pohn and Offield (1970) divided larger lunar craters into three size classes: Class I (> 45 km diameter; circular with distinct rim crenulations); Class II (20–45 km; polygonal); and Class III (8–20 km; circular). Relative ages are determined for craters within a given class by evaluating the characteristics of certain morphological components: *rays*, which disappear with increasing age; *radial ejecta patterns*, which become smooth and subdued with increasing age; *satellitic craters*, which become subdued and disappear with increasing age; *rim crest sharpness*, which decreases with increasing age; *terracing and interior radial channels*, with sharp terraces dominating the young crater walls, giving way to radial channels, which dominate the oldest crater walls; *polygonality*, which varies with crater class; and *rim texture*, which, from an initial hummocky texture, becomes smooth and finally pitted by continued cratering. The relationship of these diagnostic features to each other and to the numbered relative age scale devised by Pohn and Offield is shown in figure 4.21.

In any crater degradation scheme, several variables must be included that affect modification rates. These include: 1) rock strength; 2) volcanic materials, appearing either as initial volcanic craters of variable morphology and weathering characteristics or as fill and mantling deposits modifying impact

craters; 3) proximity aging: the modification of adjacent craters by the ejecta from a nearby crater or craters so that they appear relatively older than their true age (Pohn and Offield, 1970); and 4) seismic effects (Schultz and Gault, 1975).

Degradation of larger lunar craters can be divided into two time periods, based on differing styles and rates of crater degradation processes (Pike, 1968; Head, 1975). Comparison of lunar radiometric age scales and the relative degradation of crater morphologic features for craters larger than about 5 km diameter shows that early lunar history, prior to about 3.85–3.95 b.y. ago (Period I), is characterized by a high influx rate and by formation of large, multi-ringed basins. Later lunar history (Period II), from about 3.85–3.95 b.y. to present, is characterized by a much lower influx rate and lack of large multi-ringed basins. Craters formed throughout Period II show generally similar morphologic characteristics. Craters formed in Period I show markedly different characteristics although their residence time could not have increased more than 15 percent over the total time of Period II (fig. 4.21).

Elements of crater degradation and modification during Period I include destruction of crater exterior, rim, and wall facies and structures, decrease in crater depth, and increase in crater floor width. Examination of fresh crater geometry reveals that major changes in crater depth and floor width can occur with the addition of only minor volumes of material as crater fill. Volumes sufficient to produce these characteristic changes are readily available in the surrounding crater wall and rim deposits (Head, 1975). Depositional mechanisms associated with secondary cratering and lunar landslides are capable of moving material down the crater wall and onto the crater floor while preserving the characteristic break in slope (Howard, 1973). A prime source of crater degradation during Period I is associated with the formation of multi-ringed basins. The widespread ballistic sedimentation associated with the formation of these basins produces a near-saturation bombardment that excavates and mobilizes large volumes of local material and preferentially moves it into nearby low regions. This process appears to be of major importance in the degradation and modification of craters, in generation of interior crater fill, and in the formation of Cayley-type plains surfaces (Oberbeck et al., 1973, and Head, 1975).

The multi-ringed basin events fall within the earlier period of crater modification, with Orientale being the youngest and freshest example. The features associated with Orientale are progressively obliterated for successively older basins. Primary texture on the ejecta blanket is still visible for Imbrium, the next oldest basin, but disappears with successively

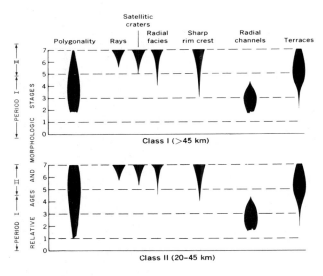

Figure 4.21. Changes in diagnostic crater features with age. The numerical values denote seven morphological stages and relative ages ranging from 0.0 (most degraded; greatest relative age) to 7.0 (least degraded; least relative age) (from Pohn and Offield, 1970). The boundary between the crater degradation periods of Head (1975) lies at about 3.85–3.95 b.y. and at about 4.8 on the Pohn and Offield scale.

Figure 4.22. An oblique view looking north over the 80-km-diameter crater Archimedes. The crater maintains a sharp rim crest. Secondaries can be seen in the foreground. Volcanism subsequent to the formation of the crater has modified its morphologic features. Lava flows have ponded inside the crater and, on the outside, have lapped over the base of the upturned rim. (AS15-1542.)

older basins. Subdued craters adjacent to a basin and blanketed by basin ejecta are present for basins as old as Nectaris. The most prominent basin elements are the last feature to be obliterated. Most of the older basins are recognized on the basis of preserved rings or parts of rings, although even these are slightly degraded by subsequent impacts (Stuart-Alexander and Howard, 1970; Hartmann and Wood, 1971; Howard et al., 1974).

An additional style of crater modification is seen in the mare regions of the moon and superficially mimics the effect of sedimentation on crater floors. Lavas flood crater floors and outer crater deposits—Archimedes, for instance (fig. 4.22). In general, crater terraces, rim deposits, etc., are unmodified because the rising lavas affect only lower topographic levels. This style contrasts with the process dominating earlier periods of lunar history.

In summary, modification of large lunar craters appears to be related primarily to an early period of high flux and to lava flooding. These processes may also be important on Mars.

Mercurian Craters

Gault et al. (1975) have shown that, although the gross morphology of fresh Mercurian craters is similar to that of fresh lunar craters, some important distinctions can be made. In particular, extensive secondary craters and chains occur relatively closer to the crater rim on Mercury than on the moon (fig. 4.23). Mercurian craters show terraces, central peaks, and central peak rings at smaller crater diameters than do lunar craters, indicating the onset of complex craters at smaller crater diameters on Mercury. Gault et al. suggest that gravitational attraction, 2.3 times greater on Mercury than on the moon, leads to occurrence of terraces, central peaks, and ringed basins at Mercurian crater diameters approximately half those for the moon. Similar gravitationally related variations should be anticipated for Martian craters.

(a)

(b)

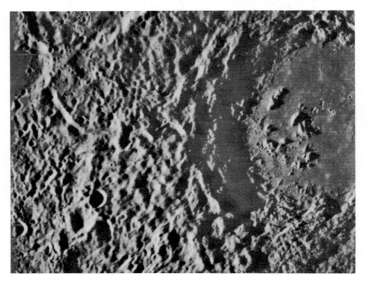

(c)

Figure 4.23. Mariner 10 views of craters and associated deposits on Mercury.

(a). A fresh Mercurian crater (crater 16) 98 km in diameter, showing extensively developed secondary crater field, radial ridges, and narrow hummocky rim facies. Terraces and central peaks are well developed. Terraces are also seen in the small fresh crater in the foreground (about 25 km diameter) (Mariner 10 frame 80) (Gault et al., 1975).

(b). Extensive development of zone of secondary craters around the 140-km-diameter Mercurian crater 12 (Mariner 10 frame 166) (Gault et al., 1975).

(c). Cluster of central peaks and hummocky rim of 120-km-diameter Mercurian crater 1 (Mariner 10 frame 27461) (Gault et al., 1975).

Martian Craters

Crater Morphology

The Mariner 9 photographs demonstrate that Mars, like the moon and Mercury, is peppered with craters. Both simple and complex crater morphologies can be discerned. Most of the smaller craters, less than about 15 km in diameter, fit into the simple crater class. They are bowl-shaped, have raised rims and smooth-appearing walls and floors (fig. 4.24). Block fields presumably exist on some of the younger craters, although the 100–300-m ground resolution of B frames precludes identification. As will be discussed in chapter 7, rayed craters are extremely rare (fig. 4.25) because eolian erosion quickly removes or modifies the thin cover of bright ray material.

Bowl-shaped craters are replaced by flat-floored craters at sizes larger than 15 km (fig. 4.26). Central peaks begin to appear at these sizes. Crater shapes vary from circular to polygonal. This trend parallels that found on Earth and the moon and, accordingly, these crater types can be termed *complex*. In complex as well as simple craters, prominent ejecta blankets and secondary crater fields (fig. 4.27) are considerably less abundant on Mars than

(a)

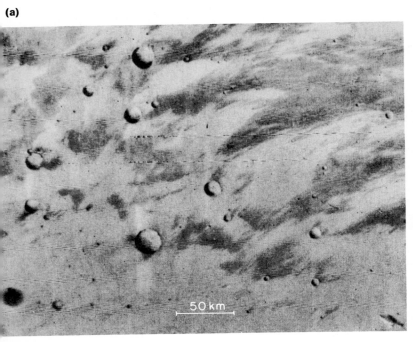

(b)

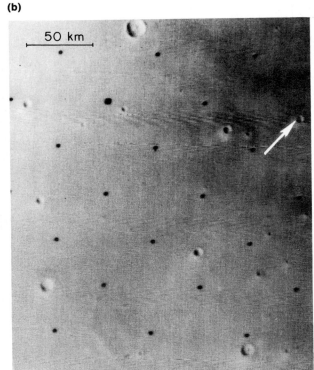

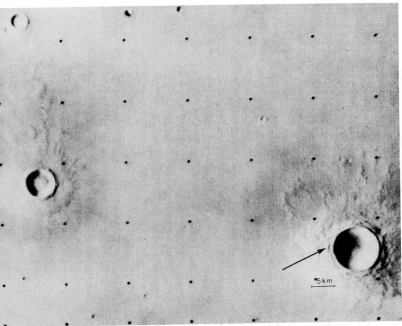

Figure 4.24 (a). Simple craters, exposed in Syrtis Major Planitia. For the most part, ejecta has probably been stripped away, but craters retain deep bowl shape, upturned rims, and circular plan view. (DAS No. 07147318, rev. 155, A camera, center at 284°W, 9°S.)

(b). Simple, bowl-shaped craters. The arrow points to a crater, which is shown at higher resolution in part (c) of the figure. (DAS No. 08838804, rev. 202, A camera, center at 231°W, 29°N.)

(c). Simple, bowl-shaped craters. Note ejecta deposits around the two larger craters. Many more craters are resolvable in this picture than in a lower resolution picture of the same region (part (b) of the figure). Undoubtedly, even more craters would be revealed with increase in picture resolution. (DAS No. 08838839, rev. 202, B camera, center at 238°W, 31°N.)

(c)

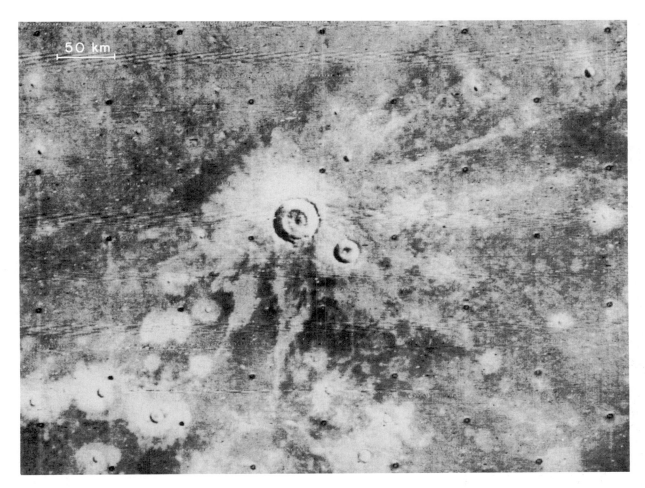

Figure 4.25. Martian bright-rayed crater. Bright haloes surround other, smaller craters. (DAS No. 13460423, rev. 675, A camera, center at 30°W, 40°N.)

Figure 4.26 (a). Martian complex craters, with flat floors and central peaks. (DAS No. 08225534, rev. 185, A camera, center at 155°W, 12°S.)

(b). A crater that is transitional between simple and complex. Polygonal walls, interior terraces, and flat floor are present, but a central peak is absent. This crater, indicated by an arrow in part (a) of the figure, is located on the floor of a larger complex crater. (DAS No. 08225499, rev. 185, B camera, center at 157°W, 14°S.)

(b)

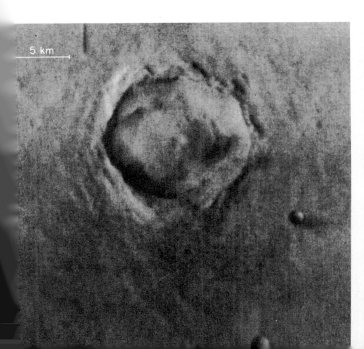

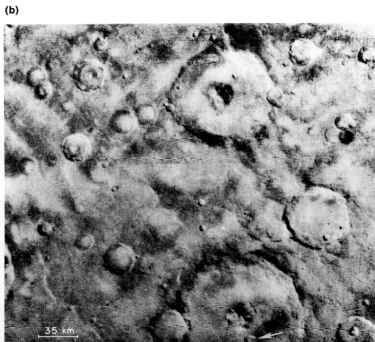

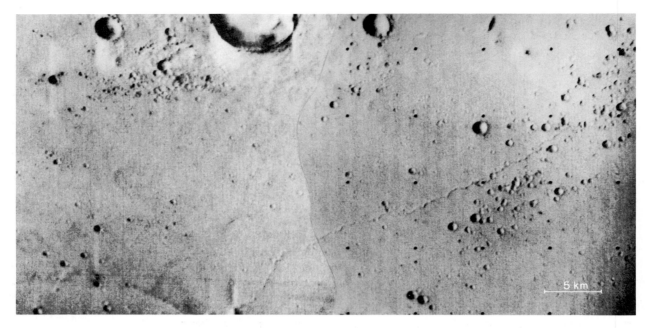

Figure 4.27. Swarms of small cup-shaped craters that may be secondaries, associated with a much larger primary impact. (DAS No. 12685958, rev. 473, B camera, center at 269°W, 6°N; DAS No. 12686028, rev. 473, B camera, center at 269°W, 6°N.)

Figure 4.28. Distribution of distinctive morphologic features according to crater size for both the moon and Mars. The diameter value separating simple and complex craters is somewhat arbitrarily chosen. In fact, a transition zone exists. All Martian measurements were made on Mariner A frames. The decrease in bowl-shaped craters at smaller diameters is a resolution effect. To a slight degree, the same effect may account for apparent decrease in central peaks and terraces.

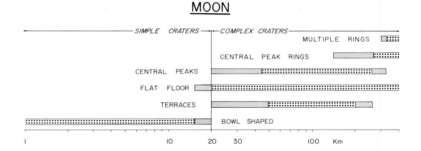

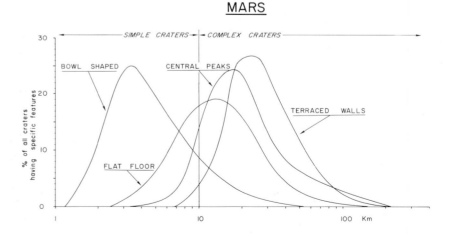

on the moon. Some of the better examples of secondary craters occur around several large craters in the polar regions (Cutts, 1975).

Although the division between simple and complex craters is observed on the moon, Mars, and Mercury, the size boundary between classes changes (fig. 4.28). On the moon it is in the vicinity of 15–25 km; on Mars it is between 5 and 15 km. Complex craters on Mercury have diameters 10 km or greater (Gault et al., 1975). As noted above, this is probably a gravity effect. As surface gravity increases, material failure occurs in craters of smaller size (Hartmann, 1972a). Surface gravity on Mars is about the same as on Mercury.

Above diameters of approximately 100 km, central peaks begin to be replaced by a central ring and, at even larger diameters, by a group of concentric rings. (The equivalent division on the moon occurs at about 200 km.) More than fifteen multi-ringed basins are identifiable on Mars (fig. 4.29). Lowell, a fresh central-ringed basin 200 km in diameter, has two well-defined rings (fig. 4.30). The outer ring is a typical crater rim. The inner ring consists of rugged peaks similar to the peak ring of the lunar crater Schrödinger (fig. 4.18). Also, there is a finely textured ejecta blanket associated with Lowell, which appears to be superimposed on rilles that intersect the adjacent terrain (Wilhelms, 1973).

The freshest multi-ringed basin found on Mars is Argyre, telescopically identifiable as a bright feature. This basin consists of an inner plains unit of roughly circular outline 900 km in diameter, which is surrounded by a rugged mountainous region (fig. 4.31). Wilhelms (1973) tentatively defines four or five rings. The innermost ring includes islands of bedrock mostly buried by plains material. The second and third rings are delineated by the prominent mountain ranges that fringe Argyre Planitia. The outer two rings consist of discontinuous, low, inward-facing scarps identifiable to the northwest of the basin. In the mountainous terrain of the inner rings, numerous radial valleys, arcuate troughs, and partly degraded crater rims are seen. An ejecta deposit with associated secondary craters similar to those associated with the younger lunar basins has not been identified around the Argyre structure.

Isidis is a more degraded basin (fig. 3.19). Its main rim, 1400 km in diameter, is partly buried by plains. To the south of the basin, isolated mountain peaks are seen (fig. 3.20). To the northwest and east, radial valleys are prominent. A large, concentric, basin-facing scarp is visible in the cratered terrain south of the basin.

Even more highly degraded are the smaller basins Schiaparelli (450 km) and Iapygia (450 km), which are occupied by plains material and whose rims consist solely of basin-facing scarps. Central ring structures are difficult to discern.

A 900-km-diameter basin, centering at 83°S, 262°W, consists only of a semicircular inward-facing scarp. In addition to being highly degraded, the basin is partly buried by polar layered deposits.

Figure 4.29. Distribution of Martian craters with diameters greater than 64 km, plotted on a Lambert equal-area base. Incomplete rings for multi-ringed basins are indicated.

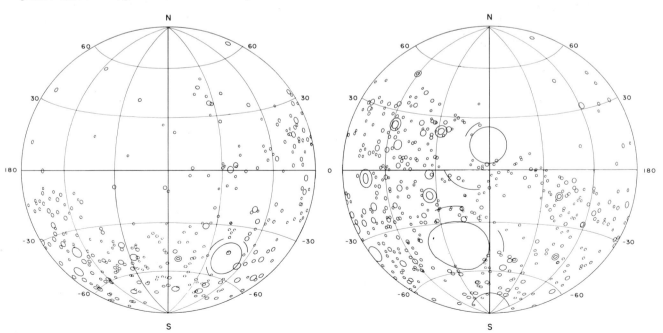

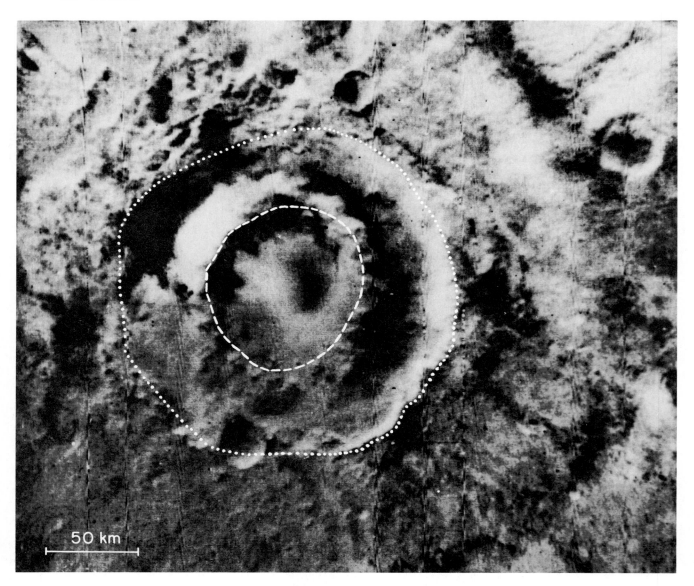

Figure 4.30. The multi-ring basin, Lowell. The crater wall is indicated by a dotted line and an interior ring by a dashed line. (DAS No. 06210438, rev. 129, A camera, center at 81°W, 53°S.)

Hellas is the largest basin structure on Mars, approximately 1600 km in diameter. Many isolated mountain peaks are found in a roughly circular outline, occurring over a large area to the east of the basin. Valleys radial to the basin are seen to the northeast and southwest. Beyond the western rim smooth concentric parallel ridges are seen. Several vaguely defined rings consisting of concentric scarps in the western region are described by Wilhelms (1973). While Hellas undoubtedly represents an original impact structure, its present morphology indicates that it has been very highly degraded.

In summary, then, Martian basins are much more degraded than lunar counterparts. Extensively eroded rings are identifiable only as isolated peaks and low scarps. Ejecta blankets are recognized inferentially by a paucity of craters in the vicinity of the basin (see crater maps of appendix C). Presumably the ejecta has covered and obscured many pre-existing craters. Plains deposits of possible volcanic origin fill the basins and extend across parts of the rim.

Crater Degradation

Martian craters appear to have formed different initial shapes and to have experienced more intense modification than lunar craters. This contrast is

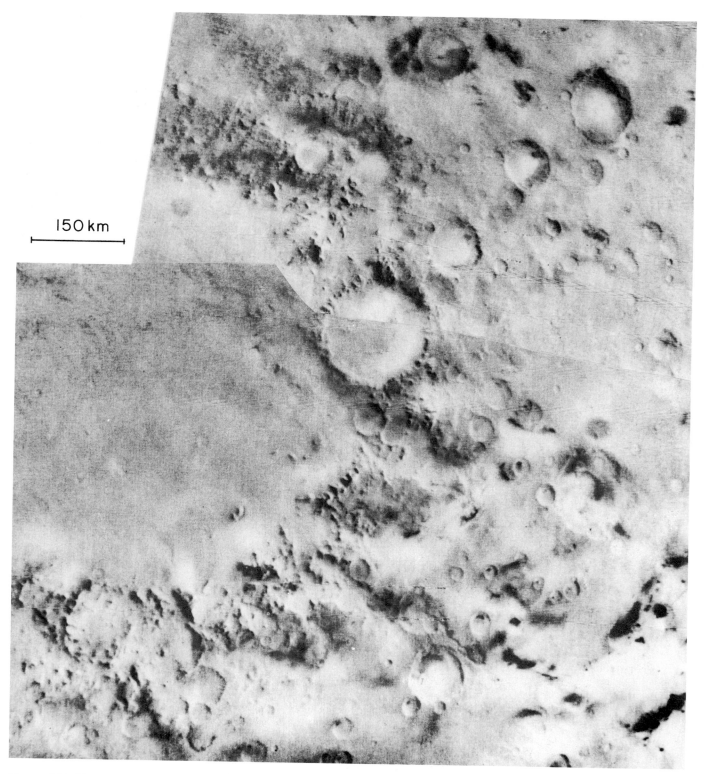

150 km

Figure 4.31. The eastern margin of the Argyre basin. Mountainous terrain borders the plains that fill the basin interior. Discontinuous concentric scarps are present at greater radial distances, particularly in the lower right of the figure. (DAS No. 05740598, rev. 116, A camera, center at 35°W, 56°S; DAS No. 05812628, rev. 118, A camera, center at 25°W, 45°S.)

shown in Pike's (1971) comparison of depth/ diameter ratios for Martian and lunar craters (fig. 4.32). Similar results are obtained by plotting the depth/diameter ratios of Martian craters as determined by Mariner 9 UV measurements (fig. 4.32). The Martian craters are uniformly shallower than fresh lunar craters for all diameters. Even relatively fresh Martian craters, with little or no eolian fill, fall in the range of highly modified lunar pre-Imbrian craters. This strongly suggests a gravity scaling effect similar to that seen on the moon and Mercury (Gault et al., 1975). Craters formed under conditions of greater surface gravity will undergo less excavation and more fallback modification. After formation they will be more susceptible to massive wall slumping by gravitational failure. Some systematic variations exist within the Martian depth/ diameter data (fig. B.2). Martian craters that are visually classified as fresh or slightly degraded have depth/diameter ratios higher than more degraded craters, suggesting that the processes of crater infilling are associated with changes in crater morphologic parameters. Some portion of crater morphologic modification may be associated with an early intense bombardment of the planet, as it is on the moon (Head, 1975). Eolian activity undoubtedly is responsible for continued modification.

A continuous sequence from fresh-appearing simple and complex craters through barely discernible crater remnants can be found on both A- and B-frame Mariner pictures (fig. 4.33). The first features to be removed are ejecta deposits. The process

is described in chapter 7. Following removal of ejecta, terraced walls are modified to hummocks and radial gullies. At the same time crater interiors are filled with smooth plains materials. At first, these processes occur without appreciable modification of crater depth or rim sharpness. Eventually, however, upraised crater rims are beveled level as infilling continues. Gullying and pitting of crater walls become prominent at this stage. Finally, craters are obliterated by infilling and complete erosion of rim and wall fragments.

This degradation sequence applies to craters in the equatorial region that apparently have not been subjected to burial by regionally disposed eolian blankets of sedimentary debris. By contrast, in polar and layered etched plains and in high-latitude areas covered with debris mantles, a sequence ranging from partial burial through complete burial to subsequent exhumation can be defined. These features are further discussed in chapter 7.

Processes besides eolian modification that could have contributed to crater degradation on Mars include: lunar-style impact erosion, water-related processes, volcanic mantling, and enhanced thermal creep due to the large daily temperature variation (> 130°C) (Sharp, 1968). In the next sections we discuss several models of Martian history that lead to divergent conclusions regarding the dominance on Mars of erosive processes previously identified on the moon. We warn the reader that the discussions are confusing because they reflect a problem that is incompletely encompassed in terms both of

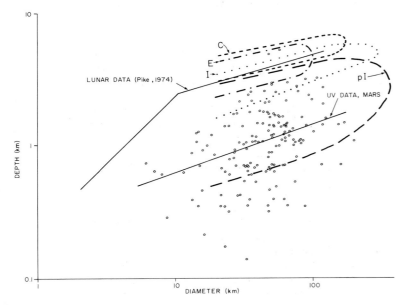

Figure 4.32. Depth-diameter relationships for lunar and Martian craters. The curve for fresh lunar craters shows a prominent inflection point at approximately 10.6 km. The reason for this is not clearly understood, but is probably related to slumping of wall materials above a critical crater size (Quaide et al., 1965). The distribution of lunar depth-diameter values shows wide scatter but, in general, measurements for Copernican (C), Eratosthenian (E), Imbrian (I), and pre-Imbrian (pI) craters fall within the indicated envelopes. This illustrates reduction of depth with increasing age (Pike, 1971). Depth-diameter values for individual craters on Mars are shown by dots. These are determined using UV spectrometer topographic profiles as drawn by Barth et al. (1973). The data are fitted to a straight line by a least-squares technique. As in the case of the moon, Martian values show wide scatter. There is no noticeable decrease in depth with increase in age, determined by visual inspection. The fact that most Martian points fall within the lunar pre-Imbrian envelope suggests that eolian infilling on Mars may be volumetrically less important than previously supposed. Gravitational modification, which would be more prominent on Mars than on the moon, can account—in large part—for the observed relationships. Eolian infilling has produced remarkably flat floors but has not greatly reduced crater depth.

data and analysis. Various models have so little in common with respect to starting assumptions and data selected for analysis that any integration—however desirable from a pedagogic point of view—is difficult.

Techniques of Crater Analysis

Diameter-Frequency Distributions

The technique most commonly used for analysis of crater populations is to plot crater diameter against a variable that measures the abundance or frequency of craters. Early investigators of the moon found that the number of craters increases geometrically as the size decreases, a consequence of the fact that the distribution of interplanetary debris is dominated by smaller-sized particles. To a first approximation, diameter-frequency distributions for lunar maria and highlands followed a power law ax^{-k}, where x is diameter and k and a are constants that must be determined from the data. This functional dependence is emphasized when logarithmic axes are used for plotting data, since the resulting distributions approximate straight lines with slopes $-k$ and intercepts equal to $\log a$ at $x = 1$ km (fig. 4.34). All discussions in this section will be based on log diameter–log frequency diagrams.

The diameter-frequency distributions used to present the data in this book are of a new type. There are three types of diameter distribution diagrams presently in the literature: "cumulative distributions" (Leighton et al., 1965; Murray et al., 1971), "logarithmic incremental distributions" (Hartmann, 1966a, b), and "incremental frequency distributions" (Chapman and Haefner, 1967). The diameter-frequency distributions used here can be defined mathematically as either the derivative with respect to diameter of the cumulative distributions or the limiting case of Chapman's incremental-frequency diagrams as the diameter increment approaches zero. The frequencies so defined are directly comparable to Chapman's frequencies in all respects, including the tendency for the slopes on a log-log frequency plot to be steeper by one unit in comparison with the cumulative or incremental distributions.

Details of the calculation of diameter-frequency distributions are given by Jones (1974b). The advantages are as follows:

1. The accuracy is constant over the entire diameter range.
2. The method produces usable distributions for very small samples.
3. There is one point plotted for every measured crater diameter.
4. The ratio of the frequencies measured at any

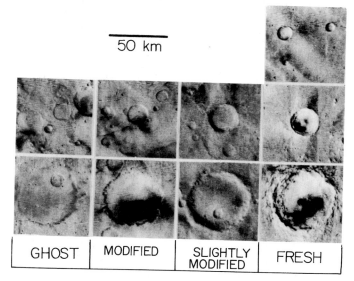

Figure 4.33. Different degradational states for Martian craters. From top to bottom, the fresh crater class consists of bowls, bowls with terraces, and flat-floored craters with terraces. The top rows for the other three classes are for craters with smooth-appearing walls, the bottom rows for classes with hummocky or furrowed walls. (From right to left and top to bottom, DAS No. 07506488, rev. 165, A camera, center at 248°W, 16°S; DAS No. 07866778, rev. 175, A camera, center at 188°W, 17°N; DAS No. 07506488, rev. 165, A camera, center at 248°W, 16°S; DAS No. 06678808, rev. 142, A camera, center at 170°W, 11°S; DAS No. 06678808, rev. 142, A camera, center at 170°W, 11°S; DAS No. 07362708, rev. 161, A camera, center at 264°W, 10°S; DAS No. 07362708, rev. 161, A camera, center at 264°W, 10°S; DAS No. 07290818, rev. 159, A camera, center at 271°W, 5°S; DAS No. 07650198, rev. 169, A camera, center at 232°W, 23°S.)

two diameters gives the relative frequency of one diameter to the other. (This point may seem trivial, but it is not true for either cumulative or incremental distributions.)

5. The final distributions allow good visual interpretation of the effects of various processes of crater formation and obliteration.

6. The distributions can be compared directly with theoretically calculated frequency distributions.

At small diameters, the "bumps" in the distributions are residual from the digitizing method and are not real (Jones, 1974a, b). The limiting digitizer accuracy of 0.01 in. produces a segregation of craters into diameter increments that, since photographs were fairly constant in scale, were preserved through the reduction programs (see appendix B). Although the bumps attract one's attention, their existence should be ignored.

As mentioned earlier, the older a surface the more

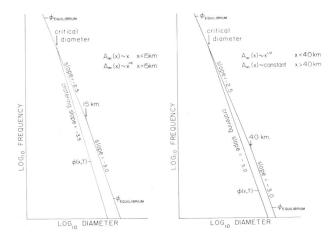

Figure 4.34. A cratering function $\phi(x,T)$ is shown at time T, displayed in a log diameter–log frequency diagram. In addition, an equilibrium distribution ϕ_{eq} is shown. The diameter dependence (and hence the two-segmented appearance) of the equilibrium function depends on the product of $A\infty(x)$ (equivalent to $A(x)$ in the text) and x^{-k}. (See equation 4 in the text.) The intersection of the equilibrium and cratering functions defines the critical diameter.

(a). The exponent of x^{-k} is chosen as -3.5. The amount of obliteration necessary to remove craters, $A\infty(x)$, is based on the depth-diameter relation for fresh lunar craters obtained by Pike (1971). It is assumed here that $A\infty(x)$ is proportional to a crater's initial depth.

(b). The exponent of x^{-k} is chosen as -3.0. The diameter dependence of $A\infty(x)$ is obtained from our data, as is explained in the text.

cratered it becomes. However, in the absence of obliteration processes, distributions remain parallel, with slope equal to $-k$. Only the value of the coefficient a increases. By "calibrating" a set of parallel curves relative to time, a set of isochrons is derived. Diameter-frequency distributions can be matched to isochrons and the length of time a surface has been exposed to impact can be obtained.

However, observed diameter-frequency distributions only approximate straight lines. The deviations from these ideal cases provide the clues for interpreting the effects of obliteration processes on a crater population. Other deviations may be due to complexities in the cratering distribution. These may include a surplus of primary or secondary craters at small diameters or a depletion of large-diameter impacting objects capable of producing large craters.

Crater Obliteration

Craters are not only formed but are also obliterated by a variety of processes. Consider an experiment in which one starts with a perfectly smooth surface and produces a very large number of nonoverlapping craters with identical diameters and morphologies. If all craters are subjected to a uniform obliteration process, they vanish at exactly the same instant. However, for all actual Martian obliterative processes, such as dust filling or wind abrasion, there is certainly a spread in time of crater disappearance. If one were to measure the lifetimes of all craters, it is reasonable to expect a binomial (or Poisson for large samples) distribution where observed crater lifetimes cluster around a central value t_e. If t_e is measured into the past, it is called the retention age (Hartmann, 1966b); if t_e is measured into the future, it is called the expected lifetime.

The retention age t_e is dependent on the obliteration rate and the amount of obliteration necessary to render a newly formed crater undetectable. This amount of obliteration, designated by $A(x)$, depends on the crater's initial diameter, morphology, and the type of obliteration process. The obliteration rate and amount of obliteration are defined so that $A(x)$ equals the obliteration rate $c(t)$ integrated over the lifetime of a crater of diameter x:

$$A(x) = \int_{T-t_e(x,T)}^{T} c(t) \cdot dt \qquad (1)$$

A crater of diameter x, formed at time $T - t_e(x,T)$ will just be disappearing at time T. That is, in a duration of time $t_e(x,T)$ the crater will have experienced enough obliteration $A(x)$ to make it undetectable. The lifetime $t_e(x,T)$ is not necessarily the age of a surface; it is only the oldest possible age of a crater of diameter x.

Physically, one can relate $A(x)$ and $c(t)$ to the experiment described above. Consider a simple obliteration process where flour is sifted onto a cratered surface. The obliteration rate $c(t)$ might be measured in centimeters of flour per minute. This rate is independent of crater diameters. The amount of obliteration required to make a crater of a specified diameter undetectable might be measured in centimeters of flour and is independent of the cratering rate. By defining $c(t)$ and $A(x)$ in this way, the time and diameter dependencies are completely separated.

The obliteration rate is thus defined, as is the cratering rate, in terms of the effects of a process rather than in terms of the process. A change in the obliteration rate changes only the expected lifetime of a crater and does not necessarily result in an instantaneous change of the observed frequency (as changing the cratering rate would).

Assuming a diameter-dependent $A(x)$ and a time-dependent obliteration rate $c(t)$, a retention age can be calculated by means of equation (1) for each diameter. The frequency $\phi(x,T)$ observed at time T and diameter x resulting from the cratering function $\phi(x,T)$ is simply

$$\phi(x,T) = x^{-k} \int_{T-t_e(x,T)}^{T} a(t) \cdot dt \qquad (2)$$

where $a(t)$ is the cratering rate. The cratering and obliteration rates may be related or independent. Computer-generated models presented later are calculated from an algorithm based on equations (1) and (2). These equations are extremely important since they can describe virtually any combination of cratering and obliteration processes. However, the functional dependence on time and diameter for $\phi(x,t)$, $A(x)$, and $c(t)$ must be specified. This specification can be done by experimentation or by theoretical modeling as exemplified in numerous papers by Marcus (1964; subsequent papers), Hartmann (1971b), Chapman et al. (1969), and Soderblom (1970).

Equilibrium Between Cratering and Obliteration

Equilibrium can exist between any processes of obliteration and cratering. Saturation equilibrium represents a special case where the obliteration process results from craters overlapping and degrading previously formed craters.

At any time T the cratering and obliteration rates are $a(T)$ and $c(T)$, respectively. The equilibrium frequency at any time T is the frequency that one would eventually observe if the cratering and obliteration rates were maintained indefinitely at constant values. From (1), one can solve for the expected lifetime $t_e(x,T)$:

$$t_e(x,T) = A(x)/c(T) \qquad (3)$$

From (3), one finds that the equilibrium frequency $\phi_e(x,T)$ is

$$\phi_e(x,T) = \left[a(T)/c(T) \right] \cdot A(x) \cdot x^{-k} \qquad (4)$$

This frequency represents the instantaneous equilibrium frequency between $a(t)$ and $c(t)$ at time T.

A necessary and sufficient condition for the existence of equilibrium at any diameter is that the crater frequency observed must equal the equilibrium frequency predicted by (4). Equilibrium need not exist over the entire diameter range. However, if equilibrium has been attained for any diameter, equilibrium must also have been attained for all smaller diameters as well, provided that smaller craters require less obliteration to disappear than large craters. Once equilibrium has been attained, changing the ratio of cratering rates to obliteration rates results in disequilibrium at all diameters.

The diameter dependence of the function $\phi_e(x,T)$ clearly depends on the product of $A(x)$ and x^{-k} (eq. 4). Two examples are plotted in figure 4.34, where the diameter dependencies of the cratering

function are $x^{-3.5}$ and x^{-3}. The equilibrium frequency distributions ϕ_{eq} are two segmented distributions with slopes of -2.5 and -3.0 and have breaks in the slopes at 15 and 40 km, respectively—these values resulting from the choice of $A(x)$ in each example.

A physical interpretation of the double-segmented equilibrium distribution arises from the observation that freshly formed craters do not have a uniform depth-diameter configuration over all diameters. Pike (1971) presents a depth/diameter relationship for fresh lunar craters. The depth is proportional to the diameter for diameters less than 15 km and proportional to the square root of the diameter for diameters greater than 15 km. Assuming that the total amount of obliteration necessary to remove a crater is directly proportional to its initial depth, one obtains the distribution shown in figure 4.34a.

One cannot necessarily assume that Martian craters follow the lunar depth-diameter relationship or that the amount of obliteration is proportional to the depth. Nevertheless, it is reasonable to expect that large Martian craters, which apparently form with an initial depth/diameter ratio shallower than that for small bowl-shaped craters, require less obliteration relative to their diameters. The amount of obliteration $A(x)$ represented in figure 4.34b appears to best match the observed Martian crater data. The value of $A(x)$ is proportional to the square root of the diameter below 40 km, and $A(x)$ is constant above 40 km. The uncertainties in the data allow for $A(x) \sim x^{0.6 \pm .2}$ below 40 km and for $A(x)$ within $x^{\pm 0.2}$ at large diameters.

The critical diameter, defined by other investigators (Öpik, 1966; Gault, 1970), is the largest diameter along the equilibrium distribution where equilibrium has been attained. This definition is applicable only if small craters erode more rapidly than large craters. Graphically, the critical diameters (fig. 4.34) occur at the intersection between the cratering distribution $\phi(x,T)$ and the equilibrium distribution. As the cratering function $\phi(x,T)$ shifts to higher frequencies at all diameters, the critical diameter moves to larger diameters. In figure 4.34 equilibrium is attained simultaneously for all diameters larger than 40 km since $A(x)$ is constant.

A change in either the cratering rate $a(t)$ or the obliteration rate $c(t)$ shifts the equilibrium frequency distribution to a new value. However, the behavior of the equilibrium frequency distribution ϕ_{eq} is simple. If the cratering (obliteration) rate is increased, each frequency along the equilibrium distribution is shifted up(down), and vice versa, the slope at each diameter remaining unchanged. If the ratio of the cratering rate to the obliteration rate is held constant, the equilibrium distribution remains unchanged.

Models of Martian Cratering History

Early Speculations

Following the limited photographic return from the Mariner 4, 6, and 7 missions, the dominant view was that Mars is enveloped by an ancient, densely cratered crust. After examining the first Mariner 4 pictures, Leighton et al. (1965) likened the Martian cratered terrain to the lunar uplands. However, the literal aspects of the comparison were immediately challenged. Witting et al. (1965), Anders and Arnold (1965), and Baldwin (1965) all concluded, first, that the Martian meteoroid flux is ten or more times greater than the lunar flux and, second, that Martian crater densities are substantially lower than those of the lunar highlands. All these authors argued that the Martian surface must have an exposure age somewhere between 300 and 800 m.y. The consensus was short-lived. Within a year Binder (1966) and Hartmann (1966b) pushed the estimated ages back beyond 2 billion years. Although they accepted a Martian meteoroid flux approximately an order of magnitude greater than the lunar flux, they argued that the Martian surface should be compared not with the saturated lunar uplands but with the maria, then estimated to have an age between 2 and 4 billion years. Since the heavily cratered region of the Martian surface had 10 to 20 times as many craters as the lunar maria, the hypothesized ages of the two surfaces fortuitously converged.

Preliminary understanding of Martian degradation history was stimulated by Öpik's (1965) observation that diameter-frequency distributions of craters display a break in slope at 20 km. He attributed this to an obliteration process that affected all craters, but below 20 km diameters it progressively erased larger numbers of craters with decreasing diameter. This is essentially the critical-diameter argument, described in a subsequent section. The interpretation was further elaborated by Hartmann (1966b). Chapman et al. (1968, 1969) offered the alternate interpretation that the break in slope was caused by observational bias associated with the relatively poor-quality Mariner 4 pictures. Although pictures from subsequent missions document the break in slope as a real phenomenon, Chapman's general concern for misleading incompleteness of the photographic record is well founded.

The diversity of interpretations based on Mariner 4 crater statistics is hardly surprising. Most investigators confined their analysis to fewer than 100 craters. Final review of specially enhanced pictures revealed 300 craters and an equal number of vaguely circular depressions. Confronted with this meager sample, the early investigators deserve credit for successfully outlining some of the models that are still being argued on the basis of much more comprehensive Mariner 9 data.

Crater Dichotomies

Mariner 6 and 7 pictures showed the crisp, small-scale detail lacking in Mariner 4 images. Within the cratered terrain, Murray et al. (1971) identified two distinct morphological classes of craters. One included small bowl-shaped craters less than 15 km in diameter. The other comprised flat-floored craters generally larger than about 15 km. These were thought to have been extensively modified, in contrast with the smaller craters, which had retained their original impact morphology. The apparently sharp distinction between the two classes led Murray et al. to postulate one or more episodes of widespread crater modification that postdated the final stages of planetary accretion but predated formation of the bowl craters. In addition, the authors noted the absence of larger, fresh, bright-rayed craters that so dominate the lunar landscape—craters like Copernicus, Aristarchus, and Tycho. This absence they tentatively attributed to a difference in the size-frequency distribution of impacting bodies, with a deficiency of larger bodies in the vicinity of Mars.

The reality of any knife-edged distinction between small fresh and large degraded craters was questioned by McGill and Wise (1972), who noted that small craters showed a progression of morphological forms from fresh, crisp bowls to barely discernible shallow remnants. Because McGill and Wise were deciphering details close to the resolution limit of the pictures, their interpretations were difficult to confirm. However, Mariner 9 pictures clearly reveal the variety of degradational stages that they first recognized and defined.

In summary, the general consensus of pre-Mariner 9 investigators was that large craters had existed for most of Martian history, possibly remnant from the end of the accretionary period. Large craters had been subjected to some process of obliteration that had removed or at least highly modified most craters below 15–20 km diameter. Present obliteration processes were thought to be less intense than previous erosion since there was a population of fresh, small craters superposed on the older degraded craters.

A First Analysis of Mariner 9 Results

The first comprehensive review of Mariner 9 crater information was by Hartmann (1973a). He observed that the two Martian satellites, Phobos (mean radius 11 km) and Deimos (mean radius 6 km), were cratered to densities approaching saturation (fig. 4.35). The densities were higher than those

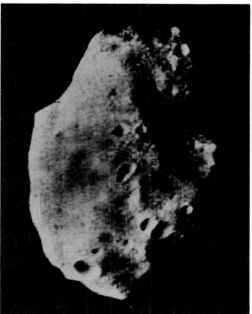

Phobos

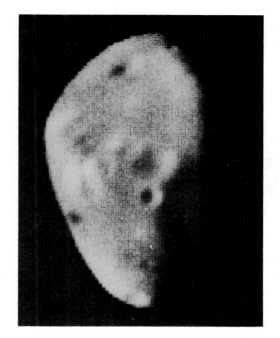

Deimos

Figure 4.35. B-frame pictures of Phobos and Deimos. Both satellites are irregularly shaped triaxial objects. The axes of Phobos are 13.5, 10.7, and 9.5 km; those of Deimos are 7.5, 6.1, and 5.5 km. The satellites have a low albedo, suggesting a basaltic or carbonaceous chondritic composition. Photometric properties are consistent with the presence of an impact-generated regolith overlying well-consolidated material (Pollack et al., 1973; Noland et al., 1973; Zellner and Capen, 1974). For discussion of the shape and topography of Martian satellites see Veverka et al. (1974a) and Duxbury (1974). (Deimos, DAS No. 06918173, rev. 149. Phobos, DAS No. 02816315, rev. 34.)

tabulated for the more densely cratered parts of the Martian surface, suggesting that the two satellites, with no atmospheres, had retained more of their crater record than had Mars.

Turning his attention to Mars, Hartmann used Mariner 9 pictures to construct improved diameter-frequency curves for the cratered terrain. He identified three linear segments of the crater distribution, each interpreted as the result of a distinct episode in Martian history. A relatively steep segment (CD in fig. 4.36) records the period of terminal accretion. During or immediately after this period there was extensive erosion, perhaps associated with a temporary dense atmosphere. This erosion reduced the number of craters below saturation values, even though the slope of the frequency curve remains parallel to the saturation curve—and Hartmann's influx curve. A segment with reduced slope (BC in fig. 4.36) was identified at the intermediate diameters of 5 to 30 km. This was interpreted to be the consequence of a long period of constant degradation, probably eolian in nature, during which craters were erased by infilling on the order of 10^{-4} cm/yr.

An equilibrium, or steady-state condition, was established between cratering and obliteration, and the break in slope at 30 km corresponds to a critical diameter. The modification would preferentially destroy smaller craters, accounting for the decrease in slope of the frequency curve. This is essentially the same observation previously made by Öpik (1965) and Hartmann (1966b) from study of Mariner 4 pictures. At diameters less than 5 km, a third segment (AB in fig. 4.36) was identified. The frequency curve steepens until it parallels the influx slope. Hartmann interpreted this as the record of a recent period of reduced erosion during which craters have retained their impact morphologies. Because there has been little or no erosion, it was reasoned that the size distribution of craters must precisely parallel the size distribution of impacting meteoroids.

Immediately after Mariner 9 completed its photographic survey of Mars, there was inevitable pressure from many scientists to translate diameter-frequency relationships into absolute ages. For example, the cautious statement that Olympus Mons

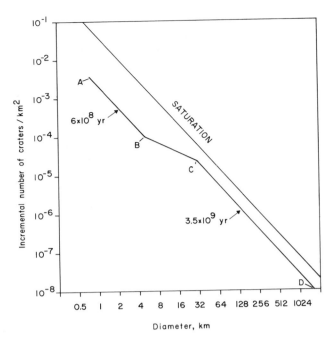

Figure 4.36. Crater distributions on different Martian surfaces, after Hartmann (1973a). For explanation, see text.

is a recently active volcano only whetted the audience's appetite: How recent? Might it be active now, a hundred years ago, a million years ago? In order for Hartmann to use his data to determine absolute ages, it was necessary for him to assume a singular Martian cratering rate. Primarily on the basis of previous studies of asteroidal distributions, he concluded that the influx rate for Mars is 10 times greater than for the moon, the latter rate having been calibrated by radiometric age determinations of lunar samples. Because the impact velocities on Mars are roughly five-sevenths those for the moon, projectiles will form proportionally smaller craters on Mars. Hence, the cratering rate on the two planets differs by a factor of approximately 6. Hartmann was fully aware of the uncertainty of his influx estimate and, in fact, specified a possible error of about 3. Any absolute ages similarly could be three times too high or too low. Putting these uncertainties to one side, Hartmann estimated that the exposure age of the heavily cratered terrain (corresponding to segment CD in fig. 4.36) was about 3.5 b.y., and that an intermediate period of erosion and deposition (corresponding to BC) continued until 600 m.y. before the present. Since then there has been little or no erosion (segment AB).

Crater distributions in the volcanic plains of the northern hemisphere yield an exposure age of 300 m.y. indicating, to Hartmann, major volcanic inundation at that time. The most recent volcanic surfaces display very few craters, and some of these may be volcanic in origin, so calculation of reliable ages is difficult. Hartmann estimated an age of 100 m.y. for the flanks of Olympus Mons, 200 m.y. for Arsia Mons, 100 m.y. for Ascraeus Mons, and 80 m.y. for Pavonis Mons. Most investigators, including Hartmann, subsequently have regarded these absolute ages as lower limits since they are based on a cratering rate that is at the high end of probable values.

Evidence for a Period of Increased Erosion

Chapman (1974) and Jones (1974a) independently demonstrated that Mars has experienced an erosional history that is not directly coupled to the cratering history. The distinctive feature of their models is the presence of a brief episode of an increased erosion rate that was independent of the cratering process. Their theses are complex, but the basic arguments are easily appreciated. Instead of looking at crater diameter-frequency curves for all craters, consider separately each of the four classes of Martian craters compiled by Arvidson (1974a) and described here in appendix B: fresh (f), slightly degraded (s), moderately degraded (m), and highly degraded (h) (table B.2).

If crater production and crater modification have both proceeded at constant rates throughout Martian history, then the diameter-frequency curve for each morphologic type should mimic the curve for all craters. Each curve will display a critical diameter below which equilibrium is attained. Equilibrium slopes for all morphologic types will be identical to the slope for all craters, although the frequencies will be decreased. This illustrates the fact that each crater spends a fixed amount of time in a youthful, mature, and old-age stage. The situation is analogous to human populations. Although each person grows older, the number of persons of particular age is constant with time—assuming that birth rate and expected lifetime also remain constant.

Martian craters do *not* show the parallel relationships just described. Different morphologic types show frequency maxima at different diameters and, as such, deviate markedly from the distribution described by all craters. The curves can best be duplicated by a model in which there is a prominent high-intensity spike in the erosional history. It is easiest to imagine instantaneous blanketing of massive proportions. All craters below a certain size will be completely obliterated. At progressively greater diameters, progressively less degraded craters will be visible (fig. 4.37). Following this intense erosional spike, a new crater population will be visible at smaller diameters, and, eventually, the total crater distribution will reflect the equilibrium associated

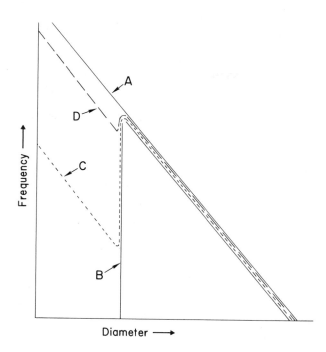

Figure 4.37. The results of instantaneous blanketing; schematically shown. At time A, prior to the blanketing event, an equilibrium distribution has been established. At time B, immediately following the blanketing, all craters smaller than a certain diameter are completely obliterated by burial. By time C a few craters have formed on the new surface. By time D the original pre-blanketing equilibrium has almost been re-established.

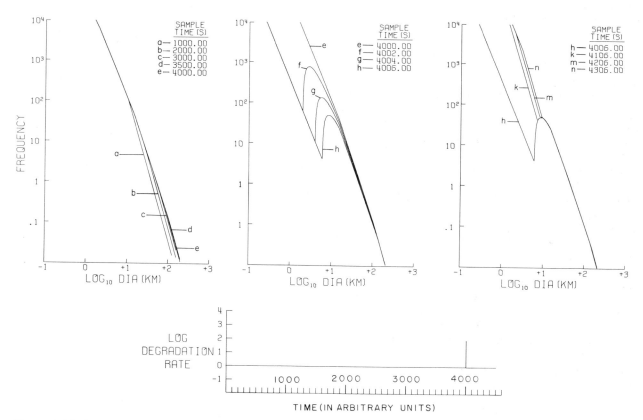

Figure 4.38. Computer-generated diameter-frequency distributions for the total (f + s + m + h) crater population. A constant cratering rate was maintained. A constant obliteration rate was maintained except for a brief increase (times e through h) to 100 times the original rate. The behavior of the distribution is described in the text.

with the current erosional rate. All traces of the intense spike will be removed.

Chapman (1974) describes the effects of different cratering rates on the age and character of the erosional event. Predictably, a higher flux rate favors a younger age for the event. If the cratering rates for Mars and moon have been approximately equal, as proposed by Soderblom et al. (1974), then the event occurred about 3.5 billion years ago. If the rates have been about 5 to 1, the event is positioned at about 1.5 billion years.

In an attempt to see if the crater data place any constraints on the cratering rate, Jones (1974a) demonstrates that, under a set of specific, plausible conditions, the observed distributions could have resulted from a specific model in which an erosional episode with an intensity 100 times the background value occurred four-fifths of the way through planetary history (fig. 4.38). When the distributions for morphologic types are compared with the actual values, a general coincidence is observed (fig. 4.39). Both Chapman and Jones examine the uniqueness of their solutions. Jones, in particular, concludes that the existence of a period of enhanced erosion is unavoidable. However, the position of that period in absolute time is uncertain. It has not been demonstrated that the relatively recent event indicated in figure 4.38 is a unique requirement. In fact, relationships between regional geologic units and consideration of models for atmospheric evolution (chapter 8) suggest a more ancient period of obliteration.

An additional problem in the analyses of Chapman and Jones is imposed by the available data. Many of the more significant details of the diameter-frequency distributions are evidenced at crater diameters less than 10 km. These data are not available in Mariner A frames. Jones, for example, determines morphological classes only for craters greater than 10 km. Below that size, all craters are classified as fresh. Analysis of Mariner B frames might assist in this resolution problem, but a systematic study will be feasible only with the return of Viking orbiter pictures.

Although the analysis of Jones is limited by picture quality, it is important to note that the observed falloffs at small diameters for different morphologic classes are *not* resolution effects. If this were the case, the logical relation would be that the positions of the maximums for the morphological classes would occur at increasing diameters for the fresh though highly degraded craters, in a sense opposite to that observed. The reason is that degraded craters with subdued features are more difficult to detect than fresh craters with well-defined features. Therefore, the observed relation of highly, moderately, and slightly degraded craters agrees with the model

of obliteration described in figure 4.38, but not with the progression predicted from resolution effects.

A Moon-like Mars

Another attempt to determine the Martian cratering rate from observed crater distributions, considered together with lunar distributions, was attempted by Soderblom et al. (1974). They used the observation that large Martian craters display a range of morphologies to conclude that intense early cratering was coupled with an early episode of intense obliteration. They do not state that the cratering process was, itself, responsible for the obliteration, only that cratering and obliteration were coupled in time. It is important to emphasize that an identical range of morphologies can be obtained from any model in which cratering and obliteration rates maintain a constant ratio.

Following their postulated period of early intense cratering and erosion, Soderblom et al. specify a decrease in the obliteration rate resulting in today's relatively low values. This decrease is identical to the one described by Hartmann in a previous section and—although the underlying interpretations are different—identical to the decrease in obliteration rate following a period of increased erosion postulated by Chapman and Jones. As erosion decreased in effectiveness, small craters forming on ancient cratered terrain retained their crisp, fresh morphologies. The first of the plains units formed shortly thereafter.

Soderblom et al. observe that the densities of 4–10 km craters on ancient cratered terrain are similar to the densities measured for lunar maria, which are radiometrically dated at approximately 3.5 billion years. A critical assumption justifying the choice of the 4–10 km range is that all craters of this size formed early in Martian history were destroyed by the early intense obliteration. Larger degraded craters are interpreted as having survived from the terminal stages of accretion. Because of their large dimensions, they have escaped complete obliteration.

Soderblom et al. argue that a correspondence in 4–10-km crater densities between Martian cratered terrain and heavily cratered plains on one hand, and lunar maria on the other hand, can scarcely be coincidental. The similarity constitutes evidence that emplacement of lunar maria and oldest Martian plains occurred at the same time on planets experiencing the same cratering fluxes.

In further support of a common history for the moon and Mars, they emphasize that both bodies display a dichotomy of terrains—heavily cratered regions and lightly cratered plains. However, in the 4–10 km diameter range, the most densely cratered plains on Mars have almost as many craters as the

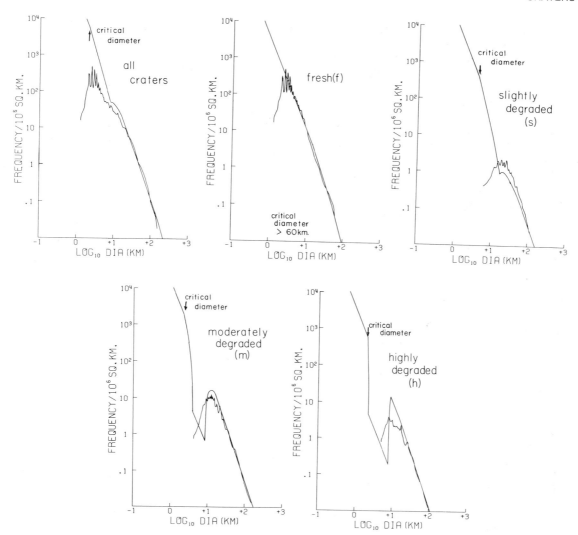

Figure 4.39. Computer-generated diameter-frequency distributions for the total crater population and for each of the four morphologic classes. The distributions shown correspond to time m in figure 4.38. Observed distributions combined from regions 1, 24, 29, and 32 (see fig. 4.40) are superimposed. Distributions for more degraded craters display troughs resulting from the increased obliteration rate. The maximums between troughs and large diameters, both observed and predicted, occur at decreasing diameters for more degraded morphologies. Although the right sides of the troughs are delineated, the left sides are missing because of the definition of all small craters as fresh. The appearance of a range of morphologies at small diameters in B frames suggests that the predicted behavior is real, at least qualitatively.

ancient cratered terrain. Other plains units display a fairly continuous range of ages, implying that these strata have been forming more-or-less continuously since the cessation of intense obliteration.

It should be emphasized that the arguments of Soderblom et al. are largely circumstantial. Many tentative similarities between the moon and Mars are mentioned, but an exclusive argument is not developed. Soderblom et al. prefer to believe that the dichotomy of Martian cratered terrain and plains can be explained in terms of a rapidly decreasing flux rate rather than a gap in time between early heavy bombardment and the beginning of plains' formation. They then accept this as evidence that cratering rates on the two bodies have been similar. However, Jones's (1974a) model reviewed in the previous section explains the crater data equally well. An erosional spike at a fairly recent time coupled with a comparably recent period of plains formation explains not only total crater abundances but also morphological statistics, even though it leaves one with the currently unpopular conclusion that cratering rates on Mars have been greatly in excess of those for the moon.

Finally, Oberbeck et al. (1975) have noted that Mercurian cratered terrain is similar to Martian cratered terrain in that both have a crater population deficient in crater sizes less than about 30 km in diameter. They also note that lunar highlands, far from basins and their associated secondary craters, show a similar deficiency. They suggest that the inner solar system was bombarded by objects that broke up under tidal fission as they approached the planets. Such a mechanism would explain the production of a crater population deficient in small crater sizes, and it would also explain the high degree of spatial clustering of large craters on Mars, the moon, and Mercury. According to this model, many of the lunar craters less than 30 km in diameter are secondaries associated with large basins. Because of higher gravity on Mercury and Mars, secondaries formed closer to basins. Crater deficiencies observed on these planetary surfaces might be due to production statistics, not to obliteration effects.

Crater Densities and Terrain Types

We return now to consideration of the terrain types described in chapter 3. Do crater densities follow consistent patterns and do they provide new insights regarding origin and age of the various terrain types? To answer these questions, Jones (1974a) selected 38 regions confined to the $\pm 50°$ latitude band (fig. 4.40). These regions were chosen to sample the terrains qualitatively identified (fig. 3.2). In addition, regions were selected with the aid of computer-plotted maps showing all craters in the 8–16 and 2–8 km diameter ranges. Within each of the regions, crater densities were determined qualitatively to be constant.

Diameter-frequency plots for the 38 regions indicate that cratered plains (pc) units, although scattered across the entire planet, have identical crater densities, and presumably identical ages (fig. 4.41). Most of the volcanic plains (pv) sampled were of one age, except that region 4 is younger and region 7 is the youngest. Both of these regions are within the Tharsis Volcanic province. Moderately cratered plains (pm) (region 16) are identical in age to the pv units.

The age relationships of the plains units can be summarized as:

$$pv_7 < pv_4 < pv_{3,5,6,9,35,37} = pm_{10} < pc_{11,12,13,28,31}$$
youngest   *oldest*

The relative frequencies for these groups are 1 : 3 : 8 : 8 : 20.

The highly cratered regions (cu) (1, 2, 14–26, 29, 30, 32, and 34) are more complicated since they record the integrated history of crater formation and obliteration over a much longer period of time. The most important single relationship demonstrated by each region is evident when the fresh craters are separated from degraded craters as shown in figure 4.42. In each case the fresh crater distribution resembles, at least qualitatively, that for the plains units. The frequency of degraded craters drops off sharply below 10 to 15 km. The slope of the frequency distributions for degraded craters larger than 15 km is the same as for fresh craters. The relative fre-

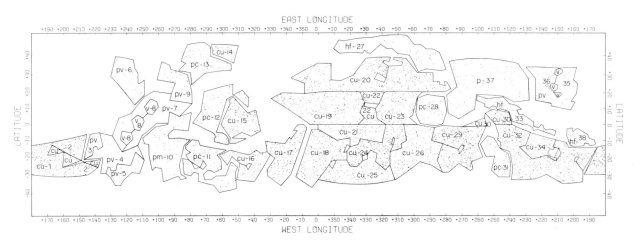

Figure 4.40. Thirty-eight regions chosen for detailed crater analysis by the criteria described in the text. All craters between 8 and 16 km in diameter are shown. The geological mapping units are modified from Carr et al. (1973).

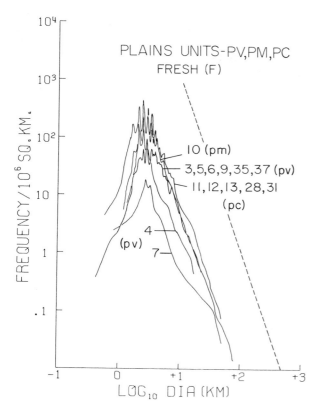

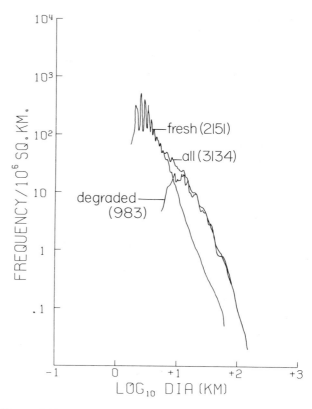

Figure 4.41. Fresh crater distributions for the various plains units. The pv and pc regions indicated in this figure, and in figure 4.40, were combined on the computer before plotting. The relative ages of the units are described in the text.

Figure 4.42. Fresh, degraded, and all craters for region 32 (3,411,998 km²) in the cratered terrain (cu). The number of craters in each sample is shown in parentheses. Similar distributions are obtained for all cu regions. The fresh distribution is nearly identical to the fresh distributions on the pc units.

quency of degraded to fresh craters at large diameters is approximately 9. Below 10 km diameter, the total crater population on the cratered terrain is roughly identical to the total population on the cratered plains (pc) units. Cu frequencies are 1.0–1.3 times as great as pc frequencies. Jones (1974a) interprets the inflection in the diameter-frequency distributions for the cratered units to be the result of a brief intense episode of increased crater obliteration, as described in the previous section. The slightly higher crater densities on some cu units relative to the pc units would at first appear to imply a time interval between the obliteration of craters on cratered units, and the formation of and subsequent cratering on the pc units. However, Jones shows that the pc units probably formed concurrently with the obliteration event. The apparent excess of craters on cu relative to pc is interpreted as a measure of the intensity of the obliteration episode. Regional variations in the crater densities within the cu units (fig. 4.43) are attributed to similar regional variations in the intensity of the obliteration episode.

Soderblom et al. (1974) independently conclude

that the Martian plains units record episodic volcanic activity throughout most of the planet's history, rather than a single volcanic event. Measuring the densities of 4–10 km craters, they define four stratigraphic units. The heavily cratered surfaces (200–250 craters/10⁶ km²) are the oldest, the small craters having formed since the end of early obliteration (3.0–3.5 b.y.). The equatorial cratered plains and polar mottled cratered plains (130–180 craters/10⁶ km²) formed next (1.8–2.4 b.y.). The Elysium volcanics (60–120 craters/10⁶ km²) were emplaced 0.8–1.6 billion years before the present. Finally, in the past 0.8 billion years, the Tharsis volcanic plains (10–70 craters/10⁶ km²) and associated structures have formed. The age estimates are obtained by assuming that lunar and Martian fluxes have been essentially identical (fig. 4.44).

Soderblom et al. also describe the distribution of 0.6–1.2 km diameter craters that are visible in Mariner B frames. They find an abrupt drop in crater density at latitudes higher than 35°S and 35°N. This

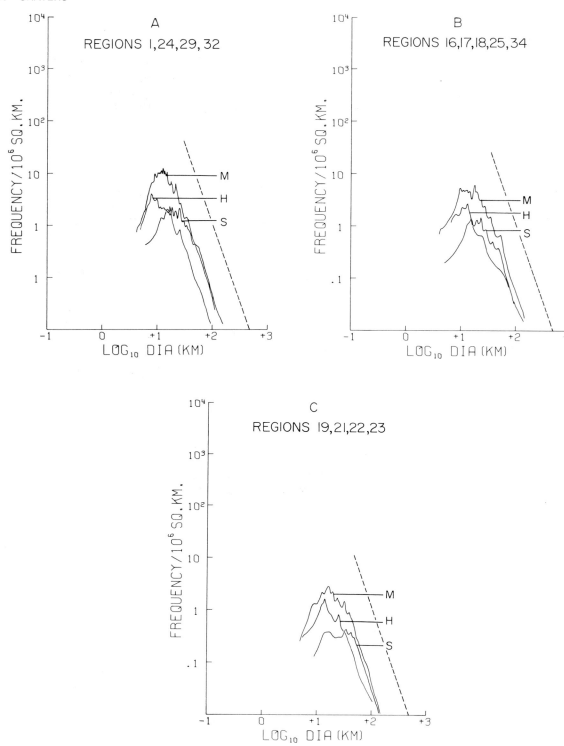

Figure 4.43. The three morphologic classes, slightly (S), moderately (M), and highly (H) degraded, for the three groups.

(a). (regions 1, 24, 29, and 32),

(b). (regions 16, 17, 18, 25, and 34), and

(c). (regions 19, 21, 22, and 23) of regions within the cratered terrain. All three groups exhibit the same characteristics. Morphologic distributions are parallel at large di-ameters with a slope of −3, and the maximums occur at progressively larger diameters for all three distributions. The sharp drop-off below the maximums for each morphologic class and the progressively increasing diameter of maximums from highly to slightly degraded craters are the major indications of an intense episode of obliteration.

The crater data are also combined by morphologic type.

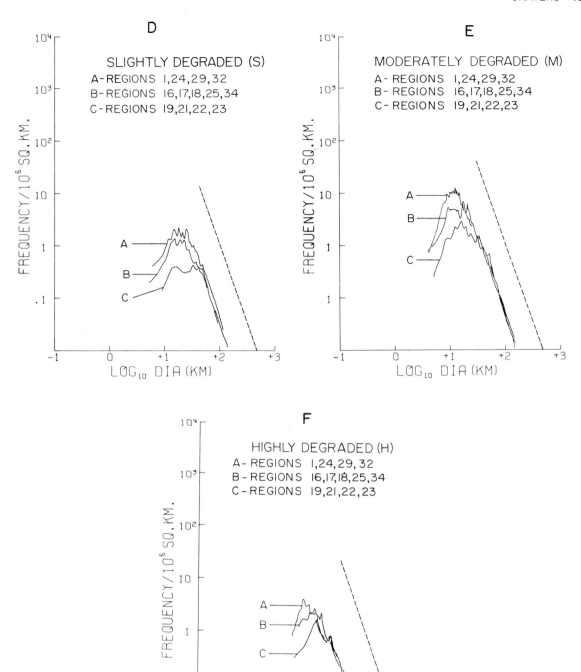

(d). Slightly degraded craters for region groupings (a), (b), and (c).

(e). Moderately degraded craters for groups (a), (b), and (c).

(f). Highly degraded craters for groups A, B, and C. The congruence of all three morphologic distributions at large diameters indicates similar equilibrium conditions in all regions prior to the obliteration event. The noncongruence below the maximums indicates different amounts of obliteration experienced among the three groups during the obliteration event described in the text. Regions included in C have suffered more degradation than those in B. Regions in A have been least affected.

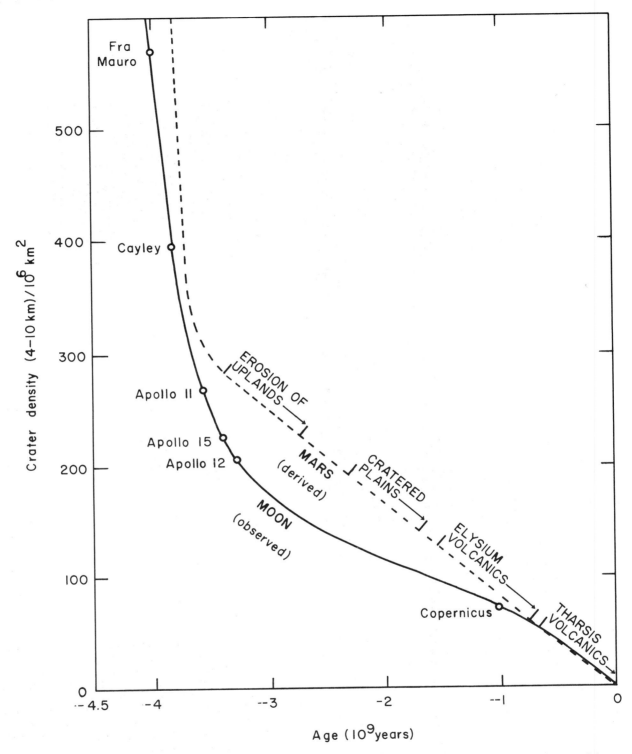

Figure 4.44. Crater densities and absolute ages for certain lunar and Martian surfaces, according to Soderblom et al. (1974). Lunar and Martian meteoroid fluxes are assumed to be approximately the same. Radiometric analysis of lunar samples establishes the scale for absolute ages.

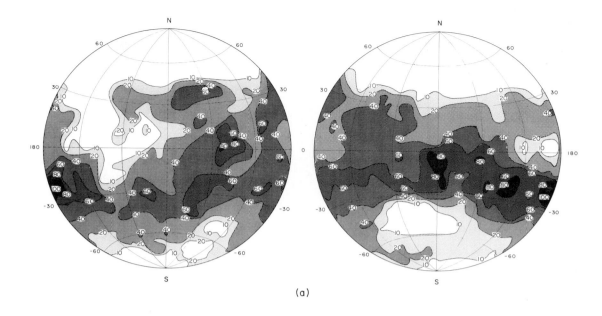

(a)

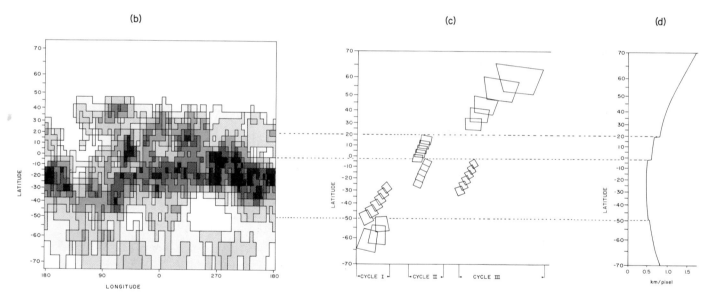

Figure 4.45 (a). Distribution of all craters between 4 and 10 km, plotted on a Lambert equatorial equal-area base. Isolines were generated by measuring crater densities within a 10° diameter search circle moved around the globe on 5° centers. For comparison with physiographic provinces refer to figure 3.2.

(b). Distribution of all craters between 4 and 10 km plotted on a Mercator base to show latitude dependence. Darker shadings indicate higher crater densities.

(c). Distribution of Mariner 9 photographic coverage,

showing the restriction of certain cycles—or groups of revolutions—to particular latitudinal bands.

(d). The average scale of A-frame photography, expressed in kilometers per pixel. (*Identification resolution* is four to six pixels. For further discussion see appendix A.) The steps in the curve represent different resolutions between mapping cycles. The 38 regions shown in figure 4.40 were segregated, where it was practical, into the 0°–50°S, 0°–20°N, and 20°–50°N latitude bands.

is attributed to depletion of craters in polar regions due to burial by the debris blanket first described by Soderblom et al. (1973b). Local variations in the equatorial regions are attributed to local topographic control of dust deposition and mantling.

The effect of photographic resolution in determining crater densities should not be underestimated. Figure 4.45 shows a crater density map for 4–10-km craters prepared from the Brown University data library. A resolution chart for mapping frames as a function of latitude is shown for comparison. Discontinuities in the density plot correspond to changes in slant ranges between successive groups of pictures. In addition, a progressive increase in resolution throughout the mission is associated with atmospheric clearing. Cycle I (rev. 100–138) covers the southern latitudes. Cycle II (rev. 139–177) covers the equatorial regions. Cycle III (rev. 178–217) covers the northern equatorial regions as well as parts of the southern equatorial regions missed by Cycle I. The southernmost coverage of Cycle III is at 35°S, exactly corresponding to the observed break in the 4–10 km crater distribution.

Global Distributions

Using the crater information in the Brown data bank, we have prepared global maps for particular crater types. A search circle of 5° radius is centered on intersections of 5° intervals of latitude and longitude. At each location the number of craters within the search circle is summed. In this way a two-dimensional array of 2664 values is obtained. (Obviously, there is increasing redundancy in the values as the geographic poles are approached.) Crater densities are contoured and the information transferred to Lambert equal area projections.

Some of the products are shown in accompanying maps. Figure 4.46 is a plot of all craters larger than 15 km in diameter. This clearly indicates the dichotomy in terrain type between ancient crust and more recent plains. As previously indicated (e.g., fig. 4.42) this distinction is most clearly delineated for craters with diameters greater than 10–15 km. Below that size, crater densities for ancient terrain and cratered plains converge.

Figure 4.47 is a plot of fresh craters greater than 8 km in diameter. This is approximately an isochron map, as is the map of all craters between 4–10 km (fig. 4.45) (Arvidson, 1974a). The close similarity between the two products, therefore, is to be expected. Using 8 km as a lower cutoff for crater diameter eliminates resolution problems that may affect the 4–10 km distributions.

A map of degraded craters is shown in figure 4.48 and the three subclasses of degraded craters are shown in figures 4.49–4.51. The patterns are subject to many interpretations. Concentrations of degraded craters (fig. 4.48) should delineate regions of most ancient crust. If all heavily cratered regions have undergone approximately the same erosional history, the distributions for three subclasses of degraded craters should mimic the distribution for all degraded craters. In fact, we have previously demonstrated that this assumption is incorrect (e.g., fig. 4.43). The distribution of highly degraded craters is particularly divergent when compared with the distribution of all degraded craters (fig. 4.52). We interpret this to mean that some regions of ancient terrain have undergone extensive mantling early in Martian history either by volcanic deposits (Wilhelms, 1974) or by sedimentary deposits. This mantling not only *reduces* the total number of degraded craters but also *increases* the relative abundance of highly degraded craters.

There are some regions of Mars that have high densities of all types of craters, fresh and degraded (fig. 4.52). Apparently these are regions that have been consistently protected from erosion. Exploring this speculation a little further we have attempted some correlations between crater densities and two pervasive global variables, albedo and elevation. The results, shown in figures 4.53 and 4.54, do not indicate any obvious trends. There is some tendency for density of degraded craters in cratered terrain (cu) to decrease as the elevation decreases. Similarly, the densities decrease as albedo increases (although a converse relationship has been described by Woronow and King, 1972). This suggests that the primitive crust is preferentially located in dark, high regions. Fresh craters in cratered terrain show an opposite trend, increasing in number in telescopically bright regions. There are at least two reasons for this relationship, both based on the assumption that the changes in brightness within cratered terrain are caused by variable amounts of intercrater plains materials. When such materials are abundant, the albedo increases. In regions of plains, small cup-shaped craters may be relatively easy to identify against a background of flat intercrater material. In some dark regions, the bright intercrater plains material and the associated craters may have been stripped by eolian erosion.

Finally we have plotted several morphologic properties for craters. Furrowed and pitted wall craters occur in a latitudinal belt, just south of the equator (fig. 4.55). This distribution is remarkably different from any others previously shown. Other writers have remarked on the concentration of channels (Sagan et al., 1973a) and furrows (Soderblom et al., 1974) in a similar latitudinal belt (fig. 3.32). We interpret all three of these features to be the

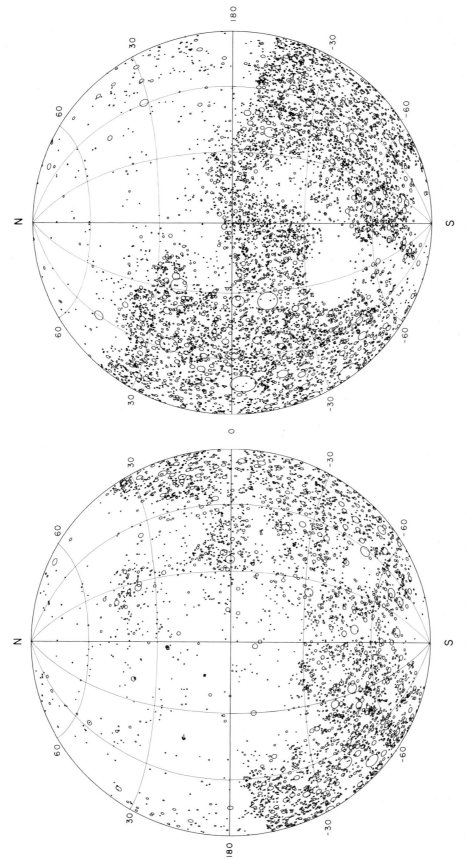

Figure 4.46. All craters greater than 15 km in diameter, plotted on a Lambert equatorial equal-area base. For comparison with physiographic provinces refer to figure 3.2.

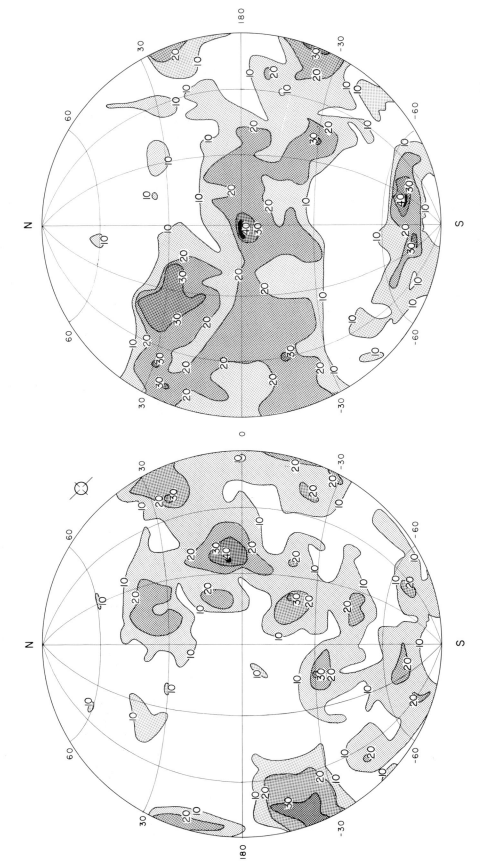

Figure 4.47. Distribution of all fresh (f) craters greater than 8 km in diameter, plotted on a Lambert equatorial equal-area base. Isolines were generated by measuring crater densities within a 10°-diameter search circle moved around the globe on 5° centers. For comparison with physiographic provinces refer to figure 3.2.

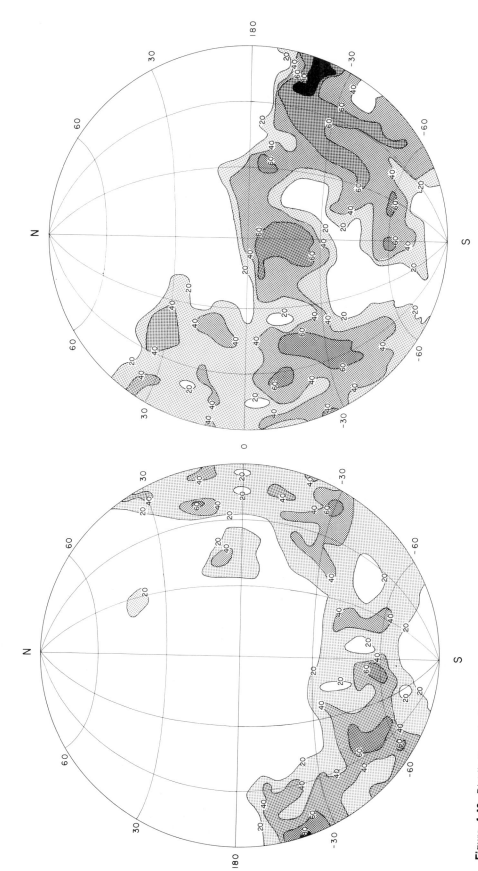

Figure 4.48. Distribution of all degraded (s + m + h) craters, plotted on a Lambert equatorial equal-area base. Isolines were generated by measuring crater densities within a 10°-diameter search circle moved around the globe on 5° centers. For comparison with physiographic provinces refer to figure 3.2.

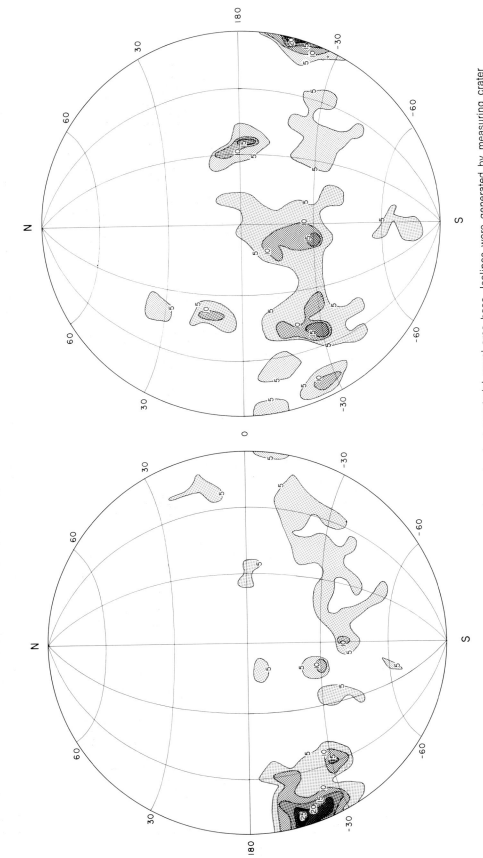

Figure 4.49. Distribution of all slightly degraded (s) craters, plotted on a Lambert equatorial equal-area base. Isolines were generated by measuring crater densities within a 10°-diameter search circle moved around the globe on 5° centers. For comparison with physiographic provinces refer to figure 3.2.

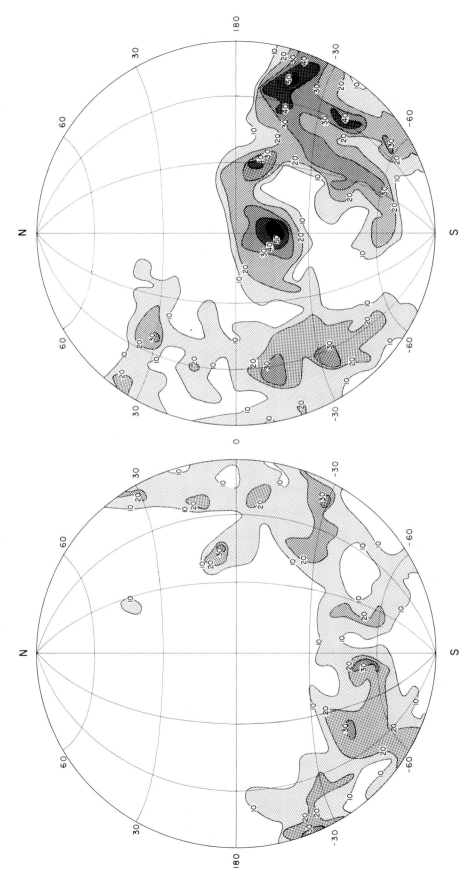

Figure 4.50. Distribution of all moderately degraded (m) craters, plotted on a Lambert equatorial equal-area base. Isolines were generated by measuring crater densities within a 10°-diameter search circle moved around the globe on 5° centers. For comparison with physiographic provinces refer to figure 3.2.

144

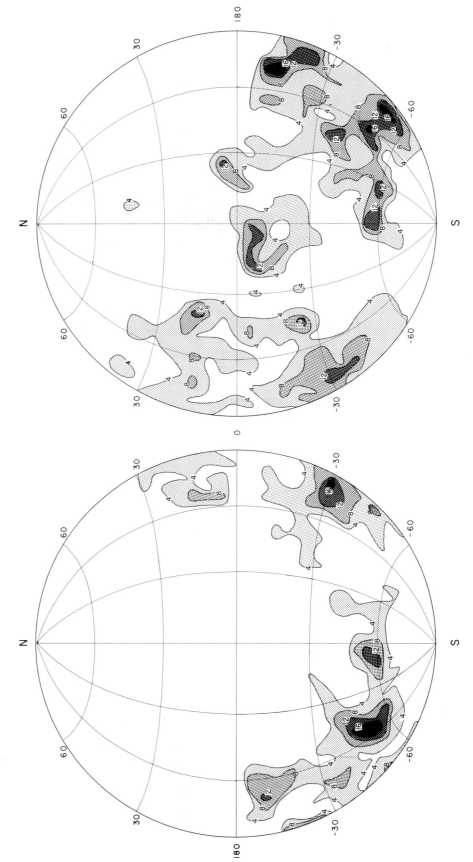

Figure 4.51. Distribution of all highly degraded (h) craters, plotted on a Lambert equatorial equal-area base. Isolines were generated by measuring crater densities within a 10°-diameter search circle moved around the globe on 5° centers. For comparison with physiographic provinces refer to figure 3.2.

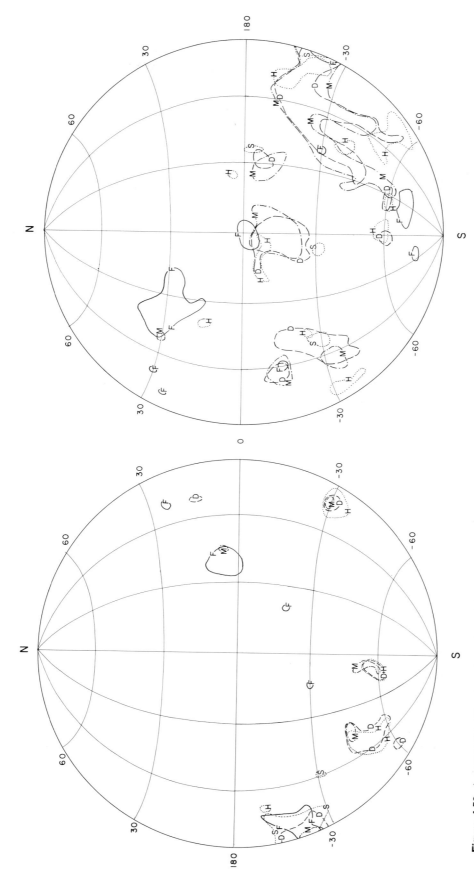

Figure 4.52. A composite map, based on data of figures 4.47–4.51, showing areas where there are concentrations of craters with various morphological properties. The following abbreviations are used: F, fresh craters; D, all degraded craters; S, slightly degraded craters; M, moderately degraded craters; H, highly degraded craters.

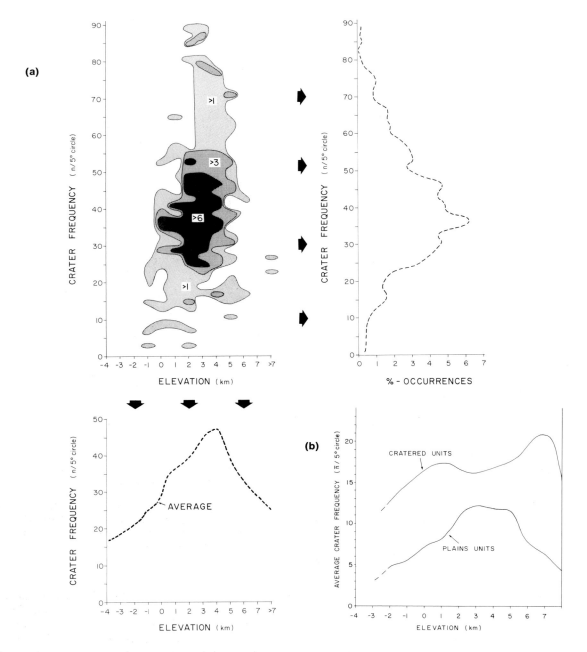

Figure 4.53 (a). Correlation of crater density with elevation within cratered terrain. Numbers of all degraded craters greater than 10 km in diameter within a 10°-diameter search circle were recorded from 60°N to 60°S on 5° centers. (The data have been normalized to eliminate the bias introduced by close spacing of points at higher latitudes.) For these same 5°-spaced centers elevation was recorded, using the data of figure 3.4. A scatter diagram reveals generally what one would expect from random combination of two normal distributions of values. Number pairs that occur most often (greater than six times) cluster close to the modes for elevation and crater density. The graph on the right shows that the distribution of crater frequency values is, in fact, unimodal. The graph at the bottom shows the variation in mean

crater frequency as a result of elevation. Since there is a great range in density values for any elevation, the significance of the average value is questionable. There is a general tendency for crater density to increase with elevation. The reverse in slope at highest elevations may result from an inadequate statistical base.

(b). Average frequency of fresh craters greater than 8 km in diameter for cratered terrain and for all plains units, combined. At very low elevations, the statistical base is inadequate. There is a general tendency for crater densities to increase with elevation. The reversal in slope for plains units above 4 km represents the presence of young volcanic units in the Tharsis region.

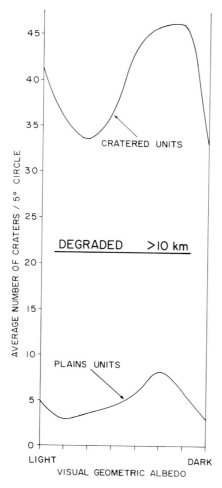

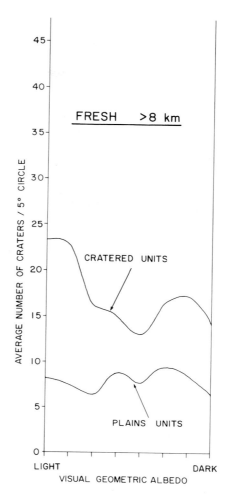

Figure 4.54. Correlation of albedo with crater density, using the same techniques described in figure 4.53. Albedo data are from de Vaucouleurs (1967) as contoured by DeWys (1968). (See fig. 1.11.) The several inflections in the curve for degraded craters within cratered terrain may be spurious, produced by an inadequate statistical base for the relatively small regions of extremely high and low albedo. The trend for fresh craters within cratered terrain to increase with telescopic brightness appears real.

result of water erosion in the equatorial regions. An alternate interpretation (Soderblom et al., 1974) is that the fluvial features are distributed across the entire planet but have been preferentially exposed in the equatorial regions by removal of overlying sediment by eolian action. Breached wall craters follow a different distribution (fig. 4.56), suggesting some process of crater erosion that operates more randomly from a geographic point of view.

The last global plot is for central peak craters (fig. 4.57). A concentration in MC 16 in the vicinity of 165°W, 30°S occurs along prominent fault zones in association with many irregularly shaped and breached craters, which appear volcanic in origin. The broad latitudinal belt between 180°W and 350°W is coincident with the boundary of cratered terrain and plains. This may mean that some process

of sapping and stripping makes the central peaks prominent (Arvidson and Coradini, 1975). Elsewhere they tend to be covered by infilling crater-wall debris, basin ejecta mantles, and eolian blankets. Along the erosional boundary between cratered terrain and plains the infilling material has been flushed out, preferentially revealing the central peaks. It is probable that the peaks there formed by impact mechanisms associated with rebound, as described earlier in this chapter.

Cordell et al. (1974) note unusual concentrations of central-peak craters in the south polar region. The same concentrations are apparent in our own data when ratios of central-peak craters to all craters are plotted. Cordell et al. suggest that preferential production mechanisms may be related to the presence of permafrost.

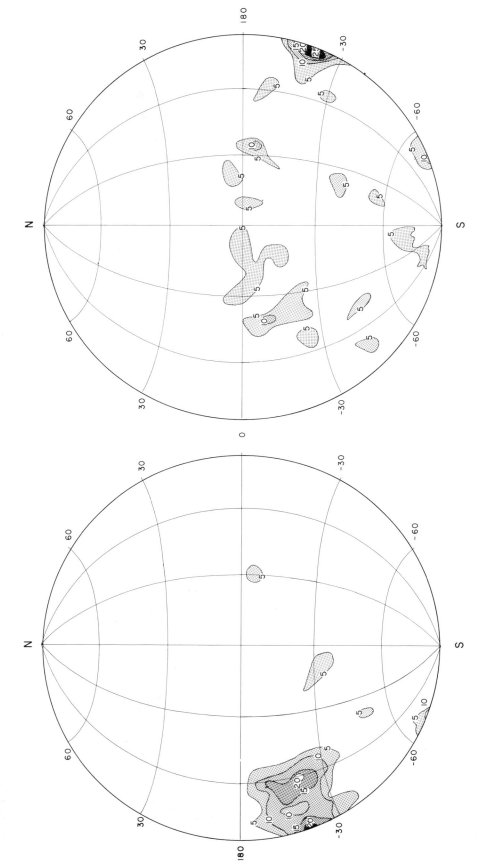

Figure 4.55. Distribution of craters with furrowed and/or pitted walls, plotted on a Lambert equatorial equal-area base. Isolines were generated by measuring crater densities within a 10°-diameter search circle moved around the globe on 5° centers. Figure 3.2 gives comparison with physiographic provinces.

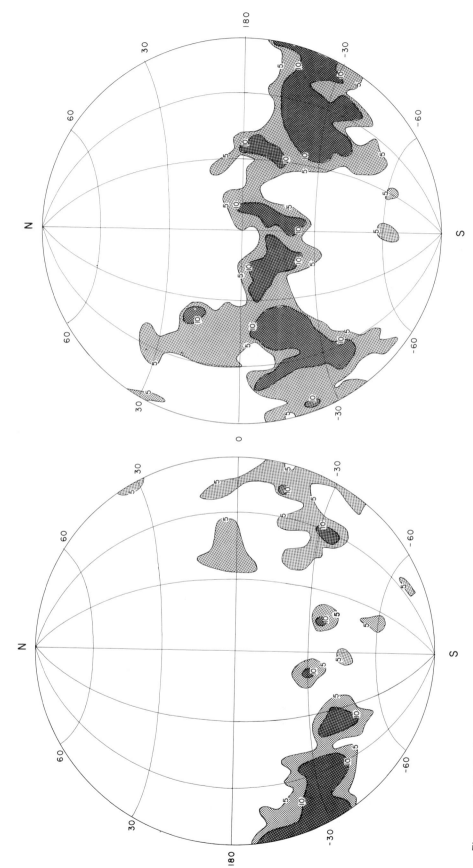

Figure 4.56. Distribution of craters with breached walls, plotted on a Lambert equatorial equal-area base. Isolines were generated by measuring crater densities within a 10°-diameter search circle moved around the globe on 5° centers. Figure 3.2 gives comparison with physiographic provinces.

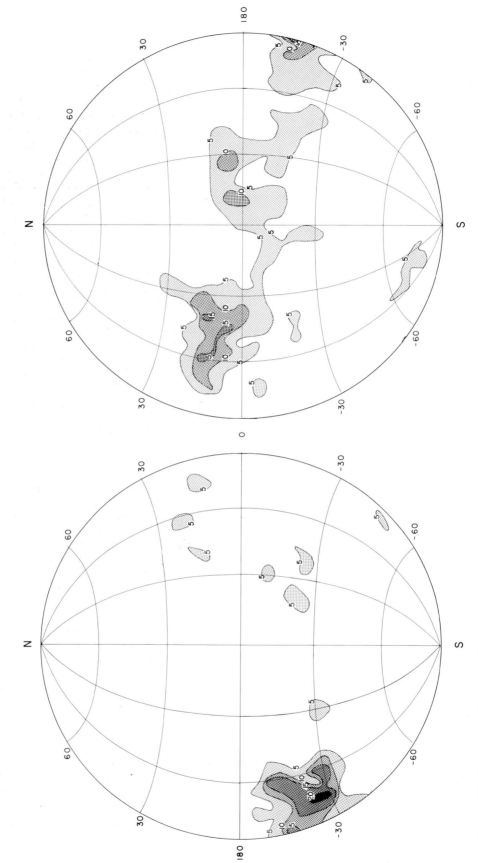

Figure 4.57. Distribution of craters with central peaks, plotted on a Lambert equatorial equal-area base. Isolines were generated by measuring crater densities within a 10°-diameter search circle moved around the globe on 5° centers. Figure 3.2 gives comparison with physiographic provinces.

5.

Volcanism

Introduction

ALTHOUGH Earth, moon, and Mars each represent special environments, they have in common at least two modifying processes: cratering by external bodies and volcanic activity from within. Before the Apollo missions to the moon even this commonality was in doubt. There was never much uncertainty about the reality of impact cratering. The only problem lay in the positive identification of scars resulting from such collisions. On Earth, the atmosphere shields the surface from smaller bodies, and larger impact structures soon are rendered obscure by erosion and by mantling with younger sediments. Nevertheless, more than fifty possible impact structures have now been identified (table 4.1; fig. 4.7). Because of the absence of an atmosphere on the moon, the probability of formation and preservation of craters is immeasurably greater. Telescopic observations early confirmed these probabilities, even though a few critics continue to delight in perpetuating the unreasonable thesis that all, or almost all, lunar craters could have been formed by internal processes. The early missions to Mars, Mariner 4 in 1964 and Mariner 6–7 in 1969, revealed a moon-like cratered landscape, thereby indicating once again the importance of impact events in modifying planetary surfaces.

Although we can quickly reach agreement on the general proposition that impact cratering is an important modifying process for all three planets, the question of internal volcanism is more troublesome. Here on Earth, of course, volcanic activity ranks among the most spectacular and destructive of all the crustal processes. One has only to think of the massive eruption of Monte Somma—the predecessor of Mt. Vesuvius—in A.D. 79. The neighboring towns of Stabiae and Pompeii were completely buried by falling pumice and ash. Nearby Herculaneum, not so affected by ash falls, was buried by swift-moving mudflows of volcanic ash liquified by torrential rains that accompanied the eruptions (Macdonald, 1972). Even more destructive was the 1902 eruption of Mont Pelée in the West Indies. Avalanches of ash and superheated gas—known as *nuées ardentes* from the French for glowing clouds—swept silently down the mountain slopes and enveloped the city of St. Pierre. All but four of the 30,000 inhabitants were killed.

For many centuries historians and archaeologists have been puzzled by Plato's legend of Atlantis, the story of an ancient island civilization that flourished one day and vanished the next. Recent excavations and oceanographic research persuasively indicate that the legend has some basis in fact. The Minoan Empire, with its base on the island of Crete, was probably destroyed in the fifteenth century B.C. by one or more violent volcanic explosions, attended by the disappearance of entire islands, subsidence of land, deposition of thick ash layers, and catastrophic flooding by tidal waves.

Volcanic eruptions not only modify or destroy pre-existing landscapes, they can literally create new land. Consider the submarine eruptions just south of Iceland, which broke through the ocean surface on the night of November 14, 1963, heralded by fiery fountains of ash and a towering pillar of steam. Sporadic outpouring of lava continued for four years, creating Surtsey Island, approximately 2 km in diameter. The island is of particular interest as a laboratory for studying the progressive patterns of plant and animal habitation on a pristine surface. In capsule form these patterns reflect the evolutionary niche-seeking patterns that have developed over millions of years.

Just as for earthquakes, many of the most violent volcanic eruptions have taken place in sparsely inhabited areas, so that their exact character and strength is difficult to establish. This was the case for the 1883 eruption of Krakatoa, an island in the Sunda Strait between Java and Sumatra. A violent eruption propelled a cloud of dust so high into the atmosphere that it was spread around the entire earth. A minimum of 10 km³ of ejecta fell within a radius of 500 km. This is five times the volume of the lavas that produced Surtsey Island. For three years following the eruption, values for solar radia-

tion measured at the Montpelier Observatory in France were reduced by 10 percent due to Krakatoan dust in the atmosphere. Thus, volcanic activity not only modifies the land directly, but also can episodically change climates by altering insolation patterns and the thermal structure of the atmosphere.

Terrestrial Volcanism

Volcanic Materials

Volcanic materials are of several types. (See Macdonald, 1972, for more details.) Most familiar are the lavas that form spectacular rivers, lakes, and fountains of molten rock. Basaltic rocks of low viscosity (high fluidity) can flow tens of kilometers along gentle slopes. As they cool, they form a solid crust with distinctive surface features. The two most common types of flows have exotic Hawaiian labels: pahoehoe and aa. Pahoehoe lavas have relatively smooth surfaces with gentle swales and hillocks. In

(a)

(b)

Figure 5.1 (a). Flank of Mauna Loa showing flow channel with levees developed in a basalt flow (from Greeley, 1973).

(b). Volcanic features on flank of Mauna Loa. Most recent flows are darkest. A small volcanic cone is visible in the upper right (from Greeley, 1973).

places, the plastic crust, dragged along by the fluid lava beneath, develops a rope-like or "piled-up" texture. Also common are pressure ridges, cracks, tilted slabs, and small cones and spires formed where the lava breaks through the upper crust. Aa lavas are characterized by extremely rough surfaces, formed by pervasive fragmentation of a highly viscous crust. The mantling fragments are angular, vesicular and spiny, resembling the clinkers or slag formed by industrial smelting. Although smooth at the scale of tens of meters, aa lavas are treacherously rough at more detailed scales. As such, they are a favored example for scientists seeking to demonstrate the pitfalls in using Mariner photos to identify a safe landing site for the Viking spacecraft.

Localized conduits often develop within solidifying flows. In the case of pahoehoe lavas these channels occur as roofed lava tunnels. Aa channels commonly are unroofed (fig. 5.1). The longer terrestrial channels extend for more than 25 km. Repeated surges of lava through an open channel often spill over the sides. As the overbank lava builds up, the channel floor is progressively raised. Eventually the channel is positioned along the elevated axis of an aggradational complex of flows. A comparable situation exists for conventional rivers such as the Mississippi. Flow deposits form natural levees on either side of the main channel. Catastrophic floods occur when the levees are breached, and the flanking plains, which actually stand lower than the channel, are inundated.

When gas-charged lava is explosively discharged, pyroclastic (fire-broken) particles of all sizes from fine ash to large blocks—some of which are picturesquely named cow-dung bombs—are thrown out along ballistic trajectories. Free-fall deposits of substantial thickness can accumulate. The 1912 Katmai, Alaska, eruption produced nearly 25 km³ of ash. Two hundred kilometers in the downwind direction, the deposits were nearly 25 cm in thickness (Wilcox, 1959).

Under restricted conditions, when gas-rich, frothy magma forms in the upper part of the volcanic vent, a mixture of gas and solid fragments overflow the volcanic lip and flow downhill as a nuée ardente. The equivalent deposits are ash flows or ignimbrites. A specialized type of gas cloud is associated with a volcanic explosion that takes place in shallow water. A doughnut-shaped cloud of steam and ash spreads laterally. The phenomenon is apparently similar to the base surge that forms following nuclear explosions.

Volcanic Landforms

Many volcanic eruptions are accompanied by the formation of mountains, but the precise size and shape of each volcano depends on many factors.

Gently sloping shield volcanoes of the Hawaiian type form when fluid basaltic lavas spread radially for great distances (fig. 5.2). Although shield volcanoes can reach large dimensions—Mauna Loa (Hawaii) is 200 km in diameter and rises more than 9 km above the sea floor—slopes seldom exceed 10°.

Composite or stratovolcanoes are formed by intercalation of ash and lava. Both because the ash is concentrated close to the vent and because the lavas are relatively viscous, slopes are steeper than for shield volcanoes. Resultant structures include conical mountains such as Mt. Fuji, Rainier, and Hood (fig. 5.3). Continuous or episodic ejection of particulate material in the absence of lava leads to formation of cinder cones with slopes at the angle of repose for disaggregated sediment (fig. 5.4). Still steeper domes and spines are formed by extrusion of highly viscous lavas that behave somewhat like toothpaste. The best known example is Mt. Pelée (fig. 5.5). During a 1903 eruption, the dome of Mt. Pelée grew vertically by as much as 23 m/day. One spine reached a height of almost 300 m in nine months.

It is generally true that viscosity of magma increases with silica content. Chiefly, this is because silica tetrahedra link together to form polymers. The association of viscosity and composition has led some photo-interpreters to assert a correlation between steepness of volcanic structures and chemical composition of the rocks. This is an uncertain speculation. Other conditions, such as rapidity of discharge, can equally well control the shape of the volcano.

Calderas, or collapse craters, commonly form at the summits of shield and composite volcanoes as well as along flank vents (fig. 5.6). It was initially thought that many of the craters were explosive in character—consistent with the image of a volcano that was literally "blowing its top." More detailed studies have shown that most calderas are, instead, the result of collapse along circular fractures following evacuation of supporting magma. Diameters reach 25 km for circular calderas and almost 100 km for elongate and irregular collapse depressions (Smith, 1966).

It is noteworthy that most calderas form atop constructional features. A volcanic crater without an associated volcanic mountain is rare. Exceptions are maars, craters which are formed by explosive release of gas that bores its way upward through overlying rock. Ejected material contains some volcanic fragments, but also fragments of wall rock, which are cored by the rising gases. Some maars, including examples in Germany and those in the Colorado Plateau, occur apart from any other volcanic features. They are simply holes in nonvolcanic strata. Indeed, a well-developed maar in Oregon has been named Hole-in-the-Ground (fig. 5.7).

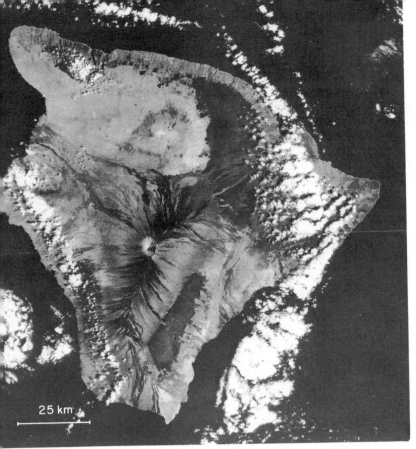

25 km

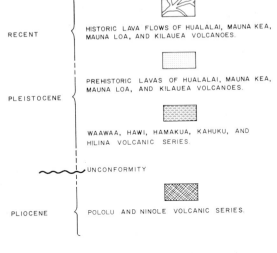

Figure 5.2 (a). Earth Resources Technology Satellite (ERTS) photomosaic of the island of Hawaii.

(b). Geologic map of Hawaii (simplified from Stearns and Macdonald, 1946).

RECENT — HISTORIC LAVA FLOWS OF HUALALAI, MAUNA KEA, MAUNA LOA, AND KILAUEA VOLCANOES.

PLEISTOCENE — PREHISTORIC LAVAS OF HUALALAI, MAUNA KEA, MAUNA LOA, AND KILAUEA VOLCANOES.

WAAWAA, HAWI, HAMAKUA, KAHUKU, AND HILINA VOLCANIC SERIES.

∼∼∼∼ UNCONFORMITY

PLIOCENE — POLOLU AND NINOLE VOLCANIC SERIES.

CINDER AND SPATTER CONES, PIT CRATERS AND FISSURE VENTS.

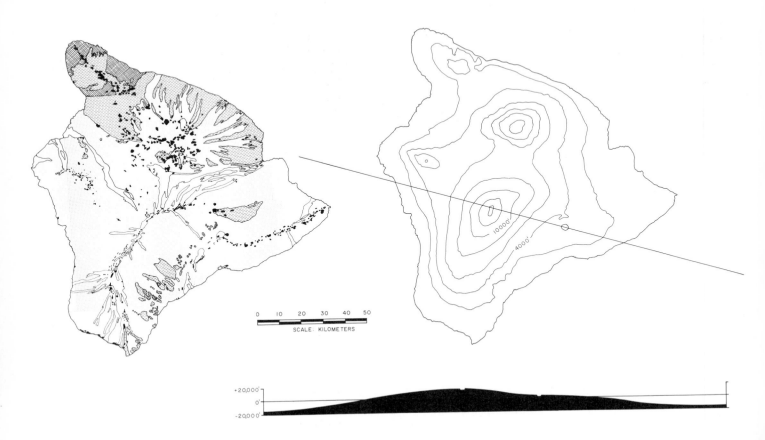

0 10 20 30 40 50
SCALE: KILOMETERS

+20,000′
0′
−20,000′

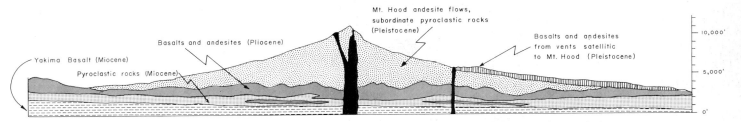

Figure 5.3. Geologic cross-section through Mt. Hood, Oregon. Horizontal and vertical scales are the same (adapted from Wise, 1969).

Figure 5.4. Paricutin volcano, Mexico. At this stage in its development the cone is about 500 m above its original base. A smaller breached cone is in the left foreground. The low rounded slopes to the right of the breached cone and at the base of the main cone are the first lava flows from Paricutin, covered by ash (from Bullard, 1962).

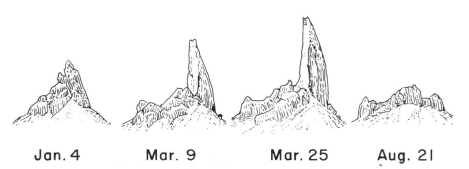

Jan. 4   Mar. 9   Mar. 25   Aug. 21

Figure 5.5. Four stages in the evolution of the Mount Pelée spine (after Lacroix, 1904).

Figure 5.6. The summit area of Mauna Loa, displaying four collapsed craters (courtesy of United States Air Force).

Figure 5.7 (a). Hole-in-the-Ground, a Pleistocene maar in Oregon. The crater is 1.6 km in diameter, and the highest rim point is 153 m above the crater floor. Basalt flows exposed in the far wall underlie the explosion tuff breccias that were emplaced during maar formation (courtesy of Oregon Department of Geology and Mineral Industries).

(b). MacDougal crater, a maar in the Pinacate volcanic field of northern Mexico. The crater is about 1.6 km in diameter and 200 m deep. Crater walls and rim are generally dark but some relatively bright tuff-breccia deposits are visible near the top of the far wall. Bright interior deposits are derived partly from erosion of tuff-breccia deposits and partly from entrapment of quartz-rich sands transported from the neighboring Sonora Desert (from Arvidson and Mutch, 1974).

(a)

(b)

In addition to volcanic mountains, there are less spectacular but volumetrically much more important accumulations of lavas termed plateau basalts because they commonly form high surfaces on continental regions. Regions of relatively recent activity are the Columbia River plateau and Snake River plains of northwestern United States as well as a North Atlantic (Thulean) province comprising Greenland, Iceland, and the Hebrides. Older occurrences include the Deccan basalts of India (Cretaceous-Eocene); basalt sequences in India, Africa, South America, and Antarctica, probably related in space and time (mid-Mesozoic); and several Precambrian lava sequences exposed on continental shields (Tyrrell, 1937).

One of the most extensive and best preserved areas of flood basalts is situated on the western margin of peninsular India (fig. 5.8). The lavas attain a maximum aggregate thickness of 3000 m, but the usual thickness is about 600 m. Individual flows average 5 m in thickness but some reach 30 m. Successive flow sheets are commonly separated by partings of ash. The ashes attest to some explosive activity, but the bulk of lavas is thought to have flowed quietly from local vents situated along major fissures. This contention is based partly on positive evidence—the monotonous horizontality of the lavas over wide areas and the occurrence of dikes around the margin of basalt exposures where erosion has removed the extrusive rocks, and partly on negative evidence—the almost complete absence of volcanic cones and craters.

The most extensive flood basalts within the United States are situated on the Columbia River plateau. The rocks are Miocene in age, the chief component of a still larger volcanic terrain that covers major parts of Washington, Oregon, and Idaho and ranges in age from Eocene to Holocene.

As in the case of the Deccan lavas, the Columbia River basalts apparently erupted through fissures. Prominent dike swarms appear in restricted areas along the southern and western margins of the plateau. Dikes within a swarm commonly trend in a single direction (Waters, 1961). A definite connection between a dike and a flow has been described by Fuller (1927). The areal restriction of dikes to the south and west suggests that flows in the northeastern corner of the plateau may have traveled more than 150 km from their source vents. The hypothesis is supported by directional features in the flows that indicate movement toward the northeast (Waters, 1961). No shield volcanoes or cinder cones have been observed, although both structures are commonly associated with the Pliocene-Holocene flows that overlie the Columbia basalt. Near the edge of the basalt plains, islands of older rock are almost completely circled by lava. The remnants of older rock are termed kipukas.

Columbia and Snake River plateau basalts have a probable volume of 300,000 km³, many times that of Mauna Loa. South American plateau basalts may be more than 200,000 cubic km in volume (Holmes, 1965). As large as these figures are, they are dwarfed by the volume of volcanic rocks that erupted beneath the oceans and are largely hidden from our view. Menard (1964) estimates that 23,900,000 cubic km are occupied by constructional forms—islands, archipelagoes, and aprons. An even greater volume of basalts has been incorporated as oceanic crust by both extrusion and intrusion at mid-

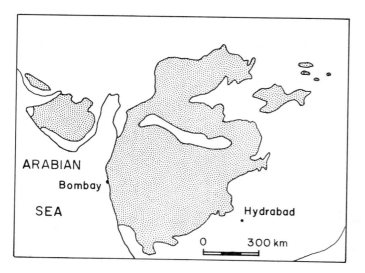

Figure 5.8. Distribution of plateau basalts in India and in northwestern United States.

ocean ridges after the megacontinent Pangaea rifted and spread apart to create the present ocean basins (Dietz, 1961).

Great thicknesses of basalt develop along rift zones where the crust has foundered. Large faults bounding down-dropped blocks provide channels for successive eruptions. A good example is afforded by the Laki, Iceland, eruption of 1783 (fig. 5.9). Initial activity took place along a set of northeast-trending faults. Lava flows subsequently spread out along several pre-existing river valleys. Another prominent rift is the Rhine graben. The faulting and associated volcanic activity apparently was the result of extension of the crust over a magma-filled arch. The Ethiopian rift is an example of contemporary plate movement, with the African plate separating from the Arabian plate. Were the seas drained, one could see the most impressive of all rift systems: the midocean ridges that form along zones of plate separation where new ocean crust is being generated.

Eroded Volcanic Landforms

Many volcanic structures on Earth have been deeply eroded. The typical surficial morphology is no longer present, and the structure can be recognized only by its internal geometry. Most commonly preserved are the dikes that connect the deep magma source to the landform on the surface. When the crust bulges upward over a magma reservoir, two dike patterns develop. Conical sheets point downward to the point where the upward stress is being applied. Radial sets of vertical dikes form as the crust fractures to accommodate a larger surface area. In the opposite situation where a subsidence caldera forms, a system of cylindrical dikes develops around the depressed plug of basement rocks.

Following deep erosion, dikes are all that remain to indicate former volcanic activity (fig. 5.10). Ring complexes—the geometrical result when a horizontal erosion plane intersects a cylinder or cone—have been identified in Scotland, Scandinavia, southwest Africa, and New England.

Larger-scale complexes of splayed dikes suggestive of a radial pattern around a regional volcanic center are present in the British Isles (fig. 5.11), New England and the Maritime Provinces, South America, and Africa. Distribution of these dikes along margins of continents once connected and now separated suggests their formation may be related to updoming, which was the precursor of plate movement (May, 1971).

Volcanism and Crustal Movement

Terrestrial volcanoes are far from randomly located. Most fringe the Pacific Ocean in what has been called a "ring of fire" (fig. 5.12). Others are in the midocean regions—Iceland and Hawaii. Rela-

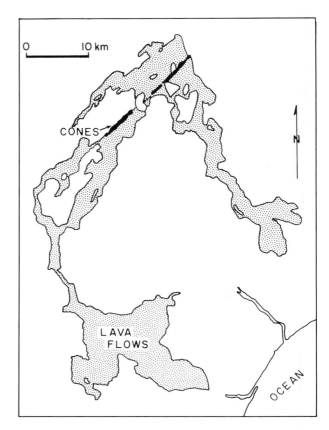

Figure 5.9. Distribution of lava flows and cones along the eruptive fissure of the Laki, Iceland, eruption of 1783 (after Thorarinsson, 1970).

tively few are situated in midcontinental regions. The correlation between volcanoes and present or ancient mountain belts has been recognized for many years. Geologists suspected that a common set of conditions gave rise to downbuckling of the crust, deposition of thick sequences of sedimentary and volcanic rocks, and subsequent compressional deformation, forming folds and thrust faults. At first it was proposed that the hot interior of the earth was sheathed by a cold, rigid crust. As the interior cooled and contracted, the crust was thought to accommodate to the smaller surface area by compressive failure, much as the skin of an apple wrinkles when the interior shrinks by dehydration. The model was physically reasonable, but failed to explain the majority of specific phenomena.

Then in the 1930s a theory of convection, first mentioned by Hopkins (1839), was elaborated by Holmes (1931), Vening Meinesz (1930), and Griggs (1939). In convection, heat is transferred by the actual movement of heated material. It contrasts with conduction, wherein heat is transferred by molecular vibrations in a rigid solid. A common

PLAN   VIEWS

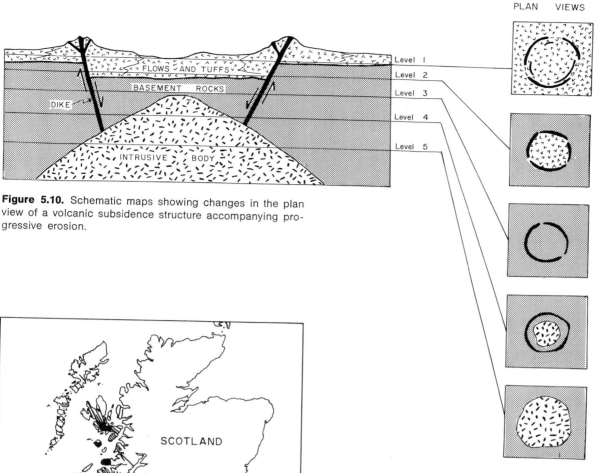

Figure 5.10. Schematic maps showing changes in the plan view of a volcanic subsidence structure accompanying progressive erosion.

Figure 5.11. Distribution of Tertiary northwest-trending dikes in the British Isles (after Ritchey et al., 1930).

example is the contrast between the movement of water in a heated saucepan—convection—and the transfer of heat through the bottom of the pan—conduction. Although one commonly thinks of convection as confined to liquids and gases, it can also occur in solids, which, subjected to high temperature and pressure over long periods of time, behave plastically. Heat generated deep within the earth, perhaps by decay of radiogenic isotopes, exceeds those amounts that can be transferred to the surface by conduction. Instead, mantle rocks move upward, displacing cooler rocks, which, in turn, sink. Ultimately, convection cells form. Between adjacent cells the crust is dragged downward to form a "tectogene," the presumed site for accumulation of sediments, volcanic activity, and compressional deformation (Kuenen, 1936). A present-day analog for these tectogenes appears to be the deep trenches that parallel island areas of the South Pacific and the West Indies (Hess, 1938; Vening Meinesz, 1930). Interestingly enough, these trenches are out of isostatic equilibrium, precisely the situation that should obtain if the crust is being actively depressed along

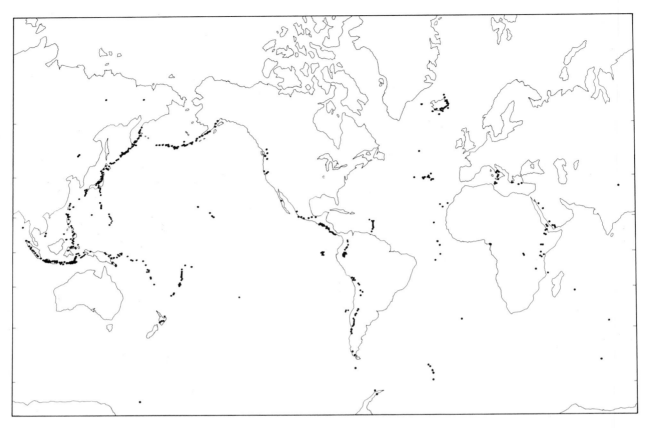

Figure 5.12. Global distribution of active and dormant volcanoes (Macdonald, 1972). Reprinted by permission of Prentice-Hall, Inc., Englewood Cliffs, N.J.

convective ramps. Were this motion to cease, the crust would presumably rebound upward, seeking its isostatic level as would a free-floating cork in a tub of water.

The tectogene model emphasized vertical motion, the subsidence of crust between two cells. However, there were some intriguing problems in horizontal or lateral arrangements of geologic elements that remained to be addressed. Chief among these was the fact that continents that are now widely separated have margins that match closely, suggesting that they were once part of a larger megacontinent that has split apart.

The theory of drifting continents was first proposed in the early part of this century by Wegener (1912). He was impressed by the complementary shapes displayed by the west coast of Africa and the east coast of South America. The geology was also complementary, with generally the same stratigraphic successions and same fossil forms on both continents. Proposals that these two land masses once constituted a supercontinent called Gondwanaland were laughed out of scientific courts on the basis that it was physically impossible for continents

to plow their way through Earth's oceanic crust.

In 1962, a classic paper by H. H. Hess shed new light on this old problem of continental drift. Hess suggested that convecting cells within the mantle have rising limbs beneath midocean ridges and that new crust is generated at those sites by extrusion of mantle material. The oceanic crust spreads laterally, riding along the crest of the convection cell. The model was bolstered by the discovery that belts of submarine lavas with reversed and normal polarity occur to either side of ocean ridges (Vine and Matthews, 1963). Earlier in 1963, Cox et al. had shown that the polarity of Earth's field has reversed repeatedly throughout geologic time. These events are "frozen" into igneous lavas by orientation of iron-rich minerals in accordance with the magnetic field at the time of their consolidation. If a sequence of lavas accumulates over a long period of time, then a paleomagnetic evolutionary sequence is also formed, not unlike the evolutionary fossil sequence in a succession of sedimentary rocks.

Vine (1966) demonstrated that the vertical paleomagnetic sequences established by Cox et al. occur symmetrically on either side of the midoceanic

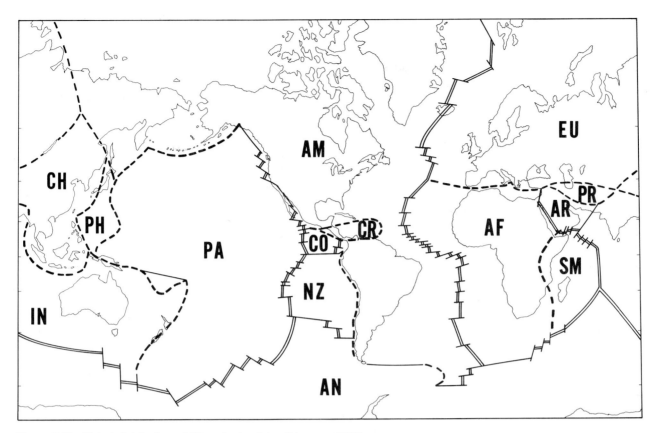

Figure 5.13. Global distribution of lithospheric plates (Morgan, 1971).

Figure 5.14. Global distribution of all earthquakes that occurred in 1972. Data supplied by NOAA Environmental Research Laboratories.

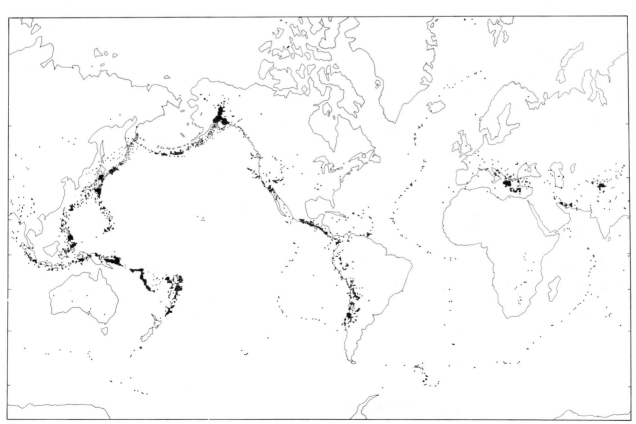

Figure 5.15. Lava flows in Mare Imbrium. These flows form a portion of the relatively young series of flows originating at the southwest edge of Imbrium and stretching 1200 km into the basin. Flows here are 10–25 km in width, average 35 m in height, and flow from lower left toward upper right. Low sun angle (about 3° in center of frame) enhances visibility of flows. (AS15-1557.)

Figure 5.16. Near-terminator photography of the Kunowsky region (eastern Oceanus Procellarum) illustrating a flow front of very low height. The flow front is approximately 3–5 m in height and is not visible on higher sun elevation angle photography. (AS14-78-10377.)

ridges, with the youngest rocks close to the ridge and the oldest rocks farthest away. The unmistakable conclusion was that ocean crust was being generated along the ridge and was moving away at a rate of approximately 2–5 cm/yr.

From ocean-floor spreading it was only a short intellectual jump to plate tectonics, the thesis that as new oceanic crust is generated, large crustal plates are displaced laterally. Where two plates collide, one is forced downward, forming deep oceanic trenches and volcanic islands along a "subduction" zone. The concept of plate tectonics is, without question, the most provocative and useful development in earth science within the present century. Provocative because it stimulates the imagination of scientist and layman alike; useful because it synthesizes a large number of structural, geological, geochemical, and geophysical observations that heretofore were unrelated and unexplained.

Our present understanding of terrestrial volcanism, then, derives from the plate tectonics model. Most volcanoes occur either along rifts between separating plates or at the juncture of colliding plates (figs. 5.13, 5.14). Admittedly, this is a descriptive explanation rather than a genetic one. The thermal and kinetic mechanisms remain speculative. The most likely candidate is convective motion, perhaps originating in the outer mantle by radioactive heating deep in the mantle where there is excess heat derived from the molten core.

Lunar Volcanic Features

The majority of features that have been associated with lunar volcanic deposits are located in the lowlands or mare regions. A review of these features, several of which occur at Apollo sites, will provide a useful background for discussion of more problematic structures preserved in the highlands.

Some of the most informative features associated with the maria are long, sinuous scarps that have been interpreted as flow fronts related to mare volcanic activity. Those in Mare Imbrium were prominent enough to have been easily detected on Earth-based telescopic photography. Apollo 15 photography provided excellent details of these flow fronts and their associated structures (fig. 5.15). Schaber (1973) has shown that these distinctive flows in Mare Imbrium emanate from a single source region in the southwest corner of that basin and represent three eruptive phases, one of which extended to distances of 1200 km and produced flow fronts with average heights of 30–35 m.

Distinctive flow fronts similar to the ones just mentioned are uncommon. However, the existence of other, more subdued flow fronts is revealed by

(a)

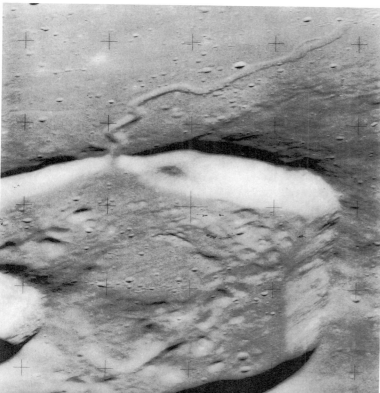

Figure 5.17 (a). View of the Apennine Mountains, looking northeast along the southeast rim of the Imbrium basin. Hadley Rille is seen in the center. The rille originates in an elongate depression at a fracture radial to Imbrium, winds along the base of the Apennine Mountains, and then turns northwest at the Apollo 15 landing site to flow out into Palus Putredinis, a portion of Mare Imbrium. The rille is over 100 km long. (AS15-139-21311.)

(b). Oblique view looking west over the 22-km crater Krieger. A sinuous rille breaches the crater wall and flows out into Oceanus Procellarum. (AS15-12480.)

(b)

Apollo photographs taken at low sun elevation angles (fig. 5.16). The presence of these sinuous scarps in areas where they are not visible on higher sun photographs suggests that degraded flow fronts may be more widespread than previously suspected.

Another prominent feature associated with maria are meandering channels known as sinuous rilles. These generally range in width from tens of meters to several kilometers, and in length from a few kilometers to several hundred kilometers. The Apollo 15 spacecraft landed at the edge of Hadley Rille, a major lunar rille on the edge of Mare Imbrium (fig. 5.17a). Although Hadley Rille occurs primarily in mare basalts, it originates in highlands adjacent to the maria. Other rilles originate in circular depressions, in large craters (fig. 5.17b), and near mare ridges. Sinuous rilles are generally interpreted as lava distributory channels or tubes (Gree-

Figure 5.18. Detail of the layering in the western wall of Hadley Rille. A distinctive bright massive unit about 10–20 m thick is seen cropping out just below the regolith. This unit contains joints that dip toward the right. A small interval of talus separates this unit from a darker, horizontally layered unit approximately 5–10 m thick. Large talus blocks appear further down into the rille. Similar basalt outcroppings were sampled by Apollo 15 astronauts Scott and Irwin on the eastern side of the rille. (AS15-12104.)

ley, 1971; Cruikshank and Wood, 1972; Murray, 1971; Carr, 1974a). Astronaut observations, photographic evidence, and petrologic analysis support a volcanic origin for most of these features (Swann et al., 1972; Howard et al., 1972) (fig. 5.18). Sinuous rilles are concentrated around the margins of the circular mare basins and in the central part of Oceanus Procellarum (Schubert et al., 1970; Murray, 1971). They often occur in clusters, such as those in the Prinz-Aristarchus region (fig. 5.19).

These clusters may signify concentrations of source vents for lunar lavas. Rilles are occasionally nested within each other, implying reactivation of a source or possibly the erosion of an initial broad valley with the channel cutting into the major "flood plain." In the case of Schröters Valley, however, the interior rille inexplicably breaches the main valley (which dead-ends a few kilometers away) and continues into Oceanus Procellarum (fig. 5.20).

Three other features: buried structures, collapse

Figure 5.19. A series of sinuous rilles surrounding the crater Prinz (lower left; about 50 km diameter). The central portion of this photograph is a regional dome, and all rilles originate in the updomed area and flow away from it (Strain and El Baz, 1975). The crater Prinz has also apparently been uplifted so that a portion of its rim escaped subsequent mare flooding. The rille on the left contains a nested inner crater and rille. (AS15-12607.)

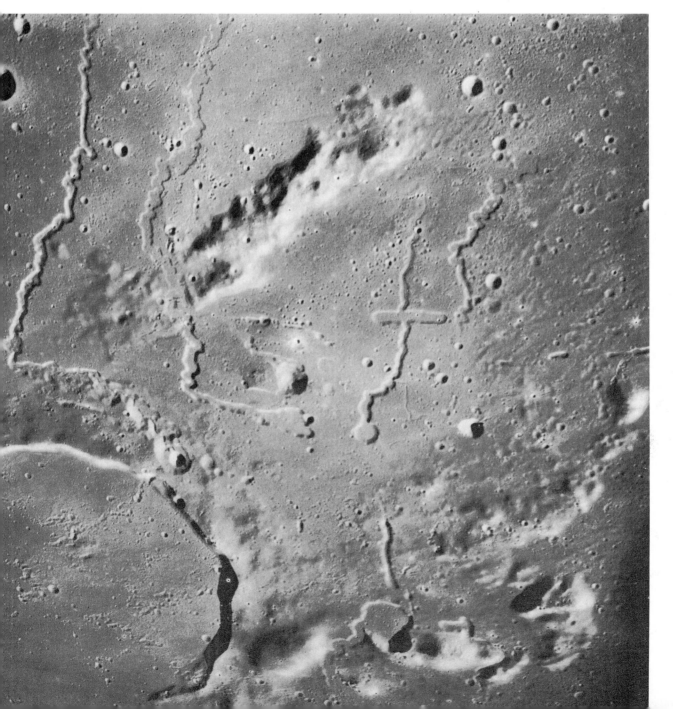

craters, and kipukas, are well developed in mare volcanic provinces, and are also common in terrestrial volcanic terrains. The floors of lunar multi-ringed basins were partially modified by postbasin pre-mare cratering. The craters Archimedes and Iridum are prime examples in the Imbrium basin. Subsequent flooding of basin floors partially or totally covered these craters. In this way many of the circular ring structures in the mare are believed to form by flooding of pre-existing craters. A 50-km-diameter ring structure south of the crater Lambert (fig. 5.21) is outlined by mare ridges and a broad smooth scarp. Abundant evidence of settling of mare lava surfaces due to shrinkage associated with cooling, or withdrawal, suggests that circular structures such as these represent draping of lavas over pre-existing structure (Cruikshank et al., 1973). Many of these mare ring structures have been attributed to shallow cylindrical igneous intrusions along ring fractures occurring over larger igneous masses

Figure 5.20. Schröters Valley, located on the Aristarchus Plateau in Oceanus Procellarum, is an excellent example of a nested sinuous rille. The large valley (up to 9 km wide) extends across the plateau, but terminates at its edge. The inner rille is extremely sinuous, flows down the center of the valley, breaches it near its abrupt end, and flows out into Oceanus Procellarum. (AS15-12628.)

(Strom, 1972). However, their abundance in the shallower maria and their partial development on craters that adjoin the highlands (Letronne and LeMonnier are examples) argue for the draping hypothesis.

Collapse craters are also common in the maria and are particularly apparent at low sun elevation angles (fig. 5.22). These rimless depressions probably formed by collapse of a lava crust as lava was evacuated from below. Some of the features may be source craters for lava flows. Finally, several types of kipukas attest to the sequential flooding of the mare regions. Figure 5.23a illustrates several characteristics of these structures. Some kipukas shown in this figure are remnants of highland material that has been flooded by early-formed lavas. Subsequent to the early flooding, subsidence of the newly deposited surface occurred. The large central kipuka formed an anchor and the mare surface settled and buckled around it, producing linear cracks at the high points. The crack at the crest of the deformed early lava is particularly well shown along the termi-

Figure 5.21. Major mare ring structure (about 50-km diameter) south of the crater Lambert. The circular ridge structure appears to be caused by the draping of lava over preexisting crater rims. (AS15-1011.)

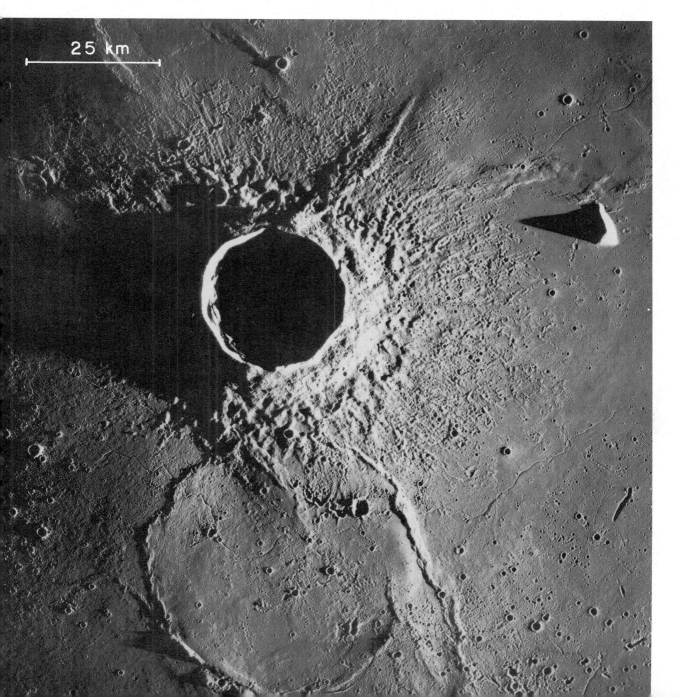

Figure 5.22. A series of collapse craters or depressions in western Mare Imbrium. Contrast the irregular shape and rimless nature of these three depressions with the sharp 2.5-km impact crater in the upper portion of the photo. These structures may have been vents or drainage features associated with the emplacement of the lava flows. (AS15-1850.)

nator in the lower sun photograph of figure 5.23b. The deformed early lava deposit became a kipuka in turn when deposition of mare lava resumed.

A number of smooth low domes with gentle convex-upward profiles (mare domes) are distributed in lunar maria. They are usually circular to slightly elliptical in plan view, less than 15 km in diameter, and have heights up to several hundred meters. Mare domes often have summit craters that are very small in relation to the total structure. They are best developed in northeast Mare Tranquillitatis and in the mare region between Kepler and Copernicus (fig. 5.24). Wilhelms and McCauley (1971) have interpreted these structures as shield volcanoes, or possibly laccoliths. Figure 5.25 compares the scale and morphology of terrestrial shield volcanoes and lunar mare domes. In general, the lunar domes are smaller than the terrestrial examples. Asymmetric terrestrial domes are usually associated with lateral rift-zone eruptions, while circular forms result from the predominance of central eruptions. The generally circular outline of most mare domes may indicate the predominance of central pipe vents over lateral rift-zone eruptions. The apparent low viscosity of many lunar lavas (Weill et al., 1971) may contribute to the low elevations of lunar mare domes.

If the lunar domes are indeed shield volcanoes, a significant difference is the absence of summit calderas, which are prominent on many terrestrial and

Figure 5.23. Lunar kipukas or steptoes form when lava flows around pre-existing features, leaving them as isolated islands. In this region of southwest Mare Imbrium near Diophantus, a succession of kipukas has formed.

(a). Lavas have flowed around projections of highland material to form highland kipukas. Much of the lava was distributed by the 600 m wide sinuous rille in the center of

the photo. Position of figure (b) is shown by lines. (AS17-23755.)

(b). A low-sun photo of the same area shows that the early mare flows have draped over the upland kipuka. Subsequent mare flooding has embayed the rille and isolated the draped mare region to form a mare kipuka. (AS17-23749.)

(a)

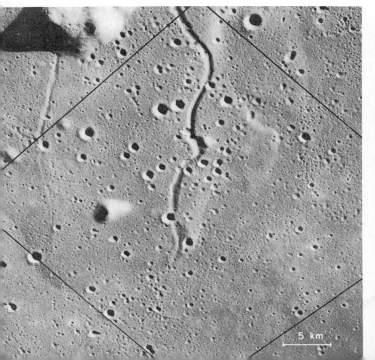

(b)

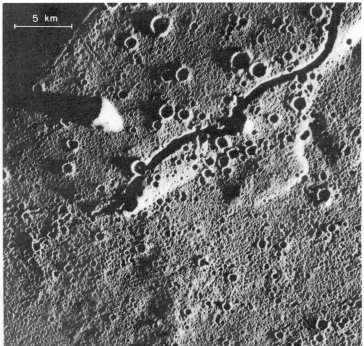

Figure 5.24. Domical features developed in the region between Kepler and Copernicus. Two low domes, about 10–
12 km in diameter, appear in the foreground of this view
looking southeastward toward Mare Cognitum. Central pits
are indicated by arrows. A sinuous rille winds around the
base of the two domes. (AS17-23739.)

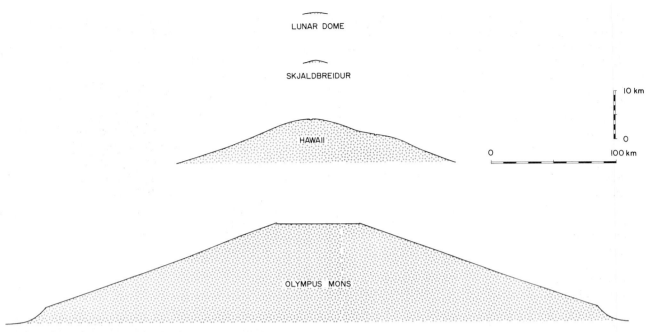

Figure 5.25. Comparative topographic profiles of the Hawaiian shield volcano Mauna Loa, an Icelandic shield volcano (Skjaldbreidur), a typical lunar mare dome such as those shown in figure 5.24 and the Martian volcano, Olympus Mons.

Figure 5.26. Model for formation of calderas illustrated by probable substructure underlying Kiluea and Mauna Loa volcanoes. Magma rises from point of origin and accumulates in a shallow reservoir. Pressure prior to eruption raises peaks. As magma evacuates and pours out on surface, collapse tends to occur because of pressure release and decrease in volume, forming caldera structures. Lack of analogous structures on the moon may mean that magma rises directly from zone of origin, rather than temporarily building up in shallow staging areas, as on Earth. Gordon A. Macdonald, *Volcanoes*, © 1972. Reprinted by permission of Prentice-Hall, Inc., Englewood Cliffs, N.J.

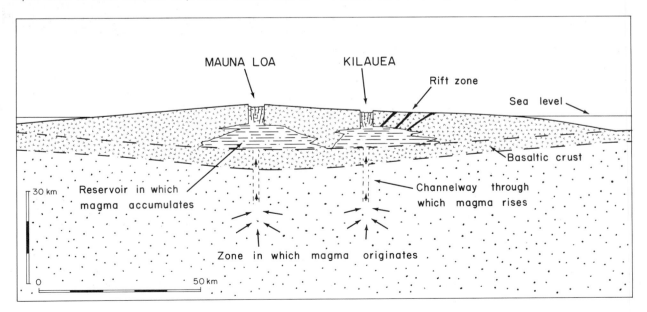

Martian shield volcanoes. Several factors are important in the formation of calderas associated with effusive eruptions of basaltic magmas, but a primary element is the evacuation of magma underneath the summit by rift eruptions and flank eruptions. Evidence for these types of eruptions is lacking on mare domes. This argues against the presence of extensive, shallow magma reservoirs in the lunar crust. Readjustment in terrestrial shallow magma reservoirs, such as those of Kilauea and Mauna Loa (fig. 5.26) seems to be responsible for much of the surface structure seen there but generally lacking on the moon. The absence of major structural deformation of large mare lava deposits suggests that there was little settling or equilibration associated with the evacuation of local magma reservoirs. Instead, it appears that deeper crustal magma sources delivered magma to the lunar surface through a rigid, and essentially passive, crust. (Even if some magma is stored close to the lunar surface, the low percentages of volatiles and low viscosity may preclude the build-up of pressures in the shield which cause tumescence and ultimate caldera collapse.)

The physical characteristics of the moon and its lack of atmosphere have important effects on the morphologic characteristics of some lunar volcanic features. If a typical terrestrial cinder cone eruption occurred on the moon, the distribution of the ejecta would be markedly affected by the lower lunar gravity and the lack of atmospheric drag on ejected particles. Instead of producing a distinctive cone, the ejecta from such an eruption would form a thin but widespread blanket of pyroclastic debris (McGetchin and Head, 1973). There is a small number of areas, most notably on the floor of the crater Alphonsus (fig. 5.27a), where distinctive patches of dark material surround low-rimmed craters. These are most likely pyroclastic deposits. A distinctive complex of domes and cones known as the Marius Hills is seen in Oceanus Procellarum (McCauley, 1969). Several structures in this complex are reminiscent of terrestrial cinder cones. However, because of the characteristics of lunar eruptions, these cones probably result from less explosive pyroclastic eruptions such as fire fountains, which deposit their ejecta at the crater to form spatter cones.

In several areas of the moon, including the Apollo 17 landing site in the Taurus-Littrow area, extensive low-albedo deposits have developed (Pieters et al., 1973; Head, 1974b; Lucchitta, 1973). Their low albedo, anomalous radar return (Zisk and Hagfors, 1970), and blanketing characteristics led to the interpretation that these areas might be cinder or pyroclastic deposits. Investigation of the dark mantle layer at the Apollo 17 site revealed that the dark

(a)

(b)

Figure 5.27. Lunar pyroclastic deposits.

(a). Dark halo craters on the floor of the crater Alphonsus. Craters are about 2–3 km in diameter. Surrounding halos represent energetic pyroclastic eruptions which are not accompanied by lava flows. (LOV-118M.)

(b). Fissure cones in the lunar Marius Hills are aligned along fractures and have associated small rilles. Pile-up of material around vents suggests lower energy eruption conditions than for the Alphonsus craters. (LOV-216, 217H.)

mantle was primarily regolith derived from underlying basalt units. Heiken et al. (1974) have identified glass droplets from the Apollo 17 soils, which form 2–18 percent of the 90–150 micron-size fraction of the soils studied. They conclude that the glass droplets are formed during fire-fountaining associated with the formation of the basaltic lavas.

Wilhelms and McCauley (1971) have described a type of crater characterized by smooth rim, lack of pronounced terraces, and absence of satellitic craters and marked radial rim deposits. These craters range from 20 to 40 km in diameter, and are generally circular. DeHon (1971) has compared two of these craters, Sabine and Ritter, to terrestrial calderas and suggested that they represent lunar resurgent caldrons. It should be noted, however, that many of these craters occur associated with or at the edge of mare deposits. In many cases, the absence of secondary craters is easily explained by postcrater flooding of mare material up to the base of the continuous ejecta blanket. A simple model of primary impact with subsequent volcanism cannot explain all occurrences of this type of crater, however. The crater Kopf (fig. 5.28) in the Orientale basin is developed in upland material adjacent to the maria and is markedly different from an adjacent crater of similar size. Crater densities suggest that Kopf is of an age similar to the adjacent mare. It exhibits a distinctive crater rim, wall, and interior, and has exterior deposits that suggest smooth mantling rather than energetic deposition of an impact ejecta blanket (fig. 5.28). These bits of evidence imply a volcanic origin. However, in a multivariate classification of craters on Earth, moon, and Mars, Pike (1974) has shown that these types of craters have the same general geometry as lunar craters of impact origin.

Attention has often been directed to lunar "volcanic complexes" (Guest, 1971). The most spectacular examples are in Oceanus Procellarum, including the Marius Hills (McCauley, 1967), the Aristarchus Plateau (Moore, 1965, 1967), and the Rumker Hills (Eggleton and Smith, 1967; Smith, 1974). These complexes display a wide variety of morphologic features such as low domes, steep domes, collapse craters, scarps, mare ridges, sinuous rilles, and cones (fig. 5.29). The concentration of these features in mare areas suggests that they are the major centers for extrusion of mare basalts. However, it should be kept in mind that the underlying plateaus (Rumker and Aristarchus in particular) are most likely highland blocks on which a thin volcanic veneer has been deposited. What we see now may be remnants of early mare flooding that may have been destroyed in low lying areas by subsequent flooding.

Mare, or wrinkle, ridges are prominent in essentially all lunar maria. They often form concentric patterns in circular mare basins and occasionally extend into the adjacent lunar highlands. In general, mare ridges are differentiated from broader arches on which they often occur (fig. 5.30) (Strom, 1972; Colton et al., 1972). Tjia (1970) noted an en echelon arrangement of mare ridges and attributed their origin to surface readjustment caused by strike-slip faulting in the basement. Strom (1972) described the broad arching associated with many ridges, and the lobate structures which are similar in appearance to flow lobes. Noting that mare ridge orientation paralleled lineaments in the highlands, Strom con-

Figure 5.28. The crater Kopf, about 40 km in diameter, is markedly different from other craters of similar size in lacking central peaks, extensive terrace development, and abundant secondary craters. Its associated features, including rilles and elongate depressions, and its differences from similar sized craters of apparent impact origin suggest that it has a volcanic origin. (LOV-195H.)

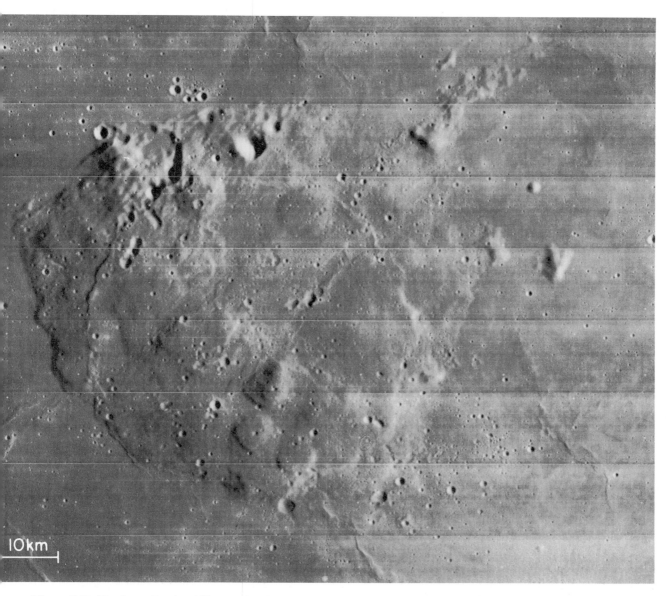

Figure 5.29. The lunar Rumker Hills, a volcanic complex in northern Oceanus Procellarum, is approximately 60 km wide and displays abundant domes and flows, which suggest that it may be a center of igneous activity. (LOV-163.)

cluded that the arches are dike, sill, or laccolith near-surface intrusions, which emerged along major sub-mare fractures. The flow-like mare ridges often seen atop the arches were interpreted as areas where lava had broken through to the surface. It is quite likely that many arches and ridges formed during the time of mare volcanism. Thus the sinuous lobate ridges may in places represent squeeze-ups of lava liberated during settling, arching, and buckling of a lava crust (Hodges, 1973). In fact, Schaber (1973) and Bryan (1973) have shown that mare ridges grew between successive flows in Mare Imbrium to such an extent

that they diverted succeeding flows. Bryan (1973) summarizes morphological relationships in Mare Imbrium and Serenitatis and concludes that mare wrinkle ridges are formed by localized compression of a relatively thin crust essentially uncoupled from underlying structure and topography. He attributes the compressional forces to the isostatic settling of a spherical surface shell represented by the mare fill. Howard and Muehlberger (1973) attribute the mare ridge in the uplands of eastern Serenitatis to thrust faulting, probably related to late anticlinal buckling of the edge of the Serenitatis basin. The ridge has

Figure 5.30. Broad arches and mare ridges developed in Mare Imbrium. The low broad arch, approximately 5–10 km wide, contains a discontinuous sinuous crestal ridge. Most of the ridges postdate the mare lava fill. (AS17-23712.)

little relationship to earlier structural trends associated with the region (Head, 1974c). The extension of the mare ridge into the highlands, therefore, appears to be tectonic, rather than volcanic in origin.

Colton et al. (1972) cite additional areas where mare ridges appear to be caused by thrusting of mare layers. As they point out, a complex origin for arches and ridges is probable, including such factors as subsidence, faulting, intrusion, and extrusion.

A classification of volcanic eruptions summarized by Macdonald (1972) is presented in table 5.1. It

is clear from the review of lunar mare volcanic features that basaltic flood and Hawaiian eruptions most closely resemble lunar mare volcanism. Table 5.2 outlines the most extensive terrestrial examples of flood basalts and compares their area to lunar basins. Many of the lunar basins are comparable in area to terrestrial flood basalt provinces. Baksi and Watkins (1973) have recently compared volcanic production rates in oceanic ridges, islands, and Columbia plateau flood basalts. Using radiometric age data, they derive a basalt production rate of 10^8 m^3/year dur-

Table 5.1. Classification of volcanic eruptions, after Macdonald (1972)

Eruption type	Physical nature of the magma	Character of explosive activity	Nature of effusive activity	Nature of dominant ejecta	Structures built around vent
Basaltic flood	Fluid	Very weak ejection of very fluid blebs; lava fountains	Voluminous wide-spreading flows of very fluid lava	Bombs and spatter; very little ash	Spatter cones and ramparts; very broad flat lava cones; broad lava plain
Hawaiian	Fluid	Weak ejection of very fluid blebs; lava fountains	Thin, often extensive flows of fluid lava	Bombs and spatter; very little ash	Spatter cones and ramparts; very broad flat lava cones
Strombolian	Moderately fluid	Weak to violent ejection of pasty fluid blebs	Thicker, less extensive flows of moderately fluid lava; flows may be absent	Bombs; cinder, small to large amounts of glassy ash	Cinder cones
Vulcanian	Viscous	Moderate to violent ejection of solid or very viscous hot fragments of new lava	Flows commonly absent; when present they are thick and stubby	Essential, glassy to lithic, blocks and ash; pumice	Ash cones, block cones, block-and-ash cones
Peléean	Viscous	Like Vulcanian, commonly with glowing avalanches	Domes and/or short very thick flows	Like Vulcanian	Ash and pumice cones; domes
Plinian (exceptionally strong Vulcanian)	Viscous	Ejection of large volumes of ash	Ash flows, small to very voluminous	Glassy ash and pumice	Widespread pumice lapilli and ash beds; generally no cone building
Rhyolitic flood	Viscous	Relatively small amounts of ash	Voluminous wide-spreading ash flows	Glassy ash and pumice	Flat plain, or broad flat shield, often with caldera

Reprinted by permission of Prentice-Hall, Inc., Englewood Cliffs, N.J.

ing a specific middle Miocene volcanic episode. According to them, this is almost three times the oceanic island production rate and four to six times the spreading midocean ridge system rate. For lunar lavas Schaber (1973) has argued for extremely high extrusion rates for Eratosthenian-age flows in Mare Imbrium, and he suggests that extreme flow lengths on the lunar surface can be accounted for by these extremely high rates, with low melt viscosity playing an important but secondary role. Therefore, high extrusion rates may be another similar characteristic of lunar and terrestrial flood basalts.

When the first lunar farside photos were returned, it was obvious that maria were absent over most of this heavily cratered half of the moon. This frontside-backside maria asymmetry led to speculations concerning varying crustal thickness and isotherm distributions. Data obtained from the Apollo orbital laser

Table 5.2. Areas of terrestrial flood basalts compared with areas within lunar basins filled with mare deposits.

Planet	Feature	Area (km^8)
Moon	Imbrium	850,000
	Serenitatis	300,000
	Eratosthenian Flows in Imbrium (Schaber, 1973)	200,000
	Humorum	110,000
	Orientale	70,000
Earth	Deccan (India)	256,000
	Columbia & Snake River (USA)	180,000

altimeter (Kaula et al., 1972) revealed an offset of the lunar center of mass from the center of figure of more than 2 km toward Earth. Kaula et al. interpreted this to indicate that the process of crustal formation yielded a thinner Earthside crust and a thicker farside crust. The thinner nearside crust enabled mare lavas to reach the surface more easily.

Radiometric dating of lunar mare samples collected during Apollo missions range from 3.83 b.y. for an Apollo 17 mare basalt (Tatsumoto et al., 1973) to 3.25 b.y. for some Apollo 12 basalts (Turner, 1971; Papanastassiou and Wasserburg, 1971a). Based on the Soderblom and Lebofsky (1972) method of crater degradation dating, Schaber (1973) places the ages of the three Eratosthenian lava pulses in Mare Imbrium at 3.0 ± 0.4, 2.7 ± 0.3, and 2.5 ± 0.3 b.y. These almost certainly represent some of the last stages of major mare volcanism. Abundant basalt clasts of the type collected from the mare surfaces are not seen in the highland breccias, suggesting that mare flooding of the type represented by present dark maria began at about the time of formation of the last major basins (Taylor, 1975). Although it is significant that no major extrusive volcanism occurred on the moon in the last 2.5 b.y., it is important to realize that the 1.3-b.y. period of mare volcanism does not represent a single phase. When compared to terrestrial flood basalt chronology (fig. 5.31), the 1.3-b.y. period emerges as a similarly complex series of events.

Morphologic features that have been attributed to highland extrusive volcanism primarily include: 1) regionally flat upland plains, 2) highland domes and mounds, 3) elongate linear ridges similar to and often extensions of mare ridges, and 4) elongate crater chains. Many of these features were thought to represent emplacement of volcanic rocks less basic in composition than mare basalts—perhaps involving remobilized lunar crust—and were thought to have occurred prior to the earliest phases of mare volcanism.

The Apollo 16 Descartes region (fig. 5.32) offers an excellent example of both the domical and plains morphology that was interpreted by most workers to be of volcanic origin. The plains units occur primarily in the floors of older craters. Plains are also distributed in irregular patches over many other intercrater areas in the highlands, and they occur at a variety of different topographic levels (Eggleton and Schaber, 1972). Similarities of plains to maria led to their interpretation as volcanic deposits, possibly of flow or pyroclastic origin (Wilhelms, 1970; Elston et al., 1972; Hodges, 1972a; and others). The domical structures include broad domes, elongate furrowed cone-like structures, and linear ridges (Wilhelms and McCauley, 1971). Wilhelms and McCauley drew attention to the similarity of these structures with those seen in terrestrial deposits of intermediate to acidic compositions (see also Trask and McCauley, 1972; Head and Goetz, 1972).

Field observations by the Apollo 16 crew contradicted most of the prior photogeologic interpretation. There was no evidence of highland volcanism. Instead, a wide variety of impact breccias were collected (Muehlberger et al., 1972). Re-examination of the regional geology led to the interpretation of the samples as ejecta deposits, related either to multi-ringed basins (Eggleton and Schaber, 1972; Hodges, 1972b; Chao et al., 1973) to local craters (Oberbeck et al., 1973), or to a combination of both (Head, 1974d). On a lunar-wide scale, it is clear that plains units are preferentially distributed around major multi-ringed basins (Soderblom and Boyce, 1972) and that, at least near the major basins, ponded ejecta deposits form smooth plains surfaces (Oberbeck et al., 1974; Head, 1973, 1974d). Recent studies of crater modification processes have shown that upland plains units, similar to Cayley plains explored on Apollo 16, may have been produced in early lunar history when an increased rate of bombardment, and formation of major basins, caused mobilization of rim and wall material and redeposition in surrounding lows (Head, 1975). Similarly, the

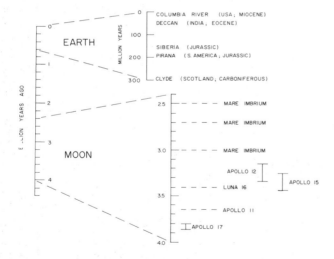

Figure 5.31. Comparison of flood basalt chronology on the moon and Earth. The three Mare Imbrium dates are for three phases of flows identified and dated by the Soderblom and Lebofsky (1972) method by Schaber (1973). The lunar period of volcanism spans about 1.4 b.y. when the youngest Mare Imbrium flows are considered. The terrestrial sequence and chronology of flood basalts (from Holmes, 1965), when compared to the moon, strongly suggests that the lunar period of volcanism represents a complex sequence of mare volcanism with its components not necessarily related closely in space and time.

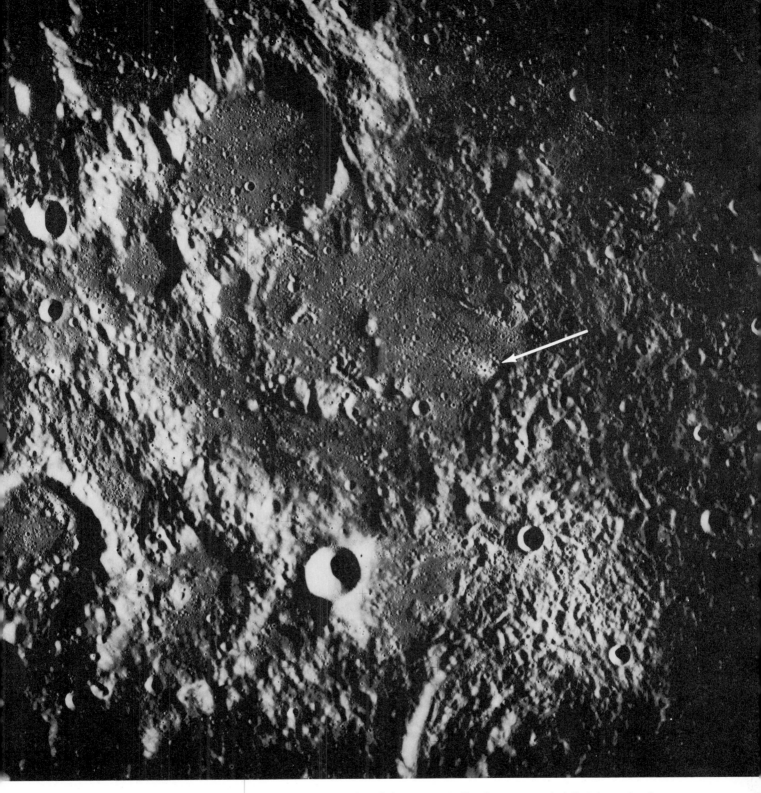

Figure 5.32. The Descartes region of the moon. The Apollo 16 site is indicated by arrow. Prior to the Apollo 16 mission, grooved and furrowed terrain to the east and south of the site was interpreted primarily as constructional volcanic landforms while the flat Cayley plains were thought to represent lava surfaces. Exploration by Apollo 16 Astronauts Young and Duke revealed terrain comprising impact brec-cias. Subsequent studies have suggested that large local craters have played an important role in the evolution of the site (Head, 1975). Note particularly the 60-km-diameter crater whose eastern rim is located near the Apollo 16 site. Impact of material from the Imbrium basin has caused extensive local mixing of material (Oberbeck et al., 1973). Picture is approximately 150 km wide. (AS16-0162.)

domes and elongate ridges and furrows have also been shown to be associated with the major multi-ringed basin deposits (Hodges, 1972b; Head, 1973). Thus a large number of deposits previously interpreted to be related to highland volcanism now appear possibly related to ejecta deposited from multi-ringed basins. The massive and cumulative influence of the lateral sedimentation associated with basin formation is only now being fully appreciated.

One area of domes appears to postdate the major multi-ringed basins but predate or be contemporaneous with the maria (fig. 5.33). Although mapped as relatively young (Wilhelms and McCauley, 1971), the Gruithuisen domes appear to be embayed by Imbrian-age mare material. The domes appear to postdate the Iridum impact, thus probably placing them somewhere between about 3.8 and 3.5 b.y. If these are volcanic, they may represent the last stage of highland volcanism of this type, and the only stage to be morphologically preserved. Malin (1974a) has

Figure 5.33. The Gruithuisen dome region at the northwest edge of Mare Imbrium. These steep-sided domes are up to 25 km in width and appear to postdate the formation of multi-ringed basins but predate or be contemporaneous with the maria (fig. 5.33). Although mapped as Iridum crater but predate the surrounding mare material. They may represent one of the few surviving examples of upland constructional volcanic features. (AS15-12718.)

suggested that this area and several others of similar age and spectral characteristics may be the surface manifestations of the pre-mare KREEP (enriched in potassium, rare-earth elements, and phosphorus) basalts.

Elongate narrow ridges similar in size and morphology to mare ridges are seen in the highlands (fig. 6.19). Because of their lobe-like morphology, these have occasionally been attributed to volcanic extrusion. Most of the highland ridges have crisp morphologies similar to mare ridges, suggesting that they are of a similar relatively young age. Although many mare ridges extend into the highlands, there appear to be no highland arches similar to those seen in the maria. Apollo 17 explored an area (Taurus-Littrow Valley) that contains a highland extension of one of these ridges. The ridge crosses the valley and winds its way through the highlands to the north. The surface trace of the ridge bears little relationship to previous structure (Head, 1974c), but there is no evidence of surface extrusion along the ridge, which was traversed by the Apollo 17 crew (Muehlberger et al., 1973). Various arguments have been advanced for the origin of this highland/mare ridge (Howard and Muehlberger, 1973; Head, 1974c; Scott, 1974a; Hodges, 1973). All authors agree that it represents some sort of fault trace unaccompanied by extrusion of lava—at least at the Apollo 17 site. Generalizing these arguments, caution should be used in citing other highland ridges as evidence of volcanism.

Crater chains are another morphologic feature suggesting volcanic activity in the highlands. Perhaps the most impressive examples of crater chains not clearly associated with multi-ringed basins are Hyginus, Davy, and Abulfeda (fig. 5.34). Of these three, Hyginus is the only crater chain that is unequivocally associated with an endogenic structure— a linear rille. Most workers interpret these structures as representing explosion or collapse craters (such as maars or calderas) associated with internal volcanic activity (McGetchin and Ullrich, 1973; Howard and Masursky, 1968). Oberbeck and Morrison (1973) attribute the formation of the Davy chain to secondary impact cratering, based on the similarity of intercrater ridges to other examples of secondary crater chains, which are more distinctly related to primary craters. If mapped crater chains interpreted to be of internal origin (Wilhelms and McCauley, 1971) are indeed of volcanic origin, they are not associated with major volcanic provinces since little of the terrain surrounding these features displays volcanic morphologies.

In summary, there is little unequivocal morphologic evidence for extensive highland volcanism, particularly in the post multi-ringed basin period of lunar history ($< \sim 3.85$ b.y.). Since observed craters on the highland surface predominantly record the last part of the early intense flux, any extensive morphologic evidence of highland volcanism prior to this time has probably been destroyed. Returned lunar samples do, however, provide evidence for pre-mare volcanism (Taylor, 1975).

Mercurian Volcanism

Landforms unequivocally identifiable as volcanic have not been seen on Mercury (Strom et al., 1975). The primary evidence for volcanism is the extensive development of plains materials (Strom et al., 1975; Trask and Guest, 1975; Murray et al., 1975). Intercrater plains represent the most extensive terrain on the planet (fig. 5.35). Origin of this unit is uncertain but it may represent volcanism related to the early differentiation of the planet (Murray et al., 1975).

A continuous plains unit (fig. 5.35), covers about 15 percent of the region viewed by Mariner 10. Most of this unit, termed smooth plains, occurs within or near the 1350-km-diameter Caloris basin. This unit is very similar to the lunar maria and is interpreted to be volcanic in origin, although other modes of formation cannot be ruled out.

Strom et al. (1975) cite the following evidence to support a volcanic interpretation: 1. large volumes of material fill older depressions and form flat surfaces; 2. much of the plains material has an albedo that contrasts with that of the immediately surrounding upland area; 3. larger volumes of plains material surround the two major basins than can be readily attributed to impacts; 4. great differences in volume of plains material occur around basins of comparable size.

Similarities between lunar and Mercurian plains units weaken the volcanic argument. Lunar light plains appear to form by a variety of non-volcanic processes.

Martian Volcanism

First Evidences

Before the flight of Mariner 9 there was no compelling reason to think that Mars had ever been volcanically active. Several authors (e.g., Fielder, 1965b; McCall, 1966) interpreted the circular structures in Mariner 4 pictures as volcanic calderas. They cited the presence of central peaks and multiple concentric scarps. However, their contentions were based less on systematic analysis than on the a priori opinion that both lunar and Martian craters *should* be volcanic.

Several telescopic observers have reported bright flares on Mars (Katterfeld, 1966; Saheki, 1955).

Figure 5.34. View of the Davy crater chain looking westward into northern Mare Nubium. The crater chain begins in the irregular crater Davy G (foreground; approximately 15 km in diameter) and stretches about 55 km to the west. Although the chain has been interpreted as a secondary crater chain (Oberbeck and Morrison, 1973), most geologists believe it is a string of maar-like volcanic craters. (AS14-73-10103.)

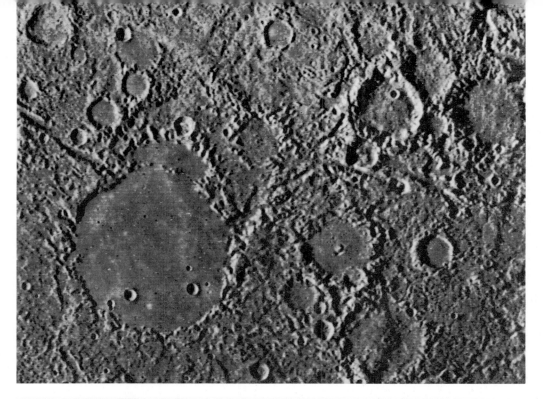

(a)

(b)

Figure 5.35. Plains units on Mercury.

(a). Smooth plains on floor of old degraded crater, about 135 km in diameter, surrounded by hilly and lineated terrain (Mariner 10 frame 27370) (Trask and Guest, 1975).

(b). View looking northward toward the Borealis Planitia (Northern Plains) surrounding a large basin (310 km diam.) near Mercury's north pole. Two types of plains are visible. Intercrater plains are present in the heavily cratered terrain in the foreground, while smooth plains dominate the background. Large craters in foreground are 75-100 km in diameter (Mariner 10 frame 152) (Trask and Guest, 1975).

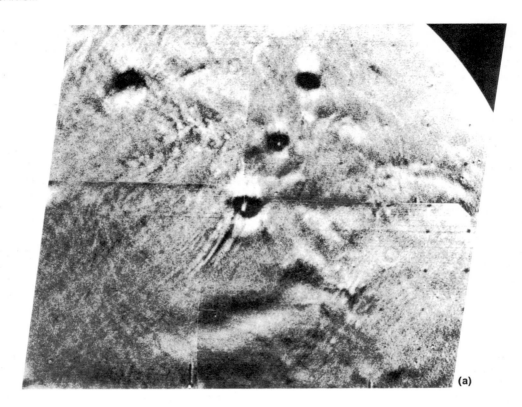

(a)

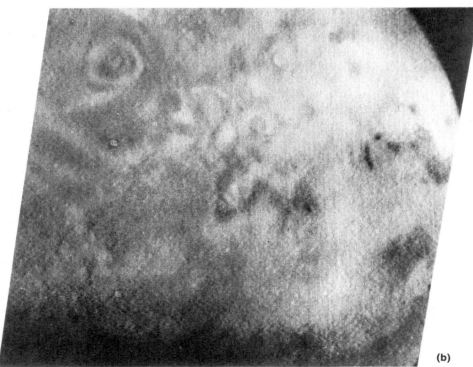

(b)

Figure 5.36 (a). A Mariner 9 photomosaic compared with **(b).** a Mariner 7 photograph. The Mariner 9 pictures were taken early in the mission. Most of the surface was obscured by dust but the summit regions of Mons Olympus, Pavonis, Ascraeus, and Arsia appear as dark spots. The volcanoes are also identifiable in the Mariner 7 picture. Olympus Mons appears as a bright annulus. The other three volcanoes are bright spots.

One of the better known sightings was in 1937 by a Japanese observer, Sizuo Mayeda. He described a bright spot, far brighter than the polar caps, which appeared near Tithonius Lacus, scintillated like a star, and disappeared after about five minutes. This flash of light, and others like it, have been attributed to eruption and radiation of volcanic gases. There have been similar reports of bright flares on the moon (Middlehurst, 1967). Because of their intrinsically random and ephemeral nature, the observations are difficult to confirm or interpret.

One comprehensive pre-Mariner volcanic theory was presented by Dean McLaughlin, an American astronomer who also played a leading role in mapping and interpreting Triassic red beds and lavas of eastern United States. Between 1954 and 1956 McLaughlin published a series of eight papers in which he attempted to explain the variable dark Martian features by a combination of volcanic and eolian processes. (See Veverka and Sagan, 1974.) In brief, he argued that dispersion of dark ash from large volcanic vents gave rise to triangularly shaped dark markings that flared in the downwind direction. Seasonal variations in wind direction accounted for redistribution of primary ash deposits, changing both the shape and albedo of telescopically observed dark markings. Although McLaughlin's ideas were summarily dismissed by most students of Mars (e.g., Kuiper, 1956), they now appear to have been unusually prophetic. Even though his identification of dark surface materials as volcanic ash is questionable, and his proposed wind circulation pattern is in error, his general thesis that a combination of volcanic and eolian processes account for many features of the Martian landscape is correct.

The development of our "space-age" knowledge regarding the volcanic history of Mars is a story full of irony. Mariners 4, 6, and 7 all photographed cratered terrain so closely resembling the lunar highlands that most scientists were quickly willing to accede that Mars was a primitive, inactive planet without internal activity of any sort. So persuasive were these arguments that serious questions were raised in some circles concerning the advisability of devoting an orbital mission to further photography of this bland lunar-like surface, let alone launching an unmanned lander dedicated to the proposition that life could exist on such a primitive body.

Whatever disappointments were associated with the first pictures taken by Mariner 9, they at least disposed of the postulate that Mars was lunar-like. Against a background of atmospheric dust, four black spots of uncertain origin were visible (fig. 5.36). One would be tempted to label them some sort of spurious or transient artifact except that they were repeatedly seen in pictures taken on successive orbits. Out of desperation as much as anything else, the narrow-angle camera was directed at these vague dark markings. Dimly first, and then with increasing certainty, a crater-like depression could be seen at the center of each spot. Was this merely a coincidence, or was one justified in stringing together a few observations to make a childishly simple conclusion?

The argument runs like this. The dark spots are bits of the Martian surface visible only because they protrude above the dust storm—that is, they are mountains. Each of the mountains has a crater at its summit. And what are the only known examples of high mountains with summit craters? Volcanoes, of course. As we said, the argument is so simple that one hesitates to advance it. Is it likely that a planet that has already been certified inactive by three previous spacecraft would choose to reveal the contrary in this spectacularly simple fashion? The answer is yes. And therein lies the irony.

Volcanic Morphologies

Three of the four volcanoes just mentioned—Ascraeus Mons, Pavonis Mons, and Arsia Mons—are aligned along the Tharsis ridge. The fourth, Olympus Mons, is located on the northwest flank of the Tharsis uplift. Olympus Mons is the most impressive of the quartet (fig. 5.37). Approximately 600 km across, its summit is 21 km higher than the surrounding plains (Davies, 1974). Its average slope, then, is between 5° and 6°. By any standard this is a prodigious mountain. On Earth, local relief in the most alpine mountain ranges is no more than 9 km. Indeed the difference in elevation between the lowest and highest point anywhere on Earth's surface is only about 19.5 km. The feature most closely resembling Olympus Mons is the Hawaiian volcano Mauna Loa, which is approximately 200 km in diameter measured at the ocean floor and stands 9 km high (fig. 5.25). This is the largest volcano known on Earth. Olympus Mons is so much larger that it clearly merits its informal title—"the largest known volcano in the solar system." Ascraeus, Pavonis, and Arsia are substantially smaller than Olympus Mons, each being about 300 km across, but are still significantly larger than Mauna Loa.

These four Martian volcanoes show certain distinctive features in common. Each is roughly conical with a summit depression. In the case of Pavonis Mons and Arsia Mons, the craters are simple circles in outline (fig. 5.38). The Olympus and Ascraeus depressions are made up of coalescing circular structures formed by successive collapse over shifting centers (figs. 5.37, 5.38).

A detailed cross section through Olympus Mons reveals a form more complex than a simple cone.

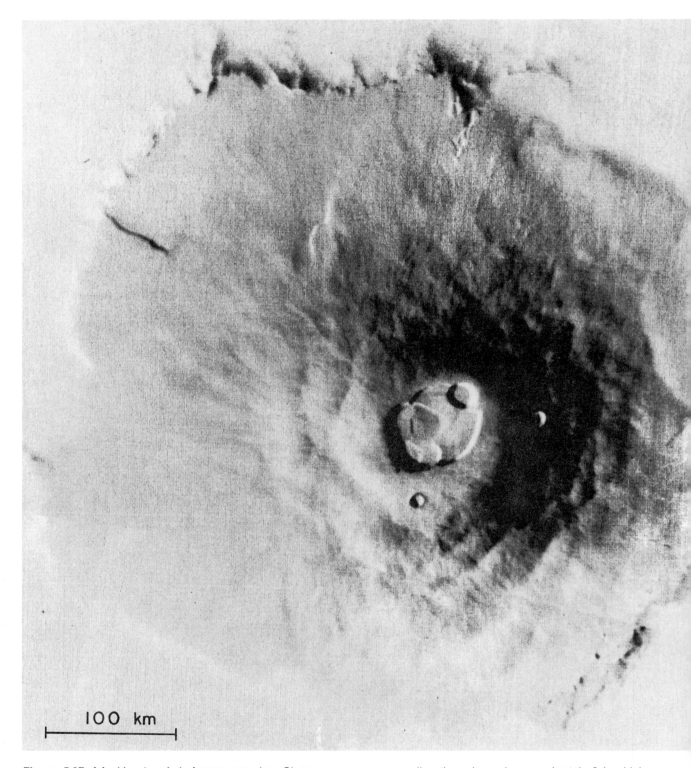

100 km

Figure 5.37 (a). Mosaic of A frames covering Olympus Mons. This volcano is the largest on Mars, 600 km in diameter and rising 21 km above its base (Davies, 1974). The scarp surrounding the volcano is approximately 2 km high (Blasius, 1973). Note the corrugated appearance of volcanic flanks. B frames resolve these features as lava flows.

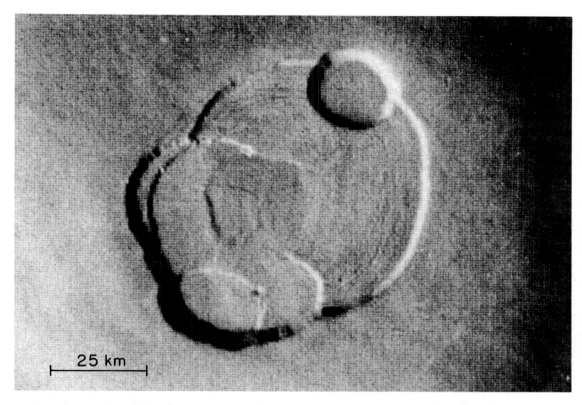

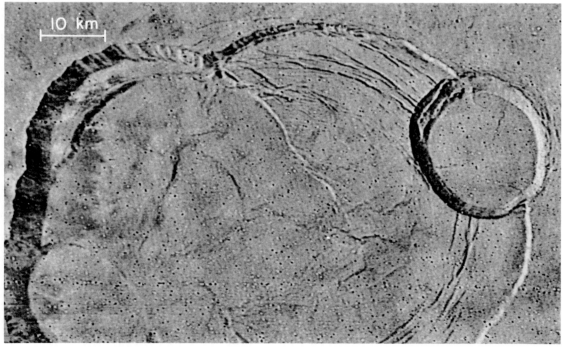

(b). Two views of summit calderas of Olympus Mons. This complex of superimposed collapse structures records a series of volcanic events. The diameter of the largest caldera is about 65 km. Special enhancements are courtesy of Karl Blasius. (DAS No. 06895388, rev. 148, A camera, center at 133°W, 15°N; DAS No. 09845439, rev. 230, B camera, center at 133°W, 18°N.)

Figure 5.38 (a). Ascraeus Mons. Several overlapping calderas are present at the summit. Concentric rings of materials with different albedo may indicate successive periods of extrusive activity. Swarms of irregular channels near the base of the volcano are possible lava channels. (DAS No. 12995262, rev. 528, A camera, center at 104°W, 10°N.)

(b). Summit caldera of Arsia Mons showing concentric fractures rimming the central lava-filled cavity. (DAS No. 02602185, rev. 28, B camera, center at 120°W, 9°S.)

(c). Pavonis Mons. This volcano is characterized by a series of concentric fractures and graben, which continue from the central caldera to the outermost flanks. (DAS No. 07111128, rev. 154, A camera, center at 113°W, 2°S.)

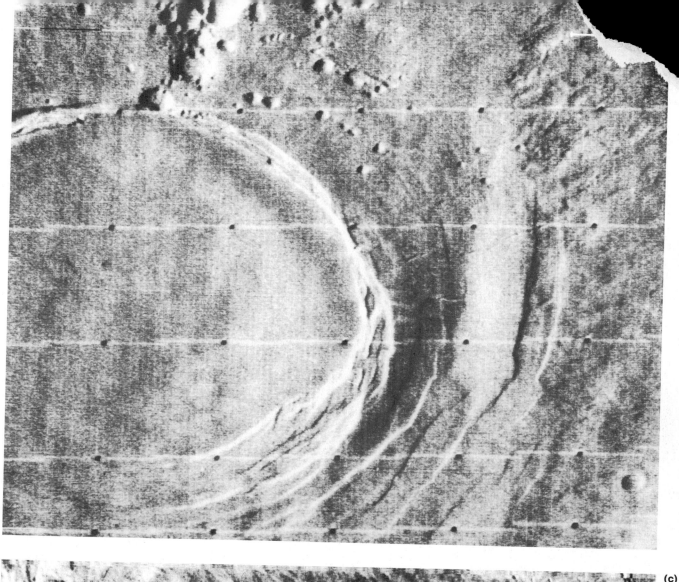

(c)

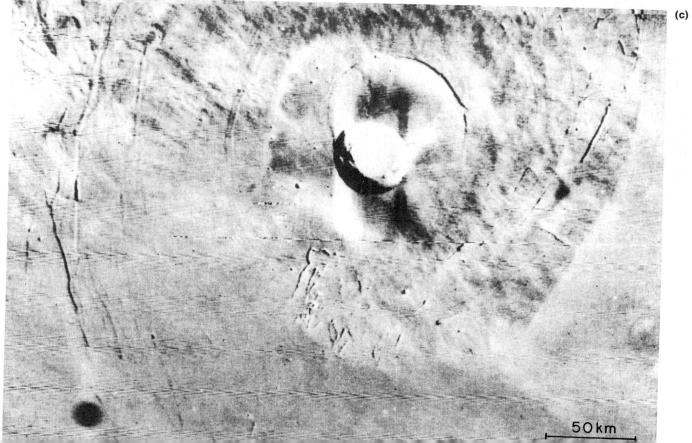

50 km

The upper slopes are steeper than the lower flanks. This two-story effect is also observed in terrestrial shield volcanoes. In the Hawaiian volcano Mauna Kea, for example, late-stage eruptions are more acidic in composition and fragmental in character than earlier ones. Stubby flows and cinder deposits form a steep-sided secondary cone astride an older shield (Macdonald, 1972).

The flanks of the Tharsis volcanoes have both radial and concentric structure. The radial corrugation is similar to texture on flanks of terrestrial shield volcanoes that are built up by successive basaltic eruptions. Individual long, narrow flows that can be identified on B-frame photographs of some of these flows indicate central channels that are analogous to lava channels and collapsed lava tubes in terrestrial situations (fig. 5.39). The concentric structures best developed on the flanks of Pavonis and Arsia Mons are cracks apparently induced by subsidence of the volcanic pile, perhaps coincident with evacuation of lava from the subsurface.

In general, the flanks of Ascraeus, Pavonis, and Arsia merge with the surrounding plains materials, the only distinction being one of surface texture. In places, however, the outer fringes of the volcanoes are marked by graben and irregular cliffs. Control by pre-existing structure is suggested, particularly for Arsia Mons, which has an outline not circular

Figure 5.39. The flanks of Olympus Mons, showing a lava channel. Arrows identify the edges of an individual lava flow. Topographic gradient is from lower right to upper left. (DAS No. 05492413, rev. 109, B camera, center at 135°W, 20°N.)

but rectilinear. Fractures tend to be oriented in two sets, one trending northeast, the other northwest.

Olympus Mons has a unique bounding feature, a cliff several kilometers high that extends around the volcano more or less continuously for 1500 km. There is evidence that the scarp is receding by slumping (fig. 5.40). However, it should be kept in mind that this kind of mass wasting activity is probably the consequence and not necessarily the cause of a scarp. It has been suggested, perhaps not very seriously, that the scarp is a wave-cut bench evidencing a former vast sea on Mars. Incidentally, any sea that lapped up on the relatively high slopes of Olympus Mons (fig. 3.4) also would cover much of the planet.

King and Riehle (1974) suggest that, at the time of its formation, Olympus Mons had a central region of lavas that was fringed by ash-flow tuffs. The tuffs were susceptible to erosion, especially in their distal regions, where they were relatively uncompacted. This tuff annulus has been removed by eolian activity and the present scarp marks the outer boundary of more resistant lava and densely compacted ash.

There is an alternate, structural explanation for the scarp. The Hawaiian volcano Mauna Loa has a number of little-noted seaward-facing scarps low on its lower flanks (Macdonald, 1972; Stearns, 1966). Called pali, they are not continuous, as is the scarp around Olympus Mons. Pali have not been well-described nor have they been satisfactorily explained. Possibly, accumulations of lava on the flanks of the volcano have loaded the crust in a region where it is not being dynamically supported by the mantle hot spot. The strength of the crust was exceeded by this load and faulting occurred. Another hypothesis is that the uplift of the volcano is greater than subsidence following eruptions. The flank faults would form during periods of central uplift. Perhaps these effects, speculative even in the case of Mauna Loa, have accounted for uplift of the central core of Olympus Mons hundreds of meters above the distal flanks.

Olympus Mons is surrounded by a ring of grooved terrain, extending almost 1000 km from the volcanic center (fig. 3.7). Parallel ridges and troughs impart a corrugated appearance to the landscape. Several features suggest a genetic relationship between the grooved terrain and Olympus Mons. The grooved terrain concentrically encircles the central shield, even though it is most prominently displayed to the north and west (fig. 3.15). Individual ridges and troughs also tend to have concentric orientations.

Carr (1973) has suggested that grooved terrain comprises the eroded roots of shield volcanoes similar to Olympus Mons, but older. McCauley (1973) has pointed out that, on a small scale, the ridges have

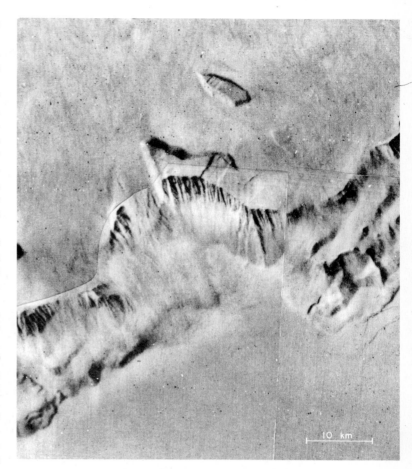

Figure 5.40. Mosaic of B frames covering the southeastern edge of the scarp surrounding Olympus Mons. This giant cliff is approximately 4.8 km high (Whitehead, 1974). The appearance of the slopes suggests mass wasting by slumping and downhill creep. Erosional gullies occur at the top of the scarp. (DAS No. 09844669, rev. 230, B camera, center at 130°W, 14°N; DAS No. 09845299, rev. 230, B camera, center at 130°W, 15°N; DAS No. 09845369, rev. 230, B camera, center at 130°W, 15°N.)

cigar shapes reminiscent of wind-sculpted ridges, or yardangs, found in terrestrial desert regions. One might argue that the topography is formed by differential erosion of volcanic lava and ash deposited on the slopes of an ancient shield volcano. However, the scale of the grooving is disproportionally large.

The grooved terrain does not extend to the shield, but is separated from it by an annulus of plains materials. Continuing the structural argument advanced for the formation of the fringing scarp, one might envision the inner, featureless ring as a strongly depressed region of the crust filled in with sediment from the bounding shield volcano. The grooved terrain, then, represents a structural hinge line where the crust has been fractured and warped downward in response to the load of Olympus Mons. The inner, featureless zone marks the region of maximum downwarping (fig. 5.41).

The four shield volcanoes just mentioned, along with a fifth in the Elysium area, Elysium Mons (fig. 5.42), constitute the largest and most spectacular Martian volcanic features. There are, however, more than twenty additional domes and calderas for which a volcanic origin is almost equally certain. Many of these are clustered in the Tharsis region. One of the better examples of a dome is Tharsis Tholus, a bulbous form with steep slopes and a steep-sided summit caldera. Concentric terraces within the caldera attest to several episodes of extrusion and collapse. Uranius Tholus is a dome with different characteristics (fig. 5.43). Numerous channels extend radially down the flanks of the dome. A larger rille-like channel issues from the central caldera and terminates in an irregular depression on the bounding volcanic plains. Uranius Tholus is among a cluster of volcanic features. One hundred kilometers to the north is a smaller dome, Ceraunius Tholus (fig. 5.43). Two hundred kilometers to the northeast is Uranius Patera, a slightly raised feature which is

dominated by a flat-floored caldera about 150 km in diameter.

Still another type of dome is displayed in the Elysium region. Hecates Tholus is a roughly textured mound about 200 km in diameter (fig. 5.44). A prominent central caldera is lacking, but a smaller, multiple-collapse depression occurs near the summit. Fine channels and crater chains radiate from this depression and extend discontinuously to the base of the dome, where they disappear beneath younger plains materials.

In addition to shields and domes, a variety of volcanic constructional features are slightly more difficult to document because they have been degraded by erosion and structural readjustment. Included here are Alba Patera of the Tharsis region, Apollinares Patera of the Elysium region, and almost all the volcanic features identified outside these two relatively young volcanic provinces. Alba Patera is unique in that it is a shield volcano without the shield (fig. 5.45). The principal feature, a complex caldera similar to those of Olympus Mons and Ascraeus Mons, is located in the midst of a generally featureless plain some 600 km across. The plain is bounded by northeast-southwest fractures that are deflected to form a structural ring that is concentric around the central caldera. Carr (1973) suggests that Alba Patera is a former shield volcano that has collapsed along arcuate fractures.

More conventional erosion is displayed by Apollinares Patera. The summit caldera has been partially breached and covered by younger deposits, and fragmentary septa hint at an original morphology like that of the Olympus caldera. The flanks of the shield are partially destroyed, but remnants display a vortex-like set of radiating grooves. The topography probably results from enlargement of original flow and channel features by eolian erosion.

Beyond the volcanic provinces of Tharsis and Elysium, a steadily enlarging group of indistinct features has been identified throughout the cratered terrain. This extends as far north as 74°, as far south as 69°. A cluster of four structures occurs around the Hellas basin, a concentration sufficient to suggest that this region was once an area of active volcanism competing in magnitude with Tharsis and Elysium. The most prominent structure occurs north of the basin at 253°W, 23°S. Early in the Mariner 9 mission it earned the name "dandelion"—radiating petals around a central disc. Later named Tyrrhenum Patera, this structure consists of a central caldera connected by a channel to a neighboring circular depression (fig. 3.9). Radiating outward are numerous troughs, some of which may be lava channels but many of which are structurally formed graben. Less striking features in the vicinity of the Hellas basin are Hadriacum Patera (fig. 5.45),

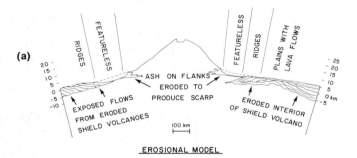

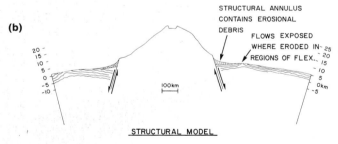

Figure 5.41. Diagrammatic cross-sections showing two models for geological relationships near Olympus Mons.

(a). According to this erosional model, grooved terrain represents the roots of an ancient volcano. Featureless terrain forms by erosion of ash on lower slopes of Olympus Mons.

(b). According to this structural model, the grooved terrain forms where pre-existing flows have been tilted along a hinge line. An annulus of smooth plains forms in a down-faulted trough.

Figure 5.42 (a). Elysium Mons, showing summit caldera and roughly textured flanks. (DAS No. 13496123, rev. 676, A camera, center at 213°W, 25°N.)

(b). The lava-filled caldera at the summit of Elysium Mons. Several overflow channels are visible. (DAS No. 13496013, rev. 676, B camera, center at 213°W, 24°N; DAS No. 13496083, rev. 676, B camera, center at 213°W, 25°N.)

(a)

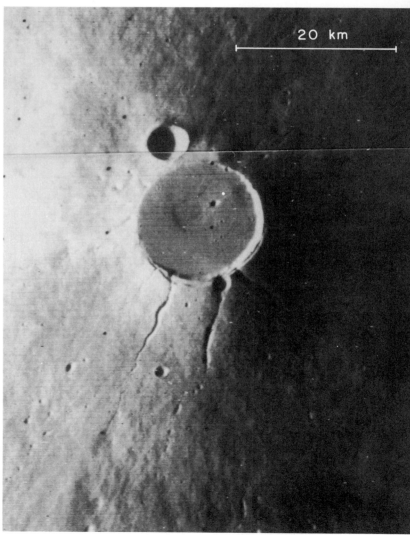

(b)

Amphitrites Patera (fig. 5.45), and an unnamed feature at 311°W, 58°S, all of which show central depressions that are the apices for sets of radial spokes. The resultant star-burst texture appears to be characteristic of ancient shield volcanoes. Erosion has been so effective that some of the central calderas are much less prominent than adjacent impact craters of the same size. Indeed, the calderas are sometimes discovered only by tracing the radial structures back to a point of common convergence.

Some extremely irregular Martian craters may be complex volcanic calderas. A prominent example is Orcus Patera, located at 182°W, 15°N, which closely resembles in shape and size the lunar crater Schiller (Scott, 1974b).

Temporal and Spatial Distribution

VOLCANOES

Having described the Martian volcanic structures, let us move on to matters of synthesis and genesis. As for Earth and the moon, we can ask four questions. Is the spatial distribution nonrandom? Is the temporal distribution nonrandom? Can one say anything about chemical/physical composition, and possible variation in composition with space and time? And finally, what does all this tell us about the thermal history of the planet?

First, spatial distribution. There is a marked twofold division of Mars into two hemispheres separated by a great circle inclined 35° to the equator. The southern hemisphere contains the cratered terrain; the northern hemisphere contains the smooth plains as well as the Tharsis and Elysium provinces. Although positive morphological evidence of volcanic deposits is lacking, it is possible that most of the surficial materials in the northern hemisphere are lavas covered by a veneer of aeolian sediments. This asymmetry is directly comparable neither with Earth, where volcanism is concentrated along plate boundaries, nor with the moon, where volcanic activity occurs in the vicinity of impact basins. True enough, lunar volcanism is restricted to the Earthside, but this hemispheric asymmetry is descriptively related to the moon's axis and period of rotation.

Because plate tectonics has proven to be such an exciting synthesizing model for Earth, some scientists wish to extend the fashionable aspects of the theory to other planets. It is difficult. There is virtually no evidence for continental fragmentation,

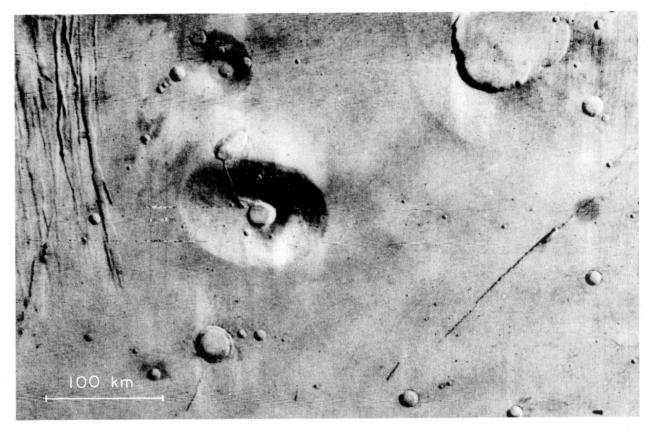

Figure 5.43. Ceraunius Tholus is situated near the center of the picture. A lava channel drains the summit caldera and terminates in a shallow depression on the adjacent plains. The smaller dome above Ceraunius Tholus is Ura- nius Tholus. The complex caldera in the upper right is Uranius Patera. (DAS No. 07183788, rev. 156, A camera, center at 97°W, 22°N.)

drift, or collision on Mars. One looks in vain for two types of features: compressional, where plates collide, and extensional, where they separate. There are some large canyons, but to measure the canyon width and cite this as the amount of plate separation is just as relevant as citing the width of the Grand Canyon in southwestern United States as the amount of plate separation in that region. The Grand Canyon is widened by erosion of valley walls, and we have no evidence at present to suggest otherwise for the Valles Marineris on Mars. This is not to deny a structural control for the canyon. Its orientation suggests that it was formed along a particularly prominent fracture zone that was part of a set radiating outward from the Tharsis uplift.

If plate movement is discounted, then must we discard the attendant proposition of convective movement in the mantle? The answer is not obvious. Hartmann (1973b) has proposed that, if the temperature gradient for Mars were less than that for Earth, slow-moving convection would displace mantle material without disrupting the crust. The only problem is that heat transfer would be similarly reduced and volcanic activity rendered less likely.

The geometric configuration of any convection currents is another problem. How many cells should one expect and how would they be positioned? If the core is small, theoretical calculations suggest that a simple two-cell system would form in the overlying mantle. As the core increases in size and the radial thickness of the mantle accordingly decreases, the number of cells increases (Runcorn, 1962). The theoretical models are usually generated for two-dimensional space. The extrapolation to three dimensions is seldom attempted. Accordingly, it is very difficult to predict what the boundaries between adjacent cells would actually look like on the surface of Mars. However, there should be some sort of a linear pattern, and this is not seen. Similarly, an expected symmetry around the pole of rotation is not observed.

There is an alternate hypothesis that draws on a recent theory of heat transfer proposed for the Hawaiian volcanoes. According to Morgan (1971),

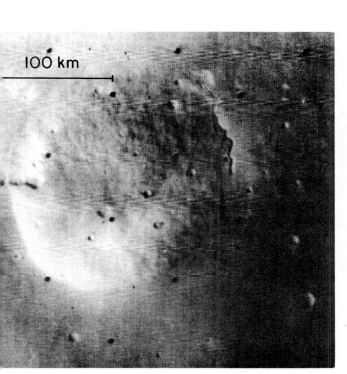

(a)

Figure 5.44 (a). Hecates Tholus, showing domical shape and rough surface texture. (DAS No. 09054544, rev. 208, A camera, center at 212°W, 30°N.)

(b). Details of Hecates Tholus. Flow channels radiate downslope from a complex summit caldera. A string of pit craters (arrows) occurs along a probable rift. Lower flanks of the dome have been eroded. Smooth plains materials have been deposited in the erosional re-entrant. On the far left, several roughly textured lobate lava flows are visible. (DAS No. 13496298, rev. 676, B camera center at 210°W, 32°N; DAS No. 13496368, rev. 676, B camera, center at 211°W, 32°N.)

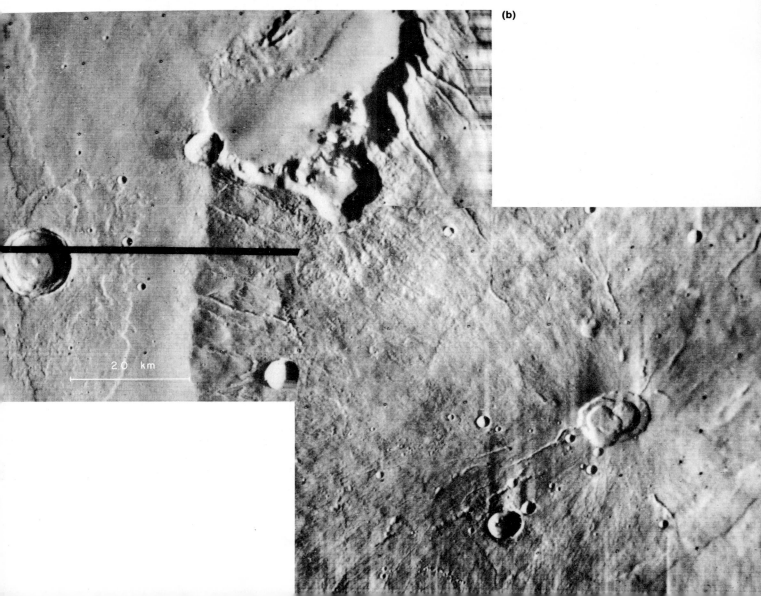

(b)

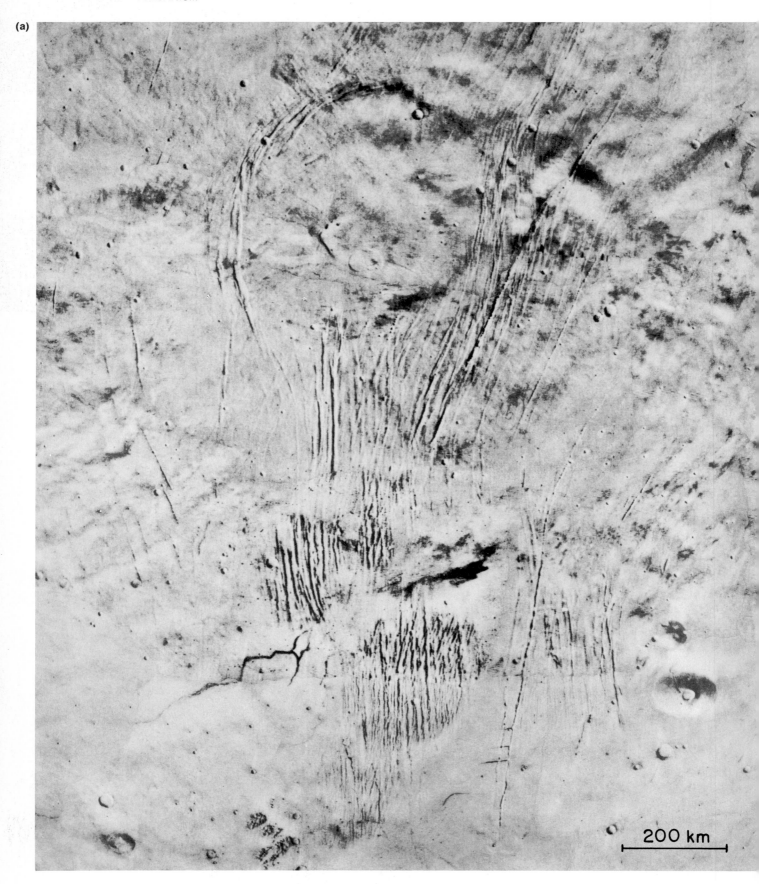

200 km

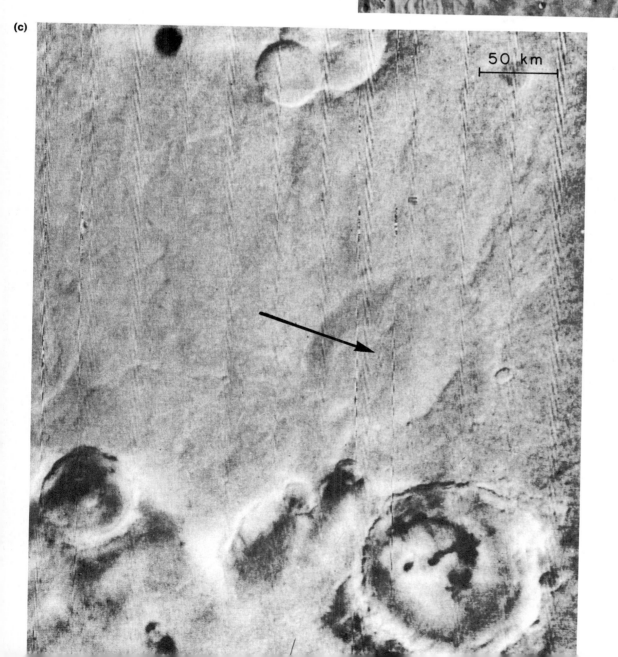

(b)

50 km

Figure 5.45 (a). A photomosaic of Mariner A frames showing Alba Patera, outlined by deflected fractures. Note that several sets of fractures are present, each with a characteristic orientation. Uranius Tholus and Ceraunius Tholus are visible in the lower right.

(b). Hadriacum Patera, a relatively old and degraded volcanic structure. Arrows point to the caldera, which is surrounded by radiating ribbons of lava. Another possible complex caldera is indicated by dashed arrows. Note the absence of raised rims which characterize other impact craters in the picture. (DAS No. 11974069, rev. 436, A camera, center at 264°W, 30°S.)

(c). Amphitrites Patera, a highly degraded volcanic structure. Arrows point to a central caldera. Radial structure is faintly visible around caldera. (DAS No. 05958578, rev. 122, A camera, center at 302°W, 60°S.)

(c)

50 km

heat is convected upward from the mantle-core interface by plumes somewhat analogous to thunderheads rising upward in the atmosphere. When the plume encounters the rigid crust, it is dissipated over a large area and the plume material returns downward in a gentle "rain." Since the plume is a point source, the volcanic activity should have a similar point stability. But then, why are the Hawaiian volcanoes arranged in a long chain? This apparent anomaly is resolved by noting that the age of the volcanoes in the ridge progressively increases as they are traced to the northwest, away from the presently active Mauna Loa. This is consistent with the movement of a crustal plate over a fixed mantle plume. On Mars the absence of a similar chain associated with Olympus Mons indicates that the crust has been fixed relative to the mantle. It could be argued that the three volcanoes along the Tharsis ridge constitute such a chain. Some support for shifting activity is lent by the morphological evidence that Ascraeus Mons appears freshest, Arsia Mons most highly degraded. By comparing the total volume of the Hawaiian chain with that of Olympus Mons and knowing the period of time over which the Hawaiian source was active, it is possible to arrive at a comparable activity period of 130 million years for Olympus Mons (Carr, 1973).

When Olympus Mons was first described, the question asked time and time again was: "How recent is it? Is it still active?" Geologists responded that it is, indeed, a young feature, no more than several hundred million years old. Soderblom et al. (1974) calculate an age of 200 m.y. based on crater densities observed on the volcanic flanks. For the person who wishes to assert the reality of a process by seeing it with his own eyes, this is hardly recent, but, in terms of the planet's lifetime—4.5 billion years—200 m.y. represents less than one twentieth of its total history.

Having determined that Olympus Mons is, if not an active volcano, at least a young one on the basis of crisp morphology and flanks unscarred by impacts, the next task is to trace volcanic activity backward in time. The first survey of Mariner pictures suggested that all volcanoes had fairly fresh morphologies and were formed over a short time span. Subsequently, as more subtle volcanic features in a variety of degradational states were identified, it became fashionable to assert that volcanic activity extended back in time, perhaps to the earliest phases of crustal formation. There are, for example, many flat intercrater areas in the most highly cratered regions. Some of the flat surfaces appear to be aggradational, the result of superposed deposits masking an older, irregular surface (Wilhelms, 1974). The infilling materials may be lavas; there is little available evidence to settle the issue either pro or con. How-

ever, the contention of "universal" volcanism can be attacked as a premature complication. If we are seeking patterns in the evolution of Mars, the conclusion that volcanic activity has occurred at many times—and in many places—is hardly a step toward simplification. For that arbitrary reason as much as any other, one should closely examine the alternate possibilities.

Fortunately, there does seem to be a more confining envelope, both in time and in space. As mentioned previously, volcanic features are not randomly located around the planet but are concentrated in three regions (fig. 5.46). The Tharsis province extends from Arsia Mons (9°S, 120°W) northeast approximately 4000 km to Kison Tholus (74°N, 359°W). The Elysium province includes a cluster of five features extending from Apollinares Patera in the south to Hecates Tholus in the north. A third province encircles the Hellas basin, including four or more degraded shield volcanoes. The freshest features appear in the Tharsis province; Elysium structures are slightly more degraded; and Hellas structures are appreciably more degraded, and presumably the oldest.

A pattern may be seen in the distribution of Martian volcanoes if one shifts the rotational pole of the planet to a position of 145°W, 55°N (figs. 5.46, 5.47). In so doing, the division between cratered terrain and plains more closely follows the equator. Volcanic features also are reasonably symmetrical with a belt around 30°–60°N latitude and clusters at both north and south poles. One would expect this sort of symmetry if internal convective motions, with attendant patterns of heat flow and volcanic activity, were controlled by rotational forces. The shifting of the pole to its present position might have been caused by outpouring of lavas and generation of a mass excess in the large volcanic province of which Olympus Mons is a part. With time, the rotational axis of the planet might have shifted to bring this protuberance into a stable equatorial position, close to its present orientation.

CRATERED PLAINS

The occurrence of Tyrrhenum Patera, a volcanic caldera, in the middle of the Hesperia Planum seems more than coincidental. Here is a correlation that suggests that the plains materials are actually superposed lavas. Two other observations lend slender support to this contention. First "mare" ridges are present throughout the plains (fig. 5.48). By analogy with the moon some of these ridges may have a volcanic origin. Unfortunately, the uncertainties mentioned previously regarding lunar mare ridges have to be extended to their Martian counterparts. Both may be compressional structures that occur in a variety of materials, including flood basalts. The sec-

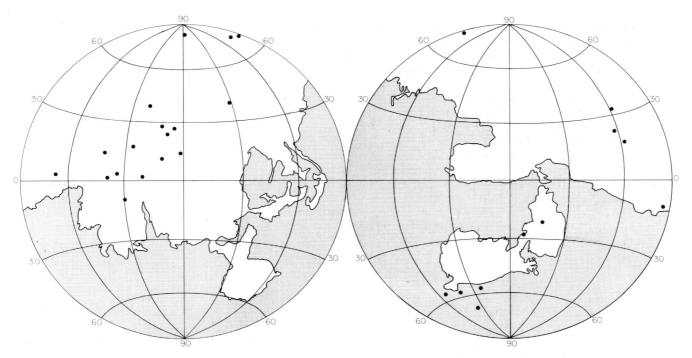

Figure 5.46. Distribution of Martian volcanic structures, shown by dots. Cratered terrain is shaded; plains are unshaded. Volcanic features occur on, or close to plains.

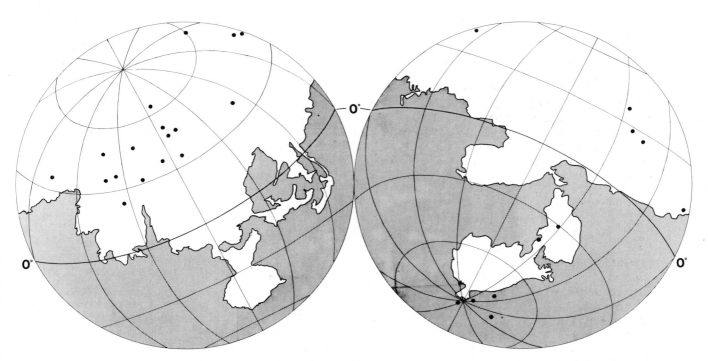

Figure 5.47. Position of Martian volcanoes, shown by circles. This is the same display given in figure 5.46 except that the pole position has been rotated to 155°W, 55°N. With reference to these shifted coordinates, the boundary between cratered terrain and plains approximately follows the equator. Most volcanoes occur in a girdle between 30°N and 60°N. A few are clustered at the south pole.

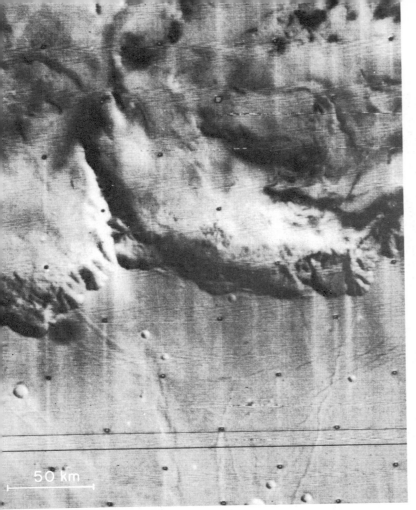

Figure 5.48. Possible dike exposed in wall of Coprates canyon. An irregular ridge on the smooth plains adjacent to the Coprates canyon may be traced nearly to the canyon rim. The low ridge extends for some 50 km onto the canyon floor. Since the canyon wall is believed to be receding by collapse and eolian removal of plateau material, the material of the ridge is evidently more resistant to erosion than the surrounding material. Competing hypotheses of origin for the ridges are 1) compressional failure of the surface, and 2) linear fissure eruptions. The evidence here tends to support the latter hypothesis. Thus the ridge protruding from the canyon wall would originally have formed as a dike. (DAS No. 07470578, rev. 164, A camera, center at 71°W, 13°S.)

ond supportive observation is that Hesperia Planum is the site for a large number of well-developed bright and dark streaks formed in the wake of craters. A situation highly favorable for generation of streaks is one in which level, dark bedrock is covered with a thin veneer of sediment. Even were there no chemical or mineralogic differences, the sediment would appear brighter than the bedrock due to multiple diffraction and reflection by the fine grains. If the bedrock of the plains was not lava but was, instead, a substantial thickness of ponded sediment, then one would not expect a brightness contrast between the bulk of the infilling sediment and the upper few centimeters or meters. Admittedly, the fact that the bedrock is relatively dark does not point exclusively to basalt. Terrestrial sedimentary rocks such as limestone and shale are frequently dark, due primarily to the incorporation of small amounts of carbonaceous material.

Other regions of plains are Solis Planum (pm) and Lunae Planum (pc), which occur to the east of the Tharsis volcanic province. Although they contain no obvious volcanic structures, some circumstantial evidence can be presented to support a volcanic origin for the plains materials themselves. First, the mere presence of large expanses of flat terrain suggests infilling. (Admittedly this argument can be extended to explain the intercrater areas in the densely cratered terrain, the only distinction being one of scale.) Second, just as for Hesperia Planum, Solis Planum and Lunae Planum are sites for prominent clusters of wind streaks (figs. 7.12, 7.20). Again, dark basaltic bedrock overlain by a thin sediment veneer is suggested.

At several points where they are cut by Valles Marineris, plains materials display horizontal structures (fig. 5.49). This is compatible with the layering that would be present in a sequence of flood basalts. Plains materials erode in a systematically polygonal fashion that is suggestive of horizontal strata dissected by a set of vertical joints or fractures. Good examples of polygonal dissection along an erosional scarp are present along the western edge of Lunae Planum (fig. 8.37). Contrast this with the complexly irregular landforms present where plains are formed at the expense of cratered terrain (cu). This distinctively serrated landscape is mapped as a separate unit: hummocky terrain, fretted (hf) (fig. 3.28).

Where the Valles Marineris cut through the moderately and densely cratered plains, the canyons are straight and regular. Further to the east where the channels are incised in cratered terrain (cu), they are irregular and discontinuous in plan view (fig. 8.44). These contrasting styles in channel shape are precisely what one would expect from development in homogeneous materials (pc) and development in heterogeneous materials (cu) that have been deformed, metamorphosed, redistributed, and generally modified by numerous large impact events.

Chemical Composition

The morphologies of Martian volcanoes supply a few clues regarding chemical composition. The resemblance of the Tharsis peaks to terrestrial shield volcanoes forcibly suggests that, just as for their terrestrial counterparts, Martian lavas are basaltic. Lava channels, which occur in association with almost all volcanic constructs, and extensive flow

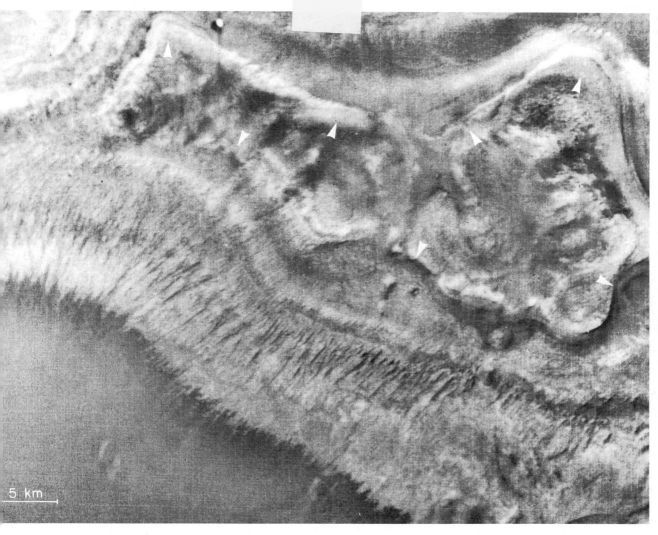

5 km

Figure 5.49. An isolated mesa (edge of top indicated by arrows) in Ganges Chasma. Layered rocks are shown in mesa walls. Rilles appear at a specific point some distance below the top of the mesa, suggesting erosion by fluid seepage (Sharp, 1973c). (DAS No. 09017619, rev. 207, B camera, center at 49°W, 8°S.)

fronts in plains materials provide additional evidence for low-viscosity basaltic flows.

The domes might be interpreted as more silicic lavas that have built steep-sided structures. However, one would feel more at ease with this interpretation if the entire spectrum of possible volcanic landforms were observed. Noticeably lacking are both strato-volcanoes and large breached structures indicative of major explosive activity (Carr, 1973). This type of igneous activity is common on Earth in mountain belts—regions of subduction of oceanic crust where rocks of calc-alkaline or andesitic composition are produced (Taylor et al., 1969). For the very reason that we have failed to see any evidence of plate tec-

tonics on Mars, we might question the uncertain morphological evidence for andesitic or rhyolitic volcanic rocks.

Evidence for the smaller-scale explosive, igneous activity responsible for basaltic cinder cones on Earth may be difficult to find on Mars. Terrestrial cinder cones are usually small, less than 3 km in diameter. Because of reduced gravity and a less dense atmosphere on Mars, the same igneous activity that forms a cinder cone on Earth would produce a shallow crater with a broad, dark halo (McGetchin et al., 1973). Thus, many explosive features on Mars may be masked by their subtle topography, as well as being susceptible to eolian erosion. Notwithstanding these difficulties, West (1974) has described some small Martian cones that may record pyroclastic activity.

Earth, Moon, Mars, and Mercury— A Comparison

When we consider volcanic morphologies and distributions on the moon, Mars, and Earth, an interesting progression toward increasing crustal activity and mobility is apparent (fig. 5.50).

One of the more remarkable aspects of the moon is that, even though Apollo sample analyses indicate that basaltic rocks are widespread, large-scale constructional features are absent. It is not sufficient to say they are so ancient that they have been destroyed by impact erosion. Some remnant of a shield volcano the size of Mauna Loa or Olympus Mons would still be identifiable, even though it formed 3.5 billion years ago. Much smaller features such as the Hadley Rille have survived with only minor modification. Experimental evidence indicates that some mare basalts formed at depths in the vicinity of 300 km (Grove et al., 1973), far below the crust-mantle boundary at about 65 km (Toksöz et al., 1972). The many kilometers that separate point of melting and point of extrusion effectively eliminate the possibility of a magma reservoir linked by a major vent to a discharge point directly overhead. Instead the magma migrates to the surface along pre-existing fractures, especially those associated with impact basins. This explains both the concentration of basalts in basins and the apparent dominance of fissure eruption. Any constructional surface topography is further inhibited by the abundance of low-silica, low-viscosity flows. The concentration of volcanic activity on the Earthside may be a consequence of a thinner crust on that side and an offset center of figure/center of mass. Lavas emplaced in a hydrostatic mode would reach the surface on Earthside before emerging on the farside (Kaula et al., 1972). There is some evidence that the depth of melting

has been increasing as the volcanic activity has become more recent. Experimental evidence indicates that early mare basalts melted at a depth of 100 km compared to several times greater depths for intermediate-age mare basalts (Hays and Walker, 1975). This progressive decrease in lower crust-upper mantle temperatures suggests a planet that experienced widespread melting during the first billion years of its development but which has been almost totally inactive for the ensuing 3.5 billion years.

Earth is at the other end of the spectrum. It experiences, at present, widespread volcanism. A relatively thin (50–100 km), highly mobile lithosphere is highly differentiated into granitic, continental and basaltic, oceanic blocks. Separation and collision of crustal plates accounts for volcanic activity with a wide range of physical type and chemical composition. In the case of Mauna Loa, magma is probably generated at depths of 50 km and stored in magma reservoirs no deeper than 5 or 6 km (Macdonald, 1972). Similar shallow magma reservoirs exist throughout the world. Such shallow chambers can be easily tapped to yield volcanoes of substantial size. With the passage of time, the structures become less prominent, either due to isostatic compensation or to planation.

Mars lies between the two extremes of Earth and moon. If one assumes that the density contrast between rock and magma was the same for Olympus Mons as for Mauna Loa, then the hydrostatic head required to discharge magma through the summit vent implies a magma chamber at a depth of 130 km (Carr, 1973). Crustal thicknesses derived from gravity information are on the order of 50 km (Phillips et al., 1973). Vogt (1974) suggests that spacing of volcanoes may be related to lithospheric thickness. Because volcanoes are more widely spaced on Mars than on Earth, the Martian lithosphere might be an order of magnitude thicker than the terrestrial lithosphere. There is no indication of plate tectonics, and the crust is apparently rigid enough to sustain large volcanic piles for millions of years.

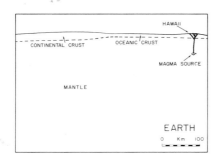 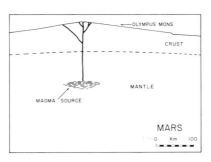 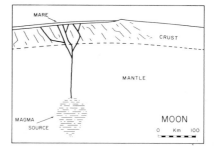

Figure 5.50. Schematic cross-sections for Earth, Mars, and the moon, showing progressively deeper magma sources for the three bodies.

Relative to the moon, a shallower depth of magma generation leads to the formation of large volcanic constructs over localized hot spots. Relative to Earth, the absence of a shifting, mobile lithosphere permits long-time activity over selected surface sites. Because the Martian crust is relatively rigid, the volcanic loads are not isostatically compensated.

The exact position of Mercury in this scheme is not yet clear. No landforms have been unequivocally identified as volcanic, yet there are widespread smooth plains that suggest the possibility of volcanism. There is no evidence of plate tectonics. Our present knowledge of Mercury suggests that crustal activity and mobility have not been important. Mercury thus appears to be similar to the moon in this comparison.

6.

Geophysics and Structure

Introduction

THIS chapter will introduce some of the important planetological aspects of Mars: planetary mode of formation, composition, early differentiation history, thermal evolution, gravity and topography, and crustal tectonics. Since Mars is a dynamically evolving planet, it is not sufficient merely to describe its present characteristics. The ultimate goal is to recreate the history of the planet from its birth to the present day.

Quantitative description of planetary history is, however, extremely difficult. Even after centuries of study, the most complete compilation and interpretation of geologic data on the evolution of our own planet, Earth, is at best a progress report. In these pages we will outline the important aspects of the problems for Mars, point out some of the gaps, but of necessity leave most of the solutions for future studies.

Planet Formation

The first scientific theory of the origin of the planets was suggested by René Descartes, formalized by Immanuel Kant, and developed in some detail by P. S. Laplace. The main feature of the hypothesis is that the planets accreted from a dense cloud of gas and dust—the nebular stage of a forming star.

Such a theory was fine for the eighteenth century because it accounted for the existence of the sun, planets, and comets. However, James Clerk Maxwell in the nineteenth century showed that a rotating ring of dispersed matter having the mass of the present solar system would remain dispersed and could not form into planets. An even more serious flaw was that the sun rotates very slowly, much more slowly than the nebular hypothesis would predict. Why should the sun have nearly 99 percent of the mass of the solar system but only 2 percent of the angular momentum? An alternate hypothesis was developed by T. C. Chamberlin and F. R. Moulton early in the twentieth century. The angular momentum problem was circumvented by postulating that the planets were formed from material ejected from the sun following a collision or a near miss by another star.

For the past three decades, the nebular accretion hypothesis has been the more seriously studied of the two competing hypotheses. This development is partly because of a number of serious objections to the various encounter hypotheses, but mostly because of the diligent analysis and support of such investigators as C. F. von Weiszäcker, H. Alfven, H. C. Urey, G. P. Kuiper, and A. G. W. Cameron. The mainstream nebular hypothesis has been made to conform with modern views of stellar evolution.

In brief summary the current nebular hypothesis has, as its main features, planets forming from a cloud of dusty gas, mostly hydrogen and helium, surrounding a protostar. The spinning cloud flattens to a thin disc, and planets condense from local eddies. The highest temperatures occur nearest the sun. The bulk composition and hence the densities of the planets are governed by temperature and composition differences in the nebula.

It is important to understand the process of planetary formation in order to construct a model of planetary thermal evolution and to determine an initial temperature profile. Such profiles have been constructed for planets by assuming symmetric accretion and by modeling the rate of accumulation of mass, balancing the energy accumulated thereby with that lost by radiation (Hanks and Anderson, 1969). The rate of heat accumulation is largely determined by the time allowed for accretion of the planet. For a planet larger than the moon, extremely rapid accretion could result in the accumulation and burial of enough heat to result in an initially molten planet. It appears reasonable that Earth could accrete in approximately 10^5 years (Lyttleton, 1973). Since Earth's magnetic field is probably associated with motions within the conducting core, short accretion times also are required to provide enough heat to form Earth's core prior to emplacement of the oldest rocks with remanent magnetism (Hanks and Anderson, 1969). It has also been argued that longer accretion times are dynamically more reasonable. High initial heat could be provided by impact of large bodies in the accretional process (Safranov, 1974). By these models the Earth may have accreted in 10^8 years. Thorough, initial degassing of the Earth has been attributed to accretional heating. Degassing is, however, independent of the rate of accretion (Fanale, 1971a).

These accretional models introduce complicating factors that have thus far been ignored in determining initial temperature profiles. For example, it has been assumed that most of the accreting material accreted in a symmetrical manner. It is also possible that the material was in orbits very near the plane of the ecliptic. Within this constraint, two models are imaginable. First, the planets may have formed by primary accretion of dust from a disc much thinner than the final dimensions of the planet. In this case the initial temperatures might be much higher at the equator, where the accretional energy was concentrated. This could result in more differentiation, and a thicker crust in the equatorial regions. The second model specifies that the dust of the solar nebula, originally taking the form of a thin disc, coalesced into a large number of lunar-size objects that subsequently collided to form the planets (Lyttleton, 1972; Safranov, 1974). Each of these moon-size protoplanets, of which some of the present planets may be larger representatives, would have accumulated from a thin local disc of nebular dust. The problem of determining an initial temperature profile in the second case seems almost insurmountable.

Ages of the Planets

The age of Earth remained an uncomfortable uncertainty until Ernest Rutherford's demonstration, in 1905, that the rate of spontaneous decay of the radioactive elements could be used to determine the age of rocks. Earlier attempts to estimate the age of Earth by the amount of time it would take to deposit all the known sedimentary rocks, for Earth to cool from a molten mass, or for the oceans to accumulate their present salts by the flow of rivers from the land seemed to allow far too little time to account for the slow process of organic evolution. With the new techniques made available by the discovery of radioactivity, it very quickly became clear that Earth was far older than had previously been supposed. Rocks nearly as old as two billion years were discovered (Holmes, 1913), and, with more analyses, older rocks were documented every few years. This trend continues to the present. Recently, rocks as old as 3.8 billion years have been found in the ancient Precambrian terrains of Greenland (Moorbath et al., 1973) and Minnesota (Goldich and Hedge, 1974).

Although the crust of Earth was shown to be very old, verification of the extreme antiquity of Earth and other planets came not from the terrestrial rocks, but from meteorites. Analysis of the lead and uranium isotopes in meteorites showed that these objects must have condensed nearly 4.6 billion years ago (Patterson, 1956). Similar analyses of rocks in Earth's crust yield an age, called the model age, of 4.6 to 4.8 billion years. Samples of rock returned from the moon yield similar ages. These ages, recording the time when the stuff of the moon became isolated and the initial ratios of lead to uranium established, should not be mistaken for the time when the particular rock, perhaps a volcanic lava, actually crystallized. With all this converging information from the moon, Earth, and meteorites, we can be confident that the inner planets formed 4.6 to 4.8 billion years ago; that is, they became closed systems with respect to the addition of lead and uranium at that time.

Composition of the Planets

The only direct evidence for bulk composition of the planets is the observed density. However, there are other observations that supply important clues. For example, the moment of inertia of a planet is governed by distribution of density with depth.

Because of observed density differences, the planets are thought to differ in bulk composition. The most dramatic differences are between the giant planets such as Saturn, which is less dense than water, and the solid inner planets, which have densities three or more times that of water. For example, Mars has a density of 3.9 g/cm³ and that of Mercury is 5.4 g/cm³. It appears that there were compositional differences in the nebula and that these resulted in compositional differences among the planets. Determining the exact nature of these differences is a difficult task, and no single hypothesis appears to be gaining primacy, although recent theoretical work by Lewis (1972, 1974b) predicts a sequence of densities for the planets closely in accord with the observed pattern.

Seismic evidence has revealed much about the interior of Earth and the moon. From the variation of seismic parameters with depth, density profiles may be constructed. Analysis of reflected and refracted waves has revealed the existence of layers of sharply contrasting seismic velocities. The continental crust of Earth has a mean thickness of about 40 km comprising rocks that are typically rich in silica (fig. 6.1). The ocean crust is thinner, with a mean thickness of about 8 km. The underlying mantle is believed to be poorer in silica and largely composed of the minerals olivine, pyroxene, and garnet. From the seismic records we also know that, at a depth of about 100 km, the mantle rocks are relatively weak and flow more easily, apparently because they are partially molten. At greater depths, between about 400 and 900 km, there are a number of discontinuities in density (Anderson, 1967; Birch, 1954). Finally, at much greater depth a liquid core slightly larger than the planet Mars encloses a central solid core.

The density discontinuities within the mantle between 400 and 900 km are of interest for theories of planetary interiors because they may represent changes in the crystal structures of the minerals at the elevated temperatures and pressures prevailing there. Such changes, ordinarily resulting in an increase in density with increasing pressure or decreasing temperature are called phase changes (fig. 6.2). Although the mantle discontinuities have long been suspected to result from phase changes, in particular the conversion of olivine to a higher density spinel structure, the details of this phase change have only recently been confirmed in the laboratory (Ringwood and Reid, 1968; Ringwood and Major, 1970; Ahrens, 1972). The spinel structure of olivine, $(Mg_{0.74} Fe_{0.26})_2SiO_4$, has also been found in a chondritic meteorite (Binns, 1970). The calculated density of this mineral, appropriately named ringwood-

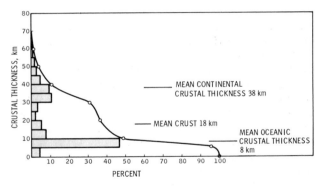

Figure 6.1. Crustal thickness distribution for Earth, showing distinct separation between continental and oceanic crust. Data from Cummings and Shiller (1971).

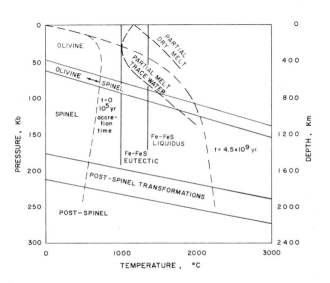

Figure 6.2. Some probable phase transformations in the thermal evolution of Mars, assuming chondritic composition. $t = 0$ is the initial thermal profile immediately after accretion. $t = 4.5 \times 10^9$ is the thermal profile 4½ b.y. after accretion. The Fe-FeS eutectic, which is generally pressure independent, marks the lowest melting temperature for any mixture of iron and iron sulfide. The liquidus is the temperature at which any mixture of these two components will be molten. This reaction facilitates migration of liquid iron to the center of the planet and formation of a core. With passage of time, and increasing mantle temperatures, the olivine to spinel and spinel to post-spinel transformations occur at increasing depth. For example, the depth of the olivine-to-spinel transformation increases from 600 km to 900 km. This would be accompanied by a 10-percent volume increase, resulting in expansion of the crust. T-P conditions for partial melting of crustal and upper mantle rocks are indicated in the figure by dashed lines. The presence of small amounts of water reduces melting temperatures. Data from Akimoto and Fujisawa (1968), Ringwood (1970), Johnston et al. (1974), Wyllie (1971, 1973), and Ringwood and Major (1970).

ite, is 3.90 g/cm³ — — or about 12 percent greater than its olivine polymorph. A similar phase change of pyroxene to the garnet structure may also occur under similar conditions. Another phase change is interpreted to take place at greater depths based on seismic data (Engdahl and Flinn, 1969). The spinel structure breaks down into its component oxides, accompanied by a further increase in density (Ming and Bassett, 1975). These phase transformations have been determined by laboratory experiments and can be expressed graphically as a function of temperature and pressure.

An interesting consequence of the phase relationships just discussed is that, if the interior temperature of a planet increases with time, then a certain volume could conceivably convert from the spinel to olivine structure with an attendant increase in volume. This might be reflected in extensional structures at the surface of a planet.

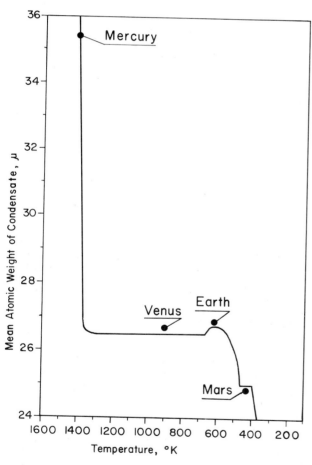

Figure 6.3. Density of condensed material in equilibrium with a solar-composition gas at 400–1600°K and 10⁻³ bars. The model values are shown by the solid line, along with the actual values for Mercury, Venus, Earth, and Mars. Simplified from Lewis (1972).

The bulk composition of Mars has been inferred from theoretical considerations and from analyses of meteorites. One theoretical approach is to assume that Earth and Mars have essentially the same bulk composition. Because Earth is larger than Mars, phase changes might result in a high density metallic core—*metallic* here referring only to the peculiar state of collapsed silicate structures. According to this argument, first enunciated by Ramsey (1948) and subsequently championed by Lyttleton (1973), such pressures are not reached in the interior of Mars, thus accounting for the difference in density between the two planets. Clearly the phase change hypothesis cannot account for the high density of Mercury—5.4 g/cm³.

According to an alternate—and more fashionable—model, Earth and Mars are composed of silicates with metallic iron cores. To account for the lower density, Mars must have a smaller proportion of iron. A model for metal/silicate fractionation in the solar system has been developed (Lewis, 1972), based on a theory of chemical differentiation in the solar nebula. This model accounts remarkably well for the observed densities of the terrestrial planets (fig. 6.3). The main features predicted for Mars are a small FeS core, and somewhat more FeO in the mantle than in Earth's. It is significant that Lewis's model predicts material of carbonaceous chondritic composition throughout the asteroid belt. Ordinary chondrites are thought to condense between Earth and Mars. These predictions appear to be confirmed by recent comparisons of the spectra of asteroids with spectra from samples of carbonaceous chondrites (Johnson and Fanale, 1973; Gaffey, 1974).

Another model featuring inhomogeneous accretion and early formation of the iron nuclei of the inner planets has been developed by Turekian and Clark (1969). In a variant of this model (Anderson, 1972), the moon forms outside the plane of the nebular disc and consequently is depleted in iron. Mars would have the composition of ordinary chondrites.

As a final explanation of density differences between planets, Ringwood (1966) assumes all the terrestrial planets to be of carbonaceous chondritic composition. Density differences are explained primarily by differences in the oxidation state of iron. Planets with lower density have a greater proportion of iron in the oxidized state in minerals such as olivine, pyroxene, spinel, and magnetite (Ringwood and Clark, 1971). This model also requires the reducing agent for the iron—carbon—to be lost as CO_2 in quantities corresponding to a significant fraction of the planetary mass.

Most of the evidence indicates that the ratio of metallic iron to silicates decreases away from the sun and that phase changes can account for some of the density discontinuities in the mantle of the Earth. Some combination of the various hypotheses just mentioned may best explain the density variations among the inner planets. Consistent with any of the proposed models, it appears reasonable to assume that Mars is approximately chondritic in composition, composed predominately of silicates with iron as a major component. The iron is not necessarily concentrated in a core.

Planetary Differentiation

Planetary differentiation, as used here, means large-scale melting and segregation of the material within a planet into chemically distinct layers. We have direct evidence for substantial differentiation from two planets—Earth and the moon. Both differentiated no later than the first billion years after formation.

Smaller-scale chemical differentiation of near-surface materials has continued on Earth up to the present (e.g., volcanism). On the moon most of this activity ceased about 3 billion years ago. The Apollo exploration program has provided much information on the early history of planets bearing directly on the time and amount of differentiation. It has long been suspected that Earth was differentiated very early, although details have been obscured by later events. On a small, airless body like the moon, both morphologic and geochemical clues to conditions existing near the time of the formation of the planets are preserved.

The Apollo samples showed that major differentiation occurred on the moon approximately 4.6 b.y. ago and no later than 4.3 b.y. ago (Papanastassiou and Wasserburg, 1972; Tera et al., 1974). This differentiation resulted in a lunar crust enriched in silica. The abundances of the elements K, Rb, U, and Th were especially enhanced. The isotopic evidence upon which this is based does not preclude the possibility that lunar material was differentiated prior to accretion. If differentiation occurred in the nebula, material may have been incorporated into planetary bodies as condensation progressed, resulting in initial chemical layering. There are convincing arguments that differentiation began in the solar nebula as soon as the first grains of dust began to crystallize out of the elemental vapor (Grossman, 1974). This concept of inhomogeneous accretion has been advanced for Earth and the moon (Turekian and Clark, 1969; Anderson and Kovach, 1972; Elsasser, 1972). Another possibility is to fractionate Earth and the moon continually as they accrete by the heat generated during very rapid accretion. There is more than enough energy released to melt an impacting projectile once a planet reaches a certain mass (Tozer, 1972). However, this critical mass is very nearly the lunar mass. Allowing for some loss by radiation, this mechanism could at best heat the outer part of the moon to no more than about 1000°C, hardly enough to result in extensive differentiation. Somewhat more heating might be achieved if the moon formed by the collision of several large protomoons (Safranov, 1974). Accretion energy may have contributed substantially more to the initial heat of Earth and Mars.

Gravitational energy is not necessarily all released at the time of accretion. If Earth were initially homogeneous, enough potential energy could be released by core formation—that is, allowing iron to melt and segregate to the center—to melt the entire body. A great amount of heat could be transferred out by convection relatively quickly.

The fact remains that the moon appears to be differentiated; the lunar surface is substantially enriched in uranium. Although some of the enrichment may have occurred in 4.0–3.0 b.y. events, the seismic observation that the moon is stratified supports the idea of global differentiation. The seismic data suggest extensive differentiation with a possibility of five zones, including a molten iron sulfide core (Nakamura et al., 1974).

The conclusion must be that the moon, and most probably any larger planet, was differentiated very early, resulting in concentration of heat-producing radioactive elements in the outer part and a complementary depletion throughout the rest of the planet. However, the differentiation mechanism remains unknown. The moon with its differentiated crust presents intriguing problems. Since the object is only marginally capable of gravitationally driven accretional heating, we cannot reject such ideas as inhomogeneous accretion or the presence of some unknown heat source—perhaps magnetic induction heating in an early solar field, decay of short-lived radioactive isotopes or fission from a large pre-existing body such as Earth.

Energy Sources

Long-lived Isotopes

On Mars, as on Earth, the long-lived isotopes appear to be the most effective fuel for the internal geologic engine. The primary heat source must be the isotopes ^{238}U, ^{235}U, ^{232}Th, and ^{40}K. The absolute abundances of these heat-producing isotopes are difficult to evaluate. As previously discussed, a chondritic composition is likely for the bulk composition of Mars. Based on density and composition, low-iron (L-group) chondrites may be the most appropriate choice. They have approximately 7 percent iron plus nickel, 1300 parts per million potassium, and 0.012 ppm uranium (Fanale and Nash, 1971). These abundances yield a K/U value of approximately

100,000. (Uranium and potassium are not separated by most igneous processes, although there may be a trend toward enrichment of potassium over uranium as differentiation proceeds. Nonetheless, the abundances of potassium and uranium are thought to vary considerably among the terrestrial planets. Therefore, the bulk ratio of the elements might be expected to be a useful parameter in the comparison of bulk compositions.)

Recent Russian gamma ray observations of Mars (Surkov and Fedoseyev, 1974) indicate uranium, thorium, and potassium abundances of approximately 2.2, 6.5, and 4×10^4 ppm, respectively, for the surface of Mars. These high abundances, similar to those found in terrestrial alkaline rocks, suggest the presence of differentiated crustal rocks. A bulk K/U value of 10,000, calculated from these data, is probably deceptively low, either because of crustal concentration or instrumental uncertainty.

The average uranium concentration of Earth or the moon cannot be determined directly. The crust contains an anomalously large amount of this element since igneous differentiation tends to concentrate uranium in phases richer in silica. Making a correction for some fossil, nonradiogenic heat and also satisfying the observed heat flow, the bulk uranium concentration of Earth is approximately 1.7×10^{-8} g/g (Conel et al., 1972). If no fossil heat is assumed, a slightly greater bulk concentration is required, 2.3×10^{-8} (Wasserburg et al., 1964). Relative to solar or chondritic abundances, both Earth and the moon are depleted in potassium, unless this element resides in unsuspected greater abundance in the interior. On Earth there is some evidence that potassium may go into a sulfide phase in an iron core (Goettel, 1972). Since the moon can have very little core to hide potassium, the average lunar crustal potassium to uranium ratio may approximate that for the whole moon.

If one assumes that the rate of heat loss measured for Earth is very nearly equal to the rate of heat production of the upper mantle and crust (Urey, 1955), there is a curious "chondritic coincidence." If Earth's upper mantle were made up of stony meteorites—which it is not—it would have approximately the same heat flow as that observed (Birch, 1958). However, it appears that Earth may have a K/U ratio of only about 5000 to 10,000 and the K/U ratio of the moon may be as low as 2000 (Fanale and Nash, 1971). Surkov and Fedoseyev (1974) argue that the moon and Earth have identical K/U values of about 3000. These contrast with the chondrite value of 100,000 just mentioned. Since, in chondrites, the contribution to heat production from potassium is comparable to the rate of heat production from uranium, it is clear that Earth must be enriched in uranium relative to chon-

drites since Earth's heat production per unit mass is the same as that of chondrites.

The observed lunar heat flow requires an even larger bulk concentration of uranium. If it is assumed that the lunar heat flow observations (Langseth et al., 1973), made at two lunar locations, are representative of the moon as a whole, then the bulk concentration of uranium in the moon (assuming K/U = 2000 and Th/U = 4) must be approximately 80×10^{-8} g/g according to Toksöz and Solomon (1973). This implies that the bulk concentration of uranium in the moon is roughly twice that of Earth. If all the uranium in the moon were concentrated in the upper part at the approximate mean abundance observed in Apollo samples, 0.5 ppm, the enriched layer would be 60–70 km thick (Surkov and Fedoseyev, 1974). Although this would represent thorough differentiation, terrestrial enrichment of uranium in a thin crust leads to much greater near-surface enrichments. A possible explanation for the apparent enrichment of the moon in uranium is that the lunar heat flow observations are not representative of the moon as a whole, and alternative explanations must be sought for the observed lunar heat flow involving special conditions at the localities of the measurements (Conel et al., 1972).

The principal conclusion from these observations, inferences, and speculations is that the inner solar system is depleted in volatile and semi-volatile elements (e.g. K) and enriched in refractory elements (e.g. U). (Volatile, in this sense, refers to the behavior of an element that is more easily vaporized at temperatures near the center of the solar nebula. Refractory elements resist vaporization at these high temperatures.) This heliocentric gradient for volatile and refractory elements is predicted by Lewis's (1972) model for the condensation of planets. By some cosmic accident, the rate of heat production per unit mass in the planets Earth and Mars may be approximately the same, although the proportions of radioactive nuclides are quite different.

Solar Energy

On Earth, water plays an important role in absorbing solar energy, which, from a geomorphological point of view, is wasted on the moon and—to a lesser degree—on Mars. Obviously, water is an important erosional agent, but it also shapes the terrestrial landscape in more subtle ways. For example, the most striking landforms of Earth, chains of fold mountains, are underlain by unusually thick sequences of water-laid sediments.

There are presently no oceans on Mars. A hydrologic cycle is not a potent force in the formation of the Martian landscape, nor are the winds, since large craters and tectonic features are preserved over long periods of time. Therefore, even if the sources of internal energy for orogenic and volcanic processes are similar on Earth and Mars, the resulting landforms are very likely to be different. We acknowledge that past conditions may have been different. As described in chapter 8, increased solar heating in the presence of large amounts of frozen volatiles may lead to dramatic atmospheric changes.

Global Topography

Topography provides vital information for the geologic study of a planet. Without it we can do little to understand fundamental processes such as crustal deformation, fluid flow on the surface, or even atmospheric circulation. As with Earth, the global topography of Mars probably reflects the crustal structure of the planet. Earth's surface consists of continents and ocean basins. A graph of the distribution of elevations clearly shows this continent–ocean basin structure (fig. 6.4). Note that the mean elevation of continental and oceanic crust is referred to the mean elevation of the boundary between continental and oceanic crust rather than to sea-level since the oceans inundate continental margins. The distinct separation of Earth's crust into elevated continents and deep ocean basins occurs because the continents comprise 30–50 km thicknesses of rocks compared to the thinner 6–8 km oceanic crust. In general, the greater the thickness of the continental crust, the greater its elevation above mean sea level. This is taken to indicate an isostatic situation, wherein less dense crustal rocks float on denser man-

tle rocks. The thicker the crust is, the higher it stands. An approximate analogy is that of an ice cube floating in water.

Sources of Data

The topographic map of Mars (fig. 3.4) has been constructed using Earth-based radar observations (Downs et al., 1971; Pettengill et al., 1971) and Mariner 9 radio occultation measurements (Kliore et al., 1972a, 1972b, 1973). Gaps in these data were filled using Mariner 9 IRIS and UVS data (Christensen, 1975). The occultation measurements give absolute radii of Mars for several hundred points scattered around the planet. The Earth-based radar measurements were made between 1965 and 1973. Most of the measured elevations are situated between 15°S and 20°N. Since the radar measurements depend to some extent on the ephemeris (position in space at a particular time) of Mars, they have been adjusted to agree with the occultation measurements. Slight errors in the calculated ephemeris cause corresponding errors in calculated elevations. Radar measurements are based on the electromagnetic travel time from a receiver on Earth to the nearest point on the surface of Mars. Generally, but not always, the nearest point is on a line between the planet's center of mass and the radar instrument. The ephemeris predicts where the center of mass of the planet is at the time the radar beam reaches the planet. By knowing the travel time for the reflected radar signal, traveling at the speed of light, and the position in space of the center of Mars, the radius can be determined. As the planet

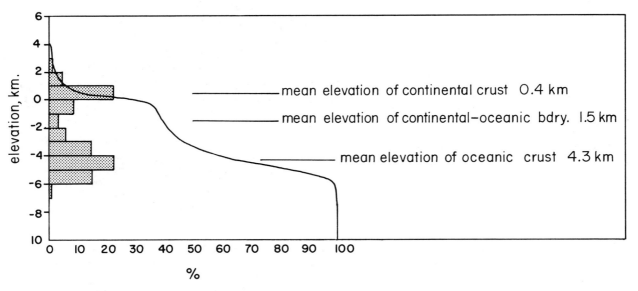

Figure 6.4. Elevations of Earth's surface. Bimodal distribution delineates continental and oceanic crust (Murray, 1888; Menard and Smith, 1966).

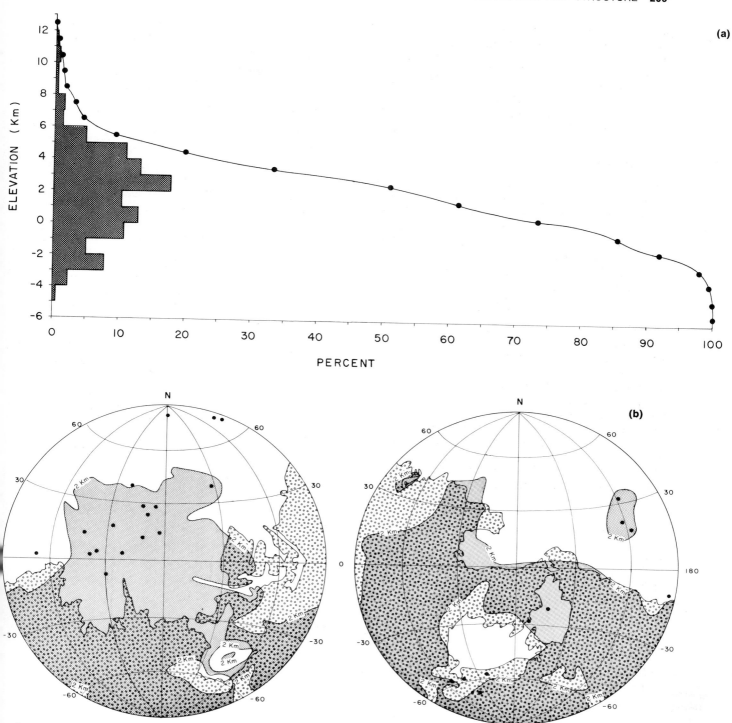

Figure 6.5 (a). Frequency distribution for Martian topography.

(b). Topographic map of Mars, divided into two regions by the 2-km elevation contour. Geographic boundary between cratered terrain (cu) and other physiographic units is shown for comparison. There is coincidence of topographic and physiographic boundary in the right hemisphere. In the left hemisphere the presence of the Tharsis uplift and Chryse trough obscure the coincidence. Volcanoes are shown by dots.

rotates, many points can be measured in a fixed latitudinal band. By completing measurements over 360° of longitude within a few days, any error of closure resulting from ephemeris drift during that short period can be corrected. Although the drift is slight over a period of weeks, the accumulated error may amount to a kilometer or more between successive oppositions. Thus, it is difficult to tie together data acquired in different years. The problem is compounded by the fact that observations during successive oppositions are generally made in different latitude bands, consistent with the apparent change in tilt of Mars viewed from Earth. The radar observations have been correlated by using Mariner 9 occultation radii that are situated near radar observations. In general, it has been found that a shift of an entire season of radar observations by a fixed amount brings them into conformity with the occultation points.

Mars Topographic Map

After a set of absolute radius measurements is assembled, they can be used to produce a contour map. On Earth, topographic maps use sea level as a reference surface. This is an arbitrary but useful device for enabling us to visualize the magnitude of elevation differences between widely separated points. There is another and better reason for using sea level as the reference figure. The surface of the oceans conforms closely to a gravitational equipotential surface. If one were to imagine Earth completely covered by water, and further imagine this global sea frozen to form an icy plain, the surface would be everywhere "level." A plumb bob would be everywhere perpendicular to the surface; a bowling ball placed anywhere on the planet would have no tendency to roll. Mostly because of the rotation of Earth, our ball of ice would not be a sphere. The radius would be many kilometers greater at the equator than at the poles. The shape of the equipotential surface very nearly represented by the surface of the oceans is termed the geoid. Although, in detail, its shape is very complex, it can be closely approximated by an ellipsoid. For Earth, an International Ellipsoid has been defined and is universally used as a reference figure. For Mars there has not been complete agreement regarding definition of a reference figure, but it appears that a simple figure whose surface would define an equipotential surface at an atmospheric pressure of approximately 6.1 millibars (the triple point pressure of water) will become the standard.

The flattening of the geoid is known to be approximately 1 part in 192. The polar radius of a hypothetical global sea on Mars is about 0.5 percent less than the equatorial radius, resulting in a slight equatorial bulge. The topographic map of Mars (fig. 3.4) was made by referring the absolute radii to an equipotential surface, approximately defined by the dynamic figure (Christensen, 1975).

The resulting topographic map is equivalent to a map of Earth, with substitution of the 6.1 mb pressure surface for sea level. Predictably, downslope is the direction in which water would flow. The coverage by Earth-based radar and Mariner 9 occultation points is incomplete. Other instruments on the Mariner 9 spacecraft collected data that have been used to fill in the gaps. These instruments (UVS and IRIS) determined the CO_2 pressure at the surface of the planet. The measurements can be directly related to the geoid in the same way that measurement of water pressure at the bottom of the ocean is a measure of the depth of the ocean floor below sea level. No prior assumptions regarding the shape of the equipotential surface are required.

The map is necessarily very general. The error is approximately 1 km, or slightly less in the equatorial regions. In some areas, especially north of about 20°N, data are sparse. Local details have been largely smoothed over. However, the global morphology is well depicted.

The most important global feature can be seen by tracing the zero contour line around the planet. The northern and southern hemispheres have markedly different mean elevations. The boundary between the two is approximately a great circle inclined about 35 degrees to the equator. The northern hemisphere is significantly lower than the southern hemisphere. This is reflected in the weakly bimodal distribution of elevations (fig. 6.5). If Mars had an ocean, it would cover the northern hemisphere, and the southern hemisphere would be a single vast continent. The margin between continent and ocean would be marked by a sharp declivity, which drops 2 or 3 km over a distance of several hundred kilometers. The scale of this slope is similar to the slopes between ocean basins and continents on Earth. The distinct separation of high and low regions on Mars, although much less pronounced than on Earth, suggests a similar division of crustal types on the two planets. The difference in elevations between the principal modes of Martian elevations is about 2 km versus nearly 5 km for Earth.

Martian Gravity Field

The shape of an irregular surface can be described mathematically. This is done by fitting the surface to a series of trigonometric functions, then summing them together. The technique is most easily seen in two dimensions (fig. 6.6). A regularly undulating surface can be approximated by a single sine function with phase (starting point), amplitude, and periodicity appropriately chosen. For more irregu-

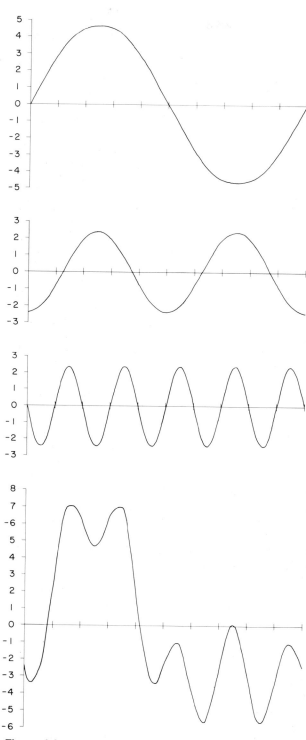

Figure 6.6. An example of harmonic analysis. The complicated curve shown at the bottom can be resolved into the three sine waves shown above, each with characteristic period, amplitude, and phase.

lar surfaces, additional sine functions are required in order to achieve a good fit. This type of representation, called harmonic analysis, has proved useful in describing and understanding many physical situations. The most instructive solutions are generally the simplest: that is, the ones in which the majority of information can be described by summation of a few sine functions. If a large number of sine functions contribute equally to the reconstruction of the observed surface, then very little is achieved by way of mathematical distillation—and ultimate understanding.

Harmonic analysis can also be applied to spherical surfaces. The significant parameter is J_{nm} where n and m describe the number of positive and negative sectors as a function of latitude and longitude, respectively. The rule of thumb is that there will be m cycles—or $2m$ sectors—and that there will be $n - m + 1$ zones, which follow parallels of latitude (fig. 6.7).

Spherical harmonic analysis permits us to describe the shape of an irregular object to any degree of exactness. In practice, because the number of terms in J_{nm} increases very rapidly as the degree (n) and order (m) increase, most fits are limited to relatively low orders. Written in the expanded form, the eighth order and degree spherical harmonic representation is a series expression with nearly a hundred terms to be summed in order to obtain the value for any particular point. The resolution of an eighth-order and degree fit is only 45° on a globe: that is, the wavelength of the smallest feature that can be depicted is about one-eighth of the circumference of the sphere. If more resolution is required, this can be achieved by increasing the order and degree and paying the computing price at a geometrically expanding rate.

The coefficient J_{00} describes a spherical planet (or point mass) where the gravity decreases as the inverse square of the distance from the center. The mean value of this coefficient is about 371 cm/sec² (gals) for Mars. The comparable value for Earth is approximately 980 gals. J_{10} effectively describes the same condition, since the gravity field of Mars is symmetric with respect to the center of mass of the planet, as is the chosen coordinate system. J_{20} (commonly referred to as J_2) describes a model in which there is no variation in longitude ($m = 0$). However, three latitudinal segments fall either above or below the mean sphere ($n - m + 1 = 3$). This is, of course, an ellipsoid of revolution. Increasingly higher values for n and m describe increasingly finer detail (fig. 6.7).

In expressing J values for Martian gravity, the J_0, J_1, and J_2 terms commonly are omitted. The resulting expansion yields a map of values from which the mean spheroid value has been subtracted.

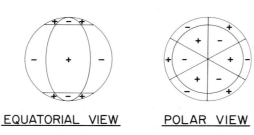

a. $J_{33} = 0.76 \times 10^{-5}$ $\lambda_{33} = 108°$ W $\Delta R = 0.9$ km

EQUATORIAL VIEW   POLAR VIEW

b. J_{53}

EQUATORIAL VIEW   POLAR VIEW

c. J_2

EQUATORIAL VIEW

Figure 6.7. Examples of spherical harmonic analysis. The spherical harmonic representation of the Martian gravity field is used to evaluate the gravity potential as a function of the areocentric coordinates. For each amplitude J_{nm} and phase λ_{nm} there is a term in the expression for the gravity potential. The contribution of each term J_{nm} and λ_{nm} can be understood by visualizing the sphere divided into positive and negative regions. There are $n - m$ zero lines along parallels of latitude, giving $n - m + 1$ zones of alternating positive and negative values. There are $2m$ zero lines along great circles of longitude, giving $2m$ sectors.

(a). For J_{33} there are six sectors and one zone.

(b). For J_{53}, schematically shown, there are six sectors and three zones.

(c). J_2, schematically shown, has three zones and one sector. The J_2 term represents most of the polar flattening.

The resulting departures from spherical values are generally termed a free-air anomaly.

The value of J_2 was initially determined from observations of the orbits of Phobos and Deimos (Woolard, 1944; and Wilkins, 1967). Revolving in orbits slightly out of the Martian equatorial plane, the satellites are subject to a slight gravitational tug as they rise above and then dip below the equatorial bulge. The result of this tug is that the nodes of the satellite orbits, the line of intersection of the orbital planes and the plane of the Martian equator, rotate slowly in a clockwise direction—opposite to the direction of revolution of the satellites. The regression of the nodes of the satellite orbits is consistent with a value for J_2 of $(1.97 \pm 0.02) \times 10^{-3}$. The accuracy of the early determination was verified by Mariner 9 investigators and somewhat refined to the presently accepted value of $(1.96 \pm 0.01) \times 10^{-3}$ (Lorell et al., 1972).

For comparison with the gravity data, the topography data may also be modeled by spherical harmonics (Christenson, 1975). The first-order coefficients in this case are related to the offset of the mean topographic figure from the center of the coordinate system. The offset of the center of figure from the center of mass is 2.5 km in the direction of 98°W, 57°S. The principal axes of the topographic figure are derived from the second-order topography harmonics, J_{22}, λ_{22} (table 6.1).

The observation of an offset of the center of figure from the center of mass by 2.5 km does not place any constraint on the interior structure of Mars. The offset is readily explained—though not uniquely so—by the difference in elevation of the two hemispheres (fig. 6.8).

Internal Structure

Our knowledge of the internal structure of Mars is largely limited to what can be inferred from its figure and its gravity field. Ideally we would like to know the axial moment of inertia, C. The moment of inertia places a constraint on the radial density profile. C is known for Earth because the sun and especially the moon exert torques on the equatorial bulge that result in precession of the axis of rotation. This allows an estimate of the dynamical ellipticity, H:

$$H = \frac{C - A}{C},$$

where C is the polar moment, and A the equatorial moment.

The coefficient J_2 is related to the difference between the principal moments of inertia:

$$J_2 = \frac{C - A}{Ma^2},$$

where M is the mass, a is the equatorial radius of the aeroid (for Earth this is called the geoid). Thus the axial moment C is:

$$C = \left(\frac{J_2}{H}\right) Ma^2.$$

For a uniform sphere, the quantity J_2/H is equal to 0.4. As mass is concentrated toward the center, the value becomes smaller. The values of this constant for Earth and the moon are 0.37 and about 0.39, respectively.

It is not possible to determine C for Mars. However, since Mars is rotating at about the same rate as Earth, it may also be nearly in hydrostatic equilibrium. In this case the moment of inertia can be derived from hydrostatic theory. The relation between the equilibrium figure and moment of inertia of a body in hydrostatic equilibrium is given by the Radau-Darwin formula for the normalized moment of inertia:

$$\frac{C}{Ma^2} = \frac{2}{3}\left[1 - \frac{2}{5}\left(\frac{5m}{2\epsilon} - 1 \right)^{1/2} \right]$$

where a is the equatorial radius of the aeroid, m is the ratio of centrifugal to gravitational acceleration at the equator

$$m = \frac{\omega^2 a}{g_e} = 4.6 \times 10^{-3}$$

and ϵ is the ellipticity. In this special case the ellipticity of the aeroid, the optical flattening, and the dynamical ellipticity are the same.

For the ellipticity, we will use the ellipticity of the aeroid, to the first order:

$$\epsilon = \frac{3}{2J_2} + \frac{1}{2m}$$
$$= 5.24 \times 10^{-3}.$$

The topographic figure is more likely to depart from the purely hydrostatic state, but the deviations will be minimized in the aeroid because of isostasy. Earth's topographic figure is slightly more elliptical than the aeroid. However, the geoid ellipticity is very nearly the same as the hydrostatic ellipticity.

Using the Radau-Darwin formula and the values of the other quantities from table 6.1 we calculate, for Mars:

$$\frac{C}{Ma^2} = \frac{J_2}{H} = 0.375.$$

This value, intermediate between comparable values

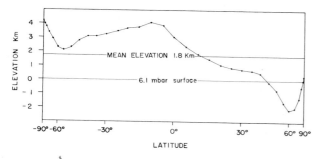

Figure 6.8. Average Martian elevation as a function of latitude.

Table 6.1. Some geophysical parameters of Mars.

Mass	6.418×10^{23} kg
Radii	a = 3399.2 km, 105°W
	b = 3394.1 km
	c = 3376.7 km

Mean equatorial radius r_e = 3396.6 km
Mean radius r_0 = 3390.0 km
Dynamical Radii (referred to mean surface)
  a = 3396
  c = 3378
Mean density   = 3.933
Equatorial gravity g_e = 370.6 gals
Dynamical ellipticity (hydrostatic approximation) $\epsilon = 5.24 \times 10^{-3}$
Offset of center of figure from center of mass   2.5 km toward 98°W, 57°S

Sidereal Period $P = 24^h37^m22^s$

Rotation rate $\omega = 7.08822 \times 10^{-5} \dfrac{\text{radians}}{\text{sec}}$

for Earth and the moon, suggests that the interior of Mars is more differentiated than the moon but less than Earth.

The visually determined equatorial radius of Mars is about 2 km greater than would be expected if it were a radially homogeneous planet in perfect hydrostatic equilibrium. Earlier estimates of this departure of the solid surface from the dynamical ellipticity were much larger, ranging from about 17 km (Loomis, 1965) to 8 km (Binder and Davis, 1973). The large values suggested significant departure of the planet from a radially symmetric body in hydrostatic equilibrium. The attendant stress differences could not be maintained, and plastic deformation of the interior would eliminate the differences in a few tens of million years. The existence of a substantial excess bulge would seem to require either internal activity such as convection, or rather large lateral variations in density—or perhaps both.

The presently observed excess bulge of only 2 km does not warrant elaborate interpretation. It appears probable that the Tharsis region has a thicker, more elevated crust (Phillips et al., 1973; Phillips and Saunders, 1975) partly in isostatic equilibrium. This may entirely account for the anomaly. The slight difference between the dynamical and optical flattening may be explained by differences in the isostatically compensated crustal thickness between the poles and equator (Lamar, 1962) or by strength in the crust (Phillips and Saunders, 1975). The contributions of each to the dynamical ellipticity may be evaluated by estimating the effect on J_2. The observed J_2 can be generalized to the following relation (Binder and Davis, 1973):

$$J_2 = J_{2H} + J_{2I} + J_{2S},$$

where J_{2H} is the hydrostatic contribution, J_{2I} is the isostatic contribution, and J_{2S} is the part supported by the strength of the crustal materials.

Binder and Davis (1973) conclude that, if the crust is in isostatic equilibrium with a core density of about 7.1 g/cm³, then the mantle is Earth-like (forsterite$_{65-80}$) and the core radius is about 1250 km (37 percent of the radius and 9–10 percent of the mass). If part of the excess bulge is isostatically uncompensated, the core could contain as much as 25 percent of the mass of the planet. These values should not be taken too literally since they are extremely sensitive to the value used for the equatorial radius. They merely illustrate the effect of crustal structure on models of the interior.

In summary, Mars lies between Earth and the moon with respect to density increase toward the interior. If part of J_2 is nonhydrostatic, then the value of ϵ in the Radau-Darwin formula is smaller, C/Ma^2 is smaller, and the core is larger.

Gravity and Topography

The comparison of global topography with the gravity field permits delineation of anomalous regions on Mars. For visual comparison, the observed gravity can be converted to equivalent topography (fig. 6.9). At first glance, gravity and topography are closely correlated (Lorell et al., 1972). This contrasts with Earth, where regional variations in observed gravity are relatively small and not correlated with topography (fig. 6.10).

On Earth the general lack of correlation between gravity and topography is a consequence of isostasy holding over large regions. The principle of isostasy is, basically, that at some depth within Earth the pressure is everywhere the same: that is, the mass of the overlying column of rock is the same. This implies that low regions must have denser rocks since their columns are shorter. The concept had its beginnings in the eighteenth century when Pierre Bouguer, surveying in the Andes on a French expedition to determine the oblateness of Earth, attempted a correction for the amount that the nearby mountains would deflect the plumb bob of his surveying instruments. Bouguer was puzzled by the fact that the deflection of the vertical seemed much smaller than the prediction, as though the total mass of the rock beneath the mountains was much less than that beneath the surrounding regions. More than a century later, after similar anomalous observations were documented by the India Survey under the direction of Sir George Everest, two major theories to explain these observations were put forth. J. H. Pratt and G. B. Airy independently developed

Figure 6.9. Elevation contours for Mars derived from gravity information and based on the assumption that all gravity anomalies are due to uncompensated surface mass excess (Sjogren et al., 1975).

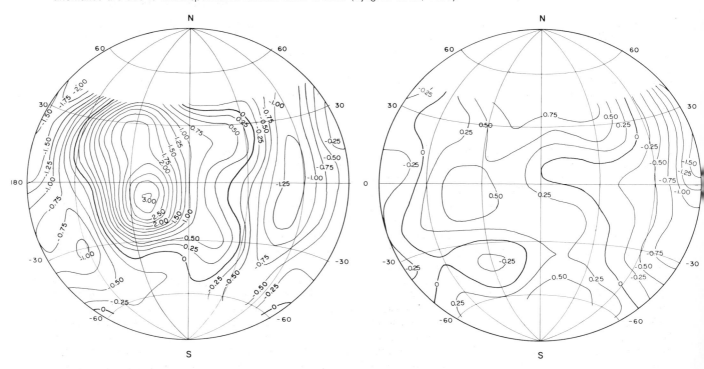

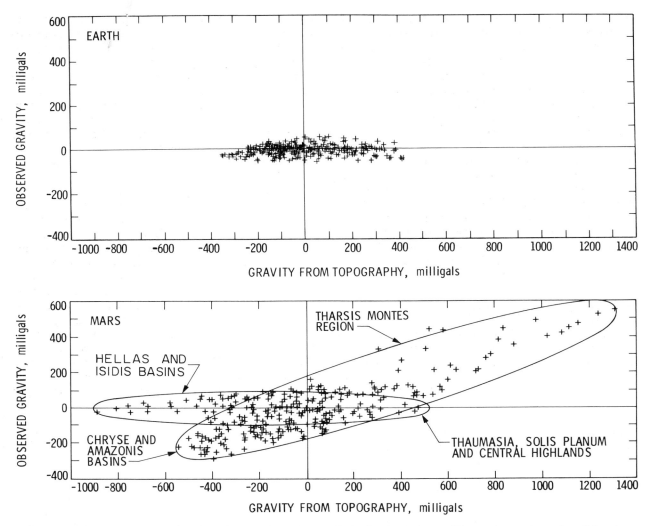

Figure 6.10. Observed gravity plotted against gravity calculated from topography. Topographic densities assumed for Earth and Mars are 2.67 gm/cm³ and 3.0 gm/cm³, respectively. For isostatic conditions, observed gravity will be close to zero (Phillips and Saunders, 1975).

explanations that remain, today, the basis for the application of gravity data to the interpretation of crustal structure. While both supposed the rocks of Earth's crust to be essentially floating in equilibrium, Pratt postulated that more elevated regions are underlain by rocks of lower density. Airy maintained that the rocks beneath mountains are essentially the same density as those elsewhere, but that there is a compensating root. The gravity effects of the two models are nearly identical, but seismic observations of greater crustal thickness beneath terrestrial mountains tend to confirm the Airy model.

The resolution of gravity and topography data on Mars is approximately 2000 km. Major features of continental scale are clearly seen in both the topography and gravity data. At similar scale, the conti-

nents and ocean basins of Earth may be seen in the topographic data, but not in the gravity variations. Because of isostatic compensation at relatively shallow depths, around 100 km, the columns of thick crustal rocks cannot be distinguished gravitationally from the thinner oceanic crust. Those slight gravity variations that *are* seen on Earth at this global scale must reflect density variations at great depth, probably in the mantle.

The data for Mars suggest either that major parts of the Martian crust are not in isostatic equilibrium or that compensation takes place at depths of 1000 km or more—deep enough so that the great distance of the compensating element from the surface reduces its effect on the observed gravity, and the topography is the only major contributor to the observed field (Phillips and Saunders, 1975). This dichotomy is illustrated by the scatter diagram (fig. 6.10) of observed gravity versus gravity predicted from topography. Points on the diagram associated with the most ancient terrains—the cratered terrain and the ancient basins—are Earth-like: that is, compensated, with no correlation of gravity with topography. The regions of volcanism on the Tharsis plateau show a marked correlation of gravity and topography, as do the adjacent low regions of Chryse and Amazonis. These regions may be as much as 40 percent uncompensated.

The Hellas basin provides evidence for the depth of compensation of the ancient terrain. If we assume a crustal density of 3.0 g/cm³, the Hellas basin and other regions of ancient terrain appear to be compensated at depths of less than 100 km (Phillips and Saunders, 1975). Assuming similar depths of partial compensation for the Tharsis region, there remains the equivalent of up to 3 km of uncompensated topography.

The positive isostatic anomalies over Tharsis and the negative anomalies over the adjacent basins, Chryse and Amazonis, follow the same linear trend on the scatter diagram (fig. 6.10). Phillips and Saunders (1975) argue that this indicates that they are the same age, have undergone the same amount of isostatic compensation, and are manifestations of the same internal process. They further show that, if the Tharsis anomaly formed any time within the past three billion years, the gravity data are consistent with viscous relaxation of the crust, and no dynamic mechanism such as convection is required

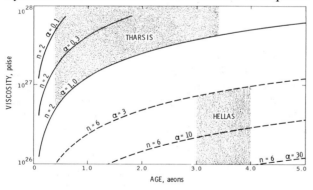

Figure 6.11. Martian viscosity plotted versus age, based on a particular model for isostatic compensation by viscous flow. Hellas topography is dominated by the sixth harmonic. Solutions for a likely range of e-folding response times (α) indicate isostatic compensation for viscosities of 10^{26} to 10^{27} poise. Tharsis is dominated by second order harmonics. The plotted range of e-folding response times and the knowledge that this structure is not isostatically compensated (fig. 6.10) suggest increasing viscosity in the outer part of the Martian crust with time (Phillips and Saunders, personal communication).

to support the anomalies (fig. 6.11). The observation that the Hellas basin is completely compensated is consistent with a model in which the interior has had a constant viscosity, since Hellas is somewhat older than the Tharsis uplift. However, it is more probable that the viscosity of the outer parts of Mars has been increasing with time, perhaps related to the loss of heat as predicted by the thermal models of Johnston et al. (1974). The viscosity history inferred is one of increase within the extreme range of 10^{25} to 10^{27} poise, corresponding to a maximum temperature decrease of about 150°C.

Crustal Structure

Terrestrial Structural Features

If, in the evolution of a planet, crustal rocks are subjected to stresses exceeding their strength, then the rocks will be permanently deformed in either a plastic or brittle manner. In the former instance, folds develop; in the latter case, faults are formed. On Earth, compression of the crust with attendant formation of major fold belts has held geologists' attention for well over a hundred years. Nowhere are these folds more spectacularly developed than in the European Alps. Sedimentary strata, once horizontal, have been deformed into cascading tiers of overturned folds with attendant thrust faults. Development of these structures implies a crustal shortening of several hundred kilometers. Geologists have been hardpressed to explain how this amount of compression could have taken place.

As mentioned in chapter 5, one of the early favored theories was that the mantle of Earth was cooling and contracting. An overlying rigid crust kept pace with this shrinking interior by doubling up along major fold belts. This model later was replaced by one of convection. Crustal rocks were pictured as being compressed between adjacent descending convection cells, something like wet laundry caught in the vise of an old-style wringer. In the past two decades the postulated vertical movements of the convection model have been supplemented by large-scale horizontal movements now believed to be associated with plate tectonics—involving continental drift and sea floor spreading. Plate tectonics serves to explain not only the compressive features of fold belts where plates collide, but also the extensional faults, or graben, that commonly occur along ocean ridges where new crust is thought to be generated (fig. 6.12). Where adjacent plates move in the same direction at unequal speeds, transform faults such as the San Andreas Fault of California form.

Almost all that we know about Earth's tectonic history derives from observations made on the surface. How much could we decipher if we had to rely solely upon high-altitude aerial photographs? The

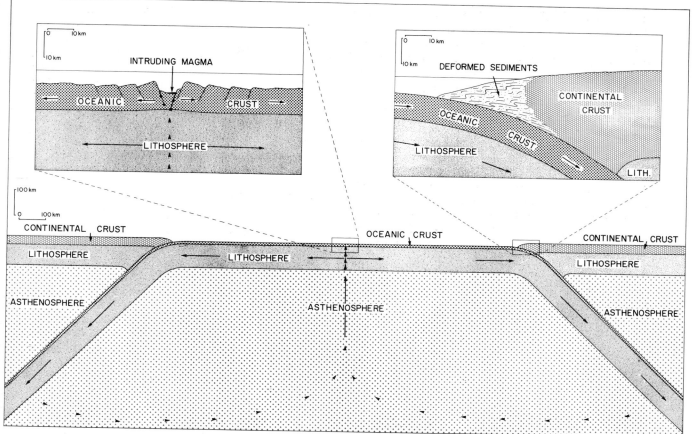

Figure 6.12. A schematic cross-section showing some of the main features of plate tectonics. Tensional features form along the ocean ridge where plates are separating. Compressional features form where oceanic crust dives below continental crust. Mantle flow takes place primarily in the asthenosphere. The lithospheric plates comprise both crust and upper mantle materials.

answer appears to be: "A lot." Folds are unmistakable, with their sinuous plan views (fig. 6.13) and arcuate terminations where they plunge beneath the surface (fig. 6.14). Fold patterns are best delineated where strata of varying resistance to erosion are involved. The highly resistant layers outline the structures. If all strata responded similarly to erosion, then the folds would not be so easily seen.

The worldwide distribution of fold belts, Paleozoic and younger, are easily delineated in aerial photographs. Their tendency to form along continental margins is remarkable. The irregular shapes of continents and their crudely matching boundaries immediately suggest the possibility that they are parts of a former megacontinent. Indeed, such a model probably would be more reasonable to a casual observer with only aerial photographs at his disposal than it would be to a geologist who devoted his life to field studies on Earth's surface.

Aerial photographs reveal other structural features that are more difficult to interpret. Many surfaces, especially those underlain by Precambrian rocks, are dissected by prominent lineaments. Some of these have sharp topographic expression, occurring either as scarps, troughs, or ridges. Many others are more subtly defined by color contrast and are difficult to correlate with features on the ground. Earth Resources Technology Satellite (ERTS—also called LANDSAT) pictures have revealed thousands of previously unrecognized lineaments (fig. 6.13).

The structural significance of the majority of lineaments is obscure. A few are associated with major faults. Others form on bedrock fractures along which there has been no movement. These are referred to as joints. Joints commonly form in three mutually perpendicular sets, a situation that is exploited in the quarrying of building stone. As widespread as joints are, their origin is poorly understood. Certainly their distribution and orientation is related to stress within crustal rocks.

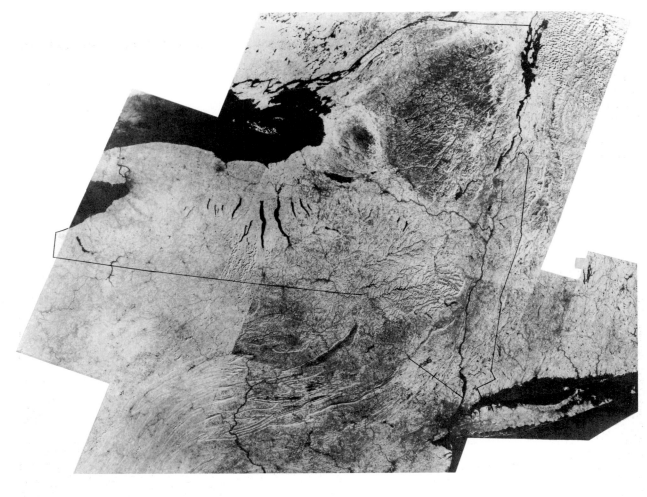

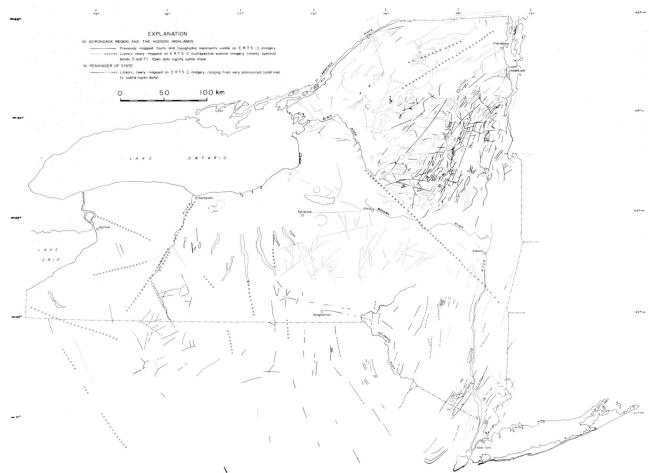

EXPLANATION

IN ADIRONDACK REGION AND THE HUDSON HIGHLANDS

——— Previously mapped faults and topographic lineaments visible on E R T S - I imagery

········· Linears newly-mapped on E R T S - I multispectral scanner imagery (mainly spectral bands 5 and 7). Open dots signify subtle linear

IN REMAINDER OF STATE

——··· Linears, newly-mapped on E R T S - I imagery, ranging from very pronounced (solid line) to subtle (open dots).

0   50   100 km

Opposite page:

Figure 6.13 (a). A mosaic of Earth Resources Technological Satellite (ERTS) pictures showing northeastern United States. Sinuous pattern in Pennsylvania is caused by differential erosion of folded rocks.

(b). A map of linear features first identified in ERTS pictures (from Isachsen, 1973).

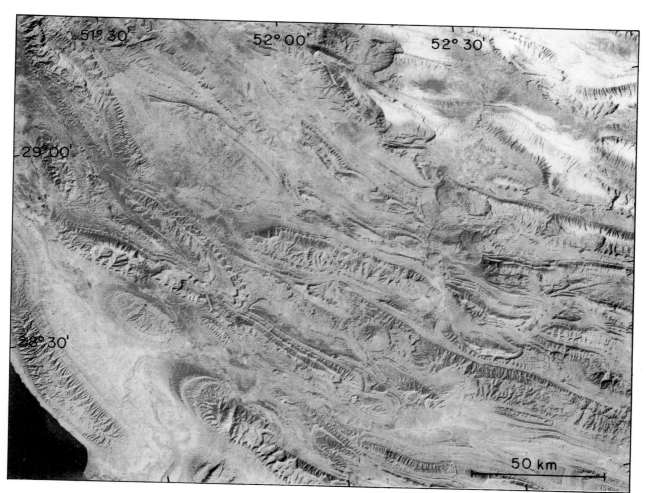

Figure 6.14. The Zagros Mountains of southern Iran. Stratified rocks have been compressed and folded to form linear domes (anticlines) and intervening structural troughs (synclines). The noses of the folds are outlined by curved outcrops. Although stream erosion has contributed to the dissection and exposure of these folded strata, the present environment is desert-like. If fold belts were present on Mars, they might display a similar pattern. This picture was taken by ERTS. The ground resolution is about 100 m, approximately the same as for a Mariner 9 B picture.

As we will shortly see, lineaments are prominent not only on Earth but also on the moon and Mars. Considering that we have such difficulty in assessing their significance on Earth, we should not be overly depressed by our inability to develop persuasive arguments for their origin on other planets.

Lunar Structural Features

A surprisingly small variety of lunar morphologic features have been attributed to structural activity. The few candidates include linear rilles, lineaments, wrinkle ridges, and mountain scarps (Baldwin, 1949, 1963; Fielder, 1965a). Linear rilles generally take the form of flat-floored, steep-walled troughs, ranging up to several kilometers in width, and tens to hundreds of kilometers in length (fig. 6.15). They

occur in both highland and mare terrain but are most prominently associated with the margins of the multi-ringed basins. They can be either curved (arcuate) or straight over long distances. The rille system in the southeast part of the Serenitatis basin (Rimae Littrow) (fig. 6.16) is typical in many respects. Here the structures are shallow linear troughs less than 3 km in width and of variable length and orientation. The flat floors and straight walls suggest that the rilles are fault-bounded, down-dropped blocks (graben). Their generally constant width at different topographic levels indicates steep dips on the bounding faults. These rilles transect both highland units and mare plains and are flooded by younger mare material. They are therefore younger than the highland units and the older lava plains, and

Figure 6.15. An oblique view of a set of lunar linear rilles transecting crater walls. The flat plains in the foreground are the floor of the crater Fra Mauro; the floor of Bonpland is seen in the upper left and Parry in lower left. The rilles are about 2 km wide. They are interpreted to be down-dropped linear blocks or graben caused by extension and faulting of the surface. High angles for bounding faults are inferred from the observation that the fault traces widen only slightly when they cross the crater walls. Post-faulting volcanic activity along the edge of the upper rille (outside the crater Bonpland) has buried a part of the rille. (AS14-73-10115.)

older than the younger mare surface at the eastern edge of Serenitatis. The Littrow graben do not precisely follow the edge of the Serenitatis basin. Several graben strike off into the uplands in a direction parallel to pre-existing lineament patterns (Head, 1974c). Therefore, the graben, in addition to reflecting basin shape, are controlled to some degree by pre-basin crustal structure.

McGill (1971) shows that certain rilles have topography characteristic of terrestrial graben. In the Littrow region, the coincidence of rilles with Serenitatis basin concentric structures suggests that sagging of the Serenitatis basin interior may have been responsible for the stress-field producing the graben. The sagging may have been caused by isostatic readjustment following early lava infill.

Other basins show similar development of linear or arcuate rilles. The best developed systems are around the Humorum, Tranquillitatis, and Imbrium basins. Other prominent systems include a generally north-trending set in the Fra Mauro region (the Parry rilles; fig. 6.15), which are radial to Imbrium, and a generally WNW trending set, which stretches across the north-central highlands into the area between Tranquillitatis and Fecunditatis.

Several larger, linear graben-like valleys are associated with the major multi-ringed basins and are oriented in a radial direction. The best known examples are the Alpine Valley (fig. 6.17) associated with the Imbrium basin and the Taurus-Littrow Valley at the southeast edge of Mare Serenitatis. These structures tend to occur where the radial directions are coincident with the lunar grid directions, discussed below.

Linear elements are abundant on the lunar surface, and their distribution and origin has long been a matter of controversy. They are often associated with major multi-ringed basins in radial and concentric orientations. As in the case of larger valleys, lineaments are enhanced in lunar grid directions suggesting that the lineaments preferentially form along pre-basin planes of weakness. Many lineaments may also be related to structural trends from ancient, subsequently obliterated basins.

Fielder (1963b), Strom (1964), and Elston et al. (1972), have discussed global lineament systems and have concluded that they represent a stress field with the maximum principal stress axis oriented north-south. Strike-slip movement allegedly has developed at 45° to the axis of maximum stress, and tension fractures have developed parallel to that axis, producing lunar-wide grid patterns. A major problem with this interpretation is the lack of widespread evidence for strike-slip dislocations that should be associated with shear. Although Fielder (1965a) has presented examples of local apparent strike-slip movement, the general lack of such evi-

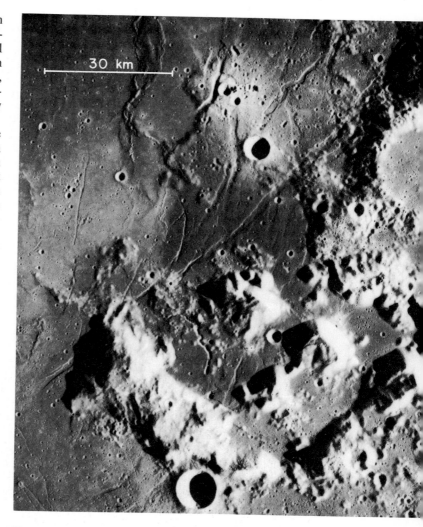

Figure 6.16. Littrow Rilles in the southeast corner of the Serenitatis basin. Most of the rilles are arrayed circumferentially to the basin along the edge of the mare deposits. However, one prominent rille begins in the southwest but continues directly northeast, following old pre-basin trends (Head, 1974c). Later mare deposits flood rilles at lower and middle left.

dence in the lineament systems is striking (Mutch, 1972). It is possible that the grid system had associated strike-slip movement but formed so early in lunar history that any evidence of lateral offset has since been destroyed or blanketed by the products of early intense bombardment. In this model, the topography of overlying highland deposits would reflect the basement fault pattern. This inheritance is quite common terrestrially where zones of Precambrian structural weakness control subsequent deformation patterns (Hoppin and Palmquist, 1965). Lack of evidence of regional strike-slip movement in the lineaments of these younger de-

Figure 6.17. View looking southwest along the Alpine Valley into northeast Mare Imbrium. The Alpine mountain range at the edge of the mare marks the position of the original Imbrium crater rim. The isolated peak at the top of the picture (Mons Pico) is a remnant of a central peak ring. The 150-km long Alpine Valley appears to have formed as a down-faulted trough in an area of lunar grid weakness. The valley was subsequently flooded and a sinuous rille formed on the floor. (LOV-120M.)

(a)

(b)

Figure 6.18. Relationships of wrinkle ridges to various units:
(a). South end of Littrow ridge, 20 km west of Littrow B crater. To the east, the dark material of the Littrow area is seen on the hilly region; younger basalts to the west lap up on this unit. The wrinkle ridge deforms both units. (AS17-2313.)

(b). North end of Littrow ridge, near LeMonnier crater, showing the ridge banked against the highlands. This suggests thrust faulting (Howard and Muehlberger, 1973). (AS17-9303.)

posits would suggest that the overlying units were influenced by the faults but not the sense of shear. This in turn would suggest that by this time—still very early in lunar history—the stresses causing global shear had relaxed (e.g., the moon receding from the vicinity of Earth), leaving an intensely sheared basement that would greatly influence subsequent lunar structure.

Stresses derived from Earth-moon tidal interaction, although particularly important billions of years ago when the moon may have been significantly closer to Earth, may still be an important factor in the activation of a previously formed lunar grid system, regardless of its origin. Recent studies of lunar seismic data (Latham et al., 1971; Hamilton,

1972; Lammlein et al., 1972) indicate that many moonquakes are triggered by tidal stresses. On Earth, regional joint systems are often found that appear to bear no relation to regional stress fields (Hodgson, 1961). It has been suggested (Kendall and Briggs, 1933; Price, 1966) that such joint systems might be fatigue phenomena resulting from repeated tidal strains. Blanchet (1957) has mapped fracture patterns in the Canadian shield that bear no relation to Precambrian tectonic trends. These patterns are also found in the overlying sedimentary cover, even in young, unconsolidated glacial sediments. Blanchet believes that the stress field was derived from tidal motion of Earth's crust.

Diverse origins have been proposed for wrinkle

ridges (mare ridges) and their analogous structures in the highlands (fig. 6.18). They may be indicators of intrusive and extrusive igneous activity (Strom, 1972); pressure ridges in flows; tension dikes result-ing from deeper strike-slip faulting (Tjia, 1970); lava lake-related surface deformation and intrusions (Hodges, 1973); or post-mare deformational structures (Baldwin, 1963; Howard and Muehlberger, 1973; Head, 1974c; Bryan, 1973). Conel (1969) presents convincing evidence for thrust faulting with accompanying horizontal displacement of at least 150 m in the case of one lunar ridge. Howard and Muehlberger have similarly interpreted some wrinkle ridges as thrust faults (fig. 6.19). The continuation of many of these features from the maria into the uplands suggests that their origin is, indeed, tectonic. There is no apparent volcanic activity associated with most of the ridges in the lunar uplands. Several large features of similar morphology on Mercury show clear evidence of overthrusting (Murray et al., 1974).

Lunar mountain scarps and ranges are almost exclusively circumferential to major lunar basins and are probably the result of crater excavation and subsequent faulting in surrounding areas (Hartmann and Wood, 1971; Wilhelms and McCauley, 1971; Head, 1974a; Howard et al., 1974). The formation of the mountains was essentially contemporaneous with the formation of the basin. Thus, there appears to be no process of mountain formation that has

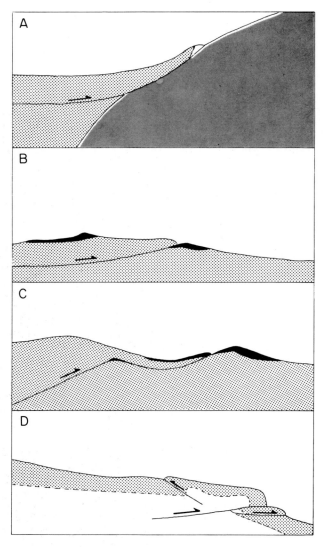

Figure 6.19. Schematic cross-sections of wrinkle ridges showing thrust-fault interpretation (Howard and Muehlberger, 1973).

(a). Ridge banked against highlands (similar to situation shown in fig. 6.18b) is interpreted as a basalt slab riding up against highlands.

(b). Cross-section of truncated crater relationship occasionally seen with wrinkle ridges. Ejecta blanket of crater is indicated as a stratigraphic marker (black).

(c). Cross-section of flow lobe into crater, a relationship often cited as evidence for lava flows. Ejecta blanket (black) is shown on the lobe that rides into crater.

(d). Origin of "backflow" features seen on many wrinkle ridges (see fig. 6.18) according to a thrust-fault interpretation.

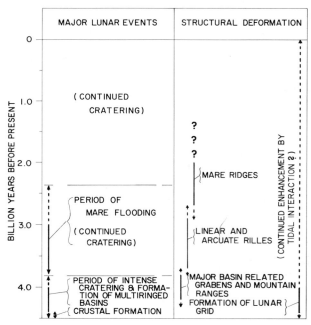

Figure 6.20. A synopsis of lunar structural history. Early activity is associated with the formation and early history of the lunar crust and the major impacts that modified it. Later activity is primarily related to modification and filling of the major basins. Virtually no major tectonic activity can be inferred for the last 2.5 b.y. on the basis of tectonic features.

been active on the moon during the last four billion years.

A summary of the chronology of lunar structural features (fig. 6.20) shows that there has been almost no tectonic activity in the past 2.5 billion years. Clearly, this contrasts with Earth. Both the ages and styles of dominant structural features are different on the two planets.

Mercurian Structural Features

The dominant tectonic feature visible on Mercury is a series of lobate scarps (fig. 6.21a), apparently indicative of major compression in the Mercurian crust occurring in the latter stages of intense bombardment. Although some of these features are similar to the wrinkle ridges developed on lunar mare surfaces, the size, morphology, and apparent global distribution of the Mercurian ridges suggest that they represent a major phase of crustal shortening (Strom et al., 1975). In contrast to Earth, Mars, and to a lesser extent, the moon, features representing tension are rare and appear to be related to the Caloris basin (fig. 6.21b).

The cause of the compressional deformation has been attributed to the presence of a large iron core (suggested by Mercury's high mean density; 5.44 g/cm³) and early chemical differentiation (Murray et al., 1975). Apparently the decrease in radius of 1–2 km suggested by the amount of compressional shortening of the crust could be caused by: a) a phase change (liquid to solid) of as little as 6 percent of an iron core making up about 50 percent of the volume of Mercury, or b) the difference in the coefficient of expansion between a large iron core and a silicate mantle (Strom et al., 1975).

Martian Structural Features

Two hemispheric asymmetries dominate the tectonic character of Mars. One, involving the Tharsis uplift and its related swarms of faults, affects most of the hemisphere between 0°W and 180°W (fig. 6.22). In contrast, the hemisphere between 180°W and 360°W is largely devoid of major tectonic features that are systematically controlled by a single mechanism (fig. 6.23). An even more striking asymmetry exists between a northern hemisphere of lowland plains and a southern hemisphere of cratered uplands.

The Tharsis Plateau, its related structural and volcanic features, and the adjacent Chryse Trough transform Mars from a largely lunar-like body into a dynamic planet. The Tharsis Plateau is the most elevated region of Mars. The relief is not due solely to the accumulation of volcanic rocks, since unmantled regions of ancient terrain form part of the plateau (Phillips et al., 1973). Structural and stratigraphic evidence show that the region was uplifted

Figure 6.21. Structural features on Mercury.

(a). Santa Maria Rupes, a prominent Mercurian scarp, cuts across center of photo from upper left to lower right. The scarp transects both old craters and intercrater plains and is typical of the large distinctive Mercurian scarps of apparent compressional nature (ranging up to 3 km in height). Area shown is 200 km across (Mariner 10 frame 27448) (Trask and Guest, 1975).

(b). Part of the floor of the Caloris basin of Mercury, showing abundant ridges and fractures. Rough terrain to the right is part of Montes Caloris, which partly surround the Caloris basin. Width of view is about 400 km (Mariner 10 frame 110) (Strom et al., 1975).

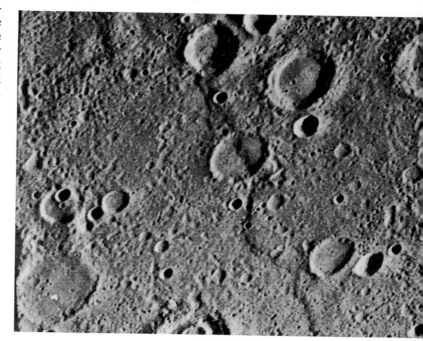

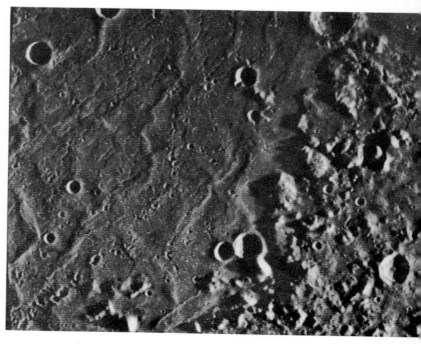

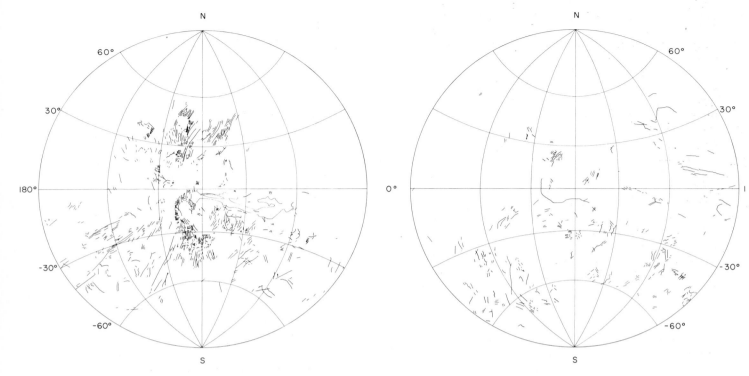

Figure 6.22. The more prominent Martian structural features, primarily fractures, plotted on a stereographic base.

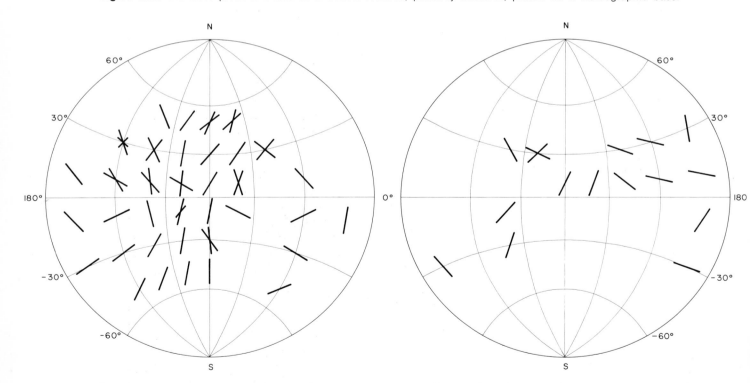

Figure 6.23. Modes for trends of Martian structural features, generally faults, plotted on a stereographic base. Data are averaged within 20° squares, between 60°N and 60°S. Modes were visually determined from rose diagrams on which all structural elements were plotted according to orientation and length. In some cases, secondary modes are identified. The absence of modes in some 20° squares indicates the absence of major structural features.

and the crust fractured before volcanism occurred. The major volcanic features have been discussed in chapters 3 and 5. The extent of volcanic deposits is shown in figure 3.2.

Adjacent to the Tharsis Plateau are exceptionally low regions. Amazonis, to the west, is a region of low plains. The Chryse Trough, to the east, is a pronounced depression extending from the north rim of the Argyre basin northward for 6000 km (fig. 6.24). The lowest part of the trough is Chryse Planitia, into which the channels that form the east end of the Valles Marineris debouch. All of the chaotic terrain on Mars and many of the major fluvial-like channels occur in the Chryse Trough. The channels all flow down the regional slope. There can be little doubt that the regional topography predates the formation of the channels and chaotic terrain.

The large-scale elevation differences that characterize the Tharsis Plateau and the depressions of Amazonis and Chryse are generally not of the type

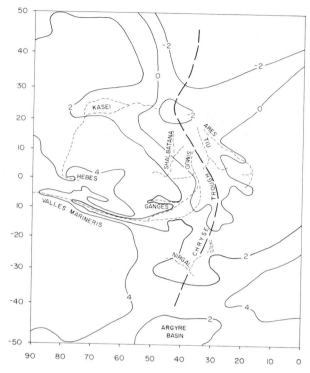

Figure 6.24. The Chryse trough. The contour interval for elevation is 2 km. The more prominent channels, identifiable in Mariner pictures, are shown by dotted lines.

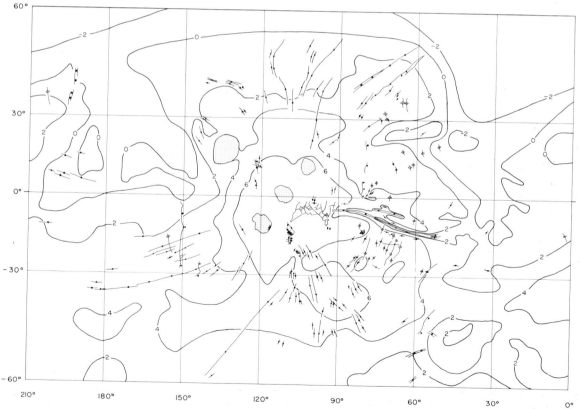

Figure 6.25. The Tharsis region, showing topography and structure. The contour interval for elevation is 2 km. Graben are indicated by solid circles, wrinkle ridges by open diamonds.

induced by terrestrial plate movements. However, large vertical displacements are known to occur on "stable" continents of Earth. In mid-Paleozoic time (Silurian), for example, the Michigan basin of the eastern United States was depressed more than one kilometer in 15 million years. Unfortunately, the analogy between Earth and Mars is weakened by our ignorance regarding the mechanisms for such large vertical movements on terrestrial continents.

The most striking features of the Tharsis Plateau are the approximately radial swarms of graben that extend to the south, west, and north of the uplift (fig. 6.25). Orientations of these structures, seldom more than a few kilometers wide and hundreds of kilometers long (figs. 6.26, 6.27), are qualitatively consistent with the stress distribution to be expected from regional doming (Carr, 1974b).

There apparently has been a long history of deformation in the plateau region, as evidenced by several sets of graben that show different orientations. The closely spaced Thaumasia and Claritas Fossae are related to the earliest uplift and fracturing. Memnonia and Sirenum Fossae formed more recently and are related to a volcanic center situated to the north of the Montes Tharsis. Consistent with this interpretation, some of the volcanic flows on the plains are displaced by Memnonia-Sirenum Fossae.

In the Alba and Tempe regions, on the northeast fringe of the Tharsis uplift, several sets of fractures occur within a small area, adding to the evidence

Opposite page:

Figure 6.27 (a). A set of graben, averaging 5 to 10 km in width. In the lower right a chain of craters is aligned along one of the fractures. Note that irregular channels, less than one km in width, trend from lower left to upper right. They are cut by fault scarps, so were apparently formed before the graben. (DAS No. 08443059, rev. 191, B camera, center at 104°W, 38°N.)

(b). Oblique view of intersecting sets of graben in the region of Claritas Fossae. (DAS No. 05635983, rev. 113, B camera, center at 110°W, 17°S.)

for a continuing tectonic history with stress patterns changing with time. Wise (1974) documents a sequence of five geologic episodes: (1) development of cratered crust, (2) formation of the volcano, Alba, (3) fracturing along northeast trends, (4) additional volcanic activity, and (5) renewed faulting in a north-south direction. The change in fault direction is attributed to an eastern migration of the Tharsis uplift. Deflection of fractures around Alba may have been a crustal response to the local load of volcanic deposits, but more likely resulted from regional extension over a region where there were inhomogeneities in the strength of crustal rocks (fig. 6.28).

Deep within the Tharsis Plateau, in a region only slightly lower than the crestline of Montes Tharsis, is a singular structural complex, the Labyrinthus Noctis. This interlocking web of broad graben has the same pattern as "turtle structures," described by Beloussov (1962). Turtle structures are reticulate faults that occur along the axial regions of domical uplifts on Earth (fig. 6.29). Conceptually they can be imagined as the combined effect of concentric and radiating components of concentrated stresses near the center of an uplift. Farther from the center, the stresses decrease in magnitude. Although the radiating fractures continue to propagate outward and relieve the relatively small stresses, the concentric faults no longer form. Thus the Labyrinthus occurs near the center of the Tharsis uplift, but only radial fractures occur farther away.

Most prominent of all the distal radial fractures are the Valles Marineris. The straight canyon walls and the alignment of pit craters along parallel fractures in cratered plains attest to a structural origin, although the canyons certainly have been enlarged by erosion. This sort of massive rifting may be analogous to activity along the East African rift valleys (fig. 3.22).

A somewhat analogous stress situation may have existed in the pre-drift megacontinent of the Americas, Europe, and Africa during the Late Triassic (May, 1971). Triassic dike swarms form a radiating pattern on the reconstructed continent. Apparently the stresses associated with a spreading center led to radiating tensile fractures and extensive volcanism (fig. 6.30). It has been suggested (Hartmann,

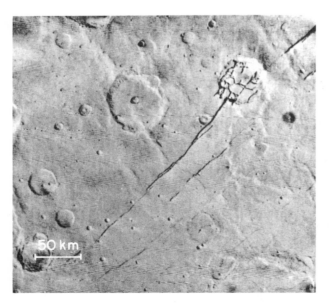

Figure 6.26. Several graben, part of a larger set (Memnonia Fossae) that extends 2000 km across the Martian surface. The irregularly fractured crater floor suggests crustal collapse and rupture over a shallow magma chamber. (DAS No. 06822588, rev. 146, A camera, center at 155°W, 20°S.)

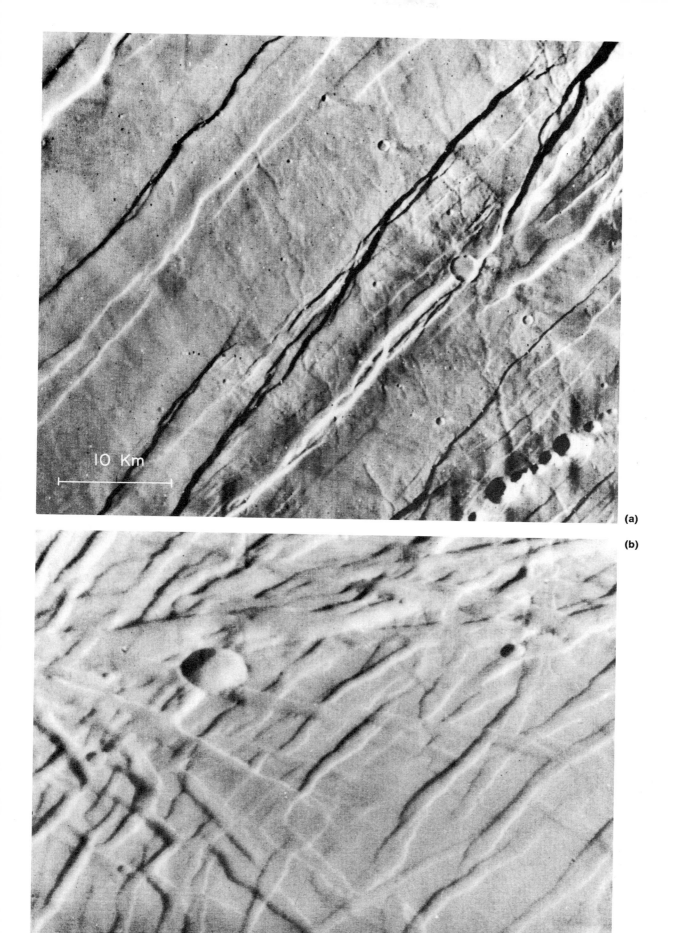

10 Km

(a)

(b)

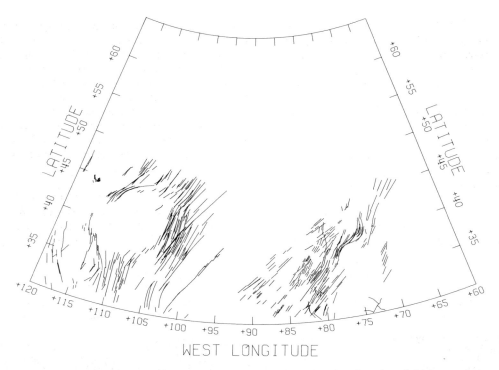

Figure 6.28. Some of the more prominent fractures in the Arcadia quadrangle. Multiple orientations indicate several periods of structural deformation. In the lower right the fractures are deflected around Alba Patera. This figure was machine-plotted using the digitizing techniques described in appendix B. A composite of similarly measured data for the entire planet is shown in figure 6.23.

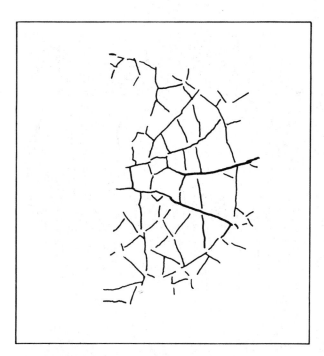

Figure 6.29. Concentric and radial faults, sometimes called turtle structure, formed on a structural dome (from Beloussov, 1962).

Figure 6.30. Triassic-Jurassic dikes in eastern North America, West Africa, and northeastern South America, with the continents restored to their relative position in the Triassic. Simplified from May (1971).

1973b; Carr, 1974b) that Mars is at an evolutionary stage where regional doming has occurred and horizontal plate movement is just beginning. Based on crater densities of affected surfaces, Carr (1974b) estimates that the doming occurred at least 1 b.y. ago.

The cratered plains to the east of the Tharsis Plateau of Mars contain linear ridges reminiscent of those on the lunar maria (fig. 3.13). Such features on the moon are widely believed to be structural in origin (e.g., Baldwin, 1963; Bryan, 1973) resulting from isostatic settling of the lunar maria and consequent crustal shortening. The distribution of the ridges on Mars, concentric to the Tharsis Plateau, is also consistent with this structural interpretation. Thus we see two types of structures on the plateau.

The earlier graben formed during uplift and the ridges formed on older volcanic plains surfaces during the later period of relaxation of the Tharsis uplift. However, as on the moon, one may also argue for a volcanic origin for these features. At one place where a ridge intersects the canyon wall, there is evidence that the ridge is not solely a surface feature (fig. 5.48). The ridge appears to be the trace of a subsurface planar body resistant to the erosion that has resulted in scarp recession along the canyon. This feature is most easily explained as an igneous dike.

In the region to the south and west of the Tharsis Plateau are a number of irregular scarps approximately perpendicular to the trends of the radiating graben (fig. 6.31). Those scarps generally face the

Figure 6.31. Irregular en echelon ridges of possible compressional origin, indicated by arrows. Graben transect some of the ridges. Youngest geologic units are plains materials that partly cover graben. (DAS No. 06894548, rev. 148, A camera, center at 147°W, 21°S; DAS No. 06894618, rev. 148, A camera, center at 145°W, 16°S; DAS No. 08297424, rev. 187, A camera, center at 151°W, 18°S; DAS No. 08297494, rev. 187, A camera, center at 149°W, 13°S.)

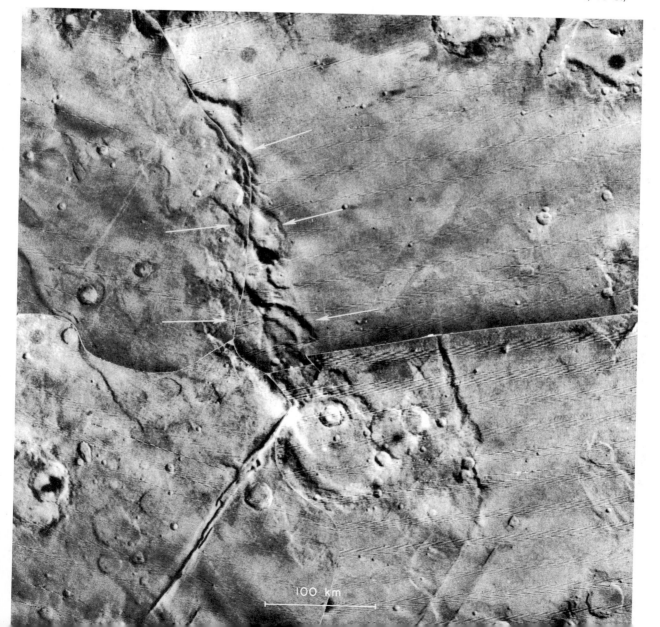

100 km

plateau and appear to be older than the graben. One interpretation of these scarps is that they are normal faults formed in the early stages of uplift of the plateau. They may be more extensive, but buried beneath the volcanic plains (fig. 6.32).

The Plains-Cratered Terrain Hemispheric Asymmetry

Except for the Hellas basin, the northern plains are the only region that lies below the zero elevation contour. The northern plains constitute an extensive, lightly to moderately cratered region. The argument that will be developed here is that these low plains occupy a region of crust that has been thoroughly broken by faults (Saunders, 1974). The process of disruption that marks the boundary be-

tween these plains and the undisturbed and highly cratered parts of the crust to the south (figs. 3.27, 3.28) has been carried to completion over much of the northern hemisphere.

The evidence for disruption along the cratered terrain-plains boundary is discussed in chapters 3 and 8. Additional evidence that the entire region of lightly cratered plains has been similarly disrupted is contained in the B frames. Many pictures show small knobs (fig. 3.33) that resemble those of the fretted terrain. This suggests the presence of a region of eroded crustal blocks, extending well beyond the fretted terrain. In addition, several B frames show extensive fracturing of the plains surface (fig. 3.33c and d).

As pointed out by Sharp (1973a), other processes

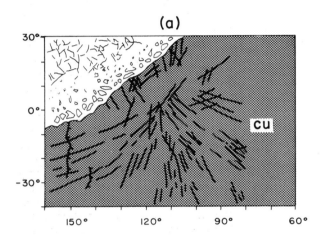

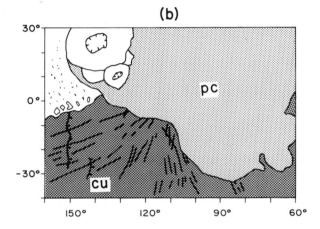

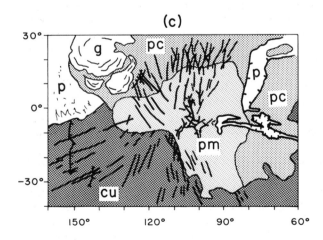

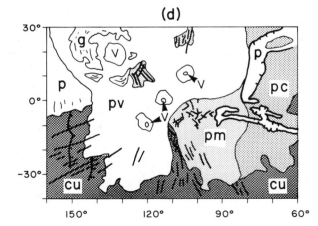

Figure 6.32. Paleomaps showing the geological evolution of the Tharsis region.

(a). Uplift and fracturing of cratered terrain.

(b). Emplacement of densely cratered plains (pc), with associated volcanoes.

(c). Emplacement of moderately cratered plains (pm), renewed uplift and fracturing, including formation of Valles Marineris. Erosion of former volcanoes to form grooved terrain.

(d). Emplacement of most recent volcanic plains (pv).

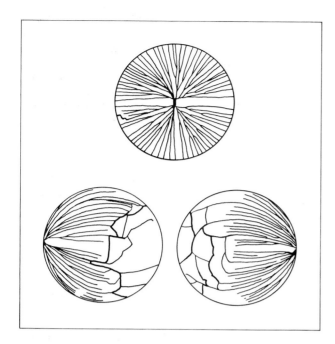

Figure 6.33. Three views of a thin spherical glass shell fractured under tension caused by expansion of the interior (from Bucher, 1924).

may contribute to the appearance of the fretted terrain and the knobby terrain. These may involve the removal of permafrost and certainly must include generation and removal of sediment by eolian activity. The reason for the tectonic disruption is unknown. As in the Tharsis Plateau and at Alba, it is highly likely that crustal extension and pervasive fracturing has occurred. The chaotic terrain and no doubt the fretted terrain has broken up along preferred directions (Wilson et al., 1973). Apparently the tilted blocks and fractures are shallow phenomena. Deeper than about 10 km, the strength of the crustal rocks is exceeded and any nonhydrostatic stresses tend to result in plastic behavior rather than fracture. Assuming that the structural blocks are more or less equidimensional, that might explain why they average about 10 km in diameter and—by implication—in thickness.

There is little or no evidence for directed horizontal movement of the Martian crust. As mentioned before, the motions of Earth's crust are primarily horizontal with an apparent balance between extension and compression. On Mars, the evidence suggests extension only. Observational evidence for plate tectonics has been cited (Sengör and Jones, 1975), but the interpretation is speculative.

One mechanism that could result in extensive rifting of the Martian crust involves phase changes in rocks of the interior. Lyttleton (1965a, b) first sug-

gested that the phase change mechanism should dominate Martian tectonics and correctly predicted that the surface of Mars would be extensively fractured. He further predicted that no compressional features—that is, fold mountains—would occur on Mars. The change in structure of $(Mg,Fe)_2SiO_4$ from the spinel to the olivine structure with increasing temperature is the most likely phase change based on the probable composition of the Martian mantle. The amount of increase can be estimated by choosing a representative thermal model and calculating the volume increase of the mantle with time (fig. 6.2). For example, if Mars has a thermal profile consistent with chondritic composition and an accretion time of 10^5 years, then its radius would have increased by approximately 20 km since its formation. This results in generation of 1.7 million square kilometers of new surface, an increase of approximately one percent.

If the planet has increased in radius and the crust has fractured to accommodate this increase, then the crustal structures must indicate the expansion. It is impossible to predict the pattern of fracturing that would result from the breakup of the brittle surface rocks as the mantle volume expands. It is sufficient to point out that the observed distribution of fractures is one of a larger set of possible solutions. In this context, there is one obvious question. Why does the crust not fracture everywhere to about the same degree? One possible answer is that the brittle part of the crust is uncoupled from the expanding region. The mantle is able to flow and, once fracturing has been initiated in one region, that region can continue to be the site of most of the deformation. The lateral motion required in the mantle is the same as the amount of radius expansion, or about 20 km in 4 billion years. This is 0.5 cm per 1000 years, a reasonable rate, since viscosities of Earth's mantle allow rates of approximately 2 cm/yr, nearly four orders of magnitude higher.

An amusing and perhaps instructive analogy illustrating preferred fracturing of one hemisphere is found in some experiments by Walter Bucher. Bucher (1924) froze water-filled spherical Christmas tree ornaments. They were suspended in supercooled brine to avoid the problems of undesired stresses at points of support. As might be anticipated, the water, upon cooling to the freezing point, expanded and fractured the thin glass shells. A surprisingly predictable fracture pattern was observed each time (fig. 6.33). An initial fracture formed, more or less at random, and propagated a short distance in both directions. The single fracture does not propagate far since it relieves stresses in only one direction. The ends split into more fractures that, in turn, splayed until the region of initial fracturing was a radiating mass of cracks. These fractures converged toward the hemisphere opposite the initial break. There is always an unbroken region opposite the first break. As with all such models, this one can only give us insight into the properties of the model itself. A planet may behave differently. To explain the low-lying northern plains by the expansion model, it is necessary only to postulate that most of the increase in surface area was restricted to the northern plains simply because the initial fracturing occurred in this region. A consequence of increasing the surface area is that the crust must become thinner. Because the crust becomes thinner, its isostatic equilibrium elevation will be lower. This is, of course, what is observed.

Martian Lineaments

In addition to the unmistakable topographic features just discussed, there are on Mars, as on Earth and the moon, many linear elements. Binder (1966) includes in this category any feature that is "sensibly straight and narrow"—even linear boundaries between surficial materials of differing brightness. Several hundred lineaments were identified in Mariner 4 pictures (Binder, 1966) and more than 5000 in Mariner 6–7 pictures (Binder and McCarthy, 1972). Equivalent analysis of Mariner 9 pictures has not been published. Using the same criteria adopted for previous studies, close to 100,000 lineaments should be identifiable in a photographic survey of the entire planet.

Lineaments identified on Mariner 4 pictures trend northwest and northeast, similar to the lunar grid. Binder (1966) suggests that the lineaments were formed by compressional stresses accompanying changes in the planet's rotational equilibrium figure. Early in its history Mars may have experienced a decrease in rotational angular velocity. The global grid tentatively identified in Mariner 4 pictures was confirmed by more numerous measurements in Mariner 6–7 pictures. Binder and McCarthy (1972) assert that the angle between northeast and northwest lineament sets increases from low to high latitudes. This is consistent with the calculated distribution of shear sets accompanying changes in planetary oblateness (Vening Meinesz, 1947). Binder and McCarthy also identified radial and approximately concentric lineaments associated with the Hellas basin and a south polar basin.

Binder's conclusions are difficult to assess. As he himself acknowledges, identification of subtle features is—by definition—hazardous. Some lineaments may be introduced by the television system or by the picture enhancement programs. Some may be artifacts of lighting. Howard and Larsen (1972) demonstrate that low-angle illumination of randomly hummocky models creates the illusion of lineaments preferentially oriented in two sets symmetrical with respect to the direction of the light. Finally, some lineaments, especially those indicated by changes in surface brightness, might be real but unrelated to structure.

Even with these cautionary remarks in mind a local correlation between lineaments and erosional landforms is beyond question. Wilson et al. (1973) document two sets of lineaments in the chaotic terrain, based on analysis of Mariner 6 pictures. One set trends slightly west of north, a second trends northeast. They are interpreted as shear fractures associated with north-northeast compression. The lineaments are present both in the chaotic terrain and in the adjacent cratered terrain. Apparently the chaos formed by dissection and slumping of prefractured bedrock. Independent analysis of the same Mariner 6 pictures by Arvidson (unpublished) yields the same descriptive results. At the very least, we have evidence that the lineaments, subtle though they may be, are similarly deciphered by several investigators, working separately.

7.

Wind

Introduction

THE POSSIBILITY that wind might be modifying the Martian surface was first mentioned by E. M. Antoniadi in 1909 but had been recognized earlier by Percival Lowell and others. Antoniadi noted that most telescopic observations of yellow clouds were made when Mars was closest to the sun (perihelion). He suggested that increased solar heating of the atmosphere during this period provided enough energy for winds to raise large amounts of surface dust. Later, analysis of more extensive cloud records by Wells (1966) and Capen (1971) confirmed that yellow cloud occurrences reach a maximum near perihelion.

Mariner 9 was injected into Mars' orbit two months after the 1971 perihelion, a time when the surface was obscured by the latter phases of one of the more intense planet-wide storms (Masursky et al., 1972). After about a month in orbit, the dust cover began to dissipate and Mariner 9 began systematically photographing the Martian surface. The pictures revealed that eolian activity is not restricted only to raising surface dust. Eolian processes have also played an important role in shaping Martian landforms. Mariner pictures contain evidence for wind-eroded terrain, terrain mantled by wind-blown deposits, and wind-related features surrounding many craters.

This chapter will review the effects of terrestrial wind erosion and deposition. Then some general physical constraints will be summarized. Finally photogeologic evidence for wind action on Mars will be analyzed. Our intent is to illustrate the nature of present eolian activity and to provide insight into the role played by wind during Martian geologic history.

The Nature of Eolian Processes on Earth

Some Sedimentological Details

Before considering theoretical and empirical evidence for eolian processes on Mars, we will discuss the work of wind on Earth as a background against which to view the Martian situation. Studies in natural environments and in the laboratory demonstrate that wind moves sediment in three modes: suspension, saltation, and impact creep (fig. 7.1). Suspension occurs when particles lifted from the surface have settling velocities that are smaller on the average than velocities of upward eddies. Small particles in suspension may be carried to great heights and over large distances. Particle transport by saltation occurs when the wind has sufficient energy to lift particles off the ground, but insufficient energy to

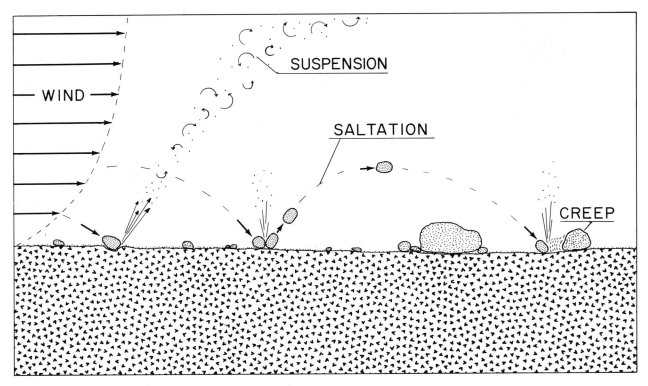

Figure 7.1. Three modes of sediment transport by wind. Small grains have settling velocities low enough to be kept aloft by upward wind eddies, and they are removed in suspension. Larger grains bounce or saltate in the direction of wind movement. The largest grains can be transported only by momentum transferred by impact of saltating grains.

keep them suspended. Grains bounce along in the direction of wind movement, imparting momentum and dislodging other grains when restriking the surface. Particles that are too large to be picked up by wind still can roll and slide in the direction of wind movement if a sufficient number of saltating grains strike the larger grains. This process is termed impact creep.

Given a sediment surface with a mixture of particle sizes, wind can move material in all three modes of transport. The net result will be a segregation of particle sizes (fig. 7.2). Suspended grains will travel with velocities nearly equivalent to wind velocities and will be carried away. Saltating grains will be removed and carried away at some fraction of wind velocity. Impact creep will transport larger particles at rates much slower than for saltation. Eventually, both suspended and saltating grains will be removed. Without a resupply from another area, the surface will become a residue of large particles—those that travel slowly by impact creep and those that are too large to be moved. When occurring at large scale, these processes tend to produce three types of sediment: a gravel or boulder lag deposit consisting of grains unable to be carried in either saltation or suspension; sand sheets or dune fields produced by deposition of saltating sand; and blanketing "loess"

dust deposits, formed when grains carried in suspension are finally deposited.

The situation on Earth is more complicated than just depicted because other processes, notably fluvial action, tend first to segregate dust and sand from larger particles. For example, during Pleistocene glaciation in North America, sand and dust were eroded from moraine deposits by glacial streams and deposited in ephemeral lakes and on plains at the terminal edges of glaciers. Winds then removed the sand and dust, producing regions of dunes near the glaciers and areas of loess deposits covering thousands of square kilometers downwind of glacial deposits (fig. 7.3).

On Earth, loess deposits blanket large regions because dust settles out of the atmosphere in a fairly uniform manner. Once deposited, the small grains form relatively cohesive sediment with a smooth surface. Atmospheric turbulence above the surface is minimal, and there is little opportunity for the grains to move further. Sand grains, on the other hand, are relatively mobile and subject to repeated transportational episodes. Winds are capable of rapidly reworking sand deposits to form a remarkable array of morphologic forms (fig. 7.4). Sand sheets, barchans, transverse, and longitudinal dunes are all common features. The particular form attained by a

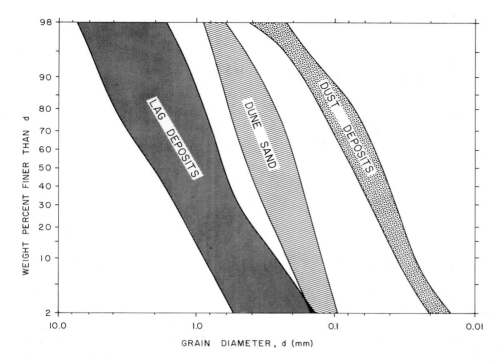

Figure 7.2. Particle size distributions for selected eolian lag, dune, and dust deposits collected from the midwest by Udden (1914). Lag sediments consist of grains too large to be moved directly by wind; dune sediments are those whose grains were carried in saltation; and dust deposits consist of particles that were carried along in suspension.

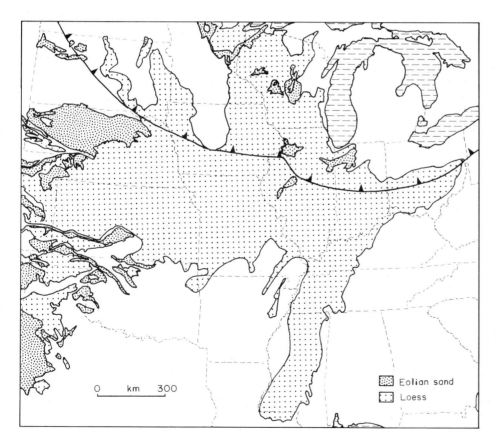

Figure 7.3. Central United States, showing areal distribution of Wisconsin-age sand and loess deposits. Barbed line indicates the southern limit of Wisconsin-age glaciation. Loess has been derived both directly from glacial outwash and from outwash of major rivers. Formation of sandy deposits is favored by abundant outwash near glaciers and by lowering of water table. Adapted from Flint (1957).

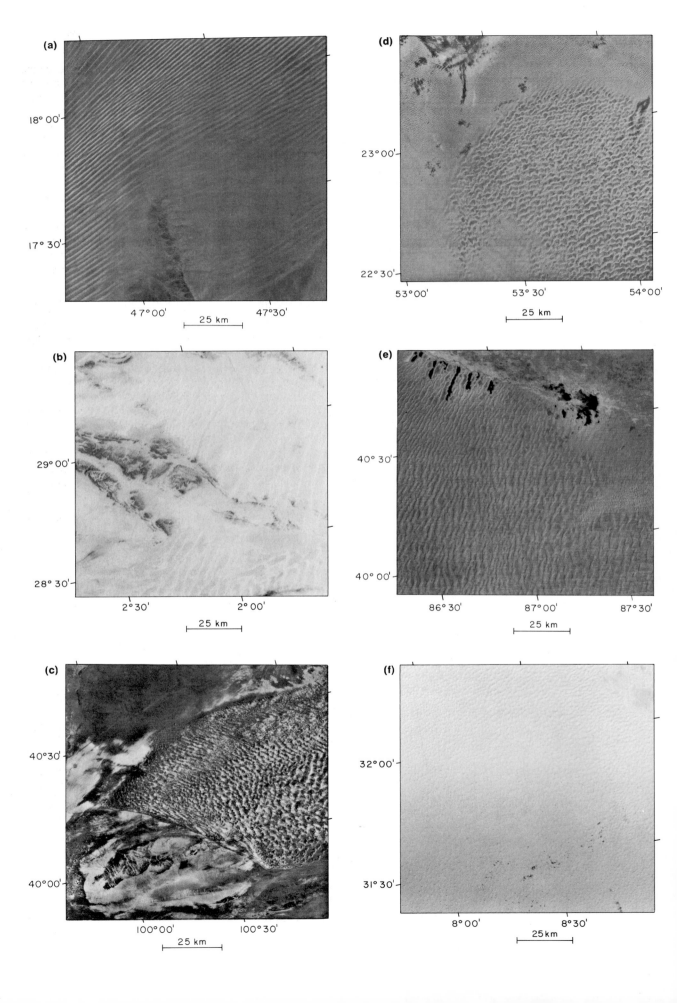

(a)

18° 00'

17° 30'

47° 00'   47°30'
          25 km

(b)

29° 00'

28° 30'

2° 30'   2° 00'
          25 km

(c)

40°30'

40°00'

100°00'   100°30'
          25 km

(d)

23° 00'

22° 30'

53°00'   53° 30'   54°00'
          25 km

(e)

40° 30'

40° 00'

86° 30'   87° 00'   87° 30'
          25 km

(f)

32° 00'

31° 30'

8° 00'   8° 30'
          25km

Opposite page:

Figure 7.4. A variety of terrestrial dune patterns, revealed in ERTS pictures. Photographs were supplied and characterized by C. S. Breed.

   (a). Compound linear dunes in the Empty Quarter, Saudi Arabia.

   (b). Star dunes aligned along the crests of linear dunes in Algeria.

   (c). Reversing crescentic dunes in the Gobi desert. Star dunes are present at the edges of the dune field. Similar patterns are revealed in Martian dune fields (fig. 7.14b).

   (d). Compound crescentic dunes in Empty Quarter, Saudi Arabia. Smaller, simple crescentic dunes occur along the margins of the dune field.

   (e). Herringbone pattern produced by complex, crescentic dunes in the Taklamakan Desert of China.

   (f). Scattered star dunes in the Erg Oriental of Algeria.

sand deposit is a complex function of sand supply, wind velocity, topography, and climate.

Terrestrial Arid Regions

On Earth, fluvial modification of landforms is far more effective than eolian modification, even in most desert regions. This seemingly contradictory situation results because: (a) rainfall and subsequent runoff, although rare occurrences in arid areas are extremely effective in the absence of vegetation, and predominate over wind in terms of total erosive capacity, and (b) the terrestrial climate is dynamic.

Long-term changes in atmospheric circulation cause shifts in locations of dry areas. Consequently, deserts are usually not exposed to rainless conditions long enough for wind to modify landforms radically.

A map of the world's deserts is shown in figure 7.5. Arid conditions result when there is a lack of atmospheric water vapor that can readily be converted to rainfall. Thermodynamic considerations demonstrate that hot air can retain more water as vapor than can cold air. Hot air rises at the equator and sinks to the surface at ±30° latitude. While descending, it heats up by compression (adiabatic heating) and by transfer of heat radiated from the surface. Any water near the surface is retained as vapor rather than dissipated as rainfall. These processes produce two zones of aridity, straddling the equator at ±30° latitude. Most of the world's major deserts fall into this band: the Sahara of Africa, the Arabian Desert in the Middle East, the Victoria Desert of Australia, the Kalahari of southern Africa, and the Atacama Desert of Peru.

Another phenomenon that occurs in high latitudes contributes to arid conditions. Near the poles, cold surface air can retain only traces of water vapor. The resulting atmosphere is relatively dry and precipitation is low, making the regions of polar ice caps zones of cold deserts.

In addition to global-scale phenomena, local climatic perturbations can lead to dry conditions. For instance, a mountain range can force air to rise and

Figure 7.5. Lambert equal-area map of the world's arid regions. Stippled areas have less than 10 inches of rainfall per year. Adapted from Petterssen (1969).

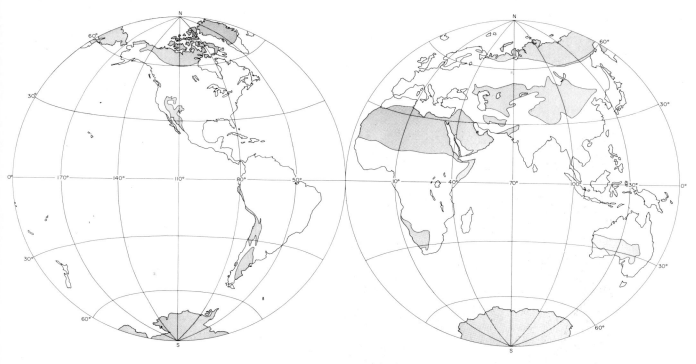

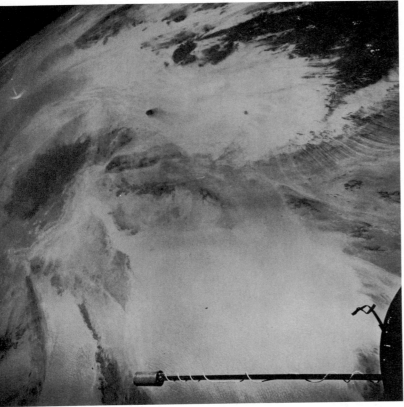

cool adiabatically, and consequently to release its vapor as precipitation. By the time the atmosphere begins to descend the leeward slopes, it is dry; adiabatic heating during descent further reduces the possibility of rainfall. When cool ocean winds blow onshore, they warm as they move over land, drawing up moisture as vapor and leading to arid conditions. Both of these phenomena play a role in determining conditions in the Peruvian-Atacama Deserts (Grolier et al., 1974). There, the high Andes Mountain belt forms a topographic barrier to moist air masses from the Amazon basin to the east. Combined with this effect, winds blowing onshore from the cool Peru current warm as they encounter coastal lands, further inhibiting rainfall.

Wind as a Sculpturing Agent

Eolian modification of pre-existing fluvial forms can be clearly seen in the Sahara Desert. Eolian sand bodies have accumulated by reworking of sedimentary deposits. They exhibit a myriad of dune morphologies and sheet-like appearances (fig. 7.6b). Etched terrains have formed where deposits have been scoured and removed by wind (fig.7.6a). Dust and silt is often picked up and blown far distances from North Africa. Fine-grained sediment, carried across the Atlantic Ocean by upper level winds, has been collected on the island of Barbados in the West Indies (Prospero et al., 1970).

Wind sculpturing can result in spectacular landforms if conditions are dry enough and if a suitable supply of material is present. The Peruvian-Atacama Desert meets the first criterion in that rainfall is less than a scant 10 mm/yr, making it one of the driest areas on Earth (Grolier et al., 1974). Combined with aridity, the Peruvian deserts are underlain by soft marine sediments that provide ideal raw materials for rapid erosion. Perhaps the most spectacular features produced there by winds are yardangs. These are sculptured mounds, aerodynamically shaped into elongate mountains resembling overturned boat hulls (fig. 7.7).

Even in wet climates, wind can be an efficient modifying agent. Many coastlines with sandy beaches have extensive dune deposits that have been blown inland by prevailing onshore winds. A good example is the coastal dune region of Oregon and Washington (Cooper, 1958).

Figure 7.6 (a). Gemini VII photograph of central Algeria. Horizontal strata have been scoured by wind to form a series of closed basins. A small barchan dune field can be seen in the center of the frame. (S65-63784.)

(b). Gemini XI photograph of Libya. Winds have reworked sands into patterns ranging from sand seas to thin sand stringers extending downwind from main sand bodies. The Mediterranean Sea is visible in the distance. (S66-54527.)

Figure 7.7. Aerial views of some Peruvian desert landforms, all taken from Grolier et al. (1974).

(a). Yardangs, sculptured from poorly consolidated marine sedimentary rocks. Streamlined shapes are the result of eolian abrasion by winds blowing from upper left to lower right.

(b). Yardangs, forming in horizontal sedimentary strata.

(c). Part of a large lee dune, forming behind a hill. The total dune length is about 5 km. The part of the dune shown here is about 1 km long, 100 m wide, and 30 m high.

(d). Sand drifts, partly covering dissected granitic terrain.

(e). Transition of a sand blanket to a field of barchan dunes.

(f). Wind-faceted hills. Attitudes of facets are controlled by joint sets in the crystalline rocks that make up the hills.

On a small scale, wind also can be an effective erosional agent, provided only that a suitable flux of saltating particles is present. Large particles and rock outcrops can be pitted and abraded by repeated impact of sand grains, forming faceted and fluted surfaces called ventifacts (fig. 7.8). In some situations, deflation of finer material and synchronous sand abrasion produce an almost continuous pavement of ventifacts, sometimes termed an armored surface.

Theory of Eolian Processes on Mars

Consider a surface composed of loose soil with wind blowing across it. At low velocities none of the particles will be moved. With increasing wind velocity, some of the particles begin rolling and bouncing in the direction of wind movement. The velocity necessary to initiate particle movement can be expressed by equating the force on a particle due to wind drag to the inertial resistance due to particle mass. When this is done, the following equation can be derived:

$$V_{*_t} = A \left(\frac{\sigma g d}{\rho} \right)^{1/2} \qquad (1)$$

where: V_{*_t} is the wind drag velocity needed to move loose particles; A is a parameter that varies with wind turbulence at the soil surface; g is the gravitational acceleration; d the particle diameter; σ the particle density; and ρ the atmosphere density.

Solutions for this equation using Martian and terrestrial values for the variables are shown in figure 7.9. Drag velocities needed to initiate particle movement are much higher on Mars because the atmospheric density is much lower. Particles that will move at lowest threshold velocities are about three times as large on Mars as on Earth, although both situations involve sand-sized grains. Smaller particles (fine sand and dust) require higher velocities because they offer too little projected area for wind drag to be effective. Larger particles (coarse sand and cobbles) are more difficult to move because of an increased inertial resistance. Thus, the particle size corresponding to the lowest drag velocities has the critical combination of enough projected area for drag to be effective and a small enough mass to offer little inertial resistance. The critical particle size also varies as a function of atmospheric density; a decrease in density shifts the critical value to coarser particle sizes.

Drag velocities can be related to wind velocities within atmospheric boundary layers by functions that describe changes in wind velocity as a function of height above surfaces. Problems arise with delineation of wind velocities on Mars because the properties of the Martian boundary layer are poorly known. The Von Karman relation will be used here as an approximation. This function describes the velocity distribution within a boundary layer for wind tunnel experiments:

$$V_z = 5.75 \, V_{*_t} \log \frac{Z}{K} \qquad (2)$$

where: V_z is the wind velocity at height Z above the surface; V_{*_t} is the drag velocity needed to move a particle; and K is the surface roughness length.

Computing the velocity at 1 m height needed to raise a 0.5 mm particle (corresponding to lowest Martian threshold velocity) yields a result of 50 m/sec for Mars and a result of 8 m/sec for Earth. Although these numbers can only be accepted as a first-order approximation, it is obvious that winds needed to initiate particle movement on a loose soil surface on Mars are extremely high relative to normal terrestrial conditions.

Once the threshold velocity is reached and exceeded, particle motion will begin. Effectively, all sizes less than the coarsest set in motion will be stirred up and carried along, the smaller ones in suspension and the larger ones in saltation. Estimating rates of travel and transport capacity at these high wind velocities is difficult because, in general, the nature of turbulent fluids is not well understood. First-order scaling of wind velocity, atmospheric density, and gravity suggest that saltating grains of

Figure 7.8. Block field on glacial till in Victoria Valley, Antarctica. Boulder surfaces have been polished and faceted by repeated impact of silt- and sand-sized grains carried along by strong winds. (From Morris et al., 1972.)

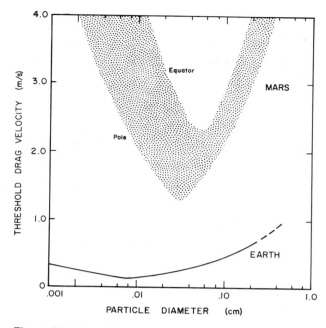

Figure 7.9. Threshold drag velocities plotted over a range of particle sizes for Mars and Earth (from Arvidson, 1972). Upper limit for Mars corresponds to equatorial afternoon (low density) and lower bound to polar night (high density). Earth data from Bagnold (1941).

a given diameter will travel about 20 times faster on Mars than on Earth (Arvidson, 1972, 1973). Sagan (1973b) has deduced that saltating particles will travel at velocities equivalent to wind velocities as they reach the apex of their flight paths. First-order derivations by him suggest extremely high rates of movement and associated rapid erosion of Martian surface materials—possibly as high as one cm/year. Presumably, rates of travel by surface creep would also be high on Mars, since the saltating load imparts momentum to grains traveling by creep.

Another parameter becomes important when discussing rates of transport for material in suspension. If particle-settling velocities are lower in magnitude than upward components of atmospheric winds, then the particles will stand a good chance of remaining aloft in suspension. To a first approximation, rates of travel will be identical to wind velocity. Settling velocities, in the absence of turbulence, can be estimated by considering that, when particles approach terminal velocities, the drag caused by atmospheric flow around the particles becomes equal to particle weight. For dust to silt size ranges this constraint leads to the following solution:

$$V_s = \frac{\sigma g d^2}{18\mu}\left(1 + \frac{2B\lambda}{d}\right) \qquad (3)$$

where: V_s is the terminal settling velocity; σ the par-

ticle density; g the gravitational acceleration; d the particle diameter; μ the atmospheric dynamic viscosity; B a constant; and λ the atmospheric molecular mean free path. Atmospheric density does not appear in this equation because the drag around the particle can be described by the fluid viscosity term, μ.

Figure 7.10 is a plot of solutions for particle terminal settling velocities under Martian and terrestrial conditions. In contrast to velocities needed to raise material from the Martian surface, vertical velocities needed to keep smaller particles suspended are relatively low. Particle sizes that would be suspended depend, of course, on the turbulence characteristics of atmospheric flow. However, it is probable that transport of fine-sized debris is an important mechanism on Mars.

In summary, theoretical deductions suggest that very high winds are needed to move and entrain loose material on Mars. High velocities also imply high rates of particle erosion and transport. When combined with relatively low settling velocities, fine particles can easily be maintained aloft. In short,

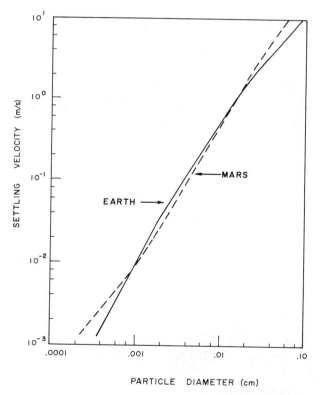

Figure 7.10. Settling velocities plotted over a range of particle sizes for Mars and Earth (from Arvidson, 1972). Note the similarity in settling velocities compared to the large difference in threshold drag velocities needed to first initiate particle movement (fig. 7.9).

Mars should have an active eolian regime if strong winds are present and if suitable particulate matter exists on the surface.

Dust Storms and High Winds on Mars

As previously noted, yellow dust clouds have been sighted on Mars by telescopic observers for almost a century. The largest storms occur shortly after perihelion, which coincides with early summer in the southern hemisphere. Due to the eccentricity of the Martian orbit, solar heat received during perihelion increases by 45 percent over that received at aphelion. Thus, considerably more energy is pumped into the atmosphere during this period, with much of it directed toward the southern latitudes. The net result seems to be an abundance of winds with velocities sufficient to entrain surface debris, especially around the perihelic subsolar latitude of 15°S. It should be noted, however, that there may be a bias introduced into telescopic cloud records that is due to favorable observing conditions during perihelic oppositions and unfavorable conditions during aphelic oppositions. During perihelic oppositions, the apparent diameter of Mars in Earth's sky is approximately twice as great as for aphelic oppositions. Thus, cloud occurrences may at least in part be clustered near perihelion because chances of observing them are greater.

Before the Mariner 9 mission there was considerable doubt about the effectiveness of eolian processes on Mars. Numerical simulation of Martian atmospheric dynamics, assuming a sphere without topo-

graphic irregularities, suggested that most winds above the boundary layer were significantly below velocities needed to erode particulate matter (Leovy and Mintz, 1969). It was not until Gierasch and Sagan (1971) demonstrated that regional slopes on Mars provide peculiarly good topography for generation of high velocity winds that some quantitative support for eolian activity was presented. Hess (1973) provided a post-Mariner 9 assessment of mechanisms to raise high winds on Mars. He concluded that a small fraction of regional winds, together with slope-generated winds and dust devils, could adequately explain the distribution of telescopically observed dust clouds.

Indirect information regarding the size of particles swept aloft during dust storms comes from a variety of sources: infrared spectrometry (Moroz and Ksanfomaliti, 1972; Pollack et al., 1974; Conrath, 1975), polarimetric measurements (Veverka et al., 1973), and trend measurements of surface contrast (Hartmann and Price, 1974). These several analyses suggest particle diameters from 0.5 to 2.0 microns, the equivalent of very fine dust.

Eolian Streaks and Splotches

When Mariner 9 first viewed Mars in November 1971, the dust storm that had begun two months earlier obscured most of the surface. A few surface details were visible on high regions, which turned out to be volcanic massifs on the Tharsis Plateau. The south polar cap was also visible, primarily because of albedo contrast between the bright ice of the cap and adjacent dark materials. Pictures in other regions, especially cratered terrain, contained vague circular bright spots, which were later documented as crater interiors (fig. 7.11). When the storm cleared and systematic photography of the Martian surface began, evidence for redistribution of surface material was widespread. Bright and dark streaks were seen to extend from many craters (fig. 7.12). In addition, a number of larger craters had dark splotches on their floors. Many of the splotches were located preferentially on the sides of craters from which streaks emanated, suggesting that splotches and streaks were formed by the same wind systems (fig. 7.13). Convincing evidence for eolian origins of dark splotches was contained in B-frame pictures that resolved a number of splotches in the higher southern latitudes into transverse dune fields, with dimensions similar to those on Earth (Cutts and Smith, 1973) (fig. 7.14).

Mariner 9 investigators repeatedly photographed a number of areas of streaks and splotches to look for changes in brightness and shapes of the features (Sagan et al., 1972; 1973b). Using a picture-differencing technique, which reveals picture-to-

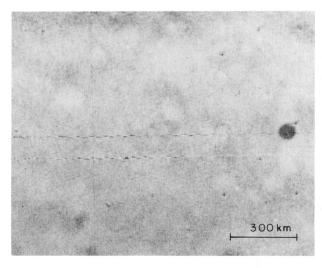

300 km

Figure 7.11. A photograph taken early in the Mariner mission showing circular bright spots. Correlation with pictures of the same region, taken after the atmosphere cleared, indicates that bright spots are crater interiors. (DAS No. 01885070, rev. 8, A camera, center at 189°W, 38°S.)

(a)

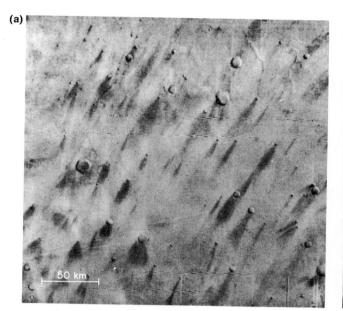

(b)

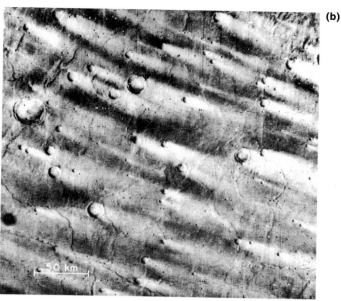

Figure 7.12 (a). Mariner 9 frame covering the Bosporus region of cratered plains. Dark streaks emanate from most of the craters. (DAS No. 06210998, rev. 129, A camera, center at 62°W, 34°S.)

(b). Mariner 9 frame of the Hesperia region of cratered plains. Bright streaks extend downwind from most craters. (DAS No. 07578308, rev. 167, A camera, center at 242°W, 23°S.)

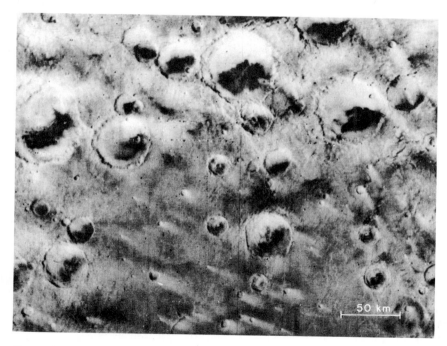

Figure 7.13. Mariner 9 frame in the Hesperia-Mare Cimmerium region. Bright streaks emanate from bowl-shaped craters, while flat-floored craters have dark downwind splotches and bright upwind floors. Note the correlation in trend between streaks and directions that bisect splotches normal to crater rims. (DAS No. 07578448, rev. 167, A camera, center at 237°W, 14°S.)

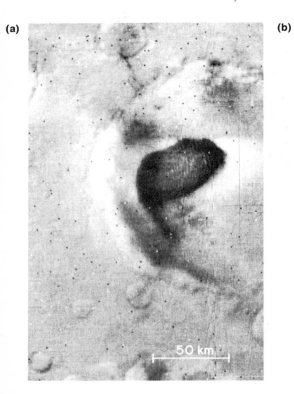

(a)

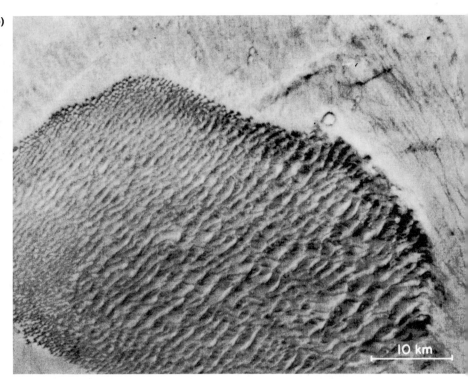

(b)

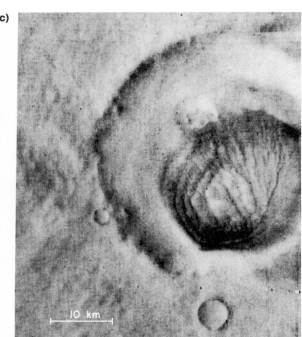

(c)

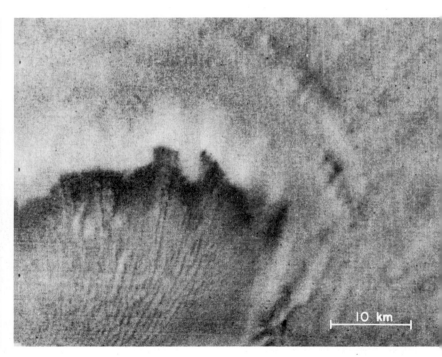

Figure 7.14 (a). A dune field in the Hellespontus region, appearing as a dark splotch at A-frame resolution. Transverse dunes are ponded against the wall of a large flat-floored crater. (DAS No. 05598988, rev. 112, A camera, center at 329°W, 45°S.)

(b). The same dune field as in (a), shown at B-frame resolution. Transverse dunes in the central part of the dune field are spaced 1–2 km apart. Smaller equidimensional dunes are visible at the margin of the dune field. Similar patterns are revealed in terrestrial dune fields (fig. 7.4c). (DAS No. 09807429, rev. 229, B camera, center at 330°W, 48°S.)

(c). Dark-floored craters with anomalous curving and concentric features that may be dune crests, but also may have a structural or volcanic origin (Cutts and Smith, 1973). (DAS No. 05921933, rev. 121, B camera, center at 206°W, 65°S.)

picture variations in brightness, they found that only dark streaks and some dark splotches changed; bright streaks showed no observable changes during the five-month monitoring period (fig. 7.15). These results suggest that, once bright streaks form, they are relatively more stable than dark features. However, even bright streaks and patches have been observed to change in the two-and-a-half-year interval

between Mariner 6–7 and Mariner 9 photography (Veverka et al., 1974b).

If one assumes that bright material is a fine-grained weathering product, then the relative stability of bright and dark streaks may be explained in part by grain size differences. Just as on Earth, fine-grained material would be difficult to move because high winds are necessary to compensate for the lack

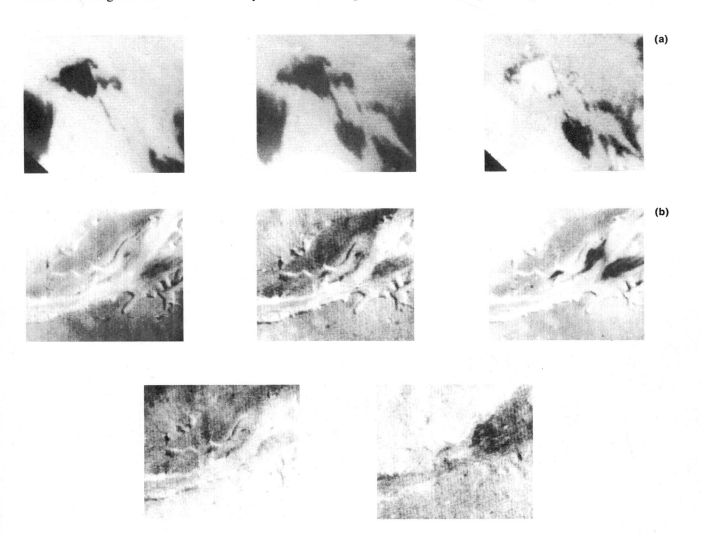

(a)

(b)

Figure 7.15. Albedo changes observed during the course of the Mariner 9 mission.

(a). Darkening of a feature shaped like a spearhead. The picture at the left was taken on rev. 99; the one in the center on rev. 126, 13 days later. The picture at the far right represents the brightness values for the left picture subtracted from those for the center picture. The darkening of the spearhead and brightening of the spear "butt" are clearly shown (from Sagan et al., 1973b; Stanford AIL picture STN 0167-050610). (DAS No. 05130443, rev. 99, B camera, center at 254°W, 71°S; DAS No. 06101693, rev. 126, B camera, center at 256°W, 71°S.)

(b). Albedo changes in Kasei Vallis. The three pictures in the top row are taken (from left to right) on rev. 125, 160, and 238. Shown below are the brightness for rev. 160 minus rev. 125, and rev. 238 minus rev. 160. During the first interval there is a slight brightening of the canyon floor, during the second interval a pronounced darkening (from Sagan et al., 1973b; Stanford AIL picture STN 0166-042911, 042912). (DAS No. 06068338, rev. 125, A camera, center at 65°W, 21°N; DAS No. 07327848, rev. 160, A camera, center at 66°W, 21°N; DAS No. 10133524, rev. 238, A camera, center at 64°W, 22°N.)

of drag on the particles. Also, fine-sized grains could settle into the interstices of rough surfaces, becoming protected from further redistribution. Thus, bright streaks on Mars may be deposits of relatively immobile dust and silt.

The relative stability of bright streaks may also be in part due to thickness of bright dust deposits. Bright streaks may record areas in which there has been substantial accumulation of bright, fine-grained material in the lee of craters. Elsewhere, bright dust is probably deposited in a thin blanket. Dark streaks could form by generation of high-velocity turbulent winds downwind of craters, quickly eroding thin dust blankets and exposing darker lag deposits or bedrock. Bright streaks would be more difficult to modify in similar time periods because a thicker deposit would have to be eroded. As we shall see, however, some dark streaks also form by streaming of dark materials out of crater floors, with deposition in a dark swath extending in a downwind direction.

One way of placing constraints on mechanisms of formation of streaks and splotches is by analysis of crater types in and around which these features form. In figure 7.16, diameter distributions for craters with streaks and splotches are plotted, along with the distribution of bowl-shaped craters and flat-floored craters. The dichotomy between streaked and splotched craters mimics the general dichotomy between bowl-shaped and flat-floored craters, suggesting that crater size and shape determine the pattern of eolian sediment accumulating in and near craters.

Bowl-shaped craters have prominent raised rims, placing them among the sharpest topographic features on the Martian surface. With high winds, they probably act as centers for downwind shedding of complex vortex tails. Consider the diagram shown in figure 7.17. If wind with velocity, V, traverses the crater and surrounding areas, then turbulence will produce two zones with velocities higher than V downwind of the crater (Greeley et al., 1974). The result should be a higher rate of particle erosion downwind of craters, eventually leading to a lag surface or exposed bedrock. In other words, a dark streak will be produced.

As discussed, bright streaks are probably areas downwind of craters where preferential deposition rather than erosion has occurred. One mechanism of accumulation would be to have dust trapped in craters during storms, only to be blown out later (Sagan et al., 1973b). An equally plausible mechanism would be to have a dust-laden atmosphere moving at low velocity (Sagan et al., 1973b; Arvidson, 1974b). Downwind turbulence would be reduced, and wind shadows would be set up behind obstacles such as bowl-shaped craters. This is a common phenomenon on Earth and leads to preferential deposition of sediment downwind of obstacles (Bagnold, 1941). On Mars, this phenomenon also may lead to accumulation of bright material behind craters.

In contrast to bowl-shaped craters, most flat-floored craters are rimless and relatively shallow. Consequently, downwind turbulence is probably considerably reduced. Thus, for flat-floored craters the major wind perturbation probably occurs when the atmosphere passes from intercrater area, over the crater wall, to the crater floor. Some fraction of transported debris probably is deposited against the

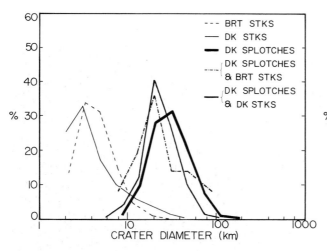
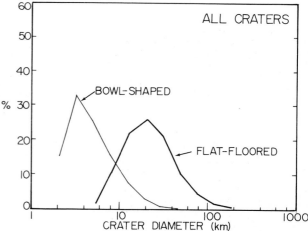

Figure 7.16 (a). Crater diameter distributions for craters with streaks and craters with splotches. Derived from Mariner 9 A frames for the whole planet. (See appendix B.)
   (b). Crater diameter distributions for bowl-shaped and flat-floored craters. Note the striking correspondence with data in (a). Derived for whole planet from Mariner 9 A frames.

Figure 7.17 (a). Schematic flow field over a raised-rim crater, showing horseshoe vortex pattern and zones of relative erosion and deposition. Wind tunnel simulations suggest that deposition occurs immediately behind downwind crater rims and erosion dominates outer areas where the vortex is active. (From Greeley et al., 1974.)

(b). Erosional depressions form in the lee of a 18-cm-diameter crater during wind-tunnel tests. Wind velocity was 420 cm/sec. Similar dark lobes are seen adjacent to a small Martian crater. (From Greeley et al., 1974.)

(c). Sand collects preferentially in the lee of a crater during wind-tunnel tests. Wind velocity was 850 cm/sec. A similar pattern is seen adjacent to a Martian 2-km crater. (From Greeley et al., 1974.)

(a)

FLOW OVER A RAISED RIM CRATER

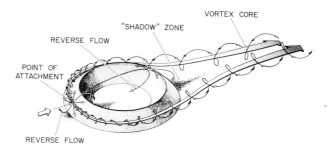

(b)

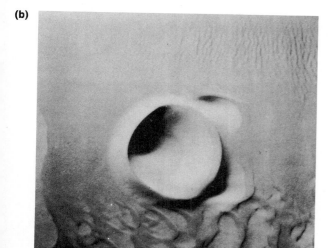

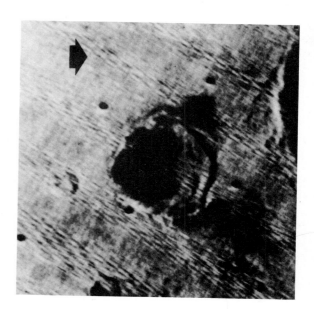

(c)

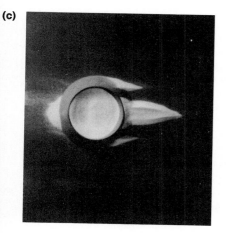

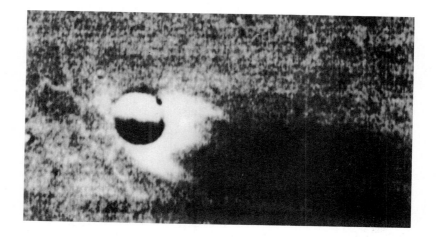

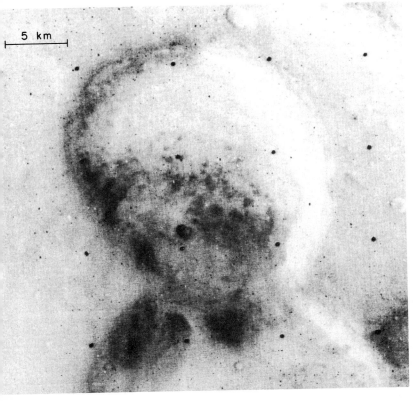

Figure 7.18. Dark splotches within craters that are feature-less at B-frame resolution. Other splotches have dune-like texture, as in figure 7.14. (DAS No. 08477009, rev. 192, B camera, center at 331°W, 38°S.)

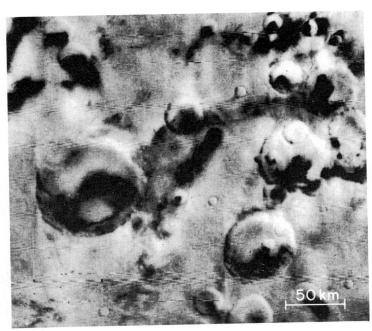

Figure 7.19. Portion of a Mariner 9 frame just southeast of the Promethei Sinus region. Most craters have dark splotches on downwind sides of floors. Several craters have dark streaks associated with the splotches, suggesting that dark material has migrated over subdued rims and walls and was then carried downwind. (DAS No. 06533768, rev. 138, A camera, center at 237°W, 61°S.)

upwind wall during this transition. This model of deposition is consistent with the observation that most flat-floored craters in areas with streaks have upwind bright deposits.

Occasionally, especially in the higher southern latitudes, dark areas on downwind sides of flat craters can be resolved into dune fields (Cutts and Smith, 1973) (fig. 7.14). In other cases, the dark areas are featureless at B-frame resolution (fig. 7.18). Sand and gravel on crater floors would tend to migrate downwind, but unlike dust, these grains would have difficulty migrating over high crater walls since they do not travel in suspension. Consequently, if winds are fairly unidirectional, accumulations should form that are ponded against downwind walls. Thus, flat-floored craters could contain dust deposits on upwind sides of the floor, and dune accumulations on downwind sides. If blanketing dust deposits are, at most, a few centimeters in thickness, as suggested by Greeley et al. (1974), then bright upwind deposits probably are extremely variable in shape and extent relative to thicker dune accumulations. If no source of sand or gravel exists on the crater floor, then downwind areas would consist of lag deposits or exposed bedrock.

One might expect that fresher (deeper, raised-rim) flat-floored craters would tend to form more dark splotches because they have profiles conducive to perturbing wind systems and trapping sediment. In fact, this is what is observed. Splotches are three times more abundant in fresher flat-floored craters than in shallow, rimless craters (Arvidson, 1974b). In very shallow craters, dune material will more easily migrate over downwind walls. Consistent with this hypothesis, there are numerous examples of dark streaks emanating from splotches in very degraded craters (fig. 7.19). In these cases, the streaks for the most part are sand or gravel deposits that were temporarily ponded in craters before migration over downwind walls.

Dust Storms and Sediment Redistribution Patterns

Both streaks and splotches are indicators of wind directions prevalent during their formation. As such, they provide data on wind systems and patterns of global sediment redistribution. These patterns are illustrated in figures 7.20 to 7.22, which are computer-drawn meridianal stereographic projections of streaks and splotches, accompanied by Lambert equal-area plots showing density of streak and splotch occurrences. The meridianal stereographic projection is a perspective projection formed conceptually by projecting a point source from the equator onto a plane tangent to the sphere at a point 180°

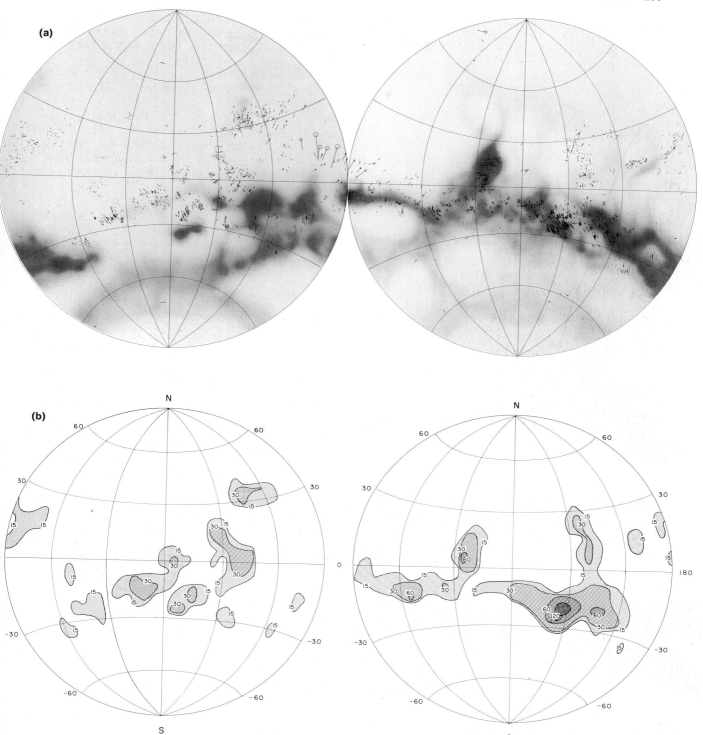

Figure 7.20 (a). Equatorial stereographic projection of craters and associated bright streaks. Craters are plotted at real map scale; streaks are plotted at 2.5 times map scale to improve visibility. Data were compiled from Mariner A frames. The background is a qualitative rendering of brightness variations, as they appeared in 1971.

(b). A Lambert equal-area projection of craters with bright streaks. Isolines were generated by measuring crater densities within a 10°-diameter search circle moved around the globe on 5° centers. Comparisons with the physiographic province map (fig. 3.2) shows that most streak concentrations are in areas of cratered plains.

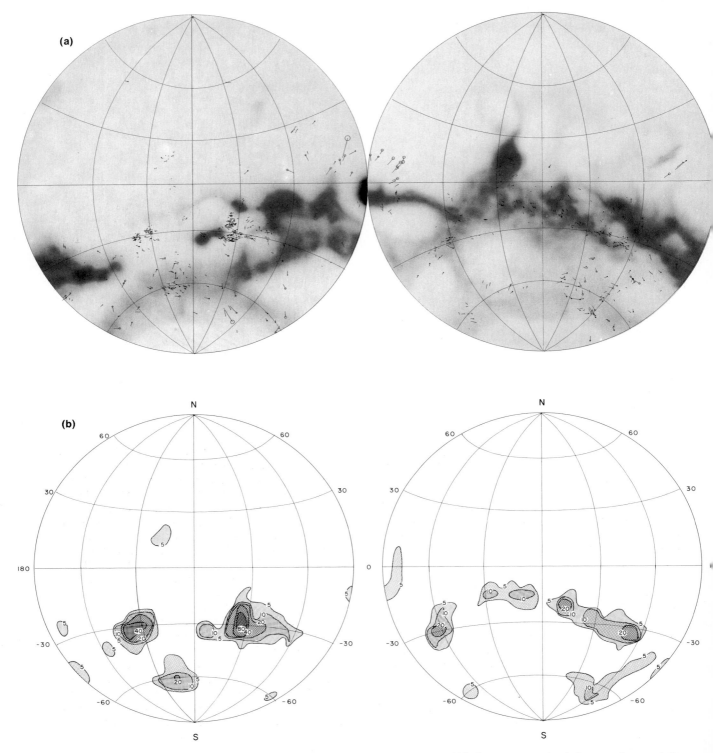

Figure 7.21 (a). Equatorial stereographic projection of craters and associated dark streaks. Craters are plotted at real map scale; streaks are plotted at 2.5 times map scale to improve visibility. Data were compiled from Mariner A frames. The background is a qualitative rendering of brightness variations as they appeared in 1971.

(b). A Lambert equal-area projection of craters with dark streaks. Isolines were generated by measuring crater densities within a 10°-diameter search circle moved around the globe on 5° centers. Concentrations occur at different positions than for bright streaks (fig. 7.20b). Comparison with physiographic province map (fig. 3.2) shows that regions of dark streak concentrations are not restricted to one terrain type. They generally occur either in cratered terrain or in cratered plains.

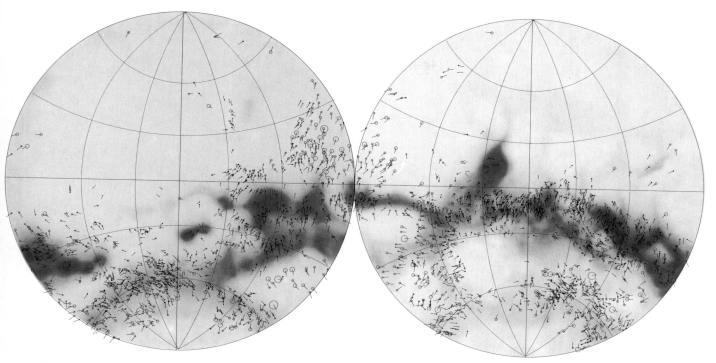

Figure 7.22. Equatorial stereographic projection of craters with dark splotches located on one side of crater floors. Craters are plotted at real scale. The trends of dark splotches are indicated by 75 km-long lines bisecting the dark splotches.

away. This projection was chosen to display streaks and splotches for several reasons. First, true angles are preserved. Second, unlike the mercator projection, where the poles are at an infinite distance from the equator, whole hemispheres can be plotted. Third, the stereographic projection, since it does include whole hemispheres, resembles the usual orthographic telescopic photographs of Mars.

A great deal of information is evident on these plots relative to spatial and trend distributions of streaks and splotches. Some of this is summarized in figure 7.23, which is a series of rose diagram plots for streaks and splotches divided into 20° latitudinal bands. There is an excellent correlation between trends of bright streaks and those of dark splotches in the northern to southern midlatitudes, a result also obtained by Sagan et al. (1973b). In the 30°–50° southern latitude band, dark splotch trends seem to include both trends of bright streaks and dark streaks. However, in the 50°–70° southern latitude band, splotch trends closely follow trends of dark streaks.

As discussed, major dust storms on Mars usually begin in the southern latitudes and then disperse to cover large areas of the planet. Often, but not always, an east to west movement is discerned in the first few weeks of the storm. The directional trend and latitudinal distribution of dark streaks suggest that some may be direct relicts of passing storms. Bright streaks may then be records of transport systems prevalent after the storm has spread to planet-wide proportions. In fact, there is a good correlation between trends derived from bright streaks and wind systems predicted from the atmospheric temperature gradient measured by the Mariner 9 IR experiment (Sagan et al., 1973b; Hanel et al., 1972). Thus, most material eroded in the southern hemisphere and dispersed over the planet seems to be returned to the southern hemisphere by a net north-to-south surface flow. This phenomenon explains why major storms continually form in the southern latitudes without completely denuding the surface of mobile dust. In addition, Hellas, which is a 1600-km diameter basin located about 7 km below the mean elevation of the surface, probably acts as a large trap for dust. When meteorological conditions are conducive, large amounts of dust probably escape from Hellas, marking the beginning of a new Martian dust storm.

In the high southern latitudes, dark streaks and splotches trend in a northwesterly direction, consistent with flow away from the south pole, modified by the Coriolis effect. The latitudinal band in which this trend occurs extends to about 30°S latitude.

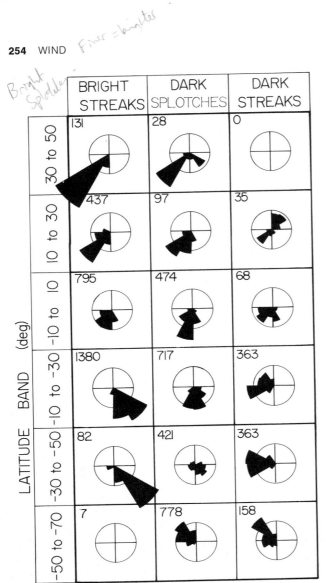

	BRIGHT STREAKS	DARK SPLOTCHES	DARK STREAKS
30 to 50	131	28	0
10 to 30	437	97	35
-10 to 10	795	474	68
-30 to -10	1380	717	363
-50 to -30	82	421	363
-70 to -50	7	778	158

LATITUDE BAND (deg)

Figure 7.23. Normalized trend distributions for bright streaks, dark splotches, and dark streaks shown in the previous three figures. Numbers in upper left of boxes represent total features used in constructing rose diagrams. Angular interval is 20°. Very few features were found above 50°N and below 70°S.

3.2), illustrate that the streaks preferentially occur in regions of cratered plains (pc and pm). This suggests that streak formation is controlled not only by atmospheric conditions but also by bedrock characteristics. The cratered plains are generally flat surfaces, which are interrupted only by the upraised rims of bowl craters. In intercrater areas, basalt strata may be overlain by a thin mantle of fine-grained, bright dust. The simple topography confines atmospheric perturbation to the lee of craters, and the albedo contrast between bedrock and sediment results in easy identification of streaks. It is possible that wind conditions are generally similar at many other localities but that streaks are not observed, either because rugged topography leads to more complicated turbulence patterns, or because there is little brightness contrast between mobile sediment and immobile sediment.

Equatorial Erosion, Polar Deposition, and the Debris Mantles

Redistribution of surface material by wind is an ongoing process on Mars. But, what about large-scale effects of erosion and deposition? Are there regions that have been subjected to massive wind excavation or burial? In contrast to Earth, has the Martian landscape been modified to any great extent by wind action?

Both the north and south Martian polar regions are covered with an accumulation of ice and wind-deposited sediment, perhaps several kilometers thick, with the southern polar deposits covering an area of about 1.5×10^6 km² and the northern polar deposits an area of approximately 1.1×10^6 km² (Cutts, 1973b). At least two groups of deposits can be resolved: (1) a massive (nonlayered) deposit that is now etched by pits and hollows and makes up the unit termed etched plains (fig. 7.24); and (2) a layered deposit that unconformably overlies the etched plains material (fig. 7.25). The layered deposit has also been eroded, forming a rolling topography with valley walls composed of benches separating individual layers or groups of layers.

Wind action has apparently also been responsible for eroding the polar material. The result has been migration of eroded debris toward the equator. In the near-polar latitudes, this process is documented by mantled and excavated terrains (fig. 7.26). A debris cover thins away from the poles to a thin-to-nonexistent cover at about ± 30° latitude, although placement of the boundary is difficult because of the obscuring effects of haze (fig. 7.27). Detailed discussion of the effect of mantling on crater populations appears in chapter 4. Suffice it here to mention that the effect is one of softening the topography,

This is also the limit of the debris mantle, which, according to Soderblom et al. (1973b), has been derived from the south polar regions. The trend displayed in the high southern latitudes may also have ramifications for the famous Martian wave of darkening. The wave starts at the pole during the spring season and advances toward the equator. The trend data presented here are consistent with the model proposed by Sagan and Pollack (1967), in which the wave of darkening is due to wind-blown removal of a thin layer of bright dust.

The equal-area density plots of figures 7.20–7.22, when compared with the geologic map of Mars (fig.

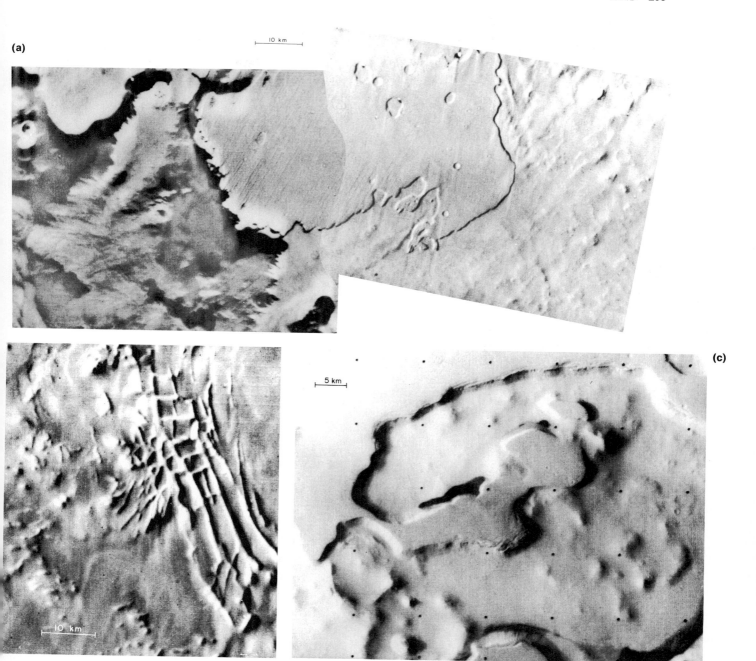

(a)

(c)

Figure 7.24. Some examples of landforms displayed in regions of etched plains.

(a). Pitted and striated layer overlying an irregular, sediment-mantled surface. The upper layer was once more extensive, but has been partly stripped away by erosion. Striae are probably formed by eolian scouring. (DAS No. 06173723, rev. 128, B camera, center at 324°W, 66°S; DAS No. 06245613, rev. 130, B camera, center at 326°W, 66°S.)

(b). Distinctive rectilinear landforms near the south pole. These regularly arranged ramparts were informally called "Inca City" during the Mariner mission. The ridges may be underlain by igneous dikes that are more resistant to erosion than are surrounding sediments. (DAS No. 08044333, rev. 180, B camera, center at 64°W, 81°S.)

(c). Detailed view of a planar stratum overlying an irregular, sediment-mantled surface. The scarp bounding the uppermost layer is receding by fracturing and slumping. (DAS No. 08763869, rev. 200, B camera, center at 358°W, 71°S.)

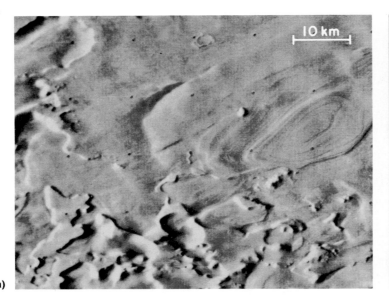

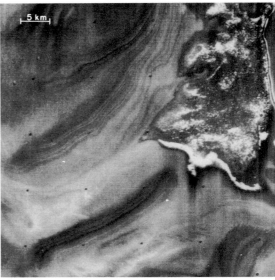

(a)

Figure 7.25. Some examples of landforms displayed in regions of layered plains.

(a). Etched plains overlain by an elliptical erosional remnant of layered plains. The individual layers stand out in topographic relief. (DAS No. 09231189, rev. 213, B camera, center at 85°W, 82°S.)

(b). Details of horizontal strata within layered terrain. Brightness differences between strata are in part caused by topographic relief, but also may be the consequence of intrinsic albedo differences. A patchy region of frost and sediment is revealed in a local depression. (DAS No. 09734909, rev. 227, B camera, center at 53°W, 83°S.)

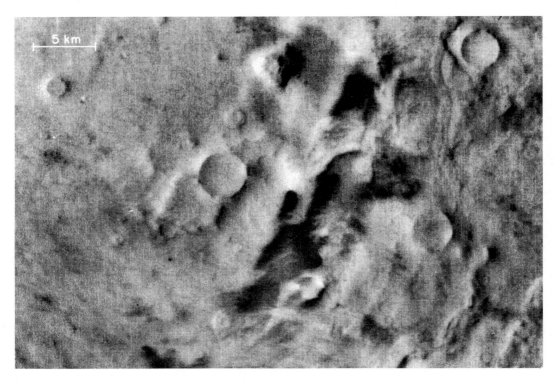

Figure 7.26. The entire region has been mantled with sediment. Subsequent erosion has stripped material between craters, but sediment within craters has been protected by crater walls. The unusual result is that the "floors" of the craters actually stand higher than the surrounding intercrater areas. (DAS No. 08764149, rev. 200, B camera, center at 321°W, 65°S.)

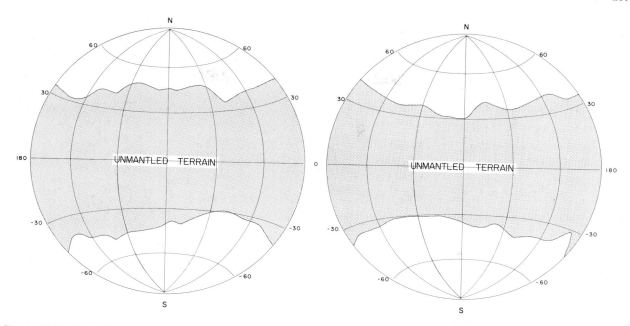

Figure 7.27. Latitudinal distribution of areas that, according to Soderblom et al. (1973b), have a cover of eolian debris eroded from polar regions. Thickness of cover (shown unshaded) probably decreases from poles to a thin-or-nonexistent cover at ± 30° latitude.

with a net decrease in the population of small bowl craters, terraces in larger craters, ejecta blankets, hummocky floors, and central peaks.

An obvious constraint for any mass redistribution model is that, if material is deposited in one place, it must have been removed from another. In other words, where did the polar sediments originate? Delineation of a particular source is difficult because the midlatitudes and equatorial regions abound with evidence for mass removal. For instance, the majority of larger flat-floored craters are shallow and rimless, and intercrater areas are generally smooth, indicating substantial erosion. The major canyons in the equatorial regions have approximately the same volume as that for the polar deposits (Cutts, 1973b). Their excavation could have been accompanied by transport of material to polar regions. The boundary between cratered terrain and most plains units on Mars is marked by an erosional escarpment. Proceeding from cratered terrain to plains, polygonal cracks, separate mesas, and finally outlines of isolated buttes are visible. The result is the formation of topographically low, intensely eroded terrain. Again, some of this material eroded from the cratered terrain may be incorporated in polar sediments. Finally, it is possible that volcanic eruptions have contributed significant material for incorporation in polar areas. Perhaps the grooved terrain adjacent to Mons Olympus represents the wind-eroded surface of an originally huge volcano. If so, some of this material may have contributed to polar deposits.

While potential sources for polar sediments can be found with relative ease, more difficult tasks are to determine (1) why the older deposits are massive, (2) why the younger ones are layered, and (3) why polar regions are today sites of erosion rather than deposition. Poles are cold traps on planets that have atmospheres. Material carried in the atmosphere would descend and be deposited at polar regions along with CO_2 and H_2O volatiles. Deposits would begin to build up and layers would form if the cyclic periods of deposition proposed by Murray et al. (1973) dominated. However, this model does not explain formation of the massive, apparently unlayered deposits underlying the layered plains. These materials can best be explained by having sediment influx rates remain constant over time. There would then be little chance for distinct layers to form.

Wind Erosion of Crater Ejecta Deposits

As discussed in chapter 4, crater statistics, when combined with cratering rate estimates, provide information on the history of surface processes. Crater ejecta deposits can be used in a similar way. Figure 7.28 shows one of the few rayed craters visible in Mariner 9 pictures. The vast majority of ejecta deposits have much more degraded appearances, probably a consequence of eolian modification. For instance, craters in figure 7.29 are surrounded by escarpments. The pedestal appearance is probably

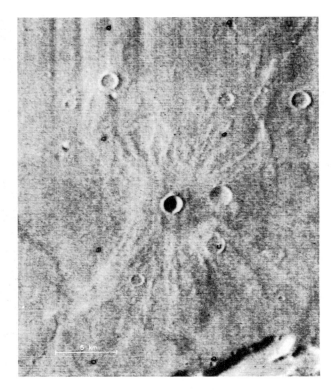

Figure 7.28. One of the few rayed craters visible on Mariner 9 pictures. The crater is surrounded by a continuous ejecta deposit that extends further outward as discrete patches and rays. Ejecta and ray material appears to be standing slightly above the surrounding terrain. (DAS No. 07975243, rev. 178, B camera, center at 343°W, 38°N.)

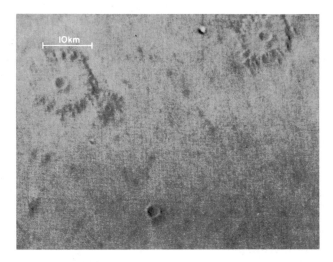

Figure 7.29. This frame contains some of the few pedestal craters found near the equator. Ejecta deposits around the craters probably act as surficial armor. Intercrater regions apparently have been lowered by wind erosion, leaving craters and ejecta on a higher topographic level. (DAS No. 12500070, rev. 458, B camera, center at 152°W, 5°N.)

due to: (a) low deflation rates for ejecta deposits, and (b) relatively high rates of removal of intercrater materials (McCauley, 1973). Figure 7.30a shows two craters surrounded by ejecta deposits, one of which exhibits a continuous blanket with a trough-ridge topography radial to the crater. The other crater is similar in size and freshness, but the ejecta has been eroded back to very near the crater rim. Figure 7.30b is an example of an ejecta deposit being eroded back toward the crater in a more irregular manner.

In general, ejecta deposits on Mars seem to degrade by erosion that begins at the edge of the continuous blanket and works in toward the crater rim. Thus, measurements of ejecta blanket and crater diameters may provide information on states of ejecta modification. For example, figure 7.31 shows that ejecta blanket diameters are larger and that pedestal craters are more abundant in higher latitudes. This trend follows the trend of increasing thickness of eolian debris mantling at higher latitudes (fig. 7.27). A plausible explanation for the correlation relates to the manner in which ejecta is eroded. In equatorial regions, craters probably form in eolian lag deposits or bedrock, excavating material along ballistic trajectories. Upon landing, the secondary projectiles excavate and mix with local material, forming ejecta blankets (Oberbeck, 1975). Subsequent scarp recession begins at the edge of the continuous blanket and works back toward the crater rim. The base of the scarp probably rests upon the pre-impact surface, and scarp recession may be maintained by eolian erosion of material from the escarpment base.

In higher latitudes, on the other hand, a significant thickness of debris must be excavated during impact. Experimental studies demonstrate that projectiles impacting into a soil overlying a firm substrate induce a complex excavation sequence (Quaide and Oberbeck, 1968). Whatever the sequence, a relatively thick and widespread debris blanket, derived directly from the primary crater or by mixing during secondary cratering, apparently is emplaced. Very little deflation has occurred on the ejecta blankets of high-latitude craters, while intercrater material apparently has been deflated to great depths. If scarp recession is the dominant ejecta erosion mechanism, then a great deal of mass must be removed from the escarpment surrounding the pedestals in order to allow the ejecta blanket to recede a given distance toward the crater center. The rate should scale inversely to the escarpment height so that pedestal craters in areas of thickest debris should erode much slower than those forming in thinner debris deposits. Thus, craters in mid- to high-latitudes that formed in debris layers could have had their ejecta blankets preserved over a fairly

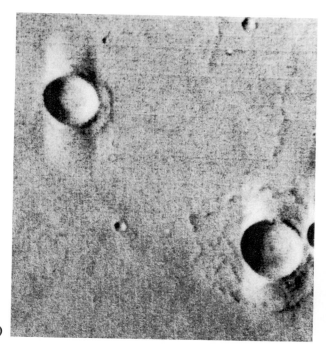

(a)

(b)

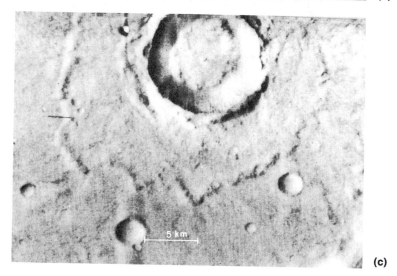

(c)

Figure 7.30 (a). A view of equatorial plains showing two craters of equal size and degree of freshness. One has an extensive ejecta blanket with concentric ridge-trough topography, in common with many lunar craters (fig. 4.11). The other crater's ejecta has been largely removed by erosion and presently exists only near the crater rim. (DAS No. 07003713, B camera, center at 298°W, 16°N.)

(b). Large crater in upper left part of frame displays an ejecta blanket that has been eroded back toward the crater rim in an irregular manner. Note that ejecta remnants are superimposed over an older crater. (DAS No. 06571953, rev. 139, B camera, center at 350°W, 17°N.)

(c). Crater with a distal rampart exposed by differential stripping of ejecta deposits. (DAS No. 06606813, rev. 140, B camera, center at 182°W, 17°S.)

Figure 7.31. Diameters of continuous ejecta blankets, normalized to crater size, and plotted against latitude. Measurements were made from B frames. Northern latitudes were not included because of the possibility of frost deposits preferentially emphasizing ejecta surfaces. Subjective judgments were made as to topographic relief between ejecta and intercrater terrain. Pedestal craters with exceptionally high relief are shown by solid circles. The two pedestals plotted for the equator are from figure 7.29. The rayed crater appears in figure 7.28.

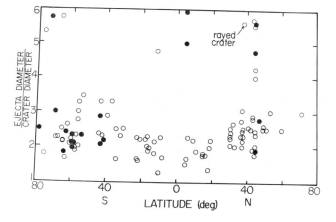

long time. Accordingly, ejecta blankets should be largest in areas of thickest debris.

A simplified model of ejecta erosion in equatorial regions can be derived if it is assumed that the rate of removal of ejecta material is controlled entirely by the rate at which wind removes material from the escarpment base (Arvidson et al., 1975b). This assumption leads to a concept known in geomorphology as parallel scarp retreat. Consider figure 7.32, a cross-section of a crater with an ejecta blanket. The rate of ejecta thickness decay away from the crater rim can be modeled by a power law. Constants A, B, and C for the function are known for nuclear blast craters on Earth and for some lunar impact craters (McGetchin et al., 1973; Arvidson et al., 1975a). The Martian situation probably requires intermediate values for the constants, since Mars is intermediate between Earth, with a high gravity field (982 cm/s²) and high atmospheric density ($\sim 10^{-3}$ gm/cm³) and the moon, with low gravity (186 cm/s²) and no appreciable atmosphere.

Assuming that no material is deflated from the ejecta surface and that the rate of removal of scarp material is a constant, equal areas of ejecta cross section in figure 7.32 will be removed in equal times. Integrating the power function from some starting ejecta diameter to some diameter after exposure to scarp recession, and then rearranging terms gives:

$$e_t = \left(e^a_{t_0} + \frac{\gamma \Delta t}{bd^c} \right)^{1/a} \qquad (4)$$

where: e_t = ejecta diameter after time $\Delta t = t - t_0$ of exposure to erosion, e_{t_0} = initial ejecta diameter, γ = area/time removed from ejecta cross-section, d = crater diameter, a, b, c = constants related to A, B, C.

Figure 7.33 is a plot of crater and associated ejecta dimensions for craters not located in the debris mantle. Lines of equal area removed from the ejecta cross-sections are, by our model assumptions, also lines recording equal times of exposure to erosion. The isochrons plotted in figure 7.33 assume a nuclear blast ejecta thickness profile. There are 18 craters larger than 6 km in diameter and less than t_2 in age. This corresponds to 2.49 × 10⁻⁵ craters/km², if it is assumed that B-frames cover one percent of the Martian surface. If it is also assumed that a lunar crater production rate of 2.55 × 10⁻⁴ craters larger than 6 km/km²/10⁹ years (Hartmann, 1973a) is applicable to Mars, then the two production rates can be equated, and t_2 can be calculated as $\sim 10^9$ years. Replacing Δt in equation (4) with this value and solving for γ gives a value of 5 × 10⁻⁵ m²/yr removed by scarp recession from nuclear-blast ejecta cross-sections. Using a lunar ejecta profile gives a value of 1 × 10⁻⁴ m²/yr. These estimates correspond to 5 × 10⁻⁴ to 10⁻³ m/yr for

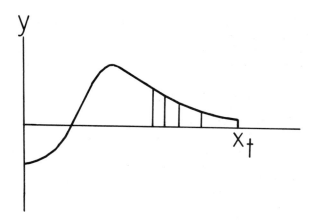

Figure 7.32. Schematic cross-section of a Martian crater with ejecta deposits. Rim uplift due to structural upwarping during impact is not shown. Decrease in ejecta thickness from the crater rim can be modeled by the power law displayed in the sketch. If ejecta erosion occurs by scarp recession and if the rate of recession is controlled by the rate at which wind removes material from the scarp base, then equal areas of ejecta cross-section would be removed in equal times if erosion rates were constant.

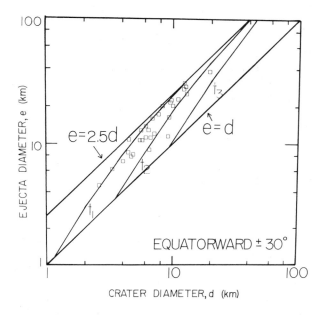

Figure 7.33. Diameters of the continuous ejecta blankets plotted as a function of crater size for craters that have not been blanketed by eolian debris. Lines corresponding to an ejecta diameter 2.5 times the crater diameter (assumed initial starting diameter of continuous blanket) and to no ejecta (i.e., ejecta diameter = crater diameter) provide limits of integration of the power function shown in figure 7.32. Lines representing t_1, t_2, and t_3 correspond to equal areas removed from ejecta cross-sections. By our model assumptions, these must also be isochrons.

scarp recession rates for a 10 m high ejecta escarpment. These results suggest that ejecta erosion rates have been low during recent Martian history. For comparison, scarp recession by bank undercutting in meanders of the Mississippi River reaches rates of 1 to 10 m/yr for similar-sized escarpments (Brunsden and Kesal, 1973).

Computed wind erosion rates using this model are necessarily order-of-magnitude estimates. We assumed that equal areas of ejecta cross-section are removed in equal times. This assumption is not strictly true, even if wind erosion rates have remained constant. An assumption was made that ejecta deposits lie on non-erodible substrate, so that scarp heights are equivalent to ejecta thickness. Most likely, scarps around ejecta, even in equatorial areas, are a combination of ejecta and some thickness of pre-existing mobile debris. Thus, the model provides a lower limit to the removal rate. This problem aside, results are significant in two respects. First, computed rates of ejecta erosion are very low and, by extrapolation, erosion rates for coarse-grained eolian lag deposits and of bedrock must be *very* low indeed. Second, pedestal craters imply high rates of intercrater debris removal and very low rates of deflation of ejecta surfaces. In sum, eolian dynamics on Mars are probably dominated by rapid redistribution of loose debris; very little new debris is being created, suggesting that most is a remnant of an earlier period of more vigorous erosion. This conclusion raises somewhat of a paradox since we said earlier that eolian erosion should be significant on Mars, even for rock surfaces. One likely explanation that resolves the contradiction is that the quantity of saltating material (sand) on Mars may be smaller than previously imagined. If so, then abrasion rates would also be considerably reduced. The lack of observable dune fields in equatorial and midlatitudes lends credence to this speculation.

Synthesis and Speculation

Even though the Martian atmosphere has a pressure of only about 5 mb at the surface, high velocity winds are capable of redistributing surface material. A global redistribution pattern is set up during large-scale storms like the one observed by Mariner 9 in 1972. Sediment seems to be eroded from the southern midlatitudes and spread as a cloud over most of the planet. However, surface winds return most of the material by a net north-south flow in the northern latitudes to southern equatorial latitudes. Winds also spiral away from the south pole, returning material northward (Arvidson, 1974b; Cutts, 1973a). The zone where these opposing trends meet is the southern midlatitudes, the belt where most large storms originate. Storms usually occur near

perihelion, which corresponds to southern summer, with the subsolar point located at 15°S. It appears that increased insolation received in and near this belt provides enough energy for winds to raise large amounts of material.

Soderblom et al. (1974) have pointed out that most dark areas on Mars are located just south of the equator (fig. 1.11). They speculate that the southern equatorial to midlatitudes have been partially stripped of bright dust during the numerous large storms that spread material over most of the planet. If not all the material is returned by surface flow, the result should be a darkening of lower southern latitudes and a net brightening of other regions. To substantiate their hypothesis, they point out that small (2–4 km) crater densities are greatest in the same latitude band in which storms are generated and most dark areas are located. They suggest that the density increase is causally related to excavation of craters buried by eolian debris. This claim is consistent with results discussed in the last section, where evidence was presented that craters are difficult objects to erode relative to softer sedimentary fill. Thus, the dominant crater degradation process on Mars today may be burial by eolian sediments. Since the subsolar point at perihelion periodically migrates back and forth from 25°S to 25°N because of orbital precession, the belt of dark areas on Mars also may migrate back and forth across the equator.

As discussed, polar regions are sites of thick deposits of eolian debris that were, at least in part, derived from erosion of equatorial and midlatitude regions. However, the polar deposits are now undergoing wind erosion, and debris is being shed toward equatorial regions by surface winds spiraling outward from the poles. Perhaps the massive deposits forming etched plains were laid down when there was a relatively dense atmosphere capable of rapid erosion of the equatorial and midlatitude regions. This may have provided a relatively constant influx of material to polar areas, forming thick, massive deposits. Layered deposits suggest periods of deposition followed by periods of nondeposition or even minor erosion. These materials may record cyclic perturbations of atmospheric pressure (and thus erosion and transport capabilities) caused by orbital precessions (Murray et al., 1973). However, it is difficult to envision the shift from layered deposition to the present polar erosion, if cyclic phenomena were based on present atmospheric conditions. More likely, layered deposits probably record dissipation of a denser atmosphere. Today, with relatively little equatorial erosion and consequently little polar deposition, winds are able to carve into the relatively soft polar deposits and transport them equatorward.

8.

Water

Introduction

OF ALL the questions associated with understanding the Martian environment, none has generated such a panoply of contradictory answers as the question of whether significant quantities of water exist, or have existed, on the planet's surface. The quest for an answer to this problem, which remains partly unresolved, has inspired considerable research and speculation, predominantly because of the apparent biological requirement for liquid water.

Because today's atmospheric conditions are characterized by low pressures and low temperatures, liquid water as a common substance on the Martian surface can be discounted. If present, water occurs only in very small quantities when frost, formed during the Martian night, melts more rapidly than it can sublimate and re-enter the atmosphere. Nevertheless, photographs returned from Mariner 9 reveal numerous features so strikingly similar to terrestrial river valleys that one is compelled to conclude either that some unknown—and currently unimaginable—process formed these features, or that sometime in the past conditions were conducive to the formation of liquid water.

Three main questions reappear in any discussion of the former importance of liquid water. First, how much H_2O is present in the Martian environment today—in the atmosphere, in the polar caps, and trapped beneath the surface? Second, how does the equilibrium between subsurface water, polar ice, and atmospheric water vapor respond to changes in planetwide and local temperatures? And third, did the apparently water-generated features form during periodic warm, wet intervals, or are they remnant from some singular ancient epoch?

Since conditions conducive to the stability of liquid water depend on the atmospheric composition, we will begin our discussion with a summary of the present knowledge of this aspect of the Martian environment. Then, because the stability of water is also dependent on pressure and temperature, we will review the phase relationships for water, carbon dioxide, and gas clathrates—especially as they are modified by changing temperature and pressure. This thermodynamic discussion will document the reasons for the present stability of CO_2 polar caps. Moreover, the discussion will demonstrate that substantial temperature changes, as well as a large addition of H_2O to the Martian environment, are prerequisites for stabilizing liquid water. Mechanisms for temperature variations will be discussed, in particular eccentricity and obliquity variations, and long-term variations in the solar luminosity. The thermodynamic discussion will show that insolation variations are insufficient to generate rainfall. Some change in the partial pressure of CO_2

may result from insolation variations, depending on the availability of CO_2 reservoirs. Increases in the partial pressure of CO_2 (essentially the total pressure) could stabilize $H_2O(l)$, if formed by subsurface melting, against rapid boiling, although rainfall is still prohibited. However, most geomorphic evidence for liquid water is consistent with the presence of an ancient dense atmosphere.

Consideration of a former dense atmosphere leads to a discussion of the *sources* and *sinks* of volatile materials, and to the more general question of the history of Martian and terrestrial volatiles. The major potential sinks for Mars include ground ice, adsorption of volatiles in the regolith, a large amount of frozen volatiles at the poles, and exospheric escape. This contrasts with Earth, where the major sinks are oceans and—rather surprisingly—minerals. Carbonates, for example, contain most of the CO_2 expelled from Earth's interior.

In the final sections of the chapter we will review geomorphological effects of terrestrial fluvial erosion and deposition, using this as background for analysis of Martian features that also may have formed by water erosion. It will be concluded that Mars experienced an episode of water erosion early in its history. This erosion was responsible for the crater obliteration event discussed in chapter 4 and possibly for the furrows described in this chapter. Ground ice then formed from downward percolation of surface water. Later, as a result of uplifting and local heating, the ice-capped ground water catastrophically burst forth to form many of the major channels. There is no geological evidence in the equatorial regions that requires a continuing alternation of "wet" and "dry" periods associated with changing orbital parameters.

Martian Atmospheric Composition

Today, as a result of Mariner spacecraft data and improved Earth-based spectroscopic observations, the model for the composition and structure of the Martian atmosphere approximately resembles the actual situation. (It was not always so!) The surface atmospheric pressure is now known to be around 6 mb (Kliore et al., 1965; Schorn and Gray, 1967), although surface topography produces variations from about 3 to 10 mb (Kliore et al., 1972b). The composition is mainly carbon dioxide (Belton et al., 1968; Carleton et al., 1969; Barker, 1967, 1971a; see also Kaplan et al., 1964). In addition, recent observations from a Soviet spacecraft, Mars 6, indicate that the inert gas argon may compose as much as several tens of percentage of the total atmospheric content. Wood (1974) has shown that an abundance of argon from 0 to 20 percent is consistent with Mariner spacecraft observations.

N_2, the major constituent of Earth's atmosphere, represents at most only a few percent of the Martian atmosphere (Dalgarno and McElroy, 1970), and probably less. This absence of nitrogen is attributed by Brinkmann (1971) and McElroy (1972) to nonthermal escape from the atmosphere resulting from photolytic dissociation of molecular nitrogen. The process of photolysis will be discussed later in connection with the evolution of the Martian atmosphere.

Minor atmospheric constituents include O_2, CO, and H_2O. Molecular oxygen, O_2, results from photo-dissociation of both CO_2 and H_2O in the upper atmosphere (McElroy, 1972). O_2 subsequently diffuses to lower altitudes, reaching a concentration of 1.3×10^{-3} that of CO_2 near the surface (Belton and Hunten, 1968; Carleton and Traub, 1972). Carbon monoxide, also formed as a dissociation product, has an abundance 8×10^{-4} that of CO_2 (Kaplan et al., 1969).

The atmospheric abundance of water vapor has been reported to average about 35 precipitable microns (Kaplan et al., 1964; Owen and Mason, 1969) with seasonal and diurnal fluctuations shifting the value from about 10 to 50 microns (Barker et al., 1970; Schorn et al., 1967, 1969; Tull, 1970; Barker, 1971b; Conrath et al., 1973; Farmer, 1971). The CO_2 content of the atmosphere is also seen to fluctuate with the season. Barker (1971b) concluded that the total atmospheric pressure varies by a factor of 2 in response to the seasonal formation and dissipation of the polar caps.

Historical "Evolution" of the Martian Atmosphere

Even with some uncertainty about the abundance of argon, our present knowledge of the Martian atmospheric composition is vastly superior to that of twenty years ago.

In his 1954 book, deVaucouleurs lists the Martian atmospheric constituents as 98.5 volume percent N_2, less than 0.1 percent O_2, about 1.2 percent Ar, and 0.25 percent CO_2. Polarimetric and photometric observations in the late 1940s and early 1950s were interpreted to indicate a total pressure of about 80 mb, compared to the currently accepted value of 6 mb. The only observed gas species was CO_2, first discovered by Kuiper (1949). By subtracting Earth's contribution to the CO_2 absorption bands seen in lunar spectra, Kuiper found that there was approximately twice as much CO_2 in an atmospheric column on Mars as on Earth. H_2O was below detectable limits with the instruments then in use. The remaining bulk of the atmosphere was believed to be spectrographically undetectable species, such as N_2, Ar, and the other rare gases. Because of its abundance in the terrestrial atmosphere, N_2 was considered the most likely of these, and therefore the predominant species in the Martian atmosphere. ^{40}Ar was independently estimated based on the expected amount of K that had decayed to ^{40}Ar. DeVaucouleurs (1954) noted that this calculation produced the same N_2/Ar ratio for both Earth and Mars, which further strengthened the likelihood of a large N_2 abundance in the Martian atmosphere.

It is interesting to note that the CO_2, H_2O, and O_2 measurements today are not drastically different from the 1954 measurements and estimates. Only the nitrogen has been eliminated, and argon is less than the CO_2 abundance. Had the correct total pressure been known, the earlier investigators would have been led to a good approximation of what is known today.

The existence of water vapor was not firmly established until 1963. Prior to this, the existence of water could only be inferred from the visual telescopic behavior of major features on the Martian surface. The polar caps were seen to expand and recede according to the seasons. Since the polar cap temperatures were at that time thought to be well above the 150°K known today, the then accepted CO_2 pressure of 0.35 mb indicated that a dry ice polar cap could never exist. Water ice seemed the logical alternative.

The only way of detecting chemical constituents in the Martian atmosphere from Earth's surface is by spectroscopic observations. This method, which works so well for measuring the composition of stellar atmospheres, is hampered by the fact that all constituents of the Martian atmosphere also exist in Earth's atmosphere. The detection of water vapor is particularly difficult since the amount of water vapor in the Martian atmosphere is extremely small relative to that in Earth's atmosphere. The Martian spectral component is largely obscured.

This problem was solved in 1963 by Spinrad et al. using a combination of methods. First, the observations were made at high-altitude desert sites to minimize the absorption due to terrestrial water vapor. Second, the combination of increased photographic film sensitivity and better spectral diffraction gratings provided improved resolution of the spectral components than had been possible in previous studies. And third, the experimenters exploited the relative motion of Earth and Mars. The resulting Doppler effect shifts the Martian absorption lines to one side of Earth's lines (fig. 8.1). Since the first detection in 1963, other observers using spectrographic measurements have verified that water vapor does exist in the Martian atmosphere, but in such small quantities that one could never expect liquid water.

The return of photographs from Mariner 4 in 1965 represented a high point in planetary exploration. However, the rugged cratered surface signaled a low point for life-on-Mars enthusiasts. There were no signs of any features that could be easily interpreted as fluvial in origin. None of the celebrated canals was verified, extinguishing the belief that intelligent life might flourish on the Martian surface.

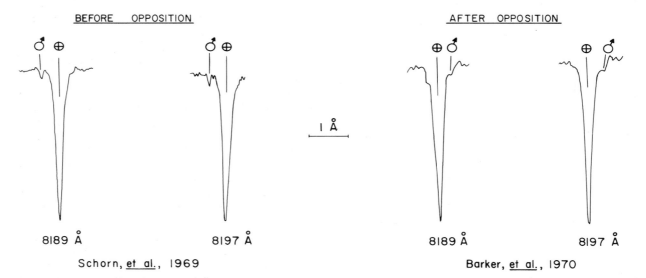

Figure 8.1. Densitometer tracings of two water absorption lines in Martian spectrograms acquired before and after opposition. Since the distance between Earth and Mars is decreasing prior to opposition, the Martian absorption lines are shifted (Doppler effect) to shorter wavelengths. After opposition, the shift is in the opposite direction. The Martian water-vapor absorption lines are dwarfed by the terrestrial water-vapor lines and, consequently, would be unobservable unless shifted due to the relative motion of Earth and Mars.

The atmospheric pressure was found to be about 6 mb, far less than the previously accepted value.

The return of photographs from Mariners 6 and 7 reaffirmed the observation that much of the Martian surface was cratered, but a variety of strange collapse features were observed in several frames. These features, now recognized as part of a larger complex of drainage basins and valleys, were interpreted at the time as resulting from a withdrawal of either permafrost or magma (Leighton et al., 1969a). The possible presence of permafrost suggested that large quantities of water might be trapped beneath the Martian surface. However, the Mariner 7 observation that the polar caps reached very low temperatures of about 150°K (Neugebauer et al., 1969, 1971) strengthened the arguments of Leighton and Murray (1966) that the polar caps consisted of CO_2 rather than H_2O.

Mariner 9 has provided the strongest evidence that substantial amounts of water in a liquid form existed at some time on the Martian surface. Many channels, some resembling braided stream channels on Earth, are scattered around the equatorial regions of the planet. The morphologies and locations of these channels, described later in this chapter, suggest more strongly than any other observations that large amounts of water have existed on the Martian surface in times past.

Thermodynamics

We have seen that the Martian atmosphere contains H_2O and CO_2. Both of these compounds form non-gas phases under present Martian atmospheric conditions. The prediction of Leighton and Murray (1966) that the seasonal polar caps are composed of carbon dioxide has been confirmed observationally (Herr and Pimentel, 1969). Other evidence suggests that at each pole there exists a permanent polar cap composed of solid H_2O.

In this section we extend the previous discussion by determining the behavior of CO_2 and H_2O under various temperatures and pressures. In particular, we explore the conditions under which liquid H_2O may exist. It should be noted that the problem has two parts. Thermodynamic calculations indicate the static conditions under which a hydrosphere may exist. Kinetic calculations indicate the situations in which ephemeral and local water may exist.

Figure 8.2 is a pressure-temperature (P-T) diagram showing the stability fields for various phases of H_2O: liquid, solid, and gas. These will be referred to as $H_2O(l)$, $H_2O(s)$, and $H_2O(g)$. In order to understand more clearly what this diagram means, let us hypothesize a planetary atmosphere composed of nothing but water vapor. Specify the pressure on the surface to be 0.1 atm., and the temperature to

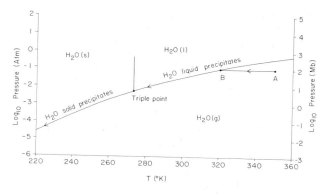

Figure 8.2. Pressure-temperature (P-T) diagram showing phase relations in the H_2O system. The path followed by a cooling atmosphere of pure $H_2O(g)$ is shown. Its behavior is described in the text. Data from Hilsenrath et al. (1955).

be 350°K. Figure 8.2 shows that point A (P = 0.1 atm., T = 350°K) falls in the $H_2O(g)$ stability field. Neither $H_2O(l)$ nor $H_2O(s)$ is stable. Suppose the atmosphere is now cooled to a lower temperature. What happens to the pressure? It moves to the left (lower T), the pressure remaining constant until reaching point B (0.1 atm., 320°K). If the temperature is further reduced, the P-T point moves into the liquid water field. This actually means that a liquid phase is beginning to form at the expense of the gas phase. Rain precipitates from the atmosphere. Since some of the atmosphere is being removed, the atmospheric pressure at the planet's surface begins to decrease. In fact, the pressure decreases so that the point in the P-T diagram exactly follows the line between the $H_2O(g)$ and $H_2O(l)$ stability fields. $H_2O(g)$ and $H_2O(l)$ are said to be in equilibrium, and the observed atmospheric pressure of $H_2O(g)$ is called the vapor pressure of $H_2O(l)$.

As the temperature is further decreased, the P-T point moves past the triple point and continues along the line between the $H_2O(g)$ and $H_2O(s)$ stability fields. At temperatures below the triple point (6.11 mb, 273.16°K), all the water on the ground freezes, and snow rather than rain falls from the atmosphere. The triple point has special significance in this diagram since it is the only point where $H_2O(s)$, $H_2O(l)$, and $H_2O(g)$ are in simultaneous equilibrium.

In our specific example, the pressure and temperature never entered the liquid or solid fields. However, during a terrestrial winter, the atmospheric pressure and temperature are well within the $H_2O(s)$ field. How can this occur? The answer is that Earth's atmosphere is not pure water vapor. Even if all the water were removed, there would still be an atmosphere of N_2, O_2, CO_2, Ar, and traces of many other species. The pressure remains close to 1 atm. and

moves into the liquid and solid fields. Also to be explained is the way the atmosphere in figure 8.2 maintained constant pressure before reaching the liquid field. Is this in conflict with the ideal gas law $PV = nRT$, which specifies that the pressure decreases as the temperature decreases? The answer is that our planetary atmosphere is not a closed box with a fixed volume. One can consider it a column of air, stretching up a great distance. The pressure at the base of the column is due to the weight of all the material in the column. As long as no H_2O precipitates, the same amount of material remains in the column while it is cooled. The pressure at the surface remains unchanged.

We have imagined a rather special atmosphere. Besides being composed of nothing but $H_2O(g)$, it was cooled uniformly. On any real planet, there are always variations over the surface. The polar regions are much colder than the equator. Water should precipitate as ice or water at the coldest points, even though neither phase can exist in warmer regions. As water or ice precipitates, material moves to the colder regions in an attempt to keep the atmospheric pressure relatively constant over the planet. In turn, more water or ice precipitates. Eventually, the atmospheric pressure at the planet's surface is much reduced, and most of the original atmosphere has collected at the polar regions. The planet's atmospheric pressure is controlled by the temperature at the coldest point on the planet.

A simple experiment illustrating this behavior consists of two boxes connected by a duct that allows $H_2O(g)$ to move freely between them. A block of ice is placed in each box. One box is cooled to 160°K, and the other to 170°K. Laboratory measurements predict that the vapor pressures in the two boxes should be 8×10^{-10} atm. and 7×10^{-9} atm. respectively at these temperatures. Since the pressures in each box must be identical, water vapor moves to the box with the lower vapor pressure and lower temperature. Since the $H_2O(s)$ in the warmer box will continue to sublimate, the final result is that all the $H_2O(s)$ is transferred to the low temperature box. Furthermore, the final pressure in both boxes will be equal to 8×10^{-10} atm., the vapor pressure of $H_2O(s)$ in the low pressure box. If the temperature in the colder box is changed, the pressure in both boxes will respond, always equalling the vapor pressure of $H_2O(s)$. The entire experiment is controlled by the coldest temperature, and the pressure of $H_2O(g)$ is said to be *buffered* by this temperature.

We mentioned that the behavior of H_2O is different if an atmosphere contains other gas species. Actually, the thermodynamic stability of H_2O itself is almost identical, but the path of a cooling atmosphere in the P-T diagram is different. If the atmosphere contains a mixture of several species, such as CO_2 and Ar in addition to H_2O, the total pressure is the sum of the partial pressures of each of these species: that is,

$$P_T = P_{H_2O} + P_{CO_2} + P_{Ar}.$$

If all the H_2O is removed, the total pressure equals the sum of the partial pressures of CO_2 and Ar alone. If all the H_2O and CO_2 are removed, the total pressure is that of Ar alone.

In addition, the partial pressure of $H_2O(g)$ in a

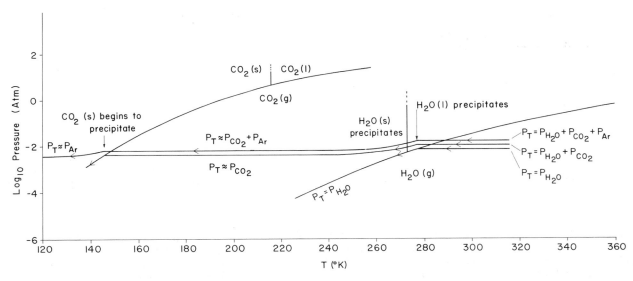

Figure 8.3. The phase relations in the H_2O-CO_2-Ar system shown in a P-T diagram. The paths followed by cooling atmospheres of pure $H_2O(g)$, a mixture of $H_2O(g)$ and $CO_2(g)$, and a mixture of $H_2O(g)$, $CO_2(g)$, and Ar are shown.

mixture of gas species when $H_2O(s)$ or $H_2O(l)$ is also present equals the vapor pressure at the specified temperature. Over the ranges of pressure and temperature found on Mars, the presence of the other gas species has a negligible effect on the vapor pressure of $H_2O(s)$ or $H_2O(l)$. Figure 8.3 compares the effect of cooling a gas of pure $H_2O(g)$, a mixture of $H_2O(g)$ and $CO_2(g)$, and a mixture of $H_2O(g)$, $CO_2(g)$, and $Ar(g)$. As each species condenses or sublimates, the pressure equals the sum of the partial pressures of the remaining species plus the vapor pressure of the non-gas phase(s).

Figure 8.4 is a generalization of figure 8.3, showing the change in total pressure as an atmosphere of any composition, but containing $H_2O(g)$, is cooled.

For example, consider an original atmosphere where $P_T = 10^{-2.7}$ atm., and $P_T/P_{H_2O} = 2$. The cooling atmosphere follows a horizontal line until it reaches the dotted line corresponding to the original P_T/P_{H_2O} ($= 2$). At this temperature, $H_2O(s)$ precipitates, reducing the atmospheric pressure. The path follows the solid line, asymptotically approaching a final pressure of 10^{-3} atm. (which equals original P_T minus original P_{H_2O}). Had the original atmosphere been at a higher original pressure (above the triple point pressure) the first phase to precipitate would have been $H_2O(l)$, changing to $H_2O(s)$ below $273°K$.

One fact is immediately apparent from examination of figure 8.4. A cooling atmosphere loses most

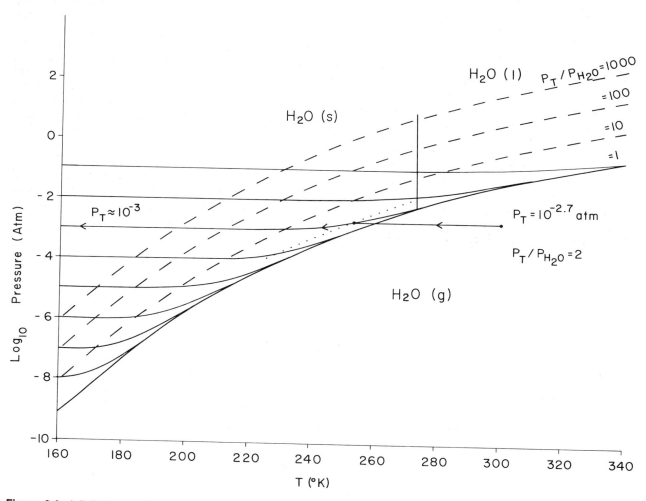

Figure 8.4. A P-T diagram that can be used to calculate the pressure drop that results when an atmosphere containing $H_2O(g)$ is cooled. Superimposed on the H_2O phase diagram are dashed lines corresponding to different ratios of total gas pressure (P_T) to the pressure of $H_2O(P_{H_2O})$, provided that $H_2O(g)$ is in equilibrium with either $H_2O(l)$ or $H_2O(s)$.

The solid lines that asymptotically approach the $H_2O(s)$-$H_2O(g)$ or $H_2O(l)$-$H_2O(g)$ equilibrium curves represent the path in a P-T diagram that an atmosphere maintaining equilibrium with $H_2O(s)$ or $H_2O(l)$ will follow as the temperature is changed. The specific example shown is described in the text.

of its water over a very small temperature range. In a 20° drop, almost all of the atmospheric water vapor precipitates. Conversely, raising the temperature from a very low original value adds a significant quantity of $H_2O(g)$ only in this restricted temperature range.

Clathrates

There is a complicating factor as the temperature is lowered. $H_2O(s)$ has the capability to contain other gas species within the interstices of the crystal structure. Such a structure is called a clathrate. Although the other gas species are physically trapped within the lattice, the resulting clathrate behaves as a chemical compound, in that it forms within a well-defined range of pressure and temperature. A mixed hydrate consists of several gas species trapped within the same H_2O framework. Since the entrapment is essentially a physical process, even inert gases such as Ar can form clathrates.

The existence of clathrates within the solar system is reasonably well established. Several investigators have compared experimentally prepared clathrates of CH_4 and NH_3 to the observed behavior of comets, concluding that clathrates of these compounds may exist in cometary material (Delsemme and Wenger, 1970; Delsemme and Miller, 1970). Carbon dioxide clathrate has been detected in ice cores

from the Antarctic. Miller (1961) suggested that, under plausible conditions, clathrates may exist on the outer planets, and on their satellites. He specifically proposed that Martian polar caps could contain a mixed hydrate of CO_2, N_2, and Ar, although the higher temperature estimate for the caps at that time led him to the conclusion that such a hydrate would be metastable. With improved temperature determinations, Miller and Smythe (1970) again suggested the existence of a CO_2 clathrate within the polar caps. They emphasized that $H_2O(s)$ and $CO_2(s)$ do not form simultaneously at the Martian polar temperatures and pressures, but either $CO_2(s)$ + clathrate or $H_2O(s)$ + clathrate is the stable assemblage (fig. 8.5).

As interesting as they are from a chemical point of view, clathrates have a very small effect on the formation of solid phases. For example, consider carbon dioxide clathrate, which has an ideal chemical formula $CO_2 \cdot 5.75H_2O$ (Takenouchi and Kennedy, 1964, 1965). Although the clathrate is stable over a large P-T range, the maximum amount of clathrate that can form is limited by the H_2O content of the atmosphere, most of which has already been condensed as $H_2O(s)$.

The ability of Delsemme and Wenger (1970) to generate an appreciable quantity of methane clathrate in their cometary investigation resulted from their particular method of synthesis. Rather than

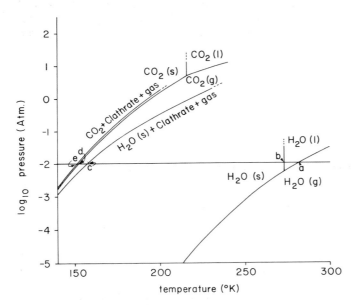
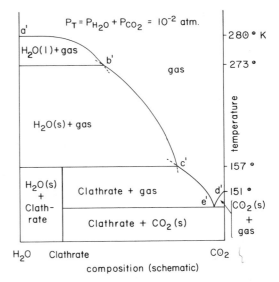

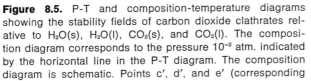

Figure 8.5. P-T and composition-temperature diagrams showing the stability fields of carbon dioxide clathrates relative to $H_2O(s)$, $H_2O(l)$, $CO_2(s)$, and $CO_2(l)$. The composition diagram corresponds to the pressure 10^{-2} atm. indicated by the horizontal line in the P-T diagram. The composition diagram is schematic. Points c′, d′, and e′ (corresponding

to points c, d, and e in the P-T diagram) are, in the actual situation, extremely close to pure CO_2 composition. Of particular significance is the absence of a $H_2O(s)$-$CO_2(s)$-gas equilibrium point, implying that, under Martian conditions, $H_2O(s)$ and $CO_2(s)$ do not coexist in equilibrium.

cooling a mixture of $CH_4(g)$ and $H_2O(g)$, they cooled a pure $CH_4(g)$ phase. They then slowly added $H_2O(g)$. The $H_2O(g)$ condensed on their condensation surface while simultaneously capturing $CH_4(g)$ within its framework. The amount of CH_4 clathrate formed was in proportion to the amount of H_2O added to the system.

We conclude that, although a small amount of carbon dioxide clathrate may form in Martian polar conditions, the quantity is severely limited by the atmospheric content of $H_2O(g)$. Clathrates will therefore be ignored in the discussions that follow.

The Martian Atmosphere

The diagrams of the previous section can be used to visualize the response of the Martian atmosphere to temperature changes. Figure 8.6 shows two boxes representative of average P_T-T ranges found on both Mars and Earth. The dashed lines in this figure, corresponding to the ratios P_T/P_{H_2O} in figure 8.4, are now relabeled to show the percentage of water in an atmosphere in equilibrium with either $H_2O(s)$ or $H_2O(l)$. The rectangular shapes of the P_T-T ranges for Earth and Mars are, of course, schematic and say nothing about what region or percentage of each planet's surface is represented by each P_T-T point. However, this is not necessary. One needs only to realize that there is a coldest and warmest point on the planet, and the pressure ranges are approximately those shown. If the atmosphere is in equilibrium with either $H_2O(s)$ or $CO_2(s)$, the atmospheric properties are buffered by the coldest temperature

and highest pressure on the planet, defined by the upper left corner of each box. The atmosphere will preferentially attempt to precipitate out at this region for the same reasons developed in figure 8.2.

Consider the "Earth-box" in figure 8.6. One might postulate that all $H_2O(g)$ should precipitate out at the coldest regions (poles). However, this is not the case because Earth's surface is covered by so much water that planetwide equilibrium is never established between the volume of condensed H_2O and the atmosphere. Clearly, if all the water precipitated as ice at Earth's polar regions, it would flow back to the equatorial regions, remelt, and eventually establish a situation where water exists in substantial quantities all around the planet. A small upset in the planetwide temperature distributions could shift the balance of materials between pole and equator.

Earth's atmosphere contains up to 5 percent water vapor, as indicated by the shaded region in the Earth-box. Locally high humidity results from evaporation over oceans and lakes, raising the $H_2O(g)$ content close to saturation. Local equilibrium conditions for precipitation of either rain or snow are attained when a cool air mass moves against a body of moist, warm air.

The "Mars-box" is significantly different. First of all, the type of local equilibrium found on Earth cannot exist because of the small amount of Martian surface water. The resulting atmospheric content of $H_2O(g)$ will be correspondingly low, implying that a shaded region of the Mars-box is always confined to a narrow band at the far left. But even if one

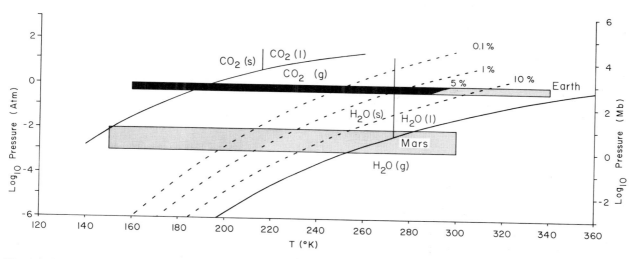

Figure 8.6. A P-T diagram showing phase relations of H_2O and CO_2. The Mars-box and Earth-box represent ranges of pressure and temperature commonly found on each planet. Although the Mars-box extends slightly into the $H_2O(l)$ stability field, $H_2O(l)$ is never stable, for reasons discussed in the text.

hypothesizes a generous 10-percent value for the atmospheric water content, the equatorial regions never reach the stability field of liquid water.

Figure 8.6 also demonstrates that the present Martian atmosphere is apparently buffered by the presence of $CO_2(s)$ at the poles. The upper left corner of the Mars-box is quite close to the $CO_2(s)$-$CO_2(g)$ equilibrium curve, exactly as one would predict if the $CO_2(g)$ pressure were controlled by the condensation of $CO_2(s)$ at the coldest point on the planet. However, as will be discussed in more length in the section on polar caps, there is disagreement as to whether this buffering is real and indicative of a large body of $CO_2(s)$, whether it is apparent and results from $CO_2(g)$ supplied by regolith adsorption/desorption, or whether the atmospheric content of $CO_2(g)$ is buffered at all.

Local Saturation and White Clouds

There may be sources that locally allow the atmosphere to become saturated with $H_2O(g)$. If the observed pressure of $H_2O(g)$ equals the vapor pressure of $H_2O(s)$ or $H_2O(l)$ at the atmospheric temperature, the atmosphere is said to be 100 percent saturated. If $P_{H_2O}^g > P_{H_2O}^{vp}$, the atmosphere is supersaturated; if $P_{H_2O}^g < P_{H_2O}^{vp}$, the atmosphere is undersaturated. Local Martian conditions of 50-percent saturation have been observed (Schorn et al., 1969). Although the amount of H_2O is still very small, it can produce the observed white clouds and possibly result in local frost conditions. One local source for Martian water is the regolith. Water adsorbed on grain surfaces is available for release to the atmosphere, a process that will be discussed later in some detail.

The major shortcoming of the P-T diagrams of the previous sections is that they say nothing about the kinetics of reactions: that is, the rate at which a process will proceed. In addition, there is no information about such variables as the latent heat of sublimation of $H_2O(g)$ from $H_2O(s)$. Since P-T diagrams describe an ideal, rate-independent equilibrium, it might be possible for local conditions to fluctuate sufficiently to generate small, but biologically and chemically significant, quantities of liquid water.

During the Mariner 9 mission, Curran et al. (1973) determined the composition of white clouds that formed over the Tharsis ridge to be $H_2O(s)$. The ice particles that composed the cloud possessed a mean radius of $2.0 \mu m$. The mass of the clouds was $5 \times 10^{-5} g/cm^2$, although local values three times this amount were measured. This is still less than the integrated mass of $H_2O(g)$ in the atmosphere column ($5 \times 10^{-3} g/cm^2$), and Curran et al. conclude that the source for the ice clouds was at-mospheric moisture, not degassed H_2O from volcanic vents.

This ice, or ice similarly formed in other locations, may fall to the surface during the Martian night. The behavior of this $H_2O(s)$ then becomes a problem of relative rates—can a layer of ice formed on the Martian surface during the night melt before it sublimates back into the atmosphere? The problem is complicated since there are numerous variables—for example, the intensity of the solar flux, the relative rates of fusion and sublimation of ice, and the effects of a small amount of dust covering an ice layer. This last effect is influenced by such variables as the change of albedo due to the overlying dust layer, the rates of diffusion of gas molecules through the dust layer, and surface interactions between the ice and soil.

In a discussion on this subject, Farmer (1973; also Sagan et al., 1968) concludes that $H_2O(l)$ may indeed form as a metastable phase under local conditions, at least in biologically important quantities. He also concludes that the white clouds seen around the planet are produced by local atmospheric condensation of H_2O formed when a saturated lower atmosphere cools in the evening. This interpretation was considered previously by many investigators. Fanale and Cannon (1971, 1974) believe that desorption of H_2O from the Martian regolith provides a local source for the formation of white clouds. Leovy (1973) calculated that the observed variation in the atmospheric water vapor content correlated with a source in the equatorial, rather than polar, region. This source he believed to be ground ice.

Although the amount of water that can be formed by this process is small, the fact that liquid water may exist at all in the upper layer of the Martian soil greatly increases the probability for the survival of Martian organisms. Also, although such small amounts of water would not produce any major morphologic features involving surface runoff, they might be significant in such processes as salt-weathering (Malin, 1974b), especially since the presence of salts lowers the freezing point of water (Ingersoll, 1970; Wade and DeWys, 1968).

Insolation Changes

The previous discussions demonstrated that the present Martian environment is not conducive to the formation of rain anywhere on the planet's surface. Although it is possible that biologically or chemically significant amounts of water may accumulate in the topsoil or on the surface, the geomorphic interpretation of the channels as a result of surface runoff forces the conclusion that past conditions have been different from present ones.

The geological evidence also suggests that more than one fluvial mechanism has operated. Several large channels, notably Mangala and Ares Valles, closely resemble catastrophically formed structures. Sharp (1973a, 1973c), Milton (1974), and Sharp and Malin (1975) postulate break-out from confined sources of underground ice melted by local heating or release of ponded water. However, the furrows observed in ancient cratered terrain (McCauley et al., 1972) appear to have resulted from surface run-off (Milton, 1973).

A major question presently being debated in the literature is whether the Martian atmosphere undergoes periodic temperature changes resulting from variations in the solar energy received. Morphological evidence for cyclic climatic changes is presented by the layered polar deposits, previously described in chapters 3 and 7. The mere existence of a succession of strata with differing morphological properties suggests some alternation between two or more depositional environments, in the same way that a terrestrial stratigraphic sequence of sandstone and limestone indicates oscillation between completely different sedimentary environments. The uncratered surface of the layered deposits suggests that they formed relatively recently, although the observation that removal of material is the only currently operative mechanism alternatively suggests that the surface may be a young erosional surface.

The possibility that periodic insolation changes might account for climatic change has been extensively debated with respect to Earth. In 1938 a Yugoslavian astronomer, M. Milankovitch (1941; Vernekar, 1972), demonstrated quantitative changes in Earth's orbital parameters over long periods of time and proposed that they might be responsible for Pleistocene changes in climate, leading to worldwide waxing and waning of ice sheets. He considered three properties: precession of the spin axis, variations in axial tilt (obliquity), and variations in orbital eccentricity. Precession of the spin axis is caused by the gravitational attraction of the sun on Earth's equatorial bulge. Like a spinning top that is dropped on the floor with its spin axis inclined to vertical, the spin axis of Earth moves conically. The time required for one complete precession cycle is 25,800 years. Because the perihelion position of the orbit is also slowly shifting in an opposite direction, the precessional cycle, viewed in space, is 21,000 years. This is sometimes called the precession of equinoxes, since the two days with periods of equal sunlight and darkness—currently March 21 and September 23—progressively change position in the orbit. This variation would be of no consequence if the orbit were circular. But because of the slight orbital eccentricity, each hemisphere will experience a 21,000-

year cycle of warmer and cooler seasons. The patterns in the two hemispheres are 180° out of phase. When the seasonal contrast in the south is great—hot summers and cool winters—the north will be experiencing relatively cool summers and warm winters.

At present, Earth's orbit is close to circular, but because of attractions between our planet, the sun, and other planets the eccentricity can reach values as high as 0.05. The amplitude and periodicity of the cycle are irregular but it tends to repeat every 90,000 years. Gravitational interactions also affect the obliquity of Earth, the angle between the spin axis and a normal to the orbital plane (ecliptic). This cycle has a periodicity of about 41,000 years and an amplitude of 2°.

One must be cautious in combining the three variables—obliquity, eccentricity, and precession—to obtain a cumulative view of changes in insolation

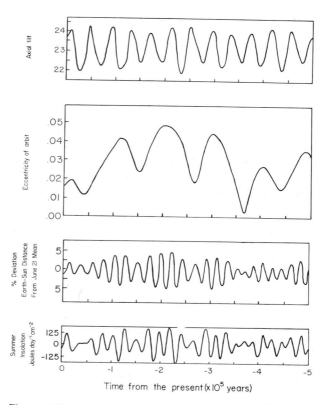

Figure 8.7. Variations in Earth's axial tilt, orbital eccentricity, and June 21 (solstice) distance from the sun over the past half-million years. The effects of these variations are shown in a summer insolation curve for 45°N. Changes are expressed as differences from the mean summer radiation of approximately 4000 joules day⁻¹ cm⁻². Insolation lows at 25,000 and 70,000 years before the present correspond to independently documented glacial advances. Adapted from Broecker and Van Donk (1970).

with time at a given latitude. As already noted, the precessional effect is not in phase for the two hemispheres. Although a change in tilt affects both hemispheres similarly, there remains a latitudinal dependence. For a given increase in tilt, there is greatest increase in insolation at the highest latitudes. Summer insolation curves for 45°N are shown in figure 8.7. There is a good correlation between the insolation peaks and the interglacial periods as determined by ratios of oxygen isotopes in deep sea cores. Similarly, the insolation lows at 25,000 and 70,000 years before the present correspond in time to the advance of Early Wisconsin and Late Wisconsin glaciers.

The same arguments that Milankovitch used to document changing insolation for Earth can be transferred to Mars (Sagan, 1971). In a comprehensive paper of unsuspected significance for climatologists Brouwer and Van Woerkom (1950) calculated the temporal changes in orbital parameters of all planets, excluding Pluto, due to gravitational interactions. In the case of Mars there are five important "changing constants": perihelion position, equinox position, orbital eccentricity, orbital plane inclination, and spin axis tilt.

The perihelion position for Mars's orbit precesses with a period of 72,000 years. The spin axis of the planet precesses with a period of 1.75×10^5 years. Although the total insolation received at the planet remains constant, both precession effects combine to influence the relative insolation at the two poles. The maximum polar radiation varies from pole to pole with a 51,000-year period (Leighton and Murray, 1966) (fig. 8.8).

The eccentricity of the Martian orbit evidences two periodic changes. Over an interval of 95×10^3 years it varies by a relatively small amount, no more than 0.06. Superimposed on this change is a larger cycle of 2×10^6 years during which the eccentricity changes from 0.14 to 0.01 (Murray et al., 1973). As a consequence, the average annual solar radiation changes. For the polar regions the maximum variation is 1 percent. As one might imagine, the insolation for the subsolar point at perihelion evi-

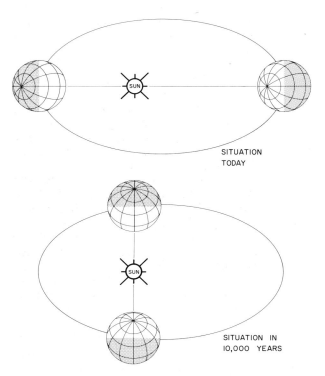

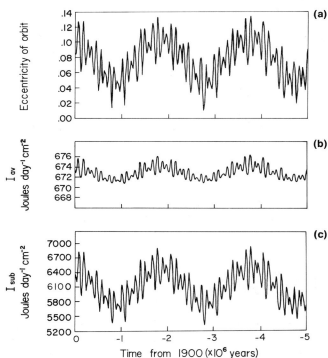

Figure 8.8. The combined effect of perihelion and spin axis precession for Mars, diagrammatically shown. At present the southern hemisphere tilts toward the sun at perihelion, so that the south polar region receives more insolation than the north pole region. In approximately 10,000 years, summer conditions will be similar at both poles. In approximately 25,000 years the north polar region will be experiencing summer at the perihelion position. Due to these precession effects volatiles in seasonally permanent ice caps might be cycled from one polar region to the other every 25,000 years.

Figure 8.9 (a). Eccentricity of Mars' orbit for the past five million years.

(b). Average annual insolation at the poles resulting from eccentricity changes.

(c). Insolation at subsolar perihelion point resulting from eccentricity changes. From Murray et al. (1973).

dences much greater variation, as much as 30 percent (fig. 8.9).

The inclination of the orbital plane undergoes two periodic variations, a 1° change with a period of 1.6×10^5 years and a 5° change with a longer period of 1.2×10^6 years. Related to this is a change in the tilt of Mars's rotational axis. It oscillates on a 1.2×10^5-year time scale (Ward, 1973). The amplitude of the oscillation also varies periodically over 1.2×10^6 years with a maximum swing from 15° to 35° (fig. 8.10). This obliquity change is particularly important in altering the amount of polar heating. Indeed, if a circular orbit is assumed, a change in obliquity from 15° to 35° approximately doubles the annual solar radiation at both poles.

As an alternative to orbital variations, Sagan et al. (1973a, also Hartmann, 1974b) suggested that long-term ($\sim 10^8$ yr) variations in the solar luminosity may result in a planet-wide heating. Clearly, this increase in the insolation would occur at all latitudes. They also suggest that a decrease in the albedo of polar volatile surface deposits resulting from dust blanketing could raise polar temperatures.

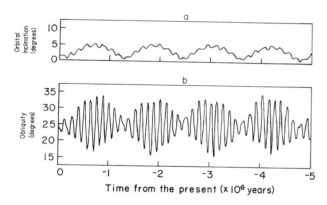

Figure 8.10 (a). Variations in inclination of orbital plane of Mars for the past five million years.

(b). Variations in the tilt of Mars' rotational axis. From Ward (1973).

Climatic Changes Due to Temperature Changes

Ward et al. (1974) calculate that the polar temperatures resulting from eccentricity- and obliquity-related insolation changes range from 130°K to 160°K. If a polar reservoir of $CO_2(s)$ of sufficient magnitude is available, they speculate that the vapor pressure of $CO_2(s)$, and therefore the atmospheric pressure, could vary between 0.3 and 30 mb, a factor of 100.

In the absence of a large polar reservoir of $CO_2(s)$, Fanale and Cannon (1974) alternatively speculate that insolation changes may release both CO_2 and H_2O adsorbed planetwide in the Martian regolith. Houck et al. (1973) present spectroscopic evidence suggesting as much as one percent by weight of bound water in surface materials. Since temperature changes dramatically alter the capacity of soil particles to adsorb volatiles, it follows that insolation changes may effect the exchange of substantial quantities of adsorbed volatiles. During physical adsorption, weak Van der Waal forces result in the condensation of the molecules of a gas phase on the surface of a solid. In the related process of chemical adsorption, the forces of attraction are stronger, resulting in the formation of a layer resembling a compound on the solid's surface. Some gas species exhibit physical adsorption over one temperature range, and chemical adsorption over another. Physical adsorption can be easily reversed (called desorption), while chemical adsorption cannot.

The amount of adsorbed volatiles depends on the exposed surface area of the solid, which for particulate material varies as the inverse of the square of the particle diameter. Thus, a finely divided Martian regolith has a larger capacity than a regolith of coarser material. However, the effective surface area is much larger than the geometric surface area one can calculate assuming spherical particles. Surface irregularities and particle porosity, present at all particle sizes, become the predominant factor for more finely divided material, increasing the expected surface area by at least an order of magnitude.

Fanale and Cannon determined experimentally that significant amounts of both $H_2O(g)$ and $CO_2(g)$ are physically adsorbed by plausible Martian regolith materials over the P-T range found on the Martian surface. Based on their results, they concluded that the capacity of the Martian regolith to store and exchange volatiles with the atmosphere may exceed that of the polar caps. Also, regolith adsorption provides a mechanism for distributing volatiles around the planet and preventing their exclusive trapping at the poles.

Figures 8.11 and 8.12 are based on the Fanale-Cannon results for $H_2O(g)$ and $CO_2(g)$ adsorption on finely divided basalts. In each figure, the x-axis is the atmospheric pressure of the pure gas phase of $H_2O(g)$ or $CO_2(g)$, and the y-axis is the experimentally measured quantity of adsorbed gas. Each of the solid lines are called adsorption isotherms. Using these figures, one can estimate the change in amount of adsorbed material resulting from changing the temperature, atmospheric pressure, or amount of available regolith. For another type of regolith material, a new set of curves must be determined experimentally.

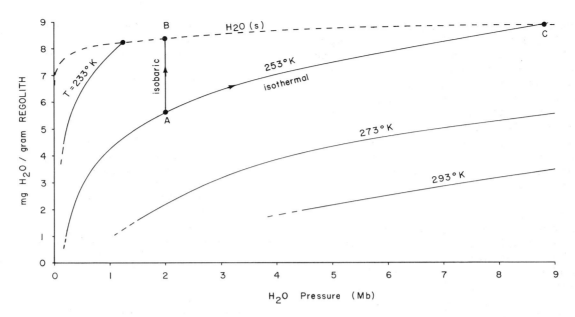

Figure 8.11. Experimental results of Fanale and Cannon (1974) for regolith adsorption of $H_2O(g)$. Their data have been replotted to demonstrate more clearly the significance of changes in temperature or pressure. Each solid line is called an adsorption isotherm, as further described in the text. Above the dashed line, $H_2O(g)$ forms $H_2O(s)$ rather than being adsorbed.

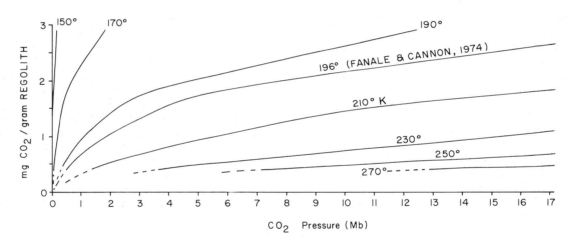

Figure 8.12. Adsorption isotherms for $CO_2(g)$ on possible Martian regolith materials, extrapolated from Fanale and Cannon (1974). For a further discussion see the text and figure 8.11.

Starting at point A in figure 8.11, one can visualize why increasing the pressure or decreasing the temperature increases the amount of adsorbed material. Two cases are of special interest. If the pressure is constant, and the temperature decreased, point A will tend to move toward point B along an isobaric line. An example of such a situation is a decreasing equatorial temperature, while the atmospheric pressure is buffered by a polar reservoir of $H_2O(s)$. However, beyond point B, $H_2O(g)$ will form $H_2O(s)$ rather than an adsorption product since the atmosphere enters the stability field of $H_2O(s)$.

At constant temperature, an increase in pressure shifts point A toward point C along an isotherm. Such a situation arises at the equator if the Martian polar temperature were raised while buffered by an adequate supply of $H_2O(s)$. As for the isobaric case, beyond point C, $H_2O(s)$ will form.

Let us consider an experiment that represents a

simplified version of the complex equilibrium between $CO_2(s)$ polar caps, CO_2 regolith adsorption, and a CO_2 Martian atmosphere. In a closed evacuated box (representing the pole), place a block of solid CO_2 (dry ice) and maintain the temperature of the box at 150°K. The dry ice will sublimate until the pressure of $CO_2(g)$ in the box equals the vapor pressure of $CO_2(s)$ at 150°K.

In a second box (representing the equator) place a sample of Martian regolith (or a suitable, less expensive substitute!) and evacuate the box of all atmosphere. Maintain the regolith at 250°K. The two boxes are now connected so that $CO_2(g)$ can move freely between the boxes. Gas will be adsorbed on the surface of the regolith material. As this occurs, dry ice will sublimate, maintaining the pressure. The $CO_2(g)$ pressure is maintained by the dry ice at 150°K, not by the regolith at 250°K. Since the total pressure in both boxes is identical, the total pressure over the regolith equals the vapor pressure of dry ice at 150°K.

Increasing the polar temperature increases the $CO_2(s)$ vapor pressure, and $CO_2(g)$ flows through the duct to the equator. Notice that changing the amount of dry ice at the poles has no effect on the movement of CO_2 through the duct, but changing the amount of regolith alters the amount of CO_2 that can be adsorbed, and material flows to or from the dry ice. This demonstrates the important property that an equilibrium is established between the polar dry ice and the equatorial regolith material. Not all CO_2 is transferred to the poles, or vice versa.

The effective regolith depth on the Martian surface is clearly an important factor in determining the amount of volatiles that could be liberated to the atmosphere. Fanale and Cannon (1974) present an intriguing illustration of the adsorbed volatile content of the Martian regolith for various regolith depths (fig. 8.13). For example, a ten-meter thick regolith raised from −110°C to −77°C liberates almost 10 grams of adsorbed $CO_2(g)$ per cm² of surface, a quantity similar to that presently observed in the Martian atmosphere.

However, although locally an important process, there are several reasons why adsorption/desorption may not influence the Martian climate as greatly as figure 8.13 might imply. First, although the process of adsorption/desorption is very rapid, the rate at which volatile atmosphere-regolith exchange occurs at any depth is limited by the process of diffusion. If there are any chemical interactions, assisted by small amounts of water in the soil, caliche-type deposits may form at shallow depths, sealing off the lower regolith from volatile exchange. Alternatively, if diffusion could occur, $H_2O(g)$ would form $H_2O(s)$ at shallow depths since the Martian subsurface re-

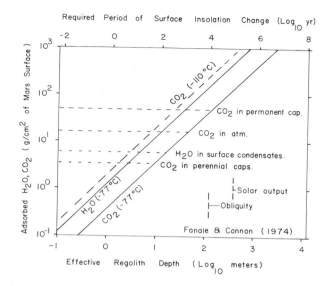

Figure 8.13. This figure, taken from Fanale and Cannon (1974), shows the amount of volatiles contained in the regolith as a function of regolith depth.

mains below freezing at all latitudes at all times. The effect would be to seal off the lower regolith, just as in the case of caliche formation.

Second, a decrease in the porosity, and therefore the adsorption capacity, occurs with depth due to compaction. The linear increase of the regolith capacity with depth shown in figure 8.13 seems unlikely. Third, although the obliquity changes described by Ward (1974) may raise polar temperatures, the average annual insolation, and therefore temperature, in the equatorial regions decreases. Because the areal extent of the equatorial regions is much greater than the polar regions, if regolith adsorption occurs, and if a large polar reservoir of $CO_2(s)$ does not exist, it is conceivable that the atmospheric $CO_2(g)$ pressure is partially buffered by the transfer of volatiles between equator and poles, rather than increasing to a high value. This was suggested by Fanale and Cannon (1974) as an explanation for the apparent buffering of the $CO_2(g)$ pressure by the annual $CO_2(s)$ polar cap.

Sagan et al. (1973a) further point out that, if any insolation variations cause the mobilization of CO_2 into the atmosphere, the possibility exists for what is called advective instability. The increased atmospheric pressure increases the heat transport capacity for the atmosphere, allowing further heat transport from equator to poles. Under certain conditions, a runaway instability exists. The planet-wide temperature rises about 30°K, and the predicted atmospheric pressure reaches 1 atmosphere before restabilizing. This can only occur, of course, if a $CO_2(s)$ reservoir of sufficient size exists to always maintain equilibrium. Ingersoll (1974), who does not believe that such a reservoir exists, argues against such changes because once such a high-pressure regime formed, it would be difficult to reverse the situation.

We have seen that there are several possible mechanisms for increasing the minimum Martian

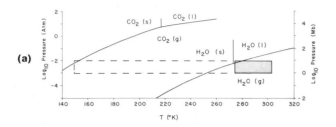

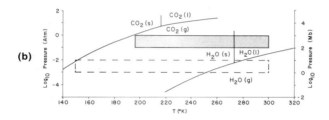

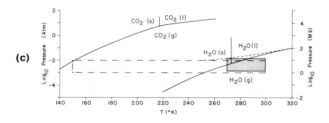

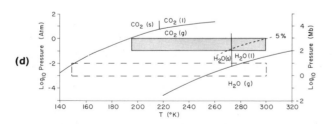

Figure 8.14. The Mars-box shown in figure 8.6 is shifted by raising the temperature of the coldest point on the planet. There are four cases depending on the polar supply of $CO_2(s)$ and $H_2O(s)$:

   (a). no $H_2O(s)$ or $CO_2(s)$;
   (b). no $H_2O(s)$, $CO_2(s)$;
   (c). $H_2O(s)$, no $CO_2(s)$;
   (d). $H_2O(s)$, $CO_2(s)$.

Although cases (b) and (d) appear to resemble the Earth-box in figure 8.6, the results are not identical, as explained in the text.

atmospheric temperature. If we assume that reservoirs of $CO_2(s)$ and $H_2O(s)$ exist so that the atmospheric pressure can respond, we can use figure 8.6 to estimate the effects of raising the minimum Martian atmospheric temperature on the thermodynamic stability of liquid water. There are four limiting cases, depending on whether sufficient polar reservoirs of $CO_2(s)$ and/or $H_2O(s)$ are available to maintain the partial pressures of $CO_2(g)$ and $H_2O(g)$ for any temperature rise.

If a polar reservoir of neither $H_2O(s)$ nor $CO_2(s)$ exists, the total atmospheric pressure must remain constant for any temperature increase. Raising the temperature (fig. 8.14a) will eventually shift the upper left corner of the Mars-box through the $H_2O(l)$ field, but the total precipitable $H_2O(l)$ remains the insignificant amount presently available in the Martian atmosphere. This amount, if condensed in one location, would form a lake about the size of Walden Pond.

If a $CO_2(s)$ reservoir exists, but a $H_2O(s)$ reservoir is absent, a small increase in the minimum temperature raises the pressure rapidly, with the upper left corner of the Mars-box following the $CO_2(s)$–$CO_2(g)$ equilibrium curve (fig. 8.14b). Although the Mars-box extends well into the $H_2O(l)$ field, the total atmospheric $H_2O(g)$ content remains insignificant since there was no addition from an $H_2O(s)$ reservoir. $H_2O(l)$ can never become stable because the maximum percentage of the partial pressure of $H_2O(g)$ relative to the vapor pressure of $H_2O(l)$ always remains very small.

If a reservoir of $H_2O(s)$ exists at the poles, but there is no $CO_2(s)$, raising the minimum temperature shifts the Mars-box to the right and to slightly higher pressures. The upper left corner of the box follows the path shown in figure 8.14c, identical to the paths defined in figure 8.4. However, in order to add a significant amount of $H_2O(g)$ into the atmosphere, the temperature at the coldest point on the planet must be raised above 250°K. Even in the final case (fig. 8.14d), with reservoirs of both $H_2O(s)$ and $CO_2(s)$, $H_2O(l)$ does not stabilize on the Martian surface unless the *minimum* temperature rises above 273°K. Although the Mars-box shifts up to the same position that was occupied by the Earth-box in figure 8.6, and $H_2O(g)$ is added to the atmosphere, the percentage of $H_2O(g)$ never becomes large enough to allow $H_2O(l)$ stability (see figure 8.4).

The fundamental difference between the Mars-box in figure 8.14d and the Earth-box in figure 8.6 is that the Martian atmospheric $H_2O(g)$ is buffered by a polar reservoir of $H_2O(s)$, while Earth's atmospheric $H_2O(g)$ is not in equilibrium with the poles. The factor that allows the terrestrial atmosphere to generate rain is its ability to locally attain saturated conditions of $H_2O(g)$ at all latitudes, even at the equator. This is a direct result of the presence of oceans and lakes, which in turn is a consequence of a vast supply of water.

We should again clarify the distinction between thermodynamically stabilizing $H_2O(l)$ and kinetically stabilizing $H_2O(l)$. The thermodynamic stabilization of $H_2O(l)$, which is necessary for rainfall, is attained only when the *partial* pressure of $H_2O(l)$ equals or exceeds the vapor pressure of $H_2O(l)$. We have shown that this condition cannot be attained during proposed Martian climatic changes. Another question altogether is whether $H_2O(l)$ can be prevented from rapidly boiling so that, if formed, it could traverse great distances over the Martian surface and form channels. This requires only that the *total* atmospheric pressure equals or exceeds the vapor pressure of $H_2O(l)$. This is the model published by Sagan et al. (1973a), who propose repeated episodes of "extensive liquid water." They speculate that a rise in planet-wide temperature resulting from insolation variations melts subsurface $H_2O(s)$ in the Martian equatorial regions. Simultaneously $CO_2(g)$ is added to the atmosphere from the poles. The increased total gas pressure, although not capable of thermodynamically stabilizing $H_2O(l)$, would at least prevent the rapid boiling of $H_2O(l)$ into the atmosphere. Consequently, water melted in the equatorial regions could flow hundreds of kilometers before vaporizing. This sequence of events has been interpreted as creating an Earth-like environment characterized by rainfall and associated erosional and biological activities. However, Sagan et al. specifically state that rainfall is not a requirement. In fact, as we have demonstrated, rainfall is impossible with only a 30°K planet-wide temperature rise.

We are thus confronted with the dilemma that the temperature changes resulting from insolation variations are apparently incapable of producing rainfall, even if one assumes "infinite" reservoirs of both $H_2O(s)$ and $CO_2(s)$. Since the actual Martian volatile reservoirs are limited in volume, we must conclude that water-eroded morphologic features did not form as a result of insolation variations. An alternative mechanism is the creation of an atmosphere. A more likely explanation is that a pre-existing dense atmosphere cooled, passing through the stability field of liquid H_2O and resulting in the precipitation of rain. Jones (1974a) speculates that the mechanism for the episode of increased surface obliteration manifested by the crater populations—as discussed in chapter 4—was the presence of liquid water. If the low Martian cratering rate postulated by Soderblom et al. (1974) and Chapman (1974) is correct, then the obliteration event occurred early

in Martian history. This is consistent with the interpretation that the furrows, confined as they are to the ancient cratered terrain, represent an ancient epoch of fluvial activity associated with an early Martian atmosphere. The presence of large amounts of equatorial ice can easily be explained as remnant from an ancient dense atmosphere. Consequently, we must consider in detail the nature of the processes of atmospheric formation early in the history of the planet.

The question of where the water came from and where it went leads to a general discussion of the sources and sinks of a planet's volatiles. Although no specific source or sink is unique, the relative importance and specific nature of each in shaping an individual planet's atmospheric history may vary dramatically. As we shall see, the source of terrestrial and Martian volatiles is nearly identical. However, the sinks are extremely different—a result of the different temperatures and masses of the two planets. Since there is more direct evidence on the history of Earth's volatiles, we shall first discuss theories of terrestrial outgassing before returning to Mars.

Sources

The volatiles present today on Earth (in the deep interior, crust, surface, and atmosphere) were almost certainly present at the end of accretion. Virtually all of these volatiles were released either upon impact of primitive objects or during subsequent heating within Earth. Other possible primary sources for Earth's volatiles are considered to be volumetrically insignificant, although they are geochemically important. There may have been an addition from the solar wind, but rare gas relative abundances argue to the contrary.

Volatiles present in the gaseous state during accretion are also insignificant on Earth. One of the most quoted observations pertinent to this issue is the observed depletion of terrestrial nonradiogenic rare gases (Ne, Ar, Kr, Xe) relative to solar abundances (e.g. Rubey, 1951). Brown (1952) interpreted this depletion as evidence that an insignificant fraction of rare gases present in the original solar nebula was retained in the accreting Earth. Since rare gases, unlike the major volatiles (e.g. H_2O, CO_2, CH_4, NH_3), do not readily form compounds, they could not have been chemically contained within the primitive material of accretion.

It is possible that sporadic additions to an atmosphere resulted from the occasional impact of asteroidal material (Rubey, 1951) or cometary material (Hunten, 1971). Comets are believed to be primarily frozen gases and clathrates of H_2O, CH_4, and

NH_3. A comet of only 100 m diameter could supply the present amount of H_2O seen in the Martian atmosphere. However, it will be apparent from a subsequent section on sinks that the amount of H_2O expected from such sources is only a small fraction of that probably present on Mars or Earth in the combined forms of vapor, water, and ice.

Although investigators agree on the above, there is still some question as to when the accreting material lost its volatiles. One extreme model involves continuous, uniform outgassing; the other extreme specifies a catastrophic loss of volatiles simultaneous with accretion (fig. 8.15). The latter model would result in a dense atmosphere early in planetary history.

Rubey (1955) argued that Earth's atmosphere has continually and uniformly accumulated, implying a constant outgassing rate. He specifically stated that it is "improbable that the present atmosphere and hydrosphere are residual from a dense primitive atmosphere." He thought that the atmospheric composition had always been rich in N_2 and CO_2. The atmospheric pressure remained fairly constant due to continuous subtraction by rock forming processes and H_2 exospheric escape, which balanced the addition by outgassing. Rubey emphasized that his theory for the gradual accumulation of volatiles at a constant outgassing rate and constant atmospheric pressure is not dependent on the early atmospheric composition.

An often cited model by Holland (1962, 1964) divides Earth's atmospheric history into three parts. Holland believed, as did Rubey, in continuous addition of volatiles over geologic time. The composition of volcanic gases, however, is supposed to have changed with time. For the first half billion years, which were thought to precede core formation, native iron was present in the mantle. Volcanic gases were in equilibrium with this iron, resulting in large amounts of CH_4 and NH_3 rather than CO_2 and N_2. After core formation and removal of iron into the core, CO_2 replaced CH_4, and N_2 replaced NH_3. This second stage lasted until the beginning of the Cambrian. Molecular oxygen, O_2, did not appear until the third stage of evolution, which continues to the present. Photosynthesis was responsible for the production of O_2. Although a biological origin of O_2 is widely accepted, Brinkmann (1969) has argued that photolytic dissociation of water vapor might provide an alternative source.

Part of the diversity of thought regarding atmospheric evolution arises from uncertainty as to whether Earth differentiated into a core, mantle, and crust during or after accretion. During the 1950s and 1960s, a cold-accretion theory was popular, in which a substantially solid Earth was heated internally by radioactive sources. The resultant melting allowed Earth to differentiate, as a result of which the radioactive material became concentrated in the crust. Additional heat was added to the interior by the gravitational energy released during core formation. Volatiles (H_2O, CO_2, CH_4, NH_3, Ar, Xe, etc.) previously contained within the interior were released by melting and migrated to the surface.

Recently, Hanks and Anderson (1969) and Anderson (1972) have demonstrated the likelihood of a hot-accretion model (see chapter 6). Melting and differentiation of Earth occurred during the accretion process, lasting no more than 500,000 years and probably less. Since differentiation occurred simultaneously with accretion, the radioactive, heat-producing elements were concentrated in the crust at that time. The internal temperatures within Earth are largely residual from the accretionary energy. The constraints placed by these two thermal evolution models have profound effects on the rate of outgassing. The cold-accretion model favors a slow outgassing after accretion; a hot-accretion model favors catastrophic degassing during accretion.

Fanale (1971a) argues that Earth developed a dense atmosphere during the accretion process. In fact, he argues that it is difficult to understand why such an atmosphere would not develop. Since accreting material is heated to extremely high temperatures on impact (and probably completely vaporized) most, if not all, of the volatiles contained in the original material should be released. Although some of the volatiles would be lost to space while accreting Earth still had a small mass, the bulk would probably be retained. This would be especially true if volatile-rich accretionary materials rep-

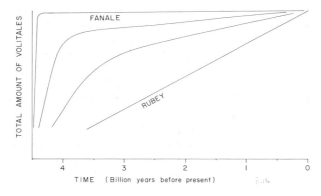

Figure 8.15. The combined amount of volatiles in Earth's atmosphere, hydrosphere, and lithosphere over geologic history, according to various models. The extreme cases are a sudden, early degassing (Fanale, 1971a) and a uniform degassing (Rubey, 1955). Intermediate paths, also shown, have been proposed as alternatives by various authors.

resented a later-stage addition to a forming planet. Clearly, this is in sharp contrast to a gradual outgassing theory. It should be noted that, while rapid accretion is sufficient to produce catastrophic early degassing, it is not essential. If Earth accreted slowly, each increment of solid material would still be heated to incandescence at the time of accretion. However, the amount of buried heat might be very low (Fanale, 1971b).

Rubey's Excess Volatiles

Rubey (1951) attempted to determine the quantity of volatiles present on Earth's surface and within the crustal rocks. He correctly argued that the present atmosphere represents only a small fraction of the total amount of volatiles expelled from Earth's interior. Enormous quantities of H_2O and dissolved CO_2 are, of course, found in the oceans. But there is a tremendous amount of H_2O and CO_2 chemically combined within crustal rocks. For example, Rubey estimated 600 times more CO_2 was contained in carbonate rocks than is seen today in the atmosphere and hydrosphere combined. He suggested that these volatiles have been released to the atmosphere throughout geologic time at a fairly constant rate. The release mechanism was thought to be outgassing from volcanic effusions.

Rubey, along with many other earlier investigators, remarked on the similarity of composition between the total volatile inventory and volcanic gases. Assuming that 0.8 percent of the magmatic gas was juvenile, as estimated by Day (1939), Rubey showed that the integrated contribution over geologic time could account for all the volatiles now contained in the atmosphere, hydrosphere, and crustal rocks.

This line of reasoning recently has been challenged by Fanale (1971a), who believes that the juvenile component in volcanic gases is close to zero. Most of the gases are recycled by crustal and upper mantle convection.

Rare Gases

The abundances of rare gases in Earth's atmosphere provide several important clues concerning the history of Earth's volatiles. As mentioned earlier, the nonradiogenic rare gases exhibit a large depletion relative to solar abundances. However, Fanale (1971a, 1971b) suggests that the appropriate standards for comparison of observed Earth rare gas abundances are not solar abundances, but the original material of accretion, which he believes to be ordinary chondrites. Ordinary chondrites resemble Earth's bulk composition and density. Fanale finds that if Earth were originally chondritic in composi-

tion, the observed rare gas abundances (except for Xe) closely match those for a completely degassed Earth. He cites this as one line of evidence that Earth completely expelled its volatiles early in its history, rather than releasing them slowly.

The Source of the Martian Atmosphere

We conclude, from the preceding sections, that there is a strong case for the early catastrophic formation of Earth's atmosphere. To conclude that Mars followed a similar history of early degassing is reasonable.

Nevertheless, one observation suggests that the Martian volatiles may not be completely residual. Rubey believed that most of Earth's volatiles formed from the crystallization of intrusive material and subsequent expulsion of volatiles. Since the volume of volcanics on Earth was thought to be small compared to intrusives, he ignored this contribution. However, Mariner 9 photographs showing the massive Martian volcanic constructs and widespread volcanic plains suggest that crystallization of extrusions might be a major source of Martian volatiles. Furthermore, since there are apparently no major recycling mechanisms for volatiles as found on Earth, there may be a large juvenile contribution to the Martian atmosphere. Let us calculate approximately how much water Mars would receive from the crystallization of the observed volcanics. The area of the planet is 4π (3390 km)^2, which is approximately 10^8 km^2. We will approximate the volume of the volcanics by assuming that 30 percent of the surface is covered by volcanic flows 2 km deep. If the water expelled during crystallization is 3 percent by volume, then 2×10^6 km^3 of H_2O, or about 20 m of water planet-wide are produced. Although this is an enormous amount compared to the present 10^{-5} m, it is small compared to the amount expected from catastrophic outgassing, about 1 km.

Furthermore, if we assume that the ratio of H_2O:CO_2 outgassed is 45, as Rubey found for Earth, we should expect about 0.5 precipitable meters of CO_2, almost exactly what is observed today. Does this coincidence imply that Mars has only received volatiles entirely as a result of slow outgassing? Not necessarily. There might be a large amount of unseen volatiles in the planet's surface and subsurface material that we have not considered.

Sinks

Just as Rubey concluded that Earth's atmosphere represents only a small part of the released terrestrial volatiles, the Martian atmosphere similarly may represent only a small fraction of Martian volatiles. Potential sinks include permanent removal (atmospheric escape, solar wind stripping), or interactive reservoirs (oceans, ground ice, ice caps, chemical combination, adsorption on surface materials).

The major terrestrial sinks are the oceans, chemical combination in sediments, and the polar caps. The Martian situation is quite different. There are no oceans. In the absence of large water bodies, widespread precipitation of carbonates is unlikely. Instead, much of the H_2O, and possibly CO_2, lies frozen at the poles. In addition, a substantial, if not predominant, quantity of both of these volatiles may exist beneath the Martian surface at all latitudes, both as ground ice and adsorbed on mineral grains. Permanent escape from the atmosphere is also important.

The availability of CO_2 is significant because climatic changes resulting from insolation variations require the episodic addition of $CO_2(g)$ into the atmosphere. The increased density of the atmosphere resulting from such an addition increases the atmospheric heat capacity, allowing the transfer of more heat from the equator to poles. The resulting unstable situation (Sagan et al., 1973a) may be capable of producing a hundredfold increase in the atmospheric pressure, *provided* that a supply of CO_2 exists.

Polar Caps

It is generally recognized that the seasonal ice caps are composed of $CO_2(s)$, as mentioned earlier. In addition, a permanent $H_2O(s)$ cap apparently exists at each pole (Murray et al., 1972). Leighton and Murray (1966), and later Murray and Malin (1973b), suggested that a permanent $CO_2(s)$ reservoir may also exist, containing several times the present atmospheric CO_2. Fanale and Cannon (1971, 1974) believe that these polar sinks represent only a small part of the volatiles compared to the amount adsorbed by the regolith. Ingersoll (1974) argues that most of the CO_2 is now present in the atmosphere. Both Fanale and Cannon, and Ingersoll, doubt the existence of a large permanent $CO_2(s)$ polar reservoir.

The existence of a permanent $H_2O(s)$ cap was inferred from the observed behavior of the shrinking polar deposits during the Mariner 9 mission. The rapidly retreating $CO_2(s)$ cap revealed a permanent cap that showed no further change. Murray et al. (1972) suggested that this cap is either a thicker $CO_2(s)$ deposit, or that it is composed of a substance other than the seasonal $CO_2(s)$ cap. They preferred the latter interpretation, the obvious conclusion being that the residual cap consisted of $H_2O(s)$, which exhibits a much lower sublimation rate under Martian polar conditions. A later observation (Kliore et al., 1973) that the south polar cap is several kilometers higher than the north polar cap led Murray and Malin (1973b) to conclude that a permanent southern $CO_2(s)$ cap would rapidly be transferred to the north, further implying that H_2O must be the composition of the permanent southern cap. Alternatively, Briggs (1974) concludes that the nature of the residual caps cannot be reliably determined with presently available data.

The amount of $H_2O(s)$ contained in surface condensates is shown by figure 8.13 to be the equivalent of 70 g/cm^2. The seasonal $CO_2(s)$ caps are only about 15 percent of the total atmospheric content. Both of these values are far lower than the expected amounts if Mars outgassed to a similar extent as Earth.

The low fraction of $CO_2(s)$ in the seasonal caps relative to $CO_2(g)$ in the atmosphere led Leighton and Murray (1966) to the conclusion that there must be a permanent $CO_2(s)$ reservoir that buffers the atmospheric pressure of $CO_2(g)$. The agreement between the observed and predicted pressures from measured temperatures and pressures (fig. 8.6) would otherwise seem to invoke the coincidence that there has been just enough CO_2 supplied to the Martian surface barely to form polar caps. This point is elaborated by Murray and Malin (1973b), who believe that they have identified a permanent $CO_2(s)$ cap near the north pole. They estimate this reservoir contains from two to five times the present atmospheric CO_2.

Ingersoll (1974) questions whether the apparent coincidence is as significant as it first appears. For example, the predicted atmospheric pressure is strongly dependent on the albedo, which is influenced by such factors as dust on the ice surface. Within a range of possible albedos from 0.61 to 0.81, the predicted vapor pressures range from 30 mb to 0.6 mb. In addition, both the presence of the polar hood, which reduces the rate of cooling of the surface, and the effects of atmospheric heat transport add uncertainty to the calculation of a predicted atmospheric pressure.

Ingersoll presents several other arguments against the presence of a permanent $CO_2(s)$ cap. The seasonal cap disappears by the beginning of the summer season. If a permanent cap of $CO_2(s)$ exists, it should also lose mass if it is in equilibrium with the atmosphere, as necessitated by the argument of Leighton and Murray. Ingersoll further points out that the observed pressure of $H_2O(g)$ should be much lower if there is a permanent $CO_2(s)$ reservoir, since such a reservoir acts as a cold trap for $H_2O(g)$. The observation that the atmospheric P_{H_2O} varies from 10 to 50 precipitable microns during the year is inconsistent with such a trap, which would tend to collect atmospheric H_2O at such a rate that one would find a much larger quantity of $H_2O(s)$ at the poles than is observed. He concludes that either a permanent $CO_2(s)$ cap is sealed from the atmosphere, in which case it cannot buffer the atmospheric pressure, or else such a reservoir does not exist.

Another explanation for the apparent buffering of the atmospheric pressure is that there may be a substantial quantity of CO_2 bound in adsorption in the Martian regolith. Fanale and Cannon (1974) suggest that the observed 6-mb pressure may be "illusory." They provide a simplified model for the interchange of volatiles between surface and atmosphere and suggest that the process of physical adsorption by the Martian regolith may be a major mechanism for the storage of Martian volatiles. For a depth of 1 km, they find that the regolith could accommodate about 40 times the present atmospheric CO_2, and 100 times the maximum estimated H_2O surface condensates. However, they also note that these calculated amounts account for only 1 percent of the H_2O and 5 percent of the CO_2 that one would expect from a Mars that has outgassed as extensively as Earth.

However, as a result of the arguments presented earlier, it is unlikely that adsorption occurs to a depth of 1 km, and these percentages are further reduced. If Mars has extensively outgassed, regolith adsorption cannot provide the primary sink.

Ground Ice or Permafrost

The conclusion we will reach is that the major sink for H_2O is beneath the surface as ground ice. We will use the term "ground ice," as opposed to "permafrost," since the strict definition of permafrost on Earth is soil or rock material that has remained below 0°C for more than 2 years without any reference to water content (Ferrians et al., 1969). By this definition, all of the Martian subsurface is presently permafrost. The presence of ground ice should be a separate issue. Nevertheless, the term permafrost in the Martian literature has come to be associated with solid H_2O.

The presence of ground ice or permafrost has been suggested by previous authors. Lederberg and Sagan (1962) suggested that local sources of heat might melt permafrost, creating local sources of water. Wade and DeWys (1968) discussed the formation of small-scale topographic manifestation of ground ice, such as patterned ground or frost mounds. However, no one seriously proposed major topographic features until they first appeared in Mariner 6 and 7 photographs. The slump features described by Sharp et al. (1971a) appear to be quite

similar to some terrestrial thermokarst features (Leighton et al., 1969a; Belcher et al., 1971; Gatto and Anderson, 1975), but the Martian features are at least ten times larger than equivalent terrestrial features.

This difference in scale does not necessarily mean that Martian features are not caused by ground ice, but it does suggest that the actual mechanisms of emplacement and decay may be very different than on Earth.

If ground ice is present below the surface, there must have been a source of H_2O, and a mechanism for placing it in a subsurface reservoir. Anderson et al. (1973) list buried glaciers, sea, lake, or river ice, buried snow, and hydroeffusions as possible sources. They suggest that, in an earlier stage of Martian development, large amounts of water may have formed surface water, possibly on the scale of a large sea. As the atmosphere cooled, this sea may have been transformed to a vast cake of ice and buried under eolian sediments.

There is no evidence that bodies of water on the scale of oceans have ever existed on the Martian surface. What is more likely is that, as occurs on Earth, water percolated through porous surface materials and reached substantial depths. Ground ice then formed as a result of the freezing of water in situ. This is exactly what one would expect to result from precipitation of an early dense H_2O-rich atmosphere. As the early atmosphere cooled, it eventually would pass through the stability field of $H_2O(l)$. Since the atmosphere was rich in $H_2O(g)$, $H_2O(l)$ would have stabilized, exactly as was described earlier in figure 8.2. Further decreasing temperatures never allowed the $H_2O(l)$ to re-enter the atmosphere, but rather favored subsurface freezing.

One might further suspect, on the basis of previous arguments, that the coldest point on a planet's surface controls the partial pressure of $H_2O(g)$, that most of the ground ice would be concentrated at the poles. However, this is not necessarily a valid conclusion.

We can dismiss atmospheric radiation cooling as the temperature-regulating mechanism because this mechanism is much too rapid. It is possible that the solar flux slowly declined early in Martian history. Although there is no direct evidence for this, the mechanism cannot be dismissed. Another alternative is that the Martian atmospheric temperature was controlled by the planet's surface temperature. A rather high surface temperature, possibly over 400°K, would clearly influence the lower atmospheric temperature. This is a conservative estimate of the surface temperature if one adopts a model requiring extensive crustal melting. Furthermore, this temperature would be nearly homogeneous at all latitudes. Water that fell to the ground would have no difficulty percolating to depth since the surface would be well above freezing. In addition, until the subsurface temperature dropped below the stability temperature of $H_2O(l)$, rain falling on the surface would have rapidly been returned to the atmosphere due to evaporation. This would explain why the crater obliteration event described in chapter 4 existed, and why it was of sufficient duration to allow the magnitude of erosion that occurred.

We conclude that it is very likely that, at least at one time, there was a substantial quantity of ground ice below the Martian surface. Furthermore, the time of its emplacement was concurrent with the crater obliteration event. This interpretation is consistent with the geologic relationships, described later, and with the thermodynamic requirements, described earlier.

However, at least one problem remains. $H_2O(g)$ easily precipitates from the Martian atmosphere. Other species, such as N_2, do not. In order to explain their absence, we must consider the final sink for Martian volatiles.

Exospheric Escape

The ability of a gas molecule to escape from the top of the Martian atmosphere depends on its velocity and the amount of atmosphere between the molecule and the vacuum of space. Clearly, a molecule must possess a velocity equal to or exceeding the escape velocity from the gravitational field of the planet. Furthermore, the molecule must be at a sufficient altitude that it can travel into space without colliding with another molecule and recoiling back toward the planet.

The distribution of velocities in a gas is predicted by the Maxwell-Boltzmann distribution. However, we will not concern ourselves with the mathematical details of escape, only with the qualitative physical interpretation. The important qualification that the Maxwell-Boltzmann distribution places on the arguments is that, at the same temperature, molecules with small molecular weight (for example, H_2: m.w. = 2) move much more rapidly than heavy molecules (CO_2: m.w. = 44, H_2O: m.w. = 18). Since smaller molecules move rapidly, they escape more easily.

Further, high temperatures induce higher velocities. The temperatures important to escape are those at the top of the atmosphere (exosphere). These were measured by Mariner 6 and 7 as 350°K (Anderson and Hord, 1971). This exospheric temperature is much lower than Earth's, which is about 2000°K. Older papers treating atmospheric escape from the Martian atmosphere (e.g., Urey, 1959)

derive a very short lifetime for gas species because of the then-assumed exospheric temperature of over 1000°K. However, when Urey's calculations are corrected for the lower observed temperature, one finds that certain important species as N_2 and H_2O cannot escape. Why, then, are they not observed? One way out is to postulate a different outgassing history for Earth and Mars. A more likely solution involves nonthermal escape mechanisms related to various photolytic reactions, resulting from the action of solar radiation.

Brinkmann (1969) concluded that photodissociation of H_2O in the terrestrial atmosphere may have produced large amounts of O_2 in the absence of biological activity. Brinkmann (1971) continued to re-examine the possibility of other photolytic reactions, and found, in particular, that N_2 could be removed from the Martian atmosphere by such processes. When N_2 absorbs solar radiation in the far ultraviolet, the molecule dissociates into two N atoms moving apart with identical velocities. The energy released during dissociation gives each N atom a boost, which, if added to the thermal motion of the molecule, can assist the N atom in leaving the Martian gravitational field. The boost must be in the direction away from the Martian surface, meaning that only one of the N atoms can escape.

Brinkmann found that, assuming an exospheric temperature of 500°K, photons of energies greater than 15.07 eV can cause the escape of N atoms. The rate is such that, if N_2 has outgassed uniformly during Martian history, 75 percent would have escaped. If N_2 formed rapidly during the early history of the planet, 90 percent would have been lost. Brinkmann concludes that the apparent absence of N_2 in the Martian atmosphere does not require a drastic alteration of the terrestrial outgassing model. An additional escape reaction

$$N_2^+ + e^- \longrightarrow N + N$$

was defined by McElroy (1972).

McElroy also discusses the effects of photolytic reactions on the escape of CO_2 and H_2O, and on the production of other minor species as CO and O_2^+ in the Martian atmosphere. The complete calculations are involved since one must take into account all possible dissociation and recombination mechanisms, the latter limiting the effectiveness of the former.

An important reaction is the photodissociation of water near the Martian surface.

$$H_2O + photon \longrightarrow OH + H.$$

The resulting hydrogen atoms recombine to form H_2, which diffuses to the upper atmosphere. The subsequent reaction

$$H_2 + photon \longrightarrow H + H$$

allows hydrogen to escape in a very short time. The result should be a buildup of oxygen, similar to the prediction of Brinkmann (1969). The observed abundance of O_2 could be attained in only 10^5 yr, implying that some mechanism for the removal of oxygen also exists.

McElroy believes that the reactions

$$CO_2^+ + e^- \longrightarrow CO + O$$

and

$$O_2^+ + e^- \longrightarrow O + O$$

provide oxygen atoms with the energy needed to escape, similar to the arguments for N presented before. Nonthermal oxygen escape regulates the loss of H_2 to space, or else the resulting imbalance would build up the concentrations of CO or O_2.

Similarly, CO_2 may be slowly removed by the reaction

$$CO^+ + e^- \longrightarrow C + O$$

where the supply of CO is produced by the reaction mentioned above. This provides a mechanism for the removal of carbon atoms from the Martian atmosphere. McElroy calculates that the CO_2 level could have been reduced by 10 percent in the history of the planet. H_2O is lost at a rate such that a surface pressure of 0.1 atm., or about 5 precipitable meters, would be lost. He concludes that, because of the observed abundances of H_2O and CO_2 today, the relative rates of CO_2 and H_2O outgassing have been comparable to Earth rates, but the absolute rates have been 1000 times less.

However, if there is a large amount of ground ice, as mentioned in the previous section, the amount of H_2O outgassed may have been larger than McElroy's estimate. This would imply either that the ratio of CO_2 to H_2O is much smaller than on Earth or that there is a mechanism for removing carbon that we have not identified. CO_2 dissolved in ground water is one possibility.

Alternatively, suppose the original atmosphere consisted of CH_4 and H_2O, as proposed by Fanale (1971b). He argues that such a composition is consistent with catastrophic degassing during planetary accretion. The fate of such an atmosphere is the production of CO_2 and the loss of H_2 resulting from the reaction

$$CH_4 + 2H_2O = 4H_2 + CO_2.$$

The number of moles of CO_2 produced equals the original number of moles of CH_4, producing the same embarrassing excess of CO_2 as in the previous model. However, the reaction above may not be the only method of controlling atmospheric composition. If nonthermal mechanisms occurred then, as now, there may be some mechanism for the loss of carbon atoms that is more efficient than merely for CO_2. It will be suggested here, without proof, that such a mechanism may have existed. Unfortunately, such common reactions as

$$CH_4 + photon \longrightarrow CH_3 + H$$

are possibly ineffective since the hydrogen atom contains almost all of the velocity. One needs a reaction that provides a carbon atom with a sufficient escape energy. Any mechanism that first requires the production of CO_2 or CO may be limited by the same factors that limit the escape of CO_2.

It might be possible that, once H_2O was trapped beneath the surface, much of the remaining Martian atmosphere was lost, resulting in a substantial period with no atmosphere on the planet. More recent volcanic activity could have produced the CO_2 observed today. There is nothing in the crater diameter-frequency data to contradict such an occurrence, and the absence of an atmosphere might help explain the apparent freshness of some of the very ancient features.

The Martian Climate, A Summary Statement

Based on our discussions of atmospheric evolution, we conclude that a good case can be made for an early dense atmosphere. Water, condensed from this atmosphere, percolated beneath the surface, becoming ground ice. More volatiles have been added by episodic volcanic activity, although these are subordinate in volume. Exospheric escape lowered original nitrogen values, but did not substantially reduce the amount of CO_2—unless a mechanism existed in early atmospheric history for removal of carbon in the form of CH_4. The volatiles presently residing in the polar caps and in the atmosphere are a small fraction of the volatiles expected from terrestrial-like outgassing. Regolith adsorption may provide a trap for some volatiles, but the volume is not great. However, the adsorption process does provide an important mechanism for observed diurnal and seasonal changes in atmospheric water concentrations.

Our favored model requires that early ground temperatures be sufficiently high to permit downward seepage of liquid water. Subsequently the ground temperatures fell below freezing, so that ground ice formed to depths of several kilometers. Still more recently, some fraction of this ground ice has been melted to form collapse and surface-runoff features.

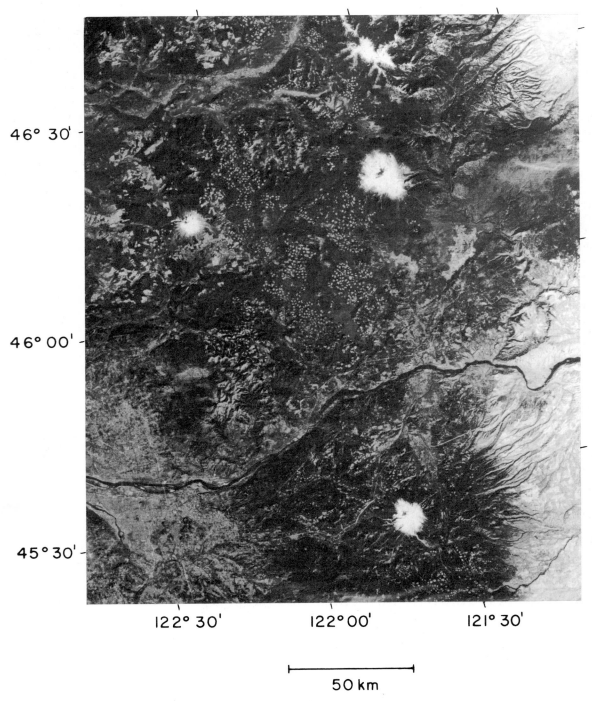

46° 30'

46° 00'

45° 30'

122° 30'   122° 00'   121° 30'

|———————————————|
50 km

Figure 8.16. ERTS picture showing a section of northern Oregon and southern Washington. The two states are separated by the Columbia River. Mt. Hood appears in the lower right, Mt. Adams in the upper right, and Mt. Saint Helens in the upper left.

A Strategy for Interplanetary Comparison of Channels

Various types of Martian depressions, particularly channels, frequently have been likened to, or contrasted with, terrestrial stream channels. Much of this discussion has lacked any conscientious attempt to present the putative terrestrial analog for examination. In some cases where "look alike" photographs *have* been displayed, major differences in scale have been ignored.

In the following sections we briefly review the sedimentological and erosional features that characterize terrestrial stream systems, in order to suggest some guidelines and establish some constraints for analysis of Martian features. A number of Earth Resources Technology Satellite (ERTS) pictures are included. Because the resolution of these pictures is approximately the same as for Mariner B frames, meaningful comparisons are facilitated.

The review of terrestrial and Martian features will demonstrate two major points: first, that many diagnostic sedimentological features of terrestrial stream systems occur at scales more detailed than Mariner pictures; and second, that the large-scale tributary patterns which characterize terrestrial rain-fed channel systems are lacking in the Martian situation.

Terrestrial Rivers

Channel Patterns

The chief agent in scouring terrestrial valleys is water, but a variety of modifying processes and products should be mentioned. Topographic irregularities can be formed by tectonic and volcanic processes. Once formed, they influence subsequent drainage patterns. Conically shaped volcanic structures lead to entrenchment of a radial drainage pattern, as on Hawaii or on some of the volcanoes of the Pacific Northwest (fig. 8.16). Down-faulted troughs or graben often are the sites for intermittent internal streams, lakes, or through-going rivers. An example is Death Valley, situated just east of the normal fault system that bounds Sierra Nevada along its eastern edge. Another example is the African rift zone, occupied by a string of narrow lakes (fig. 8.17). Uplifted mountain belts can similarly influence drainage. Even though the individual folds and thrust faults are not necessarily reflected in surface topography, the great thickness of light crustal rocks beneath the mountains causes them to stand relatively high even when they are isostatically compensated. The serpentine course of the Brahmaputra River illustrates the dramatic effects of the Himalaya Mountains. The river flows parallel to the fold belt until it reaches a point where the mountain ranges change direction and become discontinuous. There, the Brahmaputra breaks through to the Bay of Bengal, almost completely reversing direction (fig. 8.18).

Even when the initial erosional surface is perfectly level, differences in strength and orientation of underlying rocks can influence the position and shapes of stream channels. Structural fractures in bedrock commonly are present in two vertical sets, approximately at right angles. Streams flowing along these lines of weakness frequently exhibit a rectangular pattern. Where folded strata of differing resistance to erosion underlie a level plain, renewed uplift leads to the development of a trellis-like pattern.

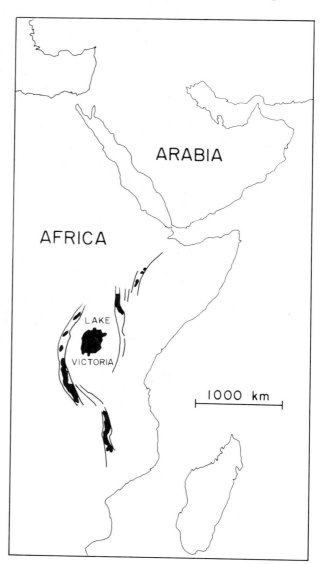

Figure 8.17. Linear lakes in central Africa, formed in great rift depressions.

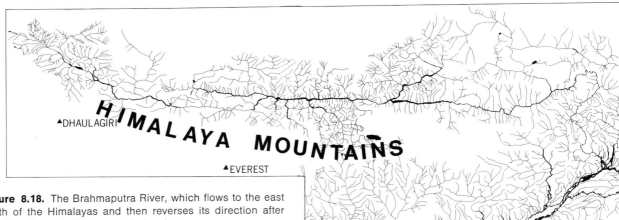

Figure 8.18. The Brahmaputra River, which flows to the east north of the Himalayas and then reverses its direction after breaking through the mountain barrier.

Tributary streams follow the strike of the folded beds, but the main channel cuts through high ridges, a phenomenon difficult to understand in terms of present topographic relief. The explanation is that the pattern of the main channel was established when the rocks had been uniformly eroded to a plain close to sea level and perhaps had been mantled by a thin veneer of sediment. The main channel followed the topographic gradient, which happened to be at right angles to the structural grain. As regional uplift took place, the elevation of the land was balanced by the ability of the stream to cut downward. There are numerous examples of this conjectural sequence of events. Most commonly cited are the Delaware and Susquehanna rivers, which cut through the folded rocks of the Appalachians to form prominent "water gaps" (fig. 8.19).

Where streams develop in a lithologically homogeneous substrate, a dendritic or veined appearance results. Although the overall pattern looks random, a certain statistical order prevails. Channel segments can be rank-ordered according to their position in the hierarchy of stream junctions. Starting at the source, first-order channels are formed. Two first-order segments join to form a second-order segment, second-order channels fuse to form a third-order segment, and so on (fig. 8.20). Numbers of stream segments in many river systems increase geometrically with decreasing rank. Similarly, stream length is a geometric function of stream order, as are stream slope and basin area. These empirical relationships permit one to make some predictions about stream development. Of equal interest is the possibility for analytical reconstruction of an ancient drainage system, which has been partially destroyed (fig. 8.20).

For those streams developed in a temperate region, where there is normal rainfall, the interstream regions are progressively weathered back and down to form rounded hills. The patterns revealed in high-altitude aerial photographs are striking. In youthful terrains, these hills are the dominant feature, with a network of ridges present where retreating slopes intersect. As erosion proceeds, the interstream regions show less topographic relief, and the meandering stream valleys are relatively prominent (fig. 8.21).

In arid regions different erosional forms develop. Occasional floods form sharp-walled arroyos. Most arroyos are discontinuous, ending in an area of ephemeral ponding rather than gaining in size and complexity as do the higher-order channels of temperate regions. Mountain fronts are faceted by steep scarps, cut back by rock slides and underground sapping. The scarps are pierced by valleys formed by intermittent streams. Where the water disgorges on the lower plain, a conical apron of debris (alluvial fan) is produced.

In some cases a river system gathers abundant

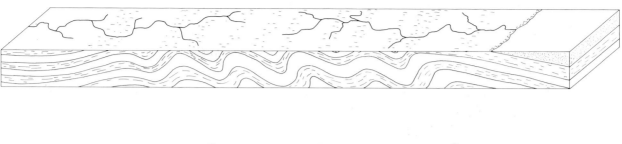

Figure 8.19. Schematic block diagrams, showing progressive development of a trellis drainage pattern analogous to that which exists in the valley and ridge province of eastern Pennsylvania.

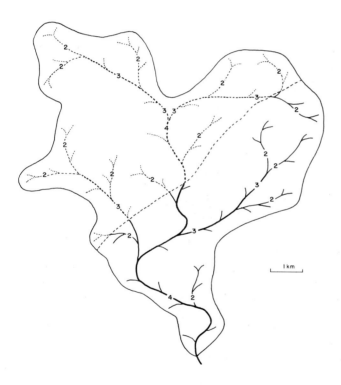

Figure 8.20. Hypothetical stream pattern. Channels shown in solid lines are visible, but those shown in dashed lines have been destroyed by erosion. Because a few of the first order stream segments are still visible, the total array can be reconstructed using the data of the accompanying table, determined by Morisawa (1962) for the Allegheny River. Numbers in parentheses are those determined by inspection of stream segments that have escaped erosion. Obviously, the exact position of eroded channels is conjectural. This type of analysis is imprecise in detail, but has general validity, based on empirical analysis of many terrestrial stream systems.

Stream order	Number of segments	Bifurcation ratio	Mean length (km)	Cumulative mean length (km)	Length ratio
I	64 (27)		.4 (.4)	.4 (.4)	
		4 (3.8)			4.2 (4.7)
2	16 (7)		1.3 (1.5)	1.7 (1.9)	
		4 (3.5)			3.1 (2.8)
3	4 (2)		3.6 (3.5)	5.3 (5.4)	
		4 (2)			3.0 (N.A.)
4	1 (1)		11 (N.A.)	16.3 (N.A.)	

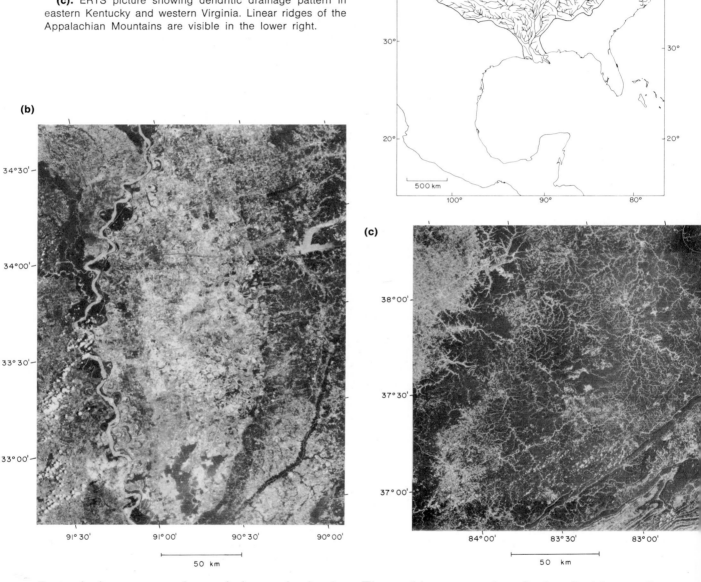

Figure 8.21 (a). The Mississippi River drainage basin. (From Matthews, 1974. Reprinted by permission of Prentice-Hall, Inc.)

(b). ERTS photomosaic showing the Mississippi River Valley, south of Memphis, Tennessee. A large number of abandoned meanders with characteristic lunate shape are visible to the east of the present river channel.

(c). ERTS picture showing dendritic drainage pattern in eastern Kentucky and western Virginia. Linear ridges of the Appalachian Mountains are visible in the lower right.

water in its upper reaches and then maintains its channel through an arid terrain. A notable example is the Colorado River, flowing out of the snowbound Rocky Mountains and cutting its Grand Canyon in the Colorado Plateau of southwestern United States. The resulting topography—flat interfluvial areas dissected by knife-edge channels—contrasts markedly to the gentle hills of temperate regions (fig. 8.22). Indeed, the plateau surfaces bordering the Grand Canyon are so little affected by stream erosion that

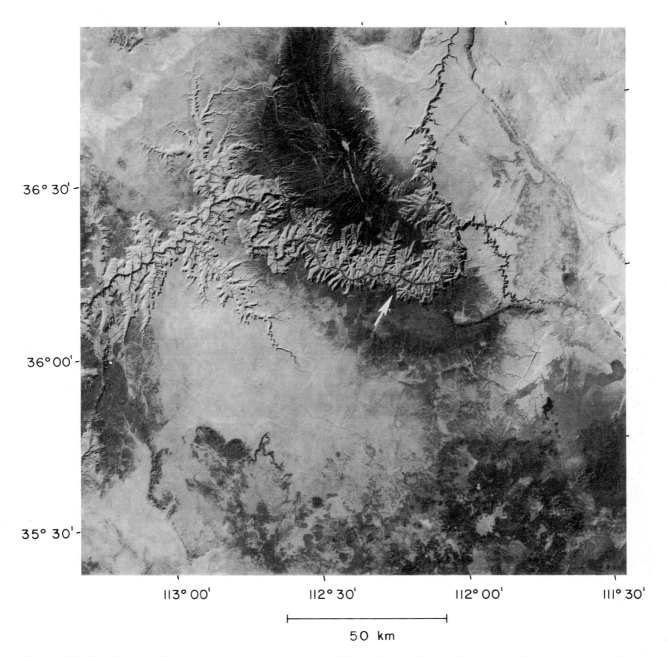

36° 30'

36° 00'

35° 30'

113° 00'   112° 30'   112° 00'   111° 30'

50 km

Figure 8.22. The Colorado River, deeply incised in the rocks of the Colorado Plateau. The town of Grand Canyon, in Grand Canyon National Park, is indicated by an arrow. This is an ERTS picture.

the presence of this great cleft is completely unsuspected by the traveler approaching from the south or north.

Valleys formed by fluvial erosion can be subsequently modified by other processes. During the recent Ice Age, glaciers occupied many mountain valleys that had previously been carved by water. Glacial erosion changed "V" profiles to "U" shapes.

Distinctive "hanging valleys" were formed where secondary glaciers flowed into the main valley. To a lesser degree eolian erosion enlarges valleys where water once flowed. More commonly, the ancient drainage pattern is partially buried by wind-driven sediment.

In rare instances, channel systems are formed by massive floods following catastrophic break-outs

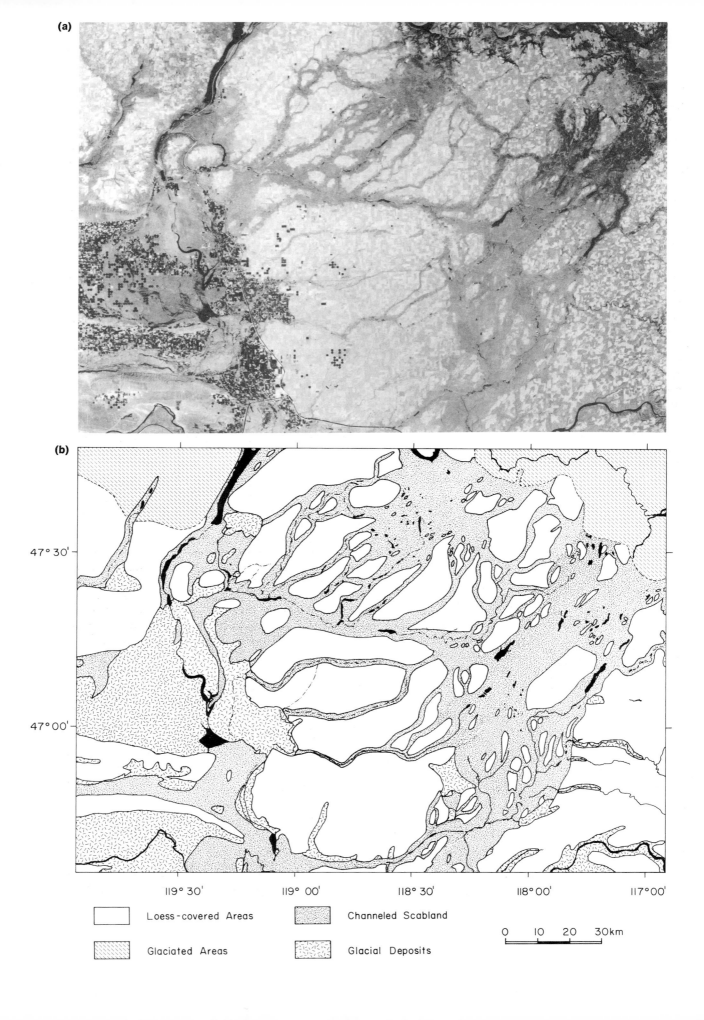

(a)

(b)

47° 30'

47° 00'

119° 30'   119° 00'   118° 30'   118° 00'   117° 00'

☐ Loess-covered Areas   ▨ Channeled Scabland

▨ Glaciated Areas   ▨ Glacial Deposits

0   10   20   30km

Opposite page:
Figure 8.23 (a). ERTS photomosaic showing the channeled scabland of eastern Washington.
   (b). Geologic features of the channeled scabland. Adapted from Bretz (1969).

from reservoirs, natural or manmade. A widely cited example occurs on the lava plateaus of eastern Washington. During Pleistocene glaciation a large lake formed along the western boundary of Montana in the vicinity of Missoula. At several times the retaining walls of the reservoir failed, either by bank collapse or by ice retreat. Water poured out to the west, coursing across a region now known as the channeled scablands. This terrain is characterized by anastomosing channels, basins carved in rock, cataract gullies, hanging valleys, coulees, giant gravel deposits, and enormous trains of sand ripples—all occurring at topographic levels high above the present drainage systems (fig. 8.23). Bretz (1923) first suggested that this collection of features could best be explained by enormous floods of short duration. Although the hypothesized deluge was initially doubted by many geologists with a bias against natural catastrophes of such epic proportions, a succession of floods is now firmly documented (e.g., Bretz, 1969; Baker, 1973).

Stream Deposits

Terrestrial streams are characterized not only by their channel geometry but also by their deposits. Two contrasting regimes are commonly found. Braided streams are often encountered where rivers emerge from mountain fronts or from glaciers, heavily laden with erosional debris. They are characterized by unusually high discharge and/or steep channel gradient. Nonsystematic erosion and deposition produces numerous channels, forming an intersecting network of channels and islands (fig. 8.24). The channel is broad with rapidly fluctuating lateral variations. Shifting the channel leads to accumulation of lensoid deposits of coarse gravel, sand, and silt. Nonsystematic variation in sediment size is present in all three dimensions: across the major channel, parallel to it, and beneath the surface. The topographic relief at any time is slight, no more than several meters. The major channel is shallow, with poorly defined banks. Sand bars extend less than a meter above the water level.

In regions where topography is subdued, and slopes are gentle, the braided pattern is replaced by a sinuous or meandering pattern. In plan view, the channel has a snake-like appearance. Over time periods on the order of a few years the meanders systematically migrate laterally and also downstream as they excavate the major valley (fig. 8.21). Erosion proceeds along the outer banks of meanders

whereas deposition occurs along the inner arc in relatively slow-moving water. Accordingly, a meander moves laterally by a combination of erosion on its outer side and deposition on its inner side. Sporadically, during times of flood, the channel is breached and shallow flood waters extend across the entire valley. Coarsest material is deposited close to the channel, and fine sediment is carried greater distances in suspension. Successive flood deposits build up natural levees of sand on either side of the main channel system. By this mechanism, coupled with deposition in the channel, the stream bottom is actually built up to a point where it stands topographically higher than the marginal parts of the valley. This situation, exemplified by the lower reaches of the Mississippi, is potentially disastrous. Once the natural levees are breached, there is catastrophic flooding in the distal parts of the valley.

The level to which a stream channel erodes is determined primarily by the base level, or water level at its lower end. Working upstream from that point, streams seek to maintain an equilibrium profile that, to some degree, is controlled by discharge and channel shape. If sea level is lowered abruptly, then the stream proceeds to establish a new equilibrium profile. As it cuts downward with new vigor, remnants of the pre-existing valley remain as terraces on either side of the valley. The Pleistocene epoch, with its accompanying changes in sea level resulting from periodic production and dissipation of polar ice, has produced sets of multiple terraces in many of the major valleys in the United States.

Most deposition in a stream system occurs not within the valleys but instead at the mouth, where rapidly flowing stream water is injected into a standing body of water. The bed load of sediment is dropped immediately; the fine-grained suspension load may be carried a short distance further before it settles to the bottom. The large-scale depositional body is called a delta—so named after the regular triangle of deposits present at the mouth of the Nile River. The Mississippi delta has a bird-foot pattern, the result of successive building outward of individual distributary channels (fig. 8.25). Deltas include many cubic kilometers of sediment. In the Mississippi delta, for example, 113 cu km have accumulated in the past 450 years. The present annual addition is about 0.25 cu km. The large volume, and especially the great thickness of deltaic sediments is made possible by the fact that they form in standing bodies of water of substantial depth. That depth essentially controls the sediment thickness. Take away the body of water, and no delta of any size can form. Instead, as we observe in arid regions, alluvial deposits spread out in fan-shaped aprons. These are formed by ephemeral streams which, at their

(a)

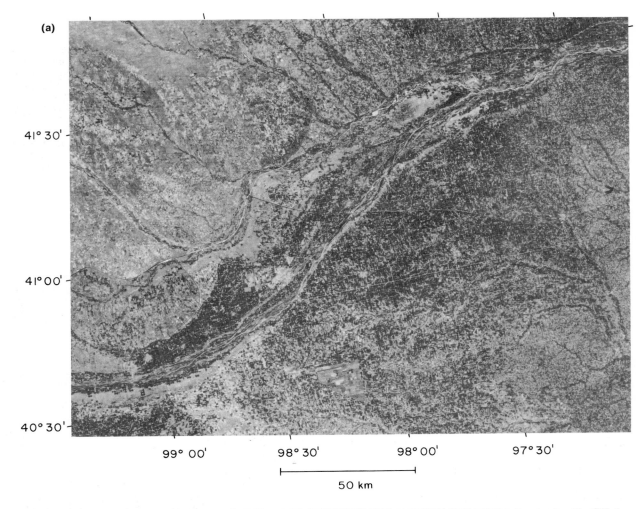

41° 30'

41° 00'

40° 30'

99° 00'   98° 30'   98° 00'   97° 30'

50 km

(b)

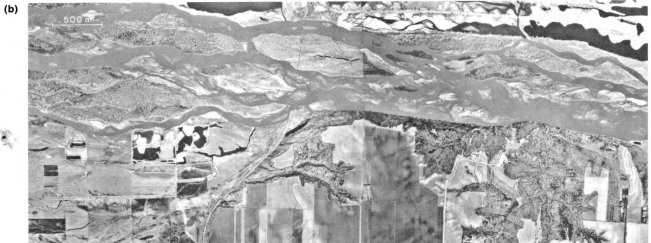

500 m

Figure 8.24 (a). ERTS photomosaic of the Platte River, Nebraska. This is a good example of a braided stream.
  (b). Aerial photograph of the Platte River between North Bend and Valley. This photograph is located directly to the east of the photomosaic shown in (a). Scale is approximately thirty times larger than that for (a). Numerous sand bars are visible in main channel.

Figure 8.25. ERTS picture showing the Mississippi River delta. The prominent finger of land is the Birdfoot delta. Plumes of sediment-laden water spread into the ocean. Inset shows the position of picture. Many former distributary channels occur in the western part of the delta. As the main course of the river has shifted, these channels have been abandoned.

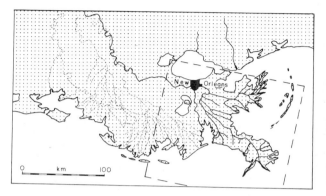

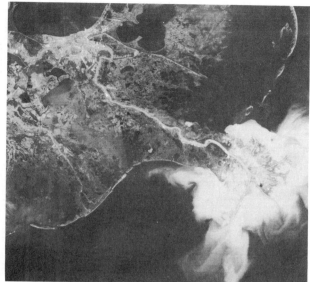

flood peaks, carry a sand and gravel load. As the flood waters reach the arid basin and rapidly dissipate, both by evaporation and percolation downward, the bed load is deposited with little reworking. Each successive flood adds to the growing cone of sediment resting against steeper mountain slopes (fig. 8.26).

In talking about stream channels, it is important to distinguish between the various morphological and sedimentological elements that we have just discussed. The channel proper, the depression that is filled with water, is seldom more than a few hundred meters wide and a few tens of meters deep, even for the largest rivers. This channel is generally emplaced in a much larger valley, which forms by oscillating migration of the channel or by a variety of lateral slope-retreat processes. The stream deposits, of particular interest to the geologist who seeks to interpret ancient sedimentary rocks by comparing them with contemporary sediments, seldom display much topographic relief. Exceptions are deltas and alluvial fans. Alternating aggradation and degradation leads to preservation of terraces along the valley walls.

Martian Depressions

First examination of Mariner 9 pictures revealed a variety of linear and irregular depressions, particularly prominent in the cratered terrain. This immediately provoked an argument as to whether or not the depressions were excavated by flowing water—

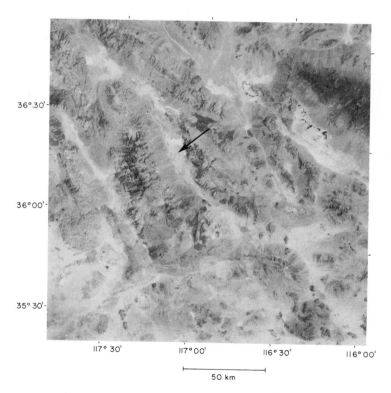

Figure 8.26. Death Valley, California, indicated by arrow. The Panamint Mountains are located to the west of Death Valley, the Amargosa Mountains to the east. A band of coalescing alluvial fans is situated between the Panamint Mountains and the axis of Death Valley. This is an ERTS picture.

an argument that continues to the present. Before entering the fray, we should try to sort through the morphological evidence. It will prove useful to divide the depressions into six categories, which are descriptively separate, even though they may be genetically linked. Those categories are canyons, chaotic terrain, fretted and knobby terrain, channels, rilles, and furrows. Most of these landforms have been briefly noted in our discussion of global geology (chapter 3). Canyons include linear depressions of probable structural origin. Chaotic terrain and fretted-knobby terrain are both related to erosion of cratered terrain along the great-circle girdle, which divides the planet into northern plains and southern cratered terrain. Channels are curvilinear depressions, somewhat resembling terrestrial stream valleys. Rilles are volcanic depressions of restricted occurrence. Furrows are something of a wastebasket category, including sets of linear, disarticulated depressions that impart a corrugated texture to the terrain.

Canyons

Troughs, as described by Sharp (1973c) include parallel, linear depressions tens of kilometers wide and hundreds of kilometers long, which, in aggregate, extend from 110°W, 10°S to 50°W, 15°S, a distance of more than 3000 km. Early in the Mariner mission these troughs were termed canyons (McCauley et al., 1972), primarily to emphasize their similarity in detail and size to the most impressive depression on Earth, the Grand Canyon of southwestern United States. In fact, the Grand Canyon is dwarfed by the Martian troughs, which, if transported to Earth, would extend all the way from Los Angeles to New York, have widths equivalent to the distance from Boston to New York, and have depths rivaling the elevation above sea level of Mt. McKinley.

In plan view, most of the individually named canyons form a continuous depression, descending in elevation from west to east. Intersecting, U-shaped, linear hollows that show strong structural control form the embryonic network of Labyrinthus Noctis. This plexus is traced into two parallel troughs, Tithonius Chasma to the north and Ius Chasma to the south. These fuse to form Melas Chasma-Coprates Chasma, which continues into somewhat shallower irregular depressions of the chaotic terrain further to the east. Ophir Chasma, which parallels the main system to the north, is a depression closed at both ends, although it has a lateral junction with Melas Chasma. Neighboring examples of closed depressions are Hebes Chasma and Juventae Chasma. These structures show a plan view similar to that of the canyons. Their emplacement parallel to Valles

Figure 8.27. Wall and floor material in the Ius Chasma. Spines of resistant wall rock are prominently exposed on the upper part of the northern wall. Talus occurs at lower elevations. Jumbled talus and collapse chaos extend all the way across the canyon. (DAS No. 10204539, rev. 240, B camera, center at 86°W, 7°S; DAS No. 10204609, rev. 240, B camera, center at 85°W, 7°S; DAS No. 10204679, rev. 240, B camera, center at 85°W, 7°S.)

Marineris suggests a common structural origin. In summary, then, the canyons form a fairly linear, continuous depression with an average slope down toward the east.

Canyon walls drop away sharply from the level upland surfaces. Slopes are estimated to be greater than 10 degrees (Sharp, 1973c). U-shaped chutes impart a scalloped appearance to the trough edge.

Sharp points out that these scallops may be formed by massive gravity slides and dry avalanches. The walls are uniformly complicated by a linear array of closely spaced spines, which, at B-frame resolution, show an anastomosing network of ridges (fig. 8.27). Some of these ridges may represent dikes of resistant igneous rock that project into the troughs (fig. 5.48), but it is difficult to explain them all in this way. For the most part they probably form where two avalanche chutes intersect.

Stubby, frond-like tributary channels occur locally, especially in the uplands south of Ius Chasma (fig. 8.28). They closely resemble the tributaries of the terrestrial Grand Canyon and, by analogy, can be attributed to underground sapping, artesian flow, or a combination of both (Lucchitta, 1974).

Sidewall erosion leads to the lateral enlargement of canyons. Where parallel canyons develop, the intervening tableland is gradually eaten away. This process is evident in the southern part of Coprates Chasma, where there is an isolated flat-topped butte, 90 km long and 35 km wide (fig. 8.29). Elsewhere in Coprates Chasma and in Tithonius Chasma, erosion has proceeded to a point where the receding slopes have intersected and formed an intertrough ridge. The continued erosion of this ridge has re-

duced its height substantially below that of the surrounding tableland, and its sculptured flanks are similar to those of trough sidewalls.

Canyon floors are of two distinctive types. Some are smooth and topographically featureless, but with prominent swirling albedo patterns. This suggests earlier infilling and contemporary movement of wind-driven sediment across the surface. Other parts of canyon floors are remarkably rough, apparently underlain by a jumbled mass of blocks. One would be tempted to identify these as debris slides except for the fact that they are not localized close to the walls. Instead they extend many kilometers across the canyon interior, their dimensions and texture dwarfing the bounding walls. Perhaps they are collapse and slump features, formed in connection with lateral retreat of canyon sidewalls and progressive enlargement of canyon floors (fig. 8.27).

Paralleling Coprates Chasma to the south is a string of pit craters about 350 km long. Other crater chains occur west of Ophir Chasma and north of Tithonius Chasma. Clearly, these depression pits are formed along a large crustal fracture. More arguable, but still defensible, is the contention that they are volcanic vents. Whatever their origin, they support the view that the canyons have formed by ero-

(b)

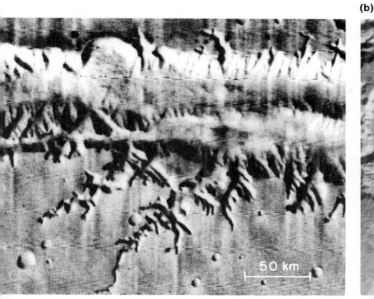

Figure 8.28 (a). Stubby frond-like tributaries along the southern margin of Ius Chasma. (DAS No. 05851968, rev. 119, A camera, center at 84°W, 8°S.)

(b). A detailed view of the Grand Canyon, Arizona, photographed by the Earth Resources Technology Satellite. There are superficial similarities with the Martian tributaries shown in (a). To the north of the Grand Canyon, valley formation is dominated by run-off stream erosion from a relatively

extensive watershed. To the south of the canyon relatively little run-off water is available. Tributary valleys appear as large amphitheaters or alcoves with walls composed of the more resistant strata. Headward erosion proceeds by plunge-pool sapping associated with intermittent run-off. These amphitheaters are unlike the spur ridges and intervening gullies shown in figure 8.27, but do resemble some of the tributary valley heads shown in (a) (Lucchitta, 1974).

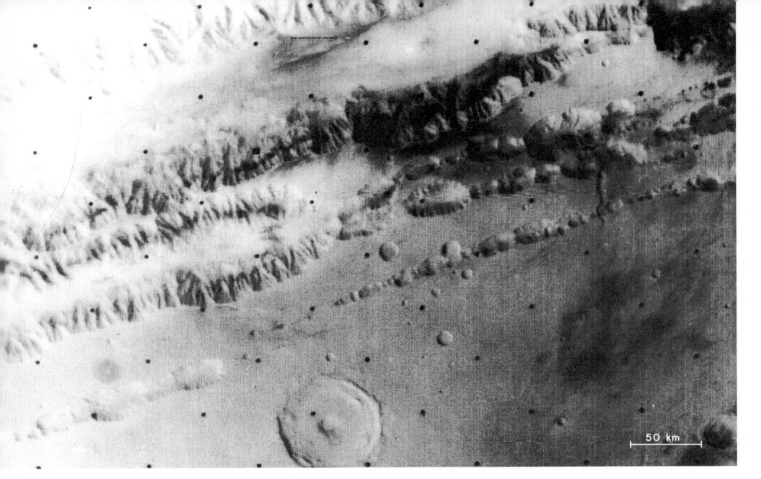

Figure 8.29. Southern boundary of Coprates Chasma. Circular and elliptical craters occur along structurally controlled fractures. As erosion proceeds, craters coalesce. Highland surfaces are first isolated to form flat-topped mesas, and ultimately are restricted to linear ridges such as the one at the left of the picture. (DAS No. 07542468, rev. 166. A camera, center at 62°W, 15°S.)

sion and enlargement of large regional fractures (Schumm, 1974) trending a little south of east, part of a radial set that surrounds the Tharsis uplift. A few scientists have suggested that the canyons may have formed by crustal spreading, with the width of the canyons indicating the amount of crustal rifting. However, the landforms described in the paragraphs above argue that the walls are receding, partly at least, by erosional mechanisms. Furthermore, any rifting along the canyons should be accompanied by transform or transverse faults at the ends, and these have not been observed (Carr, 1973; Sharp, 1973c).

Chaotic Terrain

Chaotic terrain was first identified and interpreted in Mariner 6–7 pictures. The name was selected to imply a derangement of morphological elements—irregular depressions filled with a jumble of blocks ranging in size from a few hundred meters to tens of kilometers. The tentative interpretation was that the chaos formed by removal of subsurface material, possibly ground ice, followed by large-scale collapse of overlying rock and sediment. As here defined, most of the chaotic terrain is included in an area about 2000 km on a side and centered at 30°W, 10°S. It is the middle link in a chain of negative topographic forms extending from the Tharsis uplift to the Chryse basin. To the west and upslope of the chaotic terrain are the linear troughs just described. To the north and downslope are channels that extend into the Chryse basin.

In plan view the depressions of the chaotic terrain are extremely irregular, especially by contrast with the troughs. They show no preferential elongation, parallelism, or polygonality. In short, there is no evidence of large-scale tectonic control as there is for troughs. Neither is there much evidence for through-going drainage in this cluster of chaos basins. Large axial channels are lacking. Individual chaos basins are linked by subordinant troughs.

At a more detailed level, it is apparent that there is a systematic progression of landforms as one moves from unaltered cratered terrain to the interior of a chaos-filled basin (fig. 8.30). Along such a traverse one first encounters isolated fissures and occasional breached craters. Then the fissures become

(a)

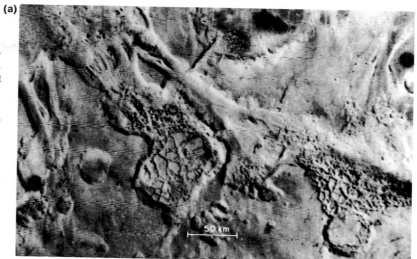

Figure 8.30 (a). Breakup of cratered terrain by fracturing and slumping. (DAS No. 07758698, rev. 172, A camera, center at 28°W, 3°N.)

(b). Detailed view of two different landforms associated with chaotic terrain. A bundle of parallel braided channels extends to the edge of a scarp. Plains to the north are dotted by small hillocks, formed by erosion and collapse of interhillock material. (DAS No. 09161399, rev. 211, B camera, center at 30°W, 5°N.)

(b)

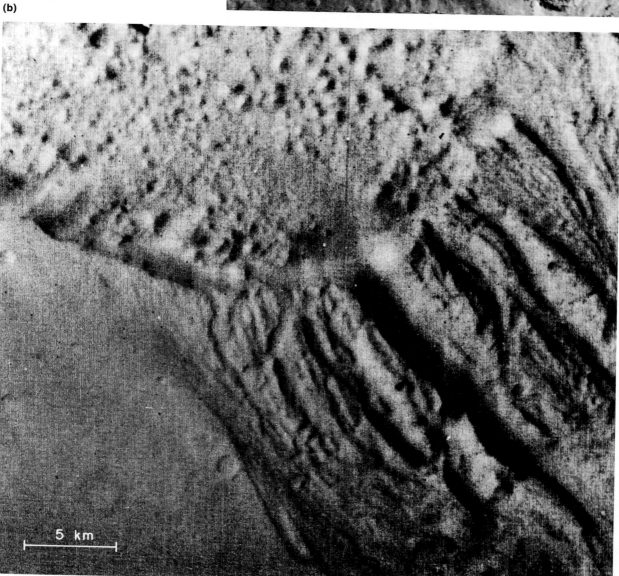

more numerous. Large, flat-topped, polygonal blocks are situated at the base of arcuate fractures. The large blocks are progressively replaced by smaller blocks and the interblock channels become areally more important. Finally, a few isolated pyramid-shaped knobs project above an otherwise level surface.

Chaos is, in a sense, a misnomer since there is some marginal order to the arrangement of blocks within "chaotic" zones. Analysis of Mariner 6–7 pictures shows a preferential orientation of all lineations (block axes, ridges, and fractures) in two directions: one northeast and a second slightly west of north (Wilson et al., 1973). This indicates some small-scale structural control.

Fretted and Knobby Terrain

Characteristically, the ancient cratered terrain in the southern hemisphere is separated from the plains to the north by a marginal zone of buttes, hummocks, and knobs. Where the flat-topped buttes are bounded by rectilinear channels, a fretted or waffle-like appearance results. Fretted terrain is especially widespread along the cratered terrain-plains contact between 180°W and 360°W. It is generally suppressed along the margins of the Tharsis uplift, although a north-south trending zone of knobby ter-

rain separates Amazonis Planitia from Elysium Planitia. An unusual variant of fretted terrain occurs along the western margin of Lunae Planum, as described in chapter 5.

A progressive change in landforms, already mentioned for chaotic terrain, is even more impressive in fretted and knobby terrain (fig. 8.31). Approaching this zone from the cratered-terrain side, one first encounters a few linear fissures. As they become more numerous, they form simple patterns of intersection. In this region degraded craters appear to be without sharp rims or steep walls. The landscape is blandly undulatory in contrast with the crisp relief present in nearby unaltered cratered terrain. Proceeding further into the fretted terrain, fissures widen and coalesce to form a maze of interconnecting linear depressions. Box canyons abound and occasional through-going drainage systems are noticeable. Next in this progressive series, depressions become so extensive that they isolate flat-topped remnants of cratered terrain. Many of these remnants appear tilted, with a prominent scarp along one side and a transitional boundary with smooth plains on the opposite side. Traced further away from the cratered terrain, the remnants become smaller in size and more widely spaced. Finally, they are present as small nubbins identifiable only in B-frame pictures.

(a)

(b)

Figure 8.31 (a). Fretted terrain, showing the progression from dissected cratered terrain (bottom) to isolated erosional knobs on plains (top). (DAS No. 08191164, rev. 184, A camera, center at 313°W, 43°N.)

(b). Detail of fretted terrain showing crater (arrow) surrounded by ejecta that mantles previously formed channels and depressions. (DAS No. 08119058, rev. 182, A camera, center at 327°W, 36°N.)

In many instances, craters are incorporated into the fretted pattern. Arcuate scarps hint at the former presence of unmodified craters. In some cases it appears that the craters provide multiple sites for concentrically retreating scarps, and that intersection or interconnection of these scarps produced the total fretted mosaic. In other instances a more-or-less complete circular rampart is present, breached only in one or two places. Here it appears that the crater walls stand as residual barriers to planation, finally to be breached and consumed by a regional process of degradation.

Within fretted and knobby terrain, features indicative of other types of depressions are observed. For example, floor units often comprise jumbled blocks, as in chaotic terrain. Occasional linear depressions, such as Nili Fossae on the eastern slopes of Syrtis Major, bespeak the same structural control evidenced by troughs. Channels, described in the following section, thread their way through fretted terrain. Good examples are contained in the Ismenius Lacus quadrangle (MC 5).

The morphological evidence that fretted terrain has formed by the "eating away" of the cratered terrain is overwhelming. The remarkable transition of landforms from smooth plains to isolated knobs, to buttes, to intersecting depressions, to occasional fissure, to unaltered cratered terrain makes the case. The scarps have remarkably uniform heights of 1–2 km. Were ground ice present, this could mark the approximate horizon at which it melted due to increasing temperature with depth. This critical geotherm might provide a planet-wide control for cliff height.

Channels

In aggregate, these features supply the best morphological evidence for the former presence of rivers on Mars. In an effort not to prejudge the conclusion, most writers have searched for a genetically noncommittal word. Channel is the consensus choice even though it has a nostalgic ambiguity, derived as it is from the Latin word, *canalis*. As used by geomorphologists, channel refers to the depression actually occupied by a stream.

Broad, meandering channels are localized in four places: south of Amazonis Planitia (Mangala Vallis, Maadim Vallis, and Al Quahira Vallis), between the chaotic terrain to the south and Chryse Planitia to the north (Shalbatana Vallis, Simud Vallis, Tiu Vallis, Ares Vallis), north of Lunae Planum (Kasei Vallis), and scattered throughout the fretted terrain in the vicinity of Deuteronilus Mensae and Nilosyrtis Mensae (Auquakuh Vallis, Huo Hsing Vallis). All of these channels are located along the cratered terrain-plains boundary, a relationship that hardly can be coincidental.

Mangala Vallis is about 350 km long. The primary channel reaches a width of 10 km. In plan view it shows incipient sinuosity (fig. 8.32). The channel is irregular in width, in places bifurcating. Although it broadens slightly toward the north, it pinches off before it reaches Amazonis Planitia. Stubby tributaries lead into the main channel.

The more provocative features of Mangala Vallis are seen within the major channel, as revealed by B-frame pictures (fig. 8.32). A braided texture is evident, conjuring up a vision of Martian floods. As Milton (1973) notes, it is the remarkable sinuosity of the landforms that triggers this intuitive impression. Further, only a fluid with substantial momentum (or speed) would generate the necessary curvatures. The overlapping braids suggest a complex flow pattern with temporal succession of events. What are the candidate fluids? One thinks immediately of water because of terrestrial experience, but are there fluids that might be equally likely in the Martian environment? Not really. Because CO_2 is the chief atmospheric and polar cap component, one might postulate its further involvement as a liquid. But, in order for the liquid to be stable, the atmospheric pressure would have to be 5 bars—or 1000 times as dense as the present atmosphere. Discharges of lava or gas-charged debris would be less improbable. Lava has some potential for melting bedrock and carving a channel (Carr, 1974a) but almost certainly not on the scale of the channels we are discussing. Gas-charged debris is even less of an erosional agent; the morphological results are almost completely aggradational.

Sharp and Malin (1975) cite Mangala Vallis as a good example of an "outflow" channel. They speculate that soil water, seeping to the surface, might have formed a large lake near the headwaters of Mangala Vallis. Episodic overflow and catastrophic failure of reservoir walls produced a drainage pattern that has many characteristics in common with the channeled scabland of northwestern United States, formed during the Spokane Flood (Baker and Milton, 1974) (fig. 8.23).

Al Quahira Vallis and Maadim Vallis resemble Mangala Vallis in dimension and form. They have the same irregularly undulose plan view. Stubby tributaries are, however, much more prominent (figs. 8.33, 8.34). Noteworthy is the termination of these two valleys well within the cratered terrain. For that reason Sharp and Malin (1975) classify Maadim Vallis as a "runoff" channel, fed by dispersed upward seepage of groundwater. Rather than debouching onto the smooth plains approximately 100 km to the north, the channels end abruptly in roughly

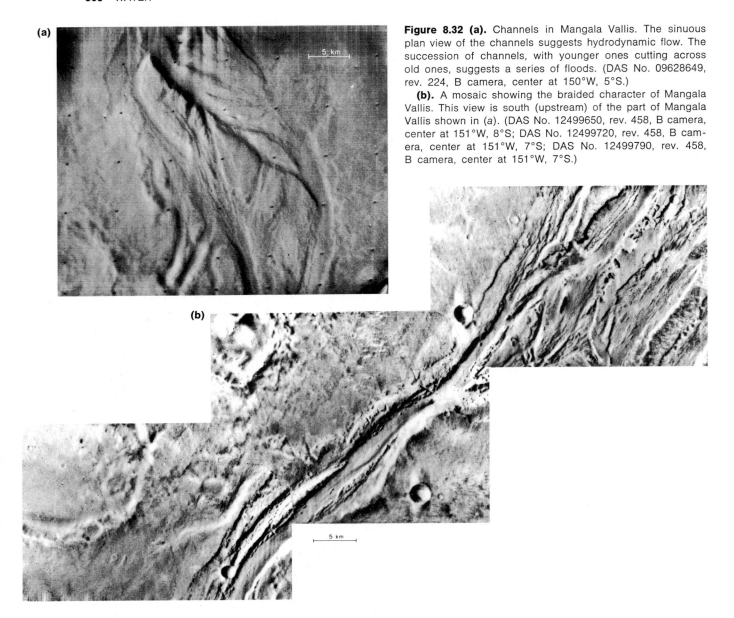

(a)

(b)

Figure 8.32 (a). Channels in Mangala Vallis. The sinuous plan view of the channels suggests hydrodynamic flow. The succession of channels, with younger ones cutting across old ones, suggests a series of floods. (DAS No. 09628649, rev. 224, B camera, center at 150°W, 5°S.)

(b). A mosaic showing the braided character of Mangala Vallis. This view is south (upstream) of the part of Mangala Vallis shown in (a). (DAS No. 12499650, rev. 458, B camera, center at 151°W, 8°S; DAS No. 12499720, rev. 458, B camera, center at 151°W, 7°S; DAS No. 12499790, rev. 458, B camera, center at 151°W, 7°S.)

circular sumps. This makes reconstruction of channel evacuation something of a problem. Where did all the eroded material go?

The channels north of the chaotic terrain are closely related to that terrain. Polygonal mesas and rounded hummocks are present in the channel floors, especially in basins that have formed at the "upstream" ends of the channels. Shalbatana Vallis and Ares Vallis have distinct lateral boundaries, delineating shallow, sinuous channels, but Simud Vallis and Tiu Vallis are indistinctly delineated, forming the western and eastern edges of a broad low plain, 400 km across, which connects the main channel of the chaotic terrain with the Chryse basin. Low, broad,

teardrop-shaped islands are situated where these channels debouch into Chryse Planitia.

Kasei Vallis, just north of Lunae Planum, is distinctive from the channels just described. It is a broad, flat-floored depression incised in cratered plains and connecting smooth plains of the Tharsis uplift to plains of the Chryse basin (figs. 8.38, 8.39). A delicate pattern of headward erosion along planes of structural weakness is revealed at the channel's western termination (fig. 8.38). Further to the east, albedo swirls and undulations that wrap around topographic obstructions suggest contemporary wind activity. Islands of stratified plains materials stand in the middle of Kasei Vallis. Exhumation of cra-

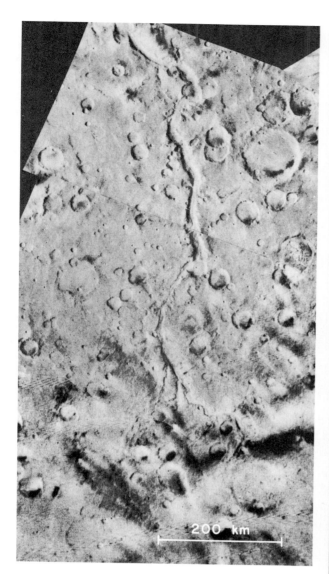

Figure 8.33. A-frame photomosaic of Maadim Vallis. Several tributaries are visible in the bottom (upstream) part of the channel system.

Figure 8.34. A-frame photomosaic of Al Quahira Vallis. This valley system is characterized by a broad flat-floored channel, general absence of tributaries, and indistinct terminations at both ends. (DAS No. 09340634, rev. 216, A camera, center at 199°W, 21°S; DAS No. 09340704, rev. 216, A camera, center at 197°W, 16°S.)

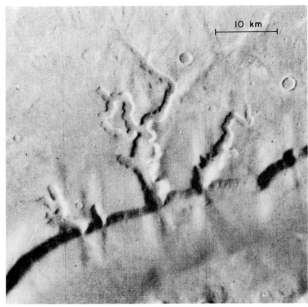

Figure 8.35. The edge of the Argyre basin. A channel cuts through the surrounding mountains. Bright and dark patches on Argyre Planitia (arrows) may be caused by sediment carried down the channel and deposited on the flat plains. (DAS No. 06497928, rev. 137, A camera, center at 48°W, 56°S.)

Figure 8.36. Stubby channels, in the fretted terrain. Parts of the channels appear to form by coalescence of circular depressions. The bright "alluvial fans" which occur where channels debouch on the plains are an artifact of picture enhancement. (DAS No. 09378189, rev. 217, B camera, center at 344°W, 37°S.)

Figure 8.37. Oblique view to the northwest, across the cratered plains of Lunae Planum. Compare with figure 8.38. (DAS No. 08873804, rev. 203, A camera, center at 72°W, 19°N.)

Figure 8.38. Headward tributaries of Kasei Vallis. Channels are formed by erosional enlargement of several sets of intersecting fractures. The large crater is surrounded by a ring of ejecta which has resisted erosion. Position of photomosaic is shown on figure 8.39. (DAS No. 13313735, rev. 667, B camera, center at 75°W, 18°N; DAS No. 13313805, rev. 667, B camera, center at 73°W, 22°N.)

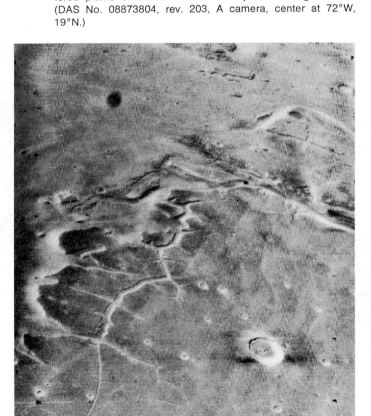

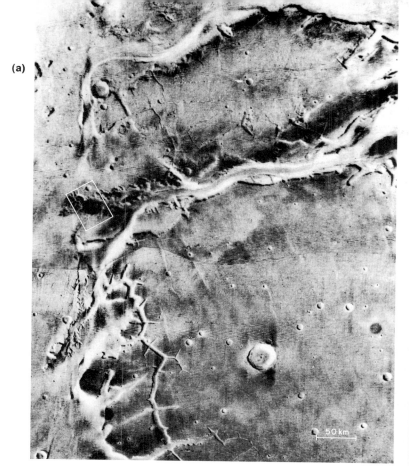

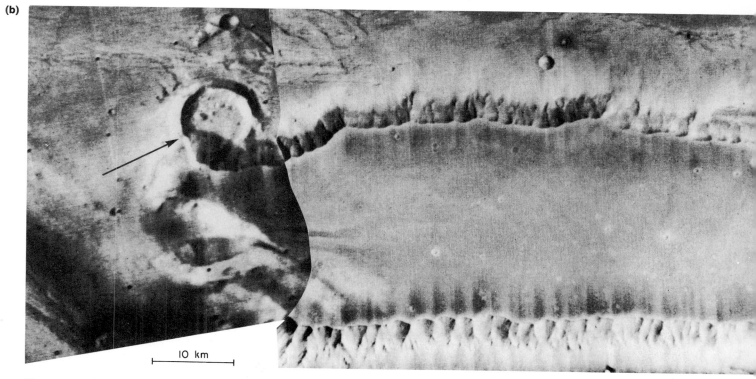

Figure 8.39 (a). Photomosaic of Lunae Planum. A large island of cratered plains appears at the top of the figure. Kasei Vallis is the broad flat-floored valley bounding the island to the south. Note the streaky and splotchy appearance of floor materials, suggesting the presence of wind-sorted sediment. The outline of figure 8.38 is shown.

(b). A flat-topped mesa in Kasei Vallis, bounded by fluted slopes. Note the large crater (arrow) which apparently has been partly exhumed by erosion of the plains materials which underlie the mesa. Baker and Milton (1974) attribute this erosion to a catastrophic flood. (DAS No. 10277409, rev. 242, B camera, center at 62°W, 24°N; DAS No. 12866208, rev. 242, B camera, center at 61°W, 24°N.)

(c). An oblique view of the same partly buried crater described in (b). (DAS No. 08945729, rev. 205, B camera, center at 62°W, 24°N.)

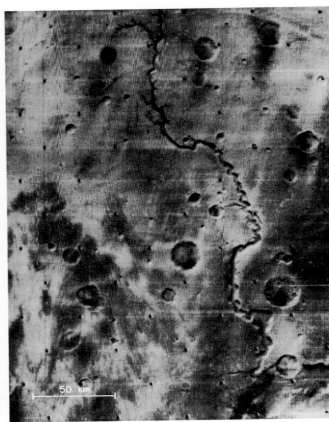

Figure 8.40 (a). Photomosaic of Nirgal Vallis, showing tributaries. Topographic gradient is from left to right.
(b). Detailed topography in Nirgal Vallis. (DAS No. 06354848, rev. 133, A camera, center at 40°W, 29°S.)

tered terrain beneath is hinted at, although not unequivocally demonstrated (fig. 8.39). Baker and Milton (1974) interpret an island within Kasei Vallis as bedrock scoured by a massive, catastrophic flood.

Two other unusual channel occurrences should be mentioned, one in an interior region of cratered plains and the second in smooth plains. Nirgal Vallis (40°W, 30°S) has a uniquely delicate plan view (fig. 8.40). Four hundred kilometers long, it displays several numerical orders of tributary branching in its western (upstream?) section, and a sharply sinuous pattern in its eastern section. The geometry of the channel meanders shows both similarities and dissimilarities with terrestrial river channel meanders (Weihaupt, 1974). Nirgal Vallis terminates in a crater that is situated in one corner of a larger scarp-bounded depression. Hephaestus Fossae (230°W, 20°N) comprise shallow northwest-trending troughs in the western part of Elysium Planitia. Teardrop-shaped islands provide clear indication of unidirectional fluid flow (fig. 8.41).

Channel Gradients

Those persons who, some years ago, chose to interpret lunar rilles as conventional stream valleys were puzzled by the fact that the channels dwindled in size as the rilles were traced downhill. Remembering this anomaly, scientists studying Martian channels awaited with considerable trepidation the correlation of inferred flow directions with topographic gradients. It now appears that the downstream direction as indicated by tributary geometry and channel width is also the downhill direction. Characteristically, the gradient is about one meter per kilometer, or one part in a thousand. More accurate determinations are difficult. The problem partly lies in differentiating elevation measurements of upland from those of channel bottom. A particular channel may form in an upland region with an average height of 4 km and flow into a basin with an average height of 2 km. But if the channel cuts through the highlands to a depth of several kilometers, then the channel slope as determined by regional elevations is incorrectly magnified.

Uncertainties other than those of measurement affect slope determinations for Martian channels. The alleged rivers flowed many years ago—millions or even billions. Since that time there may have been vertical crustal movement. For example, the Tharsis volcanic province stands anomalously high. It would not be surprising if the original channel slopes have been tectonically modified or even reversed.

Finally, we should make distinction between water level and channel bottom. When we say that a stream flows downhill, what we really specify is that the upper *water* surface has a negative slope. The base of the channel may have a different slope and in some cases, notably in parts of the lower Mississippi, the channel actually increases in elevation in the downstream direction, due largely to deposition close to the mouth. Concordantly there must be a decrease in water depth in the downstream direction.

Rilles

Numerous rilles are identified in the volcanic regions, as already discussed. They strongly resemble terrestrial lava channels and lunar rilles attributed to lava flow. Although there is little doubt that the same process accounts for Martian rilles (Carr, 1974a), a few rille-like features occur outside the Tharsis and Elysium volcanic provinces and are ambiguous on grounds both of association and form. We include here Nirgal Vallis, which is rille-like both

in its sharp sinuosity and in its linkage to a crater source—or sump. Situated on the eastern edge of the Hellas basin close to several degraded volcanic shields and to the south of Hesperia Planum are two large, sinuous channels, each about 500 km long (fig. 8.42). Their association with other volcanic features suggests a volcanic origin (Potter, 1976) even though their great size invalidates analogies with terrestrial and lunar occurrences.

Not all of the channels on the slopes of Martian

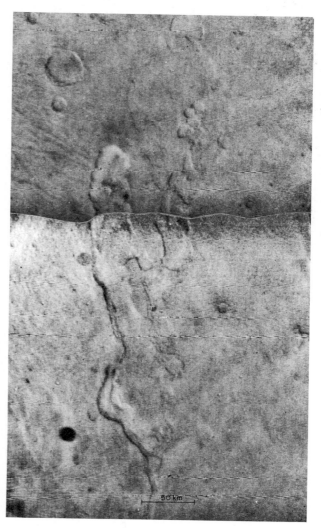

Figure 8.42. A simple, sinuous channel that heads in a quasi-circular depression on the flanks of Hadriacum Patera and descends to the southwest, toward Hellas Planitia. The occurrence of the channel in a volcanic province, its origin in a crater-like depression, and its downstream decrease in channel size all suggest that it is a lava channel. However, the size greatly exceeds that for terrestrial lava channels and lunar sinuous rilles. (DAS No. 06031168, rev. 124, A camera, center at 268°W, 37°S; DAS No. 06031238, rev. 124, A camera, center at 265°W, 32°S.)

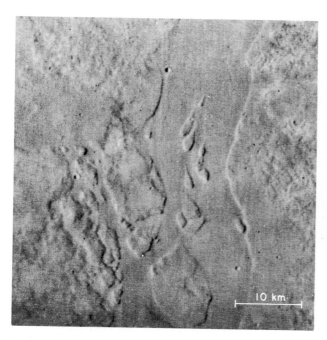

Figure 8.41. A broad, shallow channel in Elysium Planitia with tear-drop-shaped islands, almost certainly shaped by hydrodynamic or aerodynamic flow. (DAS No. 08910729, rev. 204, B camera, center at 227°W, 31°N.)

Figure 8.43. Irregularly branching furrows on the flanks of Alba Patera. Features of this sort supply some of the best evidence for surface runoff associated with rainfall. (DAS No. 08299139, rev. 187, B camera, center at 117°W, 45°N.)

shield volcanoes resemble lunar rilles and terrestrial lava channels. For example, gullies on the flanks of Alba Patera (fig. 8.43) have an irregular branching pattern. They somewhat resemble the furrows described in the next section.

Age of Channels

At first glance the Martian channels look remarkably fresh, sharply incised in the more degraded cratered terrain. However, more extensive analysis suggests otherwise. A few of the channels have large superposed craters. This suggests that some channels were formed at least 10^8 years ago (Hartmann, 1974a). In the general region of Deuteronilus Mensae and Protonilus Mensae and in the region surrounding Orcus Patera, the hummocky-fretted terrain contains a significant number of craters that, on morphological basis, appear to postdate the formation of the erosional troughs, hummocks, and mesas (fig. 8.31b) (Arvidson and Coradini, 1975). These crater densities are similar to densities on cratered plains (pc), suggesting that most of the erosion responsible for fretting and channeling ceased at approximately the same time that cratered plains were emplaced.

There is another way in which the age of the channels can be estimated by their relationship to plains units. Almost all channels are restricted to cratered terrain. Subordinate occurrences of small rilles, furrows, and channels in the Tharsis and Elysium volcanic provinces can be attributed to localized volcanic activity that may have remobilized ground ice or released juvenile volatiles. In large part, then, channel formation predates emplacement of northern plains units.

It is untenable to argue that the predominant constituent of the plains units is detritus formed by progressive eating away of the cratered terrain along the planet-wide contact between cratered terrain and plains. Partly covered and dissected inliers of cratered terrain are widespread to the north of this contact, so it is more reasonable to assume that the plains units mantle an older cratered terrain. As previously noted in chapters 3 and 5, a volcanic origin for much of the plains material is likely. In certain regions, notably south of Amazonis Planitia (fig. 8.46) and east of Acidalium Planitia (fig. 8.45) the plains encroach on pre-existing cratered and channeled terrain. These observations reinforce the previous conclusion that the plains units—volcanic in origin—were laid down after the formation of a planet-wide erosional girdle.

Plains show variable numbers of superposed craters from region to region, and their age is accordingly variable. In general, they have ages in excess of 1 b.y. (Soderblom et al., 1974).

As documented in chapter 4, the Martian landscape has probably been modified by a great erosional event. The age of this event is uncertain but, assuming approximately equal lunar and Martian fluxes (Soderblom et al., 1974), it could have occurred about 3.5 b.y. ago (Chapman, 1974; Arvidson, 1974a). From that time forward, relatively little erosion has occurred. It is reasonable to assume that most of the fluvial features we now observe date back to that great erosional event.

Furrowed Terrain

Superposed on the ancient cratered terrain in the equatorial region of Mars are numerous closely spaced, weakly sinuous furrows. Widths range from 2 to 10 km, lengths from 100 to 1000 km. The result is a wormy texture easily identifiable in A-frame pictures (fig. 3.31). The furrows occur over a sinuous band between 15°N and 45°S (fig. 3.32). If the north pole were relocated at 110°W, 75°N, then the band of furrows would follow the 15°S parallel.

The origin of furrows is uncertain, but some investigators (e.g., Milton, 1973) believe they provide the best available evidence for erosion by rainfall. They appear to head along crater rim crests and to flow down both the interior walls and the exterior rim deposits. Adjacent furrows are approximately parallel, with both joins and divisions in the downstream direction. The same reticulate pattern is observed in drainage systems of terrestrial arid regions (Milton, 1973).

The rainfall or climatic argument is strengthened by postulating a pole shift and thereby placing the furrows close to the equator at the time of their formation (Saunders, 1973). This is not quite so unlikely as might appear at first glance. Consider the Tharsis bulge, which is centered close to the equator, a position that happens to be one of rotational stability. If the bulge had formed north of the rotational equator, precessional motions and internal dissipation of energy would ultimately lead to a shift in the axis of rotation and the placement of the Tharsis bulge in the equatorial plane. This phenomenon, termed polar wandering, has been invoked by Murray and Malin (1973a) to account for the eccentric distribution of layered terrain at the poles. The shift necessary to center layered deposits on the rotational axis is only 8°. However, the shift necessary to explain furrowed deposits is 15°. Therefore, if the layered deposits near the pole do reflect polar wandering, it is likely that they reflect the final motion of a much larger adjustment of the spin axis.

An alternate explanation for the furrowed terrain has been put forward by Soderblom et al. (1974). They are impressed with the relative abundance of furrows in the vicinity of 20°S, the same latitudinal band in which the south polar debris blanket terminates and where dark albedo features are telescopically most prominent. This also happens to be the subsolar region at perihelion, the region of hot summers and presumed atmospheric turbulence with attendant dust storms. Connecting all these observations, Soderblom et al. conclude that the latitudes in the vicinity of 20°S are being exhumed by eolian activity that is stripping the polar deposits, exposing the dark bedrock, and laying bare stream channels. They argue that these same features are available for exhumation at other latitudes. As the Martian spin axis precesses and the perihelion subsolar point drifts north across the equator to a limit of 25°N, the zone of eolian stripping will similarly drift.

Groundwater vs. Rainwater

If one concedes that the channels and related features on Mars are formed by water, then the principal remaining question is whether the water was supplied from below (groundwater) or above (rain).

The evidence for Earth-like, rain-fed channel systems on Mars is not compelling. We have previously noted the intricately dendritic network of tributaries that characterize terrestrial streams in humid areas of high rainfall. Tributaries for Martian channels are largely absent. Where present, they are stubby and blunt, not at all like the delicate terrestrial examples. One might argue that the Martian tributaries have either been destroyed by erosion or have been mantled by sediment.

Comparison of Martian channels and terrestrial dendritic stream channels is vulnerable to the criticism of bias. Channels in arid and semi-arid regions take on different patterns. Large-volume surface runoff associated with episodic rain storms inhibits the formation of complex tributary networks. Arroyos, with sharply incised primary channels and irregular secondary digitations, more closely resemble Martian channels than do the humid-climate dendritic channel systems. Arroyos, however, are relatively small features. Considered on the same regional scale that the Martian channels are displayed, even arid-climate channels exhibit several orders of tributaries.

Martian channels show little evidence of sediment dispersal at their mouths. There are neither alluvial fans nor deltas. (A possibly exceptional alluvial fan is shown in figure 8.35.) Indeed, most of the channels terminate or are sharply constricted before they reach the lowland plains. Again, there is a negative explanation. One can speculate that dispersal fans, once present, have either been destroyed by erosion or have been mantled by sediment.

Martian channels are many times larger than terrestrial channels, defined strictly as the depression occupied by flowing water. Although it may be qualitatively instructive to talk about the "braided stream" character of Mangala Vallis, it should be recognized that meter-size terrestrial sedimentary features are being improperly compared to kilometer-size Martian features.

In summary then, very few of the morphological and sedimentary features that characterize terrestrial streams are identifiable in Martian channels. Although the differences can be explained away, the remaining positive arguments are very few.

Evidence for groundwater sources is more persuasive. Most of it already has been presented. The chaotic terrain shows widespread evidence of collapse and slumping. Many channels are "box canyons," indicating both underground sources and sinks. The absence of tributaries and the relatively great width of most channels is consistent with underground seepage. Channels often incorporate craters and, in the hummocky-fretted terrain, there is evidence that craters are themselves loci for ponded artesian water. Outward erosion of crater walls fuses individual circular basins into complex

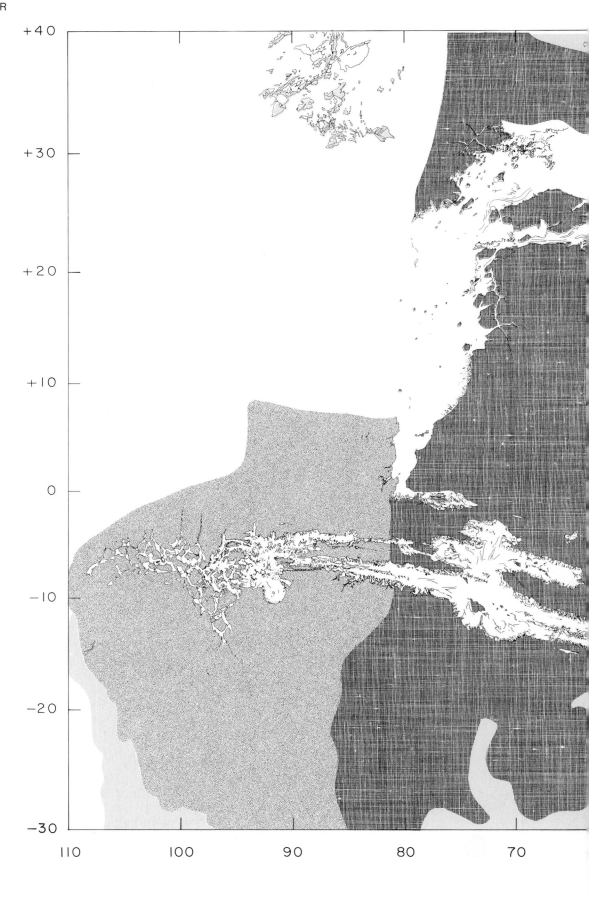

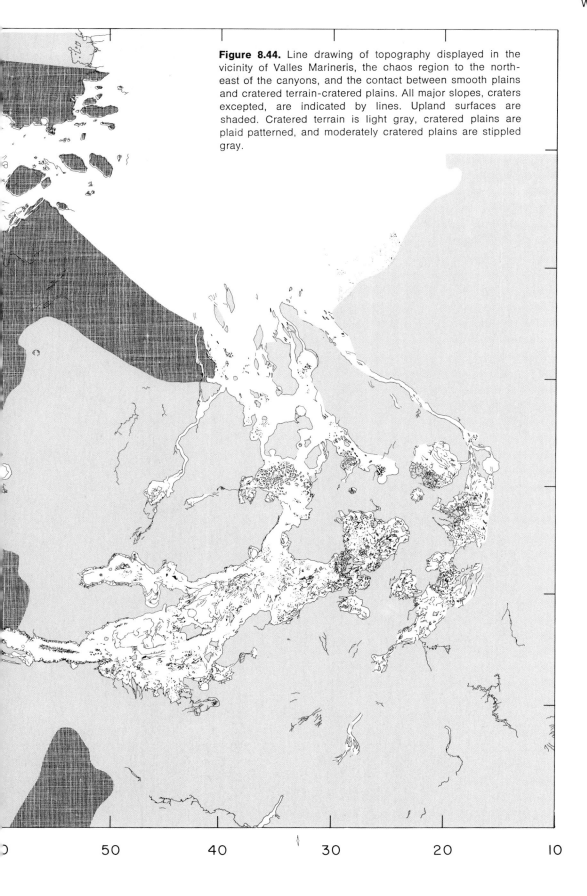

Figure 8.44. Line drawing of topography displayed in the vicinity of Valles Marineris, the chaos region to the northeast of the canyons, and the contact between smooth plains and cratered terrain-cratered plains. All major slopes, craters excepted, are indicated by lines. Upland surfaces are shaded. Cratered terrain is light gray, cratered plains are plaid patterned, and moderately cratered plains are stippled gray.

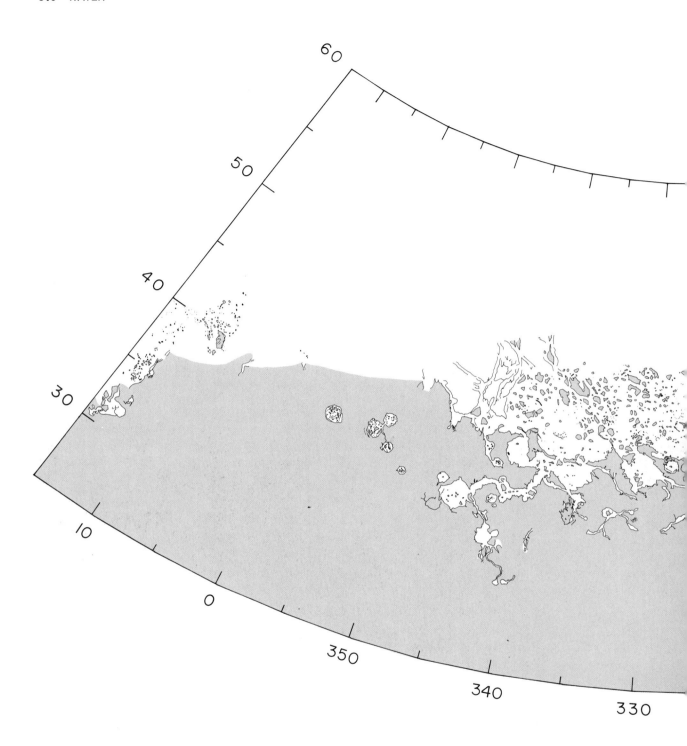

Figure 8.45. Line drawing of topography displayed along the contact between cratered terrain and plains, in the vicinity of hummocky-fretted terrain. All major slopes, craters excepted, are indicated by lines. Upland surfaces are shaded.

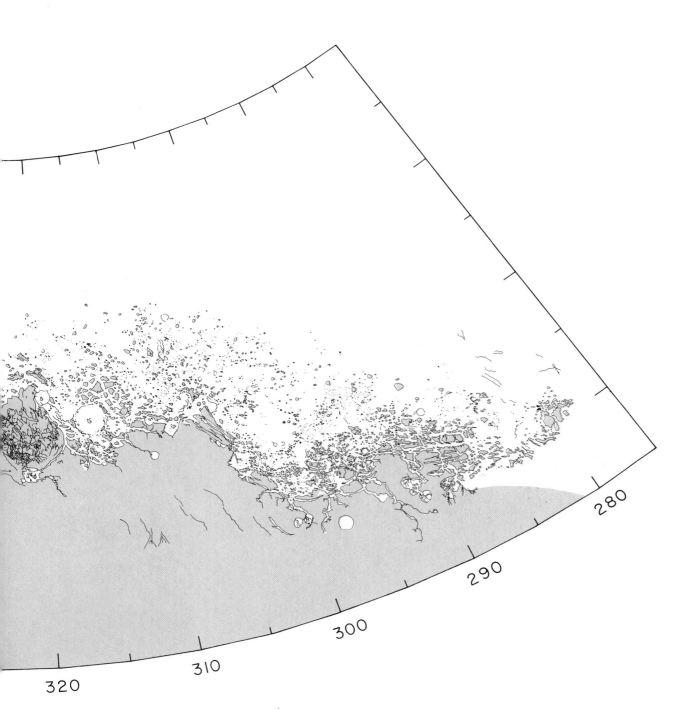

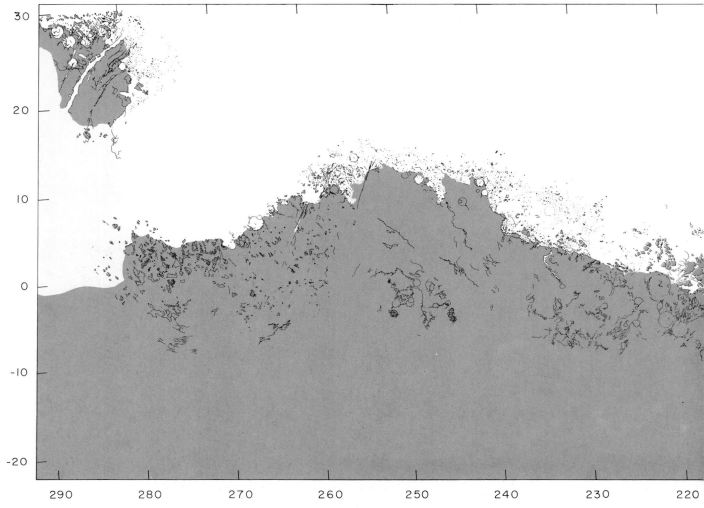

Figure 8.46. Line drawing of topography displayed along the contact between cratered terrain and plains. All major slopes, craters excepted, are indicated by lines. Upland surfaces are shaded. The center of the Isidis basin is at 275°W. The lightly shaded region to the west of Isidis Planitia is Syrtis Major Planitia. Al Quahira Vallis is located at 200°W, Maadim Vallis at 182°W, and Mangala Vallis at 155°W.

depressions, bounded in large part by fragments of crater walls.

The Plains-Cratered Terrain Erosional Contact

Most investigators of Martian channels and related features have confined their attention to piecemeal qualitative analysis of particular photographs. With the exception of Pieri's (1975) consideration of furrow distribution, little study has been given to the overall pattern of erosional forms. As a first step in that direction, we have prepared rudimentary topographic charts of the erosional features along the planet-wide great-circle contact between plains and cratered terrain (figs. 8.44–8.47). Well-defined slopes between negative and positive features, excepting craters, are represented by lines. Topographically high regions are shaded. The large-scale pattern of landforms, especially, erosional features, is shown more clearly and completely in these maps than in photomosaics.

Several remarkably different patterns can be identified along the global contact, and each one of these is associated with particular erosional, sedimentary, and structural controls. The region extending from 110°W to 10°W (fig. 8.44) includes Valles Marineris to the west and chaotic terrain to the east. Valles Marineris and related troughs and erosional scarps emplaced in cratered plains are linear and polygonal, reflecting the homogeneous character of

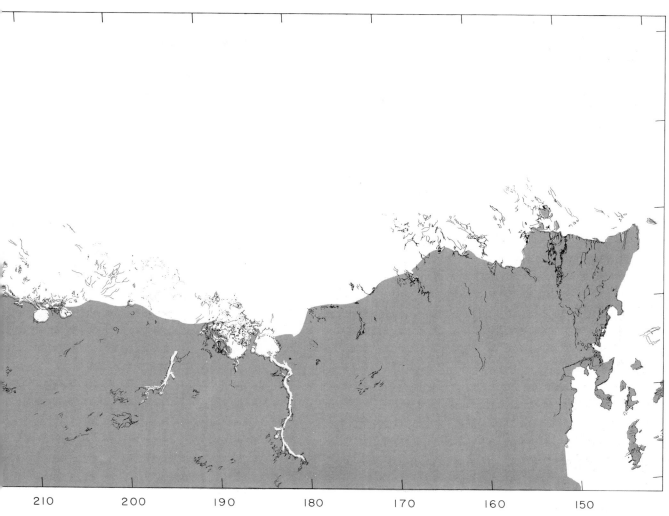

2IO	2OO	I9O	I8O	I7O	I6O	I5O

the volcanic strata that constitute the cratered plains. By contrast, chaos depressions and associated channels are irregular and discontinuous. Through-going drainage is not well displayed.

An additional distinction between Valles Marineris and chaotic terrain can be made with respect to wall types. Regularly spaced wall ridges are well developed in the former but largely absent in the latter. Again, this is probably related to erosional competence of homogeneous volcanic rocks versus heterogeneous regolith formed by impact processes.

As the global contact is traced from 10°W to 280°W (fig. 8.45), two additional patterns are displayed. Between 10°W and 350°W the contact is indistinct, most likely because volcanic and eolian deposits of relatively recent age have lapped up on the previously formed escarpment. From 350°W to 280°W, in the area mapped as hummocky-fretted terrain (fig. 3.2), the progressive breakup of the cratered terrain is graphically illustrated. As mentioned previously, disarticulated channels in the south give way to articulated channels and progressively smaller islands of cratered terrain to the north. Breached craters contribute to the general appearance of the erosional scarps. In the vicinity of 0°, chaotic-like material forms within individual craters. Around 350°W, craters coalesce to create irregularly sinuous channels. Further to the east, crater rims are visible as remnants in highly dissected terrain.

Between 290°W and 150°W (fig. 8.46) the contact between plains and chaotic terrain generally is sharply defined. In the vicinity of 290°W, arcuate fractures, related to the Isidis basin, control the pattern of depressions. South of this region cratered plains are in contact with plains, the contact between the two subdued by overlying eolian deposits.

Between 280°W and 180°W the contact is marked by a fretted zone, but one that is significantly narrower than that present between 350°W and 280°W.

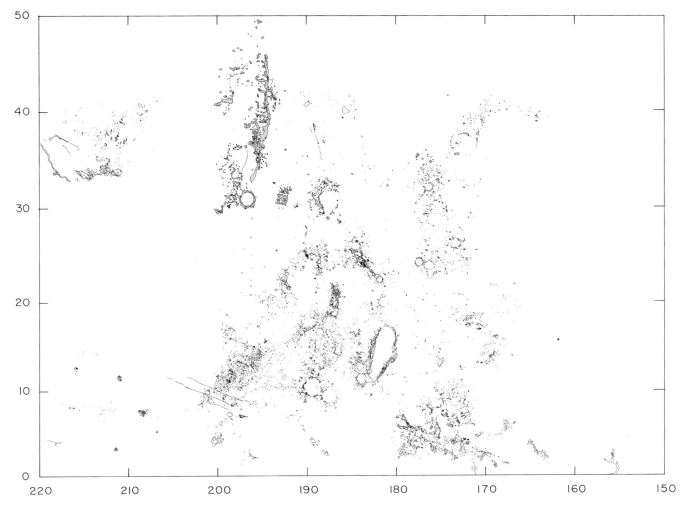

Figure 8.47. Line drawing of topography displayed in the general region of Elysium Planitia. On the physiographic map (fig. 3.2) this is mapped as hummocky-knobby terrain. All major slopes, including those associated with partly buried and degraded craters are indicated by lines. Upland surfaces are shaded. In general, the topographic relief of this region is thought to be inherited from ancient cratered terrain, largely covered by smooth plains.

Cratered terrain continues in relatively unaltered condition, except for wide-spread furrows, to within 100 km of contact with northern plains. Two prominent channels, Al Quahira Vallis and Maadim Vallis, occur at 200°W and 185°W, respectively.

Between 180°W and 150°W, the boundary between cratered terrain and plains is remarkably sharp and simple. Volcanic plains appear to lap up on a pre-existing erosional front. Mangala Vallis is situated at 155°W. In the vicinity of 150°W the trend of the contact changes abruptly from east-west to north-south, reflecting the bulge of the Tharsis uplift, situated to the east.

An additional area of eroded, cratered terrain is situated between 220°W and 160°W in Elysium Planitia (fig. 8.47). This region, lying slightly to the north of the planet-wide contact we have been discussing, includes a highly eroded cratered surface that is almost totally engulfed by more recent volcanic deposits. Some remnant crater walls are visible. The dominant features are fields of very small hummocks and furrows.

A schematic representation of all the terrains we have just discussed is shown in figure 8.48. This global overview demonstrates that the contact between plains and cratered terrain, fundamental though it is, should not be treated as a single entity. Varying conditions of erosion, sedimentary onlap, and structural control account for distinctive topographies.

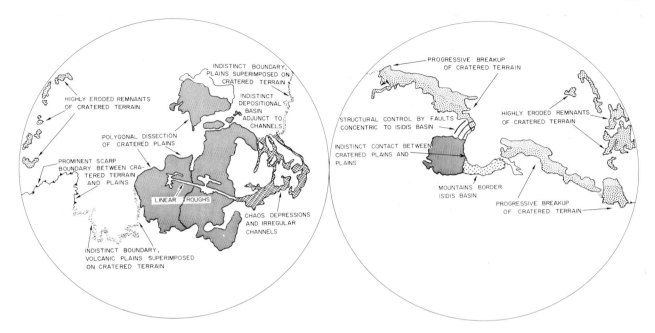

Figure 8.48. Generalized geologic map of Mars showing the different processes that have shaped the landforms along the planet-wide girdle between southern cratered terrain and northern plains.

9.

A Summary of Martian Geologic History

DRAWING ON the data and interpretations of previous chapters it is possible to construct a sequence of events outlining the geologic history of Mars. The evolutionary model is illustrated by figure 9.1, which contains five paleogeologic maps of Mars, showing the progressive development of the surface.

The planet formed by accretion of many smaller objects over a relatively short time, probably a few hundred thousands of years. During the first billion years of Martian history, the meteoroid-asteroid flux decreased to a level that has remained generally constant for the past 3.5 b.y. The final stages of heavy bombardment that occurred during the first billion years produced a densely cratered crust (fig. 9.1a). Similar surfaces occur in the lunar and Mercurian highlands. On Earth the same type of surface probably was formerly present, but has been destroyed by remobilization of crustal materials.

Shortly after the accretionary phase, planet-wide differentiation occurred—driven largely by the kinetic energy of accretion—with formation of crust, mantle, and core. If the crust varied in thickness over the globe (Phillips et al., 1973) and if the more ancient regions were essentially in isostatic equilibrium (Phillips and Saunders, 1975) then it might be expected that elevation and crustal thickness would be closely correlated. Regions of thicker crust would stand at higher elevations. It is postulated that the topographic division of the planet into a northern topographically low hemisphere and a southern topographically high hemisphere reflects this type of crustal variation, thin in the north and thick in the south.

Figure 9.1b depicts the tectonic breakup of the crust by two processes: pervasive fracturing of thin crust and radial faulting around the Tharsis uplift. Evidence for disruption of the northern crust is present along the boundary between cratered terrain and plains where the cratered terrain has been broken up to form the fretted terrain. The same fretted pattern can be seen considerable distances north of the plains-cratered terrain contact, where remnants of ancient crust are incompletely covered by plains. This suggests that fracturing and modification of crust was not restricted to the zone where it is now best displayed.

A possible mechanism for break-up of the crust has been previously outlined. This involves expansion of the mantle as the olivine-to-spinel structure phase change migrates downward with increasing internal temperatures. Regions of thin crust would be weakest and hence most susceptible to rifting over the swelling mantle. The originally thin crust would be made even thinner by this extension, and subsequent isostatic adjustments would cause it to stand at progressively lower elevations.

The Tharsis uplift is apparently an ancient fea-

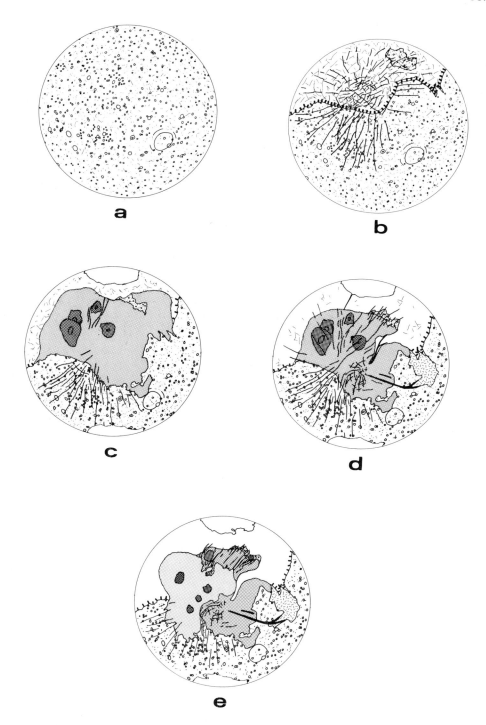

Figure 9.1. Schematic paleomaps showing geological evolution of Mars. The eastern hemisphere, from 0°W to 180°W, is depicted. The sequence of events is further described in the text.

(a). Last stage of early heavy bombardment.

(b). Uplift of proto-Tharsis region, with radial extension. Fracturing of crust throughout northern hemisphere.

(c). Widespread emplacement of volcanic plains.

(d). Renewed uplift and radial extension in the Tharsis region. Partial mantling of fractured cratered terrain throughout the northern hemisphere.

(e). Emplacement of relatively recent volcanic plains, with formation of volcanoes. Virtually complete mantling of fractured cratered terrain in northern hemisphere with volcanic and/or eolian deposits.

ture, surely much older than the volcanic rocks that cap the region. Radial fractures extend well into the cratered terrain and are exposed beneath cratered plains (pc), establishing the Tharsis uplift as an event older than emplacement of plains. It has not been possible to determine age relationships between fracturing of northern crust and radial fracturing around Tharsis.

The driving force for the Tharsis uplift is largely speculative. Plume-like convection has been suggested (Hartmann, 1973b). The set of paleomaps in figure 9.1 conforms to the suggestion that the Tharsis uplift formed at northern latitudes (fig. 9.1b). Subsequently, the relative orientations of the rotational axis and crust changed so that Tharsis was shifted into the equatorial plane (fig. 9.1c). Alternatively, broad buckling in the equatorial region might be related to differentiation that changed the moment of inertia, the rotation rate, and the amount of dynamical flattening.

It is postulated that a dense atmosphere formed early in Martian history, probably coincident with accretion. The early dense atmosphere was maintained by extensive outgassing that continued immediately after accretion. Initially, the atmosphere was relatively warm so that it held large amounts of water vapor. Condensation was favored by decreasing surface temperatures. Atmospheric losses ultimately predominated as outgassing declined, and a greenhouse condition, conducive to rainfall, could no longer be maintained. During this early period of extensive rainfall, substantial erosion occurred. Run-off furrows formed throughout the cratered terrain. As long as the top few kilometers of the crust had temperatures above the freezing point of water, rainwater percolated into the soil to depths of several kilometers. Subsurface flow led to concentration in topographically low regions, especially the Chryse trough.

With a small drop in ground temperature, ground ice formed at shallow depths. As the thickness of this ice layer increased, confining pressure for water trapped beneath the ice may also have increased. This may have created artesian situations on a regional scale. Local melting of ground ice, perhaps related to the onset of volcanic activity, and artesian release of ground water, led to collapse features in what is now the chaotic terrain. Massive floods formed some of the large channels, particularly in the equatorial zone along the previously established slope between the high southern and low northern hemisphere. With continued cooling of the atmosphere most of the remaining H_2O and CO_2 migrated to the polar caps and to subsurface reservoirs where it remains to the present as ice and possibly as chemically bound $(OH)^-$ in goethite and in clays.

This great erosional event probably occurred during, or just before, the emplacement of cratered plains (pc) (Jones, 1974). It is responsible for many of the erosional features that are now visible along the contact between cratered terrain and plains.

In summary, figure 9.1b records a particularly active period in the evolution of Mars. The northern crust was pervasively disrupted; the Tharsis uplift was initiated; a primitive atmosphere condensed; and extensive erosion occurred. These events cannot be clearly separated in time and they might be genetically related in ways not presently understood.

At the end of the great erosional event, volcanic activity continued with emplacement of the cratered plains (fig. 9.1c). The lower northern hemisphere was engulfed by deposits that, in places, lapped up on the previously formed tectonic-erosional escarpment between the two hemispheres. However, in some places the escarpment cuts through cratered plains, suggesting that the break-up of the northern hemisphere is essentially contemporaneous with emplacement of the cratered plains. Ancient volcanoes associated with emplacement of cratered plains are preserved as incomplete annuli of grooved terrain.

Widespread volcanic activity of approximately similar age occurred on the moon, and possibly on Mercury as well, forming the lunar maria and the Mercurian plains. However, in contrast to the situation for the moon and Mercury, volcanic activity on Mars has continued intermittently to relatively recent times. Volcanic deposits of the Tharsis region were emplaced a few hundred million years before the present. Also in contrast to the moon and Mercury, numerous large and unambiguous volcanoes are displayed on Mars.

Figure 9.1d records continuing uplift of the Tharsis region, with generation of more radial faults. The structural trough, Valles Marineris, formed at this time, cutting through both volcanic plains and cratered terrain. Along this relatively youthful fracture zone, there was release of subsurface water with formation of associated channels. Further development of chaotic terrain occurred at this time.

Finally, recent volcanism has mantled the Tharsis uplift with relatively unfractured lavas (fig. 9.1e). The four volcanoes—Olympus, Ascraeus, Arsia, and Pavonis—belong to this period. Much of the northern lowlands also have been mantled, either by volcanic plains or eolian deposits. On Map 9.1e recent lavas are shown only in those regions where B frames contain diagnostic flow fronts.

The volcanic evolution of Mars places it between Earth and the moon. On the moon, volcanic activity was confined to an early stage of planetary evolution, with the younger magmas apparently generated at progressively greater depths, coincident with a general decrease in upper mantle temperature. On

Earth volcanic activity continues to the present with no sign of an end. In an active environment of plate tectonics, magma is generated at depths of a few tens of kilometers. On Mars, the crustal thickness and depth of melting is intermediate between Earth and the moon. The thermal environment is insufficient to generate crustal plate movement, but adequate to fuel widespread volcanism.

The structural features of Mars differ from those of the moon, Mercury, and Earth. The moon is largely devoid of compressional structural features. Linear valleys, apparently graben, are generally related to impact basins. The Mercurian landscape is dominated by huge ridge systems that may have formed during an early stage of crustal contraction. Only on Earth is there persuasive evidence for plate tectonics, with attendant zones of separation and subduction. Great rift systems are observed, along with mountainous fold belts.

During relatively recent times eolian activity has played a dominant role in shaping the Martian landscape. It is probable that a great amount of material has been eroded from equatorial regions and has been deposited at the poles. The polar layered deposits are of uncertain age, but undoubtedly record some climatic change—ancient or recent—that led to episodic precipitation of H_2O and CO_2 frost, intermixed with dust. More or less arbitrarily we show the polar deposits first appearing in figure 9.1c, tentatively related to early atmospheric changes. Subsequent to their formation, the polar deposits have been eroded by surface winds. Mantles of debris have spread equatorward to approximately $\pm$ 30° latitude (Soderblom et al., 1973b). Smaller craters have been covered and larger craters partly buried.

At lower latitudes some depressions, such as the Hellas basin, have trapped substantial amounts of wind-blown dust. Elsewhere a thin veneer of mobile sediments has been swept across volcanic bedrock. The shifting pattern of dark basalt and brighter sediment accounts for some of the streaks and splotches that appear in association with many craters.

In summary, Mars appears to be intermediate in both size and state of geologic evolution between the Earth, and the moon and Mercury. The ancient cratered surface so evident and well-preserved on the moon and Mercury, is preserved to a lesser extent on Mars and eradicated on Earth. In part, this preservation is due to the lack of an atmosphere on the moon and Mercury. The atmosphere of Mars, while allowing for eolian modification up to the present, has not supported a continuous hydrologic cycle such as that which dominates the surface of Earth. An additional factor in the preservation of the ancient cratered terrain is the lack of extensive compressional deformation, associated with plate tectonics and continental drift on Earth. On the moon, extensive surface volcanic activity ceased about three billion years ago, and the youngest plains units on Mercury appear of similar age (Murray et al., 1975). On Mars, volcanic activity continued past that time, perhaps extending close to the present.

APPENDIX A

Guide to the Use of Mariner Images

Introduction

PLANETARY photointerpretation is beginning to acquire the status of an established discipline. Therefore, it is necessary for the investigator to gain some familiarity with the techniques for acquiring and enhancing pictures. We will discuss, in particular, the operation of the Mariner 9 television camera system and the techniques for interpreting Mariner pictures. Since television cameras have been used almost exclusively as the imaging device on all unmanned spacecraft—Lunar Orbiter with its film system is a notable exception—our remarks have applicability extending beyond the Mariner 9 mission. There are a number of sources of additional information, an excellent one being Cutts (1974).

Instrument Description

The Mariner 9 cameras are essentially similar to those used in commercial television (Masursky et al., 1970). The system is entirely electronic, capitalizing on the use of an image tube, the vidicon. The configuration is similar to that of a conventional film camera with a photoconductive target substituted for photographic film. The scene is recorded on the target by opening and closing a shutter. The inside face of the target acquires an electrical charge proportional to the intensity of light. As the target is scanned with a narrow-beam focused electron gun, electrons are supplied to the target, creating variations in the output. The entire image is recorded on the target plate but is read line by line by the electron gun. Electrical signal strengths are recorded as the electron beam scans the target. Variations of signal strength can be related to light intensity at each point on the target. Finally, the vidicon signals are converted to a digital form and telemetered back to Earth.

Each picture comprises 700 lines. Each digitized line contains 832 picture elements, or pixels. For each of these pixels 512 levels of brightness can be measured by nine-bit digitization of the signal returning to Earth. In brief, then, the pictures we examine are mosaics of 582,400 points with varying brightness. Since a pixel contains nine binary bits, each photograph represents a total of more than five million bits of information.

The entire camera consists of optical elements, filters, and shutter in addition to the vidicon. The wide-angle A camera has a 50-mm focal length lens and an eight-position filter wheel. Filter passbands in the orange, blue, and green parts of the visible spectrum permit acquisition of color data, especially if the same region is imaged in all three spectral bands under similar illumination conditions. Details of filter spectral transmittance may be found in Snyder (1971). Unfortunately, since the filter wheel mechanism failed early in the Mariner 9 mission, the majority of pictures were taken through a polarizing filter. The narrow-angle B camera has a 500-mm focal length lens with a fixed haze filter designed to block out the blue end of the spectrum. The shutters for both cameras are positioned near the vidicon target.

When one understands how the camera works, some of its shortcomings can be appreciated. The optics introduce very little distortion, but the vidicon tube has some frustrating liabilities. It is difficult to construct a target that is uniform over its entire surface. During the mission, the left side of the Mariner camera became somewhat more sensitive relative to the right side. (This sensitization may have resulted from similar lighting geometry throughout the mission. Most of the pictures were taken near the evening terminator, causing the left side of the image to be brighter than the right.) Target response also varies as a function of temperature and time. For these reasons, precise photometric correlations are impossible. Another serious problem arises because the electron beam is unable to discharge completely the scanned surface, even after several repeated scans. The result is a residual image observable in successive pictures, something like a double exposure in conventional photography. Another problem, geometric distortion of the image, results from deflection of the electron beam arising from various sources, including variations in target conductivity. Accordingly, the amount of deflection varies with light level in the scene. Electrical noise, both random and coherent, can be superposed on the image-related signal for any numbers of reasons, including faulty circuit design. Finally, blemishes on the conductive target and dust shadows on the lens must be recognized so they do not lead to faulty interpretations. The removal of all these effects is termed decalibration.

Description of Image Data

The Mariner 9 pictures were transmitted to Earth in digital form. Usually, 31 to 33 frames were recorded on a tape recorder aboard the orbiter during a single revolution about Mars. The entire tape load was then transmitted to the Deep Space Network on Earth. Data were again recorded on tape and relayed to the JPL Space Flight Operations Facility (SFOF) over high speed data lines. The first version of the pictures was then produced by the Mariner 9 Mission Test Computer/Mission Test Video System (MTC/MTVS) (Levinthal et al., 1973). Basically, the computer assembles the data in picture format,

and the video system prepares a permanent file copy. Several versions of each frame were produced (fig. A.1). The raw picture approximately duplicates what the camera actually recorded—a very flat low-contrast scene with geometric and shading distortions. The second version includes rudimentary photometric corrections for camera shading. Two other versions were produced to enhance topographic details. These are a horizontal high-pass-filter (HPF) and a vertical-automatic-gain-control (VAGC) version. Early in the mission, one or the other of these versions was used, but later frames were modified by both filter transformations. These versions are "stretched," a digital process that enhances contrast. For most users, the stretched versions are the most attractive and meaningful. A comprehensive label appears beside each MTVS picture (fig. A.2). The label data were preliminary and must always be verified. All of the MTC/MTVS versions were recorded on 70-mm film from which positive transparencies were made for distribution to experimenters and to the National Space Science Data Center (NSSDC). The NSSDC is responsible for public distribution of the data.

Decalibration

Useful as the MTVS versions of the images are for displaying the surface detail of Mars, further processing is required to allow photometric or geometric measurements. Especially important is the decalibration previously mentioned—removal of photometric distortions, geometric distortions, and residual images. The digital computer processing to produce a final archival version of the images was done by the JPL Image Processing Laboratory (IPL). The final data set is called the Television Reduced Data Record (TV RDR, or simply RDR). Technically, the RDR exists as magnetic tape containing the numerical data for each frame in a form such that each picture element is represented by a data number (DN) proportional to the luminance in the corresponding point in the scene. Data on the magnetic tape have been converted to photographic prints in two enhanced versions. One is given a linear contrast stretch that portrays correct relative brightness differences throughout the frame; the other is high-pass filtered and given a contrast stretch that emphasizes topographic detail (fig. A.3).

The RDR pixel format differs from that of the MTVS in that the pixels have been squared and scaled so that 1 mm on the vidicon tube equals 75 pixels in the RDR picture. The resultant format includes 800 lines, each with 950 samples, or pixels. The original DN range of 9 bits is retained; each pixel may have a DN value of 0 (black) through 511 (saturated white). For photometric work the

Figure A.1. Real-time MTC/MTVS versions of a Mariner 9 image of an area transitional between plains and cratered terrain south of the Elysium region. The raw picture (a) shows no detail. The shading-corrected version (b) has some of the photometric distortion removed and shows the scene enhanced, but with the relative brightness preserved across the scene. The extreme change of luminance from left to right occurs because the picture is taken in the afternoon when the sun is in the western sky; curvature of the planet leads to variable lighting conditions across the image—brightest in the west, darkest in the east. The elongate dark marking at the top is the eclipse shadow (penumbral) of the natural satellite, Deimos. In the filtered version (c), the effects of sun angle are removed by subtracting from the shading-corrected version a smoothed version of the image, and then stretching. This procedure introduces some artifacts seen as vertical lines. The other round dark spots, about 30 pixels in diameter, are the shadows of dust spots on the faceplate of the vidicon. (DAS No. 07507188, MTVS ROLL/FILE 4196-70, 71, 72.)

(b)

(a)

(c)

actual numbers from the RDR tapes should be used in preference to the grey levels recorded on the film products.

Residual Image

The formation of a residual image is a complex, poorly understood phenomenon. Basically, the response characteristics of the target depend partly on the response to preceding images. As a consequence, an image may contain a faint overprint of a preceding image. This is especially noticeable when the preceding image contained sharp, high-contrast boundaries such as the bright edge of the planet adjacent to dark space. Craters are particularly troublesome when they appear as residual images. A sharp, fresh crater may have a subtle residual image that looks precisely like a highly degraded and blanketed crater.

The removal of residual images is complicated and imperfect. It is assumed that the residual contribution is only from the directly preceding image. Although a necessary simplification, this assumption is clearly invalid. Frames taken as the spacecraft approached the planet contain residual images from all preceding frames. Residual images of the limb persisted through the first frames of orbital photography. Residual images are removed by subtracting a small part of the preceding image. If the subtraction is excessive, an equally objectional "negative" residual image may be produced.

Geometric Distortion

The raw frames contain large amounts of distortion, which can be removed by reference to a grid of 111 reseau marks (the black dots in figure A.1) placed on the vidicon target. The positions of the

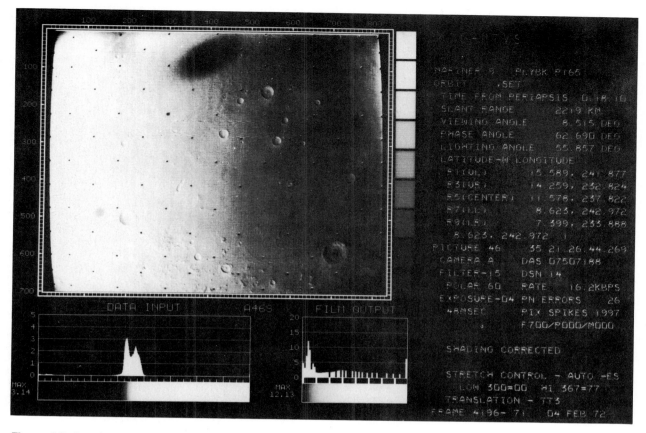

Figure A.2. Standard format for MTVS picture, data block, and histograms of input and output data numbers. The data block is reprinted with explanatory notes.

MTC-MTVS

	MARINER 9 PLYBK P165	orbit number
	ORBIT____,SET	not used
	TIME FROM PERIAPSIS 0.18.10	
	SLANT RANGE 2219 KM	
	VIEWING ANGLE 8.515 DEG	angle at center of picture from vertical to spacecraft
	PHASE ANGLE 62.690 DEG	angle at center of picture between Sun and spacecraft
	LIGHTING ANGLE 55.857 DEG	Sun elevation measured from vertical
	LATITUDE–W LONGITUDE	
reseau no	R1 (UL) 15.589,241.877	
	R3 (UR) 14.259,232.824	
	R5 (CENTER) 11.578,237.822	
	R7 (LL) 8.623,242.972	
	R9 (LR) 7.399,233.888	
	8.623,242.972 1	not used
DAS picture count (not used)	PICTURE 46 35 21.26.44.269	Earth received time (day-hr-min-sec)
	CAMERA A DAS 07507188	DAS number of first line
filter position no	FILTER–15 DSN 14	tracking station
filter (polarizing a 60°)	POLAR 60 RATE 16.2KBPS	telemetry bit rate
shutter speed code	EXPOSURE–04 PN ERRORS 26	pseudo-noise errors (loss of code indicating the end of a line)
shutter speed	48MSEC PIX SPIKES 1997	noise spikes
	F700/P000/M000	full lines/part lines/missing lines
image processing parameters	SHADING CORRECTED STRETCH CONTROL — AUTO — ES LOW 300=00 HI 367=77 TRANSLATION — 773	
	FRAME 4196 — 71 04 FEB 72	processing date

(a)

(b)

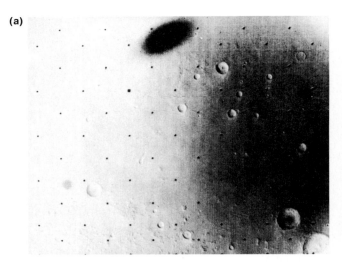

Figure A.3. RDR versions of the same Mariner 9 pictures shown in figures A.1 and A.2. The albedo version (*a*) has been given a linear stretch to preserve relative scene bright- ness across the image. The filtered version (*b*) has been processed to emphasize topographic detail. (IPL Roll Nos. 7350 and 1631.)

reseaus were accurately determined by preflight measurements. During mission operations a slight shift of the entire set was noted. This is attributed to camera operation outside Earth's magnetic field. In addition, at higher light values there is a slight upward shift of reseaus relative to the image. This is chiefly because the unscanned part of the target carries a charge that can deflect the electron beam by as much as 3 pixels.

Noise

Two kinds of noise must be removed, coherent and random. Coherent noise arises chiefly from two sources. Faint vertical bars with an amplitude of about 2 DN (not to be confused with vertical lines produced by vertical high-pass filtering) resulted from harmonic effects between the carrier frequency and the spacecraft power frequency. This gives a pervasive, but generally unobtrusive, texture to the images. A larger effect producing a horizontal pattern with an amplitude of 5 to 30 DN occurred when ultraviolet spectrometer measurements were made. Mechanical vibrations of the UVS mirror were transmitted through the scan platform. Using appropriate digital filters, it is possible to eliminate much of the coherent noise.

Random noise produces a salt and pepper pattern on the image, known more familiarly as "snow" in the early days of home television. These errors in individual pixels are produced by preamplifier noise in the camera system and by telemetry bit error (loss in transmission). The snow can be removed by arbitrarily assigning the dropped pixels brightness values similar to those of neighboring pixels, but this is only a cosmetic improvement. No additional data are displayed.

Line dropouts are significant defects related to noise. These generally arise from telemetry errors in the spacecraft to ground link. However, many lines in the MTVS frames were lost in the ground link. Most of these were subsequently recovered and appear in the RDR frames.

Blemishes

A number of dark patches and small bright specks have been identified on Mariner images. The large dark patches are about 30 pixels across. They are probably produced by shadows of dust particles on the faceplate. The bright spots are smaller, probably defects in the target. Table A.1 lists the more significant blemishes. They have caused embarrassment for some casual investigators since, in some images, they look like faint craters or protuberances. Unattractive though they are, the dark spots do have some value. They can be used to estimate contrast variation in the pictures since their amplitude is approximately known.

A series of concentric arcs appear in the corners of some stretched pictures, superficially resembling part of a natural circular feature. In fact, the circular pattern is produced by internal focusing of the electron beam.

RDR Labeling

The image processing steps applied to each picture appear on the picture label (fig. A.4). A description

		A Camera—MTVS Images	
		Dust Shadows	
Line	*Sample*	*Size, pixels*	*Amplitude*
120 (155)	165 (195)	30	2%
204 (285)	760 (865)	30	3%
485 (550)	130 (155)	30	6%
		B Camera	
		Dust Shadows	
425 (475)	645 (750)	40	< ½%
535 (590)	275 (320)	40	1%
690 (755)	330 (395)	50	1½%
		Black and White Spots	
50 (70)	320 (355)	15	+60 DN, −30 DN
187	195	10	+20 DN, −20 DN
240	148	10	+30 DN, −20 DN
118	725	10	+20 DN, −20 DN
322	770	10	+10 DN, −15 DN
522	758	10	+20 DN, −15 DN

Table A.1. Blemishes on Mariner 9 Vidicon (modified from Thorpe, 1972). RDR locations are in parentheses. In addition to those listed, there are about a half-dozen blemishes, several pixels across, on the A frames. Under certain conditions, these might be mistaken for small craters. The best check on small features is to look at the same location in several frames to verify that the feature is on Mars and not on the vidicon. The B-frame vidicon defects are listed in the table because they are large and occur as black and white spots identical to small, sharp craters.

Figure A.4. A typical picture label for an RDR picture. This version is intended to show topography to best advantage. The caldera of Pavonis Mons is shown in the picture.

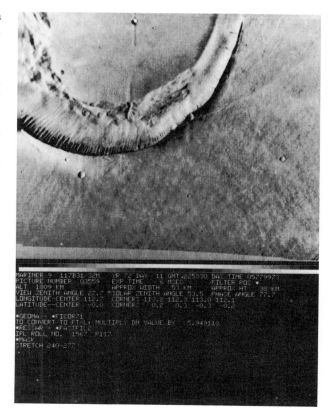

of the processing is given by Seidman et al. (1973). In general, the following sequence of labels on a picture identifies the RDR version.

"GEOMA" is the first step in decalibration. This program transforms the image to correct geometry, removing geometric distortions produced in the camera.

"RESRED71, CORRECTED FOR RESIDUAL IMAGE" appears on pictures that were processed for residual image; not all frames could be treated.

"FICOR71" is the photometric correction program. The message "TO CONVERT TO FT-L, MULTIPLY DN VALUE BY . . ." is added to allow conversion of RDR data numbers to luminance in footlamberts.

"RESSAR" removes the reseau marks and fills in by interpolation.

"FASTFIL2" is a high-pass filter program that enhances detail and suppresses low frequency in the picture.

"IPL ROLL NO . . ." label indicates the permanent roll number by which the negative is filed in the JPL Science Data Library or at NSSDC. "P . . ." is the revolution or orbit number.

"MASK" is the program step that creates the grey-scale, pixel grid, and label.

"STRETCH xxx-xxx" indicates a linear contrast stretch. The two numbers separated by a hyphen are the original data numbers that have become black and white respectively in the enhanced picture. For example a rather low contrast picture has nearly all the data numbers clustered around some intermediate value. The parameters of STRETCH 246-269 would assign the value 0 to DN's 246 and less in the original image and the value 511 to DN's 269 and greater. This "stretches" a contrast range of only 23 DN to a full 512. This process brings out details in otherwise very flat-appearing images. An image with a narrow histogram of DN values has low contrast; after stretching, it has a broad histogram. The stretch parameters must be carefully chosen to in-clude most of the original histogram. Otherwise, information is lost. If the scene has extremely low contrast, with the DN values ranging over only 5 or 6 values, the stretch produces noticeable contouring or density steps in the enhanced image.

Numerous other enhancement programs may be indicated in the labels of specially processed images. The rectified and scaled pictures produced for the U.S. Geological Survey quadrangle mosaics occur as one of the standard projections: Mercator, ortho-graphic, Lambert conformal, or polar stereographic (fig. A.5). These are made from RDR data. The nine-bit data are truncated to eight bits (256 grey levels), and they are projected in a program called SUPERMAP on many of the earlier labels, and later, more prosaically, MAP2.

Image Identification

Critical information in picture labels should always be checked against another data source, ideally the SEDR.

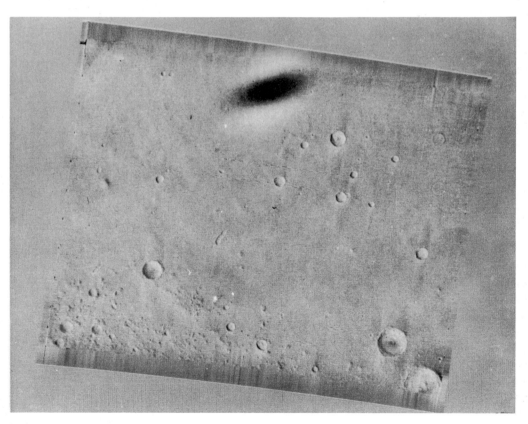

Figure A.5. Mercator version of picture in figures A.1, A.2, and A.3. North is to the top. The scale of the image can be determined precisely from equations for the Mercator projection.

Each Mariner 9 television image has a "unique" number, the DAS time, associated with it. This number is attached to each line of the image by the Data Automation Subsystem (DAS) as the target is scanned and digitized. The DAS time is an eight-digit integer incremented by one count every 1.2 seconds. Since it takes about 42 seconds to scan a frame, successive frames will be separated by no less than 35 DAS counts. In fact, three different DAS numbers may appear for a single frame. In the Supplementary Experiment Data Record (SEDR) the DAS time when the camera was shuttered is listed. This is five counts ahead of the time on the RDR version because the number listed on the picture is the time when the first line of the frame was acquired. On the MTVS picture, the DAS time is that of the first line received on Earth. This may be many counts higher than the RDR time if some of the first lines were lost in transmission.

Resolution and Scale

Resolution is a property that can be described in a number of ways. There is no completely satisfactory definition, since factors such as the experience of the interpreter may determine the effective resolution of an image. A list of the physical factors that control resolution includes target contrast, system resolving power, atmosphere, illumination, image motion, and the viewing geometry.

The resolving power of a system is simply the ability to detect two closely spaced point targets. For a lens, the diffraction-limited resolving power can be expressed by the relation:

$$a = \frac{F\lambda}{D},$$

where a is the distance two objects are separated in the image, F is the focal length of the lens, λ is the wavelength of the light, and D is the diameter of the lens.

Optical resolution is frequently stated as optical line pairs per mm (lpm), a spatial frequency. In terms of image separation,

$$k_0 = 1/a,$$

where k_0 is the limiting spatial frequency and a is the width of a line plus a space.

Film resolution is normally expressed in terms of line pairs. The resolution of film is ultimately limited by grain size and other film properties. A very high resolution film can resolve better than 400 lpm depending on target contrast.

In determining the resolving power of a vidicon tube, the scanning process complicates the matter. One of the limiting factors is the diameter of the scanning spot. To reproduce an optical pair, consisting of a line and space, it takes at least 2 tv lines. However, in imaging a line target, it is possible that, with 2 tv lines per optical line, nothing will be resolved if the tv lines fall midway between optical lines. The convention usually adopted is that the spacing between pairs of optical lines is equal to $2\sqrt{2}$ tv lines. The Mariner 9 cameras scan at 70 tv lines per mm corresponding to optical resolution of about 25 lpm for very high contrast targets. Although this is outstanding for a vidicon, it is an order of magnitude worse than for film.

Table A.2. Specifications for Mariner 9 cameras.

	Camera A	Camera B
Focal length, mm	52.267 ± 0.006	500.636 ± 0.036
Sensor dimensions, mm (within outer reseaus)	12.343 × 9.561	12.342 × 9.545
Angular field, deg	13.56 × 10.50	1.413 × 1.092
MTVS pixel dimensions, μ radians	284 × 261	29.6 × 27.2
RDR pixel dimensions, μ radians	255 × 255	26.6 × 26.6
MTVS picture elements per frame	832 × 700	832 × 700
RDR picture elements per frame	950 × 800	950 × 800

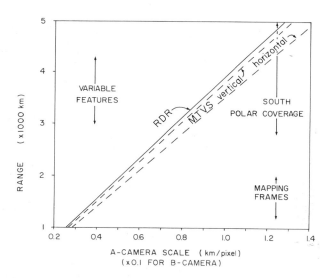

Figure A.6. Resolution of Mariner 9 pictures and approximate variation of slant range for different types of picture coverage. The two lines for MTVS pictures represent the rectangular horizontal and vertical dimensions of the pixels. Although the lines diverge at greater slant range, the ratio between the two dimensions remains constant.

A common method of expressing resolution is as *ground resolution*. This is the distance on the ground corresponding to the smallest distance resolvable in the image. For the Mariner 9 camera this would be the projected size of a pixel on the ground. This is a convenient way of measuring features since they can be easily expressed in terms of pixels and then converted to meters or kilometers. Ground dimensions of pixels may be obtained approximately by the following:

$$d = S\theta.$$

where d is the ground dimension in km, S is the slant range in km, and θ is the angular field of a pixel in radians (table A.2 and fig. A.6). The slant ranges to 9 points in each picture are given in the Mariner SEDR. Values on the individual prints are for the center of the image.

Measurements on RDR prints are more reliable than on MTVS versions because camera distortions have been removed. The pixel scale on MTVS pictures depends on whether they are measured horizontally or vertically (fig. A.6). RDR images have been reprocessed to square pixels spaced at 75 pixels per mm on the vidicon sensor. On the RDR pictures, the approximate height and width of the picture is given on the label. For MTVS pictures, a good rule of thumb for determining the approximate width in kilometers is to divide the slant range by four or forty for A and B frames, respectively. For example, a typical mapping frame taken at a slant range of 1600 km is about 400 km across. For accurate measurements, it is best to determine the pixel dimensions of the object by actual measurement. It is convenient to use the pixel grid on the margins of each photograph for this purpose.

A somewhat subjective measure of resolution is called *identification resolution*. For example, experience with planetary images has shown that most craters can be identified as such if they are four to six pixels in diameter, a dimension that is equivalent to two optical lines. Reasonably enough, the characteristic crater pattern includes a bright sunlit wall and a dark wall facing away from the sun, essentially the same pattern as two optical lines. Some small craters might be seen, depending on their contrast and the way pixels are distributed, but many other craters will be undetected since the pixels average adjacent small dark and bright regions.

A final and very significant factor in image resolution is atmospheric clarity. Particulate and molecular constituents of the atmosphere absorb and scatter light, reducing contrast. Turbulence in the atmosphere reduces resolution further by causing apparent motion of objects in the image. For these reasons, Earth-based telescopes with lenses larger than 30 inches in diameter are made exclusively for stellar studies, where capability to detect very faint objects is required but spatial resolution is less important.

Brightness, Shadows, and Topography

Many of the more spectacular lunar pictures are taken near the terminator, where subtle topographic protuberances and depressions are accentuated by dark, crisp shadows. With this lunar situation in mind, it is understandable that some persons examining Mariner pictures have identified dark slopes as shadowed areas. Were this actually the case, it would be a powerful research tool. Shadows could be used in determination of local relief.

Several lines of evidence suggest that most of the dark slopes are *not* shadows. The steepest slopes in shadow should have angular gradients complementary to the lighting angle. In terrestrial situations, unconsolidated material cannot maintain slopes in excess of 30°. For many Mariner pictures the lighting angle is 60° or less from vertical. Any alleged regions in shadow would have to be associated with unnaturally steep cliffs. Another demonstration of the lack of shadow is to look at the lower left blemish in an A frame. This region is darker than it should be by about 6 percent. In other words, if the region is fairly homogeneous, this spot is 6-percent darker than its surroundings. The upper spot is 3-percent darker. Frequently these are the darkest areas in the image although they are much brighter than the 10 or so DN to be expected in true geometric shadow. This can be verified by DN listings of particular regions on a picture, a tedious and expensive undertaking.

Although true shadows are infrequent in the Mariner 9 images, the topography is well expressed by the enhanced versions of the pictures. As with the low-sun lunar pictures, however, the stretched pictures of Mars give an impression of far more pronounced relief than the rather gentle slopes that actually exist on the surface.

Obtaining Mariner Data

The organization responsible for public distribution of spacecraft data is the National Space Science Data Center (NSSDC) at Goddard Space Flight Center in Greenbelt, Maryland. The following summarizes the Mariner 9 data available from NSSDC. For a more complete list, the reader is referred to the current NSSDC Mariner 9 Data Announcement Bulletin.

MTC/MTVS pictures in all versions.
RDR pictures in two enhanced versions.
Rectified and scaled pictures.

Picture indexes arranged by latitude, longitude, features, roll and file numbers, DAS time, or revolution.
SEDR on microfilm.
Special mosaics.
MTC/MTVS microfiche and users' guide.
RDR microfiche and users' guide.

Data Records

Most of the Mariner data of interest to potential investigators are available at NSSDC. However, many of the more specialized and intermediate data sets are in archival storage in the JPL Science Data Team library (Holmlund, 1973). The TVS EDR contains the raw data for the television pictures.

The MTC/MTVS real-time pictures are kept in the science data library as 70-mm duplicate negatives, strip contact prints, and 8 × 10 prints of all versions. Also in the library are prints of all frames processed by IPL and by the Artificial Intelligence Laboratory at Stanford University (Levinthal et al., 1973). These have been specially processed according to specific experimenter requests. The MTC/MTVS images are also available on microfiche, along with a users' guide. Information on spacecraft alti-tude, camera pointing directions, picture corner co-ordinates, slant range, etc., are catalogued in the Supplementary Experiment Data Record (SEDR). Listings of the SEDR are available from the NSSDC. A comprehensive review of picture coverage and sequence design is contained in Koskela et al. (1972) and Koskela (1973).

Maps

The United States Geological Survey has prepared a series of 1:5,000,000 scale quadrangle maps of Mars using Mercator projections in the equatorial region, Lambert conformal projections at higher latitudes and polar stereographic projections at the poles. (See figure C.1 for locations and names.) The 1 to 5,000,-000 series will contain 30 maps. A number of larger scale maps are also being made at scales of 1:1,000,-000 and 1:250,000 for Viking landing site studies and for various science studies. The 1:5,000,000 series will include controlled photomosaics, and shaded relief versions. The latter will be formally published by the Survey. A global 1:25,000,000 scale map has been compiled showing topography, albedo features and official nomenclature.

APPENDIX B

Approaches to Data Gathering and Interpretation

Description and Interpretation

THE INTERPRETATION of landforms on Mars is fundamentally a problem in geomorphology. As in terrestrial geomorphology we have two choices. We can describe the landforms according to their physical attributes, or we can classify them genetically according to the processes responsible for their formation. The latter is more intellectually stimulating, but also more hazardous. One has only to think of the influential American geographer, W. M. Davis, who classified landscapes organically as going through stages of "youth," "maturity," and "old age." Davis was a master of deduction: describing the process of slope retreat without making detailed field measurements of slope profiles, discoursing on the geology of Bermuda without ever visiting the island. In ordering the natural world, he argued logically from a few hypothetical assumptions. Looking back, it is easy for us to criticize the results, but many of Davis' ideas were useful and—at a certain qualitative level—correct.

An objective descriptive approach to landform analysis appears more "scientific." Nonetheless there are numerous pitfalls. The criteria for classification may be irrelevant to matters of genesis. To that extent the divisions are uninformative. For example, it would be easy to classify terrestrial rocks by color. However, sedimentary, igneous, and metamorphic rocks of dramatically different origin may have the same color. Indeed, a more useful first-order division employs this genetic trio: sedimentary, igneous, and metamorphic.

Even if descriptive form is related to genetic process, the correlation may not be unique. Certain types of volcanic and impact craters have similar appearance. Lava channels resemble water-cut channels in many respects.

One of the more self-conscious attempts to separate description and interpretation is recorded in the suite of lunar geologic maps produced by the U.S. Geological Survey (Wilhelms, 1970; Wilhelms and McCauley, 1971). Each of the map units is identified by purely morphologic attributes. Interpretations are presented separately. The expectation is that, even if the interpretations prove invalid, the descriptions and distributions of map units will have continuing validity. However, in actual practice, maintaining the distinction between description and interpretation is difficult. Even though the final maps consistently demonstrate separation, the author of each map is often tempted to disregard the dichotomy as he analyzes the photographic data on which the map is based. Knowing that the majority of craters form by impact, he is predisposed to identify ejecta deposits around craters and basins. Inevitably he speculates about the stratigraphic succession of rock units, and then is inclined to map distributions of material units that are consistent with the stratigraphic model. To a greater or lesser degree he can discipline himself against these interpretive forays, but he can never completely eliminate them.

Working Hypotheses, Multiple and Single

In 1890 T. C. Chamberlin published a paper entitled "The Method of Multiple Working Hypotheses." This article, widely read by each new generation of scientists, has become a generally accepted model for objective scientific inquiry. Chamberlin argues that, when we collect and evaluate data, we should have prominently in mind *all* of the hypotheses that might satisfactorily explain the data. As work proceeds we can then narrow the suite of viable hypotheses, but only after each one has received a fair trial.

Chamberlin's approach is extremely attractive. It emphasizes the objectivity of scientific inquiry as well as the requirement for accurate description and careful experimentation. At the same time it encourages the investigator to speculate. The synthesizing, process-related models are aways close at hand. One is not committed to endless and exhaustive data collection, unrelieved by any intellectually stimulating conclusions.

This approach is peculiarly tempting in the current scientific environment, which places a premium both on rapid and frequent publication and on the regular generation of "important" ideas. Investigators can postulate a stimulating hypothesis, present a few relevant data, and conclude that the data either support or contradict the hypothesis. Either outcome can be touted as a significant advance in knowledge. Subsequent papers provide continuous assessment of the candidate hypotheses in the light of a few more facts. The overall result is something like the method of multiple working hypotheses on the installment plan.

Cynics have pointed out that Chamberlin's method can be even more grievously misused. For some investigators, multiple working hypotheses become multiple working prejudices. A narrow choice of hypotheses—slender either in number or diversity—predetermines the outcome. But these criticisms are diversionary. Although seldom attainable, the method of inquiry recommended by Chamberlin is a worthy goal.

Data Now—Hypotheses Later

It is not always possible to consider simultaneously form and process, description and interpretation, facts and hypotheses. In areas of science that are in an early stage of development, often coupled to new instrumental and experimental capabilities, the emphasis must be centered on description. Some practitioners of exploratory science turn this liability into a badge of honor, scornfully dismissing anyone who has the temerity to marshal the few available data into a cohesive model.

Although we do not go so far as to wear it as a badge of honor, we do assert that planetary photogeology is in a preliminary descriptive phase. A small part of the photogeologic data has been extracted, let alone classified or interpreted. Admittedly, much of the cream has been skimmed off. Mariner 9 pictures of Mars reveal spectacular features—volcanic shields, canyons, layered polar deposits, wind streaks—that yield to rapid, qualitative analysis, much of it proceeding by analogy. (For example, a Martian feature looks like a terrestrial feature. Therefore the origins are similar.) Once this first wave of analysis is complete, further elucidation is difficult without consideration of additional data. Whereas the first analyses are usually qualitative and intuitive, the second generation of hypotheses tend to be more quantitatively documented.

Anyone moving into this second stage of photointerpretation may be led up some blind alleys. In the case of Mars there are so many unresolved questions about the physical characteristics and evolution of surface materials that any set of working hypotheses is almost certain to be incomplete. One runs the risk of selecting a currently popular hypothesis or model, collecting data to test that model, and then discovering that the model is incomplete or incorrect, and that the data are in a form unsuitable for further analysis.

A specific example will illustrate the point. Assume that an investigator suspects that crater density is related to elevation. To test this hypothesis he decides to measure all craters along belts of constant elevation. This is a laborious task, but the investigator perseveres, only to be faced with ambiguous results. Some pictures have higher resolution than others. Some regions are obscured by atmospheric haze. Crater densities are seen to change as a function of terrain type. As a final indignity, a revised elevation map is published, showing a different set of contours. If the investigator is dogged, he can make a new set of measurements. Resolution problems can be avoided by electing to measure craters with larger diameters. Measurements for different terrain categories can be separated for separate analysis. New regions can be examined, in accord with the more recent elevation calculations. If a correlation between elevation and crater density is discovered, the exercise can be counted a success—albeit a time-consuming success. If no correlation is found, it is unlikely that the crater data will be in a form useful for testing any other hypothesis. (Our own preliminary and ambiguous correlations between crater density and elevation are contained in figure 4.53.)

An obvious solution to this problem, in theory at least, is to extract all the significant data from the pictures during one comprehensive survey and to store that data in an easily accessible format. As a first step in that direction, we have measured the size, morphology, and position of approximately 50,000 craters observed on Mariner A frames. The information is stored in a computer so that it is available for combination in a myriad of formats. Instead of being bound in obsolescence to a premature hypothesis, the data can be repeatedly assembled for appropriate tests of each new model.

The Brown University Method

Classification Techniques

Table B.1 lists the different combinations of crater parameters that were recorded for A-frame craters. When tabulating data, one of the ten combinations listed in table B.1 was selected for each crater. Then the appropriate information was recorded and the reduced data were stored for each crater, using the disk crater number as a code indicating what set of parameters had been measured. Crater combinations 1 and 2 were designed to accommodate small bowl-shaped craters that were close to the resolution limit of the picture and appeared to have uniformly smooth rims, walls, and floors. For larger craters, morphological parameters were tabulated as a 9-element vector, where the elements were integers

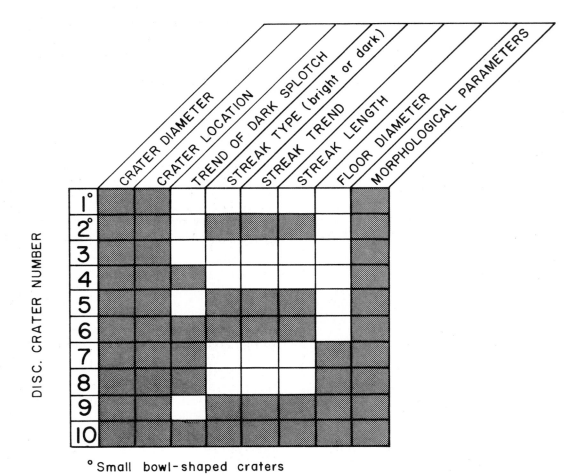

° Small bowl-shaped craters

Table B.1. Format used to record and store A-frame data. Each row represents a combination of parameters that can be recorded for a crater. Dark blocks indicate that a parameter is included in a given row. Each crater is placed in one of the 10 rows before being digitized. Then the appropriate data are digitized and stored using the disk crater number as a key to what has been tabulated for that crater.

representing descriptive values for crater geometries and features (table B.2). The morphological classification was designed to be as objective as possible by keeping divisions within each category simple and well-defined. We found that most Martian craters could be quickly assigned values for each category.

Data were recorded using a Hewlett-Packard digitizer, programmable calculator, and a tape output device. The digitizer consisted of a magnetic board with a fixed orthogonal grid and a hand-held digital cursor. The digital cursor could be placed anywhere on the board and, on command from the operator, would transmit its X-Y position to the calculator.

Figure B.1 shows conceptually how crater data were tabulated using the digitizer. Material was recorded using the following scheme:

(1) The picture was taped to the digitizer board and the origin was defined as the lower left-hand corner of the picture.

(2) The length and width of the picture were digitized and stored in the calculator, along with a number identifying the picture.

(3) Two rows of boxes were placed beneath the picture. The top row corresponded to the range of disk crater numbers listed in table B.1. The bottom row represented values for the morphological classification (table B.2).

(4) A crater was selected for measurement and classified according to table B.1. The calculator was then informed of the decision by digitizing a point within the box with the disk crater number representing the chosen combination.

(5) Crater dimensional parameters were then digitized. Small bowl-shaped crater diameters were recorded by digitizing 2 points on opposite sides of the crater rim. Larger crater diameters were recorded by digitizing 3 points on the rim crest and, if appropriate, on the contact between the wall and the floor. Streaks were recorded by digitizing a point at the downwind end of the streak. Dark patches were recorded by digitizing a point on the crater rim at the position that bisected the patch in a direction normal to the rim.

(6) For disk crater number 1, the calculator was then ready to accept another crater combination box and another set of measurements. For crater number 2, a point was digitized in one of the morphological value boxes to indicate if the streak was bright or dark. For other craters, the calculator then

A FRAME

CRATER MORPHOLOGICAL PARAMETERS

	RIM UPLIFT	RIM TO FLOOR DEPTH	RIM CONTINUITY	RIM OUTLINE	FLOOR SHAPE	FLOOR ROUGH- NESS	CENTRAL PEAK, PIT	WALL TYPE	ASSOC. OTHER CRATERS
0	RIM NOT RAISED	EXTREMELY SHALLOW	CONTINUOUS	POLYGONAL	FLAT	SMOOTH	ABSENT	SMOOTH	SINGLE CRATER
1	RAISED RIM	SHALLOW	BREACHED	QUASI- POLYGONAL	NOT FLAT	CRACKED	CENTRAL PEAK(S)	TERRACED	DOUBLET MEMBER
2		DEEP	DISCONT.: SUPERIMP. CRATERS	CIRCULAR		HUMMOCKY	CENTRAL PIT(S)	HUMMOCKY	CHAIN MEMBER
3				CALDERA				FURROWED	CLUSTER MEMBER
4								PITTED	

Table B. 2. Crater morphologic classification for A frames. With the exception of small bowls, every crater has been classified using this format. Results have been stored as an element vector for each crater.

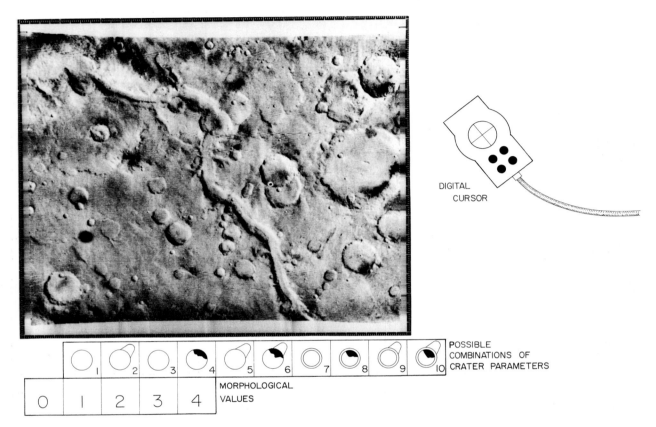

SET-UP FOR DIGITIZING CRATERS
ON A-FRAMES

DIGITAL
CURSOR

POSSIBLE
COMBINATIONS OF
CRATER PARAMETERS

MORPHOLOGICAL
VALUES

Figure B.1. Schematic illustration showing how crater data are recorded, using the digitizer. Consider the plane of the paper to be the digitizing board. The digital cursor is a hand-held instrument that can be placed at any point on the board and, on command by the operator, will transmit the x,y coordinates of that point in inches to a programmable calculator. Craters are digitized by first digitizing a point in one of the crater parameter boxes. This informs the calculator what set of data is to be recorded. Then, points are digitized relative to crater spatial parameters. Finally, morphological values are recorded by digitizing points in appropriate morphological value boxes.

accepted points digitized within the morphological boxes as integer values for the morphological classification and streak type.

(7) For each crater, the calculator would then punch onto tape the disk crater number, picture number, picture length and width, coordinates of crater parameters and, if appropriate, a string of morphological values.

The data were then fed into a large computer system. Picture coordinates were converted to latitude and longitude points with an algorithm using spacecraft slant range, the locations of the picture center and corners, and the length and width of the picture. Crater diameters, locations, and the other parameters were computed from this information. The string of morphological values was converted to vector form. Each crater was then stored on computer disk file with a disk crater number, picture number, dimensional parameters and, if appropriate, a morphological vector.

Validity of Morphological Classification

For much of our analysis, craters have been divided into four morphological classes: fresh, slightly degraded, moderately degraded, and highly degraded (Arvidson, 1974a). The subdivisions are based on floor shape, presence of terraced walls, and rim-to-floor depth, determined visually (table B.3).

	CLASS CHARACTERISTICS	% OF ALL CRATERS	% WITH RAISED RIMS	% WITH CENTRAL PEAKS	% WITH HUMMOCKY FLOORS
FRESH	DEEP, BOWL-SHAPED, WITH SMOOTH RIM, WALLS, AND FLOOR	73	100	2	—
	DEEP, BOWL-SHAPED WITH PROMINENT TERRACES OCCUPYING WALLS	1	100	80	60
	DEEP, FLAT-FLOORED, WITH PROMINENT TERRACES OCCUPYING WALLS	1	87	37	29
SLIGHTLY DEGRADED	DEEP, FLAT-FLOORED, WITH SMOOTH WALLS	3	72	9	4
	DEEP, FLAT-FLOORED, WITH HUMMOCKY WALLS. TERRACING HIGHLY MODIFIED OR ABSENT	3	67	10	9
MODERATELY DEGRADED	SHALLOW, FLAT-FLOORED WITH SMOOTH WALLS	13	5	0.7	1
	SHALLOW, FLAT-FLOORED, WITH HUMMOCKY OR FURROWED WALLS	3	9	2	3
HIGHLY DEGRADED	EXTREMELY SHALLOW, FLAT-FLOORED, WITH SMOOTH WALLS	4	2	0.2	0.6
	EXTREMELY SHALLOW, FLAT-FLOORED, WITH FURROWED WALLS	1	2	—	0.7

Table B.3. Morphologic classification of craters visible on Mariner 9 A frames (1 to 3 km ground resolution). Numerical data were obtained from analysis of the ± 30° latitudinal band covering cratered terrain.

This classification is somewhat arbitrary since there is a continuous range of degradation states for most crater sizes. However, data-bank statistics on partitioning of raised rims, hummocky floors, and central peaks listed in table B.3 demonstrate that the four classes have fairly well-defined characteristics. Fresh craters have raised rims. Central peaks and hummocky floors are commonly present. The majority of slightly degraded craters have raised rims, but the quantity of central peaks and hummocky floors is significantly reduced. This supports the characterization of fresh craters as nearly unmodified impacts, and slightly degraded craters as maintaining a generally fresh shape, but with some degradation of rims, walls, and floors. Statistics for moderately and highly degraded craters support an interpretation of major modification. Craters in both classes are predominantly rimless, have smooth floors, and lack central peaks.

Crater rim crest-to-floor diameter ratios for samples from the data bank for the four classes are plotted in figure B.2. These data were obtained by picking at random 30 craters in each class that fall within 40 to 60 km in diameter. To a first approximation, fresh (deeper) craters should have larger ratios than degraded (shallower) craters. Figure B.2 shows that fresh craters cover a range of ratios, but in general are skewed toward large values. Slightly degraded craters also cover a range of values, but are skewed toward lower ratios. Moderately and highly degraded craters both cover narrow ranges that are skewed toward low values, with moderately degraded craters peaking at larger ratios than highly degraded craters. Again, these results suggest that the degradation classes provide fairly good divisions of a range of crater morphologies.

As previously implied, there is a bias in our morphological classification that is imposed by the resolution of A frames. Small craters that are close to the limit of detection are arbitrarily classified as

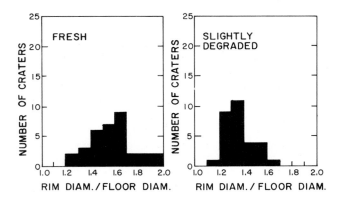

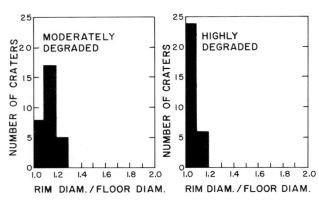

Figure B.2. Plots of rim-to-floor diameter ratios for 40- to 60-km craters within 30° of the equator. The spread for fresh and slightly modified craters is probably due to a combination of analytical error in measuring subtle boundaries between walls and floors and to variation in initial impact geometries. The ratios trend toward lower values as degradation—and thus relative age—increases.

fresh. Most of these craters do, in fact, appear to be relatively unmodified bowls, but there is always the possibility that a range of distinctive degradation features exists below the resolution limit. Indeed, this contention is an integral part of Jones's (1974a) model for crater degradation described in chapter 4.

Data Formats

The crater information that is stored in the computer can be displayed in a variety of ways. Craters of any size range or morphological type can be machine-plotted at any scale, using any map projection (e.g., appendix C, fig. 4.46). The products are particularly useful for comparison with photomosaics and geologic maps.

Crater densities can be calculated and contour maps generated, using a moving search circle routine (e.g., figs. 4.47–4.51). Quantitative information obtained in this way can be compared with other global parameters such as terrain type, elevation,

and albedo (figs. 4.53, 4.54). Crater densities can be calculated for any region, regular or irregular (fig. 4.40). Diameter-frequency distributions can be graphically presented according to any of several methods: cumulative distributions, incremental distributions, and incremental frequency distributions (e.g., figs. 4.42, 4.43).

A virtually limitless number of specific relationships can be explored. Assume that one wishes to examine the constraints of central peak distribution. He can plot the global distribution of central peaks, or examine their number within particular types of terrain. He can show peak distribution as a function of crater diameter or degradational class. More exotic relationships can be economically explored. For example, a global density map of the ratio values between central peak craters and fresh craters is a simple computational program for the computer, although a prohibitively time-consuming one if done by hand.

With a little additional programming, derivative properties can be determined. For example, characteristic ejecta dimensions as a function of crater type and size can be theoretically calculated. Planet-wide values for composite ejecta thickness, based on consideration of all observed craters, can then be machine calculated and plotted.

In this book, particularly in chapter 4 and appendix C, we have presented some of the more standard data displays. Much more remains to be done. We welcome the suggestions of readers with insights more profound and imaginations more unbridled than our own.

APPENDIX C

Crater Maps

Introduction

IN THIS appendix we present geologic sketch maps of Mars together with maps showing craters observable in Mariner 9 A frames. Crater densities provide useful clues concerning geologic processes and history. As discussed in previous chapters, most craters have been formed by impact. The majority has been subjected to erosion by wind, and possibly by water. Many craters are probably blanketed by eolian and volcanic deposits. Accordingly, crater densities do not directly record impact fluxes, or ages. Instead, they provide a measure of the combined effect of crater production and obliteration.

We have found it informative to divide craters into several classes according to amount of degradation (table B.2). In compiling the maps of this appendix we have reduced the classes to two: fresh and degraded. Fresh craters are deep, have raised rims, and retain other impact features such as terraces, hummocky floors, and central peaks. Degraded craters, which include slightly, moderately, and highly degraded craters of table B.2, range in modification from removal of terraces, peaks, and floor texture, to substantial erosion of crater rims and filling of interiors.

The division between fresh and degraded craters is genetically significant. Fresh craters have formed subsequent to an erosional episode, the character, duration, and age of which is discussed in chapter 4. To a first approximation, fresh crater densities in nonmantled regions record the age of all surfaces formed *after* this major obliteration event. A population of large (diameters measuring tens of kilometers) degraded craters indicates the presence of ancient crust that has been variously modified. Wilhelms (1974) notes that volcanic, plateau-forming materials occur throughout the Martian cratered terrain. These deposits are older than those of the cratered plains and may be closely related to the formation of original Martian crust in the first several hundred million years of planetary history. In regions where plateau materials are abundant, numbers of degraded craters are correspondingly reduced by burial and superposition.

The surface of Mars has been divided into thirty quadrangles, in keeping with the format approved by the International Astronomical Union and commonly used by all mapping agencies (Batson, 1973). From 30°N to 30°S, Mercator projections were employed for two bands of eight quadrangles each. In the latitude bands 30°N–65°N and 30°S–65°S, twelve quadrangles were plotted on Lambert conformable bases. The two polar regions, northward of 65°N and southward of 65°S, were plotted on polar stereographic bases. The layout and nomenclature for quadrangles is shown in figure C.1.

We have chosen to display groups of quadrangles together in order to facilitate regional comparisons. For each quadrangle four maps are presented: a geologic map, a plot of all craters, a plot of fresh craters, and a plot of degraded centers. Geologic maps were constructed using the data of figure 3.2. Crater maps were computer plotted from the data bank of Mariner 9 A frames. The method is further described in appendix B. Positions of most craters are in slight error relative to the control net established by Davies and Arthur (1973). In rare instances these errors are as much as a degree in latitude and longitude. Generally, they are much less.

Polar Maps
MC 1, 30 (figs. C.2, C.3)

The polar regions are shown in figures C.2 and C.3. At the south pole, etched plains (ep) and layered deposits (ld) overlie cratered terrain (cu). A remnant of a large basin is partly buried beneath ep and ld. At the north pole, ep and ld overlie plains (p). A cluster of volcanic domes appears in the vicinity of O°W, 75°N.

Highest fresh crater densities occur on cu. The ld units unconformably overlie ep units at both poles (Soderblom et al., 1973a). Consistent with this relationship, ld units have relatively fewer fresh craters.

There is a general paucity of fresh craters at high latitudes (greater than about ± 40°) relative to equatorial regions. There are at least three reasons, two observational and one substantial. Higher latitudes were partly obscured by dust in the south and by remnants of the polar hood in the north. Secondly, polar regions were photographed from relatively high altitudes. The spatial resolution of pictures was reduced from that of equatorial pictures by a factor of two or more (fig. 4.45). However, decrease in crater density is not entirely an observational anomaly. Sediment clearly has been eroded from ld and

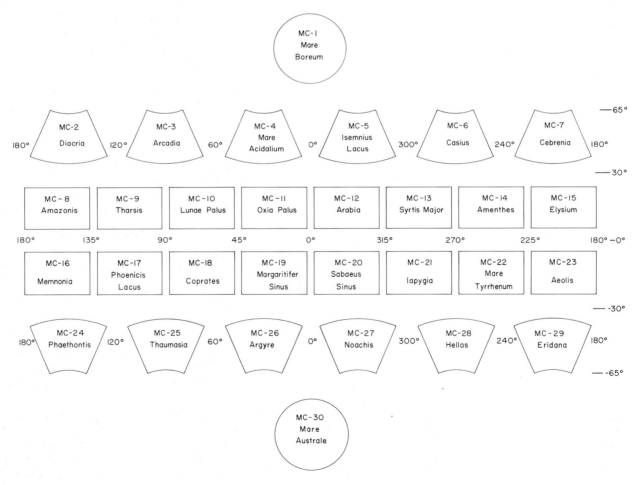

Figure C.1. Designations for the thirty 1:5,000,000 quadrangles which cover the entire surface of Mars.

ep deposits and has been swept equatorward. This has created debris blankets that extend equatorward to about ± 30° latitude, burying many of the smaller craters and mantling floors and central peaks in larger craters (Soderblom et al., 1973b).

Northern Hemisphere Lambert Maps

This set of six quadrangles is shown in figures C.4–C.6. Remnants of cratered terrain (cu) appear in the southern parts of most quadrangles. Undivided plains (p) border cu to the north and cover most of the map region. In the northern half of the quadrangles, the plains have a distinctively mottled and blurred appearance. This may be caused either by atmospheric obscuration or by the presence of large amounts of material shed from the polar regions. Common to all northern hemisphere fresh crater maps is a decrease in crater density moving from south to north. As discussed, this is predominantly the effect of progressively greater burial and mantling by wind-blown debris as the polar regions are approached.

MC 2, 3 (fig. C.4)

This region contains the northern part of the Tharsis volcanic province. A part of the annulus of grooved terrain surrounding Olympus Mons occurs in the vicinity of 140°W, 30°N. Alba Patera, located at 110°W, 40°N, is a complex of overlapping calderas surrounded by radiating lava flows. Unlike the large volcanoes to the south, these calderas are not situated in the center of a high volcanic massif. However, the appearance of white clouds in several pictures covering the eastern part of the structure does suggest that Alba is regionally a topographic high. Major structural graben, Alba Fossae and Tantalus Fossae, trend north toward Alba and are deflected around the caldera structure. A degraded volcanic feature with central caldera and radial texture occurs at 60°W, 42°N.

Scattered throughout the plains are clusters of knobs, presumably erosional remnants of older terrains. They are especially prominent in the southwest quadrant of MC 2, where they are mapped as hummocky terrain, knobby (hk).

MC 4, 5 (fig. C.5)

Break-up and erosion of cratered terrain (cu) is well displayed in these quadrangles. In the south cu is dissected by sinuous channels and broad, flat basins. Traced northward, this terrain is cut into large blocks (hf), sometimes with the formation of wide channels. North of this zone, the blocks are reduced in size until only clusters of knobs are left.

The large multi-ringed crater, Lyot, is located at 330°W, 50°N.

MC 6, 7 (fig. C.6)

These quadrangles are underlain primarily by plains (p). A sharp boundary separates mottled plains in the north from smooth plains in the south. The northern tip of the Elysium volcanic province occurs in MC 7. The volcanic dome, Hecates Tholus, is located at 210°W, 32°N. Along longitude 195°W a northern finger of cratered terrain, now almost completely destroyed and mapped as hummocky terrain, knobby (hk) separates the Elysium volcanic province from the Tharsis volcanic province. This region, known as the Phlegra Montes, has steep eastward-facing scarps, perhaps caused by normal faulting prior to erosive dissection and superposition by plains materials.

Equatorial Mercator Maps

The equatorial belt, from 30°N to 30°S (figs. C.7–C.10) contains many of the spectacular landforms and terrain types documented during the Mariner 9 mission. The exception to this general statement are the polar units—etched plains and layered plains—which occur nowhere in the equatorial zone. Similarly, the canyons, volcanoes, and regions of chaos that characterize the equatorial zone are either absent or subordinate in the polar landscapes. This dichotomy suggests two distinctive geologic cycles operating on Mars, one at high latitudes and the other at low latitudes.

MC 8, 9, 16, 17 (fig. C.7)

This region contains the many and varied landforms of the Tharsis volcanic province: the shield volcanoes Olympus Mons, Pavonis Mons, Ascraeus Mons, and Arsia Mons; the domes Tharsis Tholus, Ceraunius Tholus, and Uranius Tholus; and the calderas Biblis Patera, Pavonis Patera, and Uranius Patera. Olympus Mons is surrounded by grooved terrain that may mark the site of an older volcanic shield, now eroded (Carr, 1973). Densely cratered terrain (cu) occurs in MC 16. To the north and east it is buried by encroaching plains. The northern contact between plains and cratered terrain is, in places, a prominent scarp with small channels running directly onto the plains. A major valley system, Mangala Vallis, is emplaced in the cratered terrain in the vicinity of 155°W, 5°S, and descends to the north.

A swarm of graben, Memnonia Fossae, occur primarily in MC 16 and trend northeast along the extension of the line formed by the three volcanoes, Tharsis Montes. These extension fractures probably formed by upward bulging of the crust along a northeast trending axis. An intricate network of linear channels that form the headward extension of Valles Marineris is incised in moderately cratered plains (pm) in the northeast quadrant of MC 17. The channels are of structural origin, reflecting a complex pattern of extension close to the Tharsis ridge. They have been enlarged by subsequent erosion.

The fresh crater distributions document the relatively young age of the volcanic plains. Materials surrounding Olympus Mons are least cratered and, on that basis, youngest.

MC 10, 11, 18, 19 (fig. C.8)

These quadrangles are dominated by giant canyon systems. The fracture-controlled rifts of Valles Marineris trend eastward through MC 18. These canyons connect with irregularly sinuous valleys that flow to the north, debouching into the southern part of Chryse Planitia in the northwest quadrant of MC 11. The more prominent valleys (or valles) are Shalbatana, Simud, Tiu, and Ares. Chaos-filled depressions are associated with these valleys.

Moderately cratered plains (pm) and densely cratered plains (pc) are extensively exposed in MC 10 and 18. They comprise volcanic strata that overlie cratered terrain (cu) but are younger than volcanic plains (pv). These age distinctions are shown by fresh crater densities that are highest on cu, intermediate on pc and pm, and lowest on pv and p. Cratered plains show widespread wind streaks in the lee of bowl-shaped craters.

In MC 10, erosion of pc occurs along a north-south trending scarp that represents the eastern fault-line scarp of a broad graben. The western scarp is exposed at the southern boundary of the quadrangle near 82°W. Elsewhere, the west side of the graben is almost completely covered by younger volcanic plains. Cratered plains are being eroded back to the east by headward erosion along a coalescing network of linear valleys. This type of angular headward erosion contrasts with irregular dissection and modification of old cratered terrain elsewhere—for example, in MC 11. The former situation suggests break-up of regularly layered volcanic strata. The latter situation suggests erosion of inhomogeneous and complexly deformed materials.

A particularly spectacular example of polygonal erosion of cratered plains is provided by Kasei Vallis, in the north-central part of MC 10. A system of linear tributaries fuses into a larger flat-floored canyon, which descends to the east, toward Chryse Planitia. A secondary channel parallels the primary one some distance to the north, isolating a large remnant of polygonally fractured, densely cratered plains. Floor materials within the canyons have ir-

regularly blotchy albedo, suggestive of wind-worked dust and sand sheets.

MC 12, 13, 20, 21 (fig. C.9)

These quadrangles comprise mostly cratered terrain (cu). Several large multi-ring basins are identifiable. The most prominent is the Isidis basin, located along the eastern margin of MC 13 and filled with the relatively young materials that form Isidis Planitia. A discontinuous ring of mountains can be identified around the basin. Particularly prominent are rugged peaks along the southern rim of Isidis Planitia. A second region of mountainous terrain in the northeast part of the quadrangle contains graben structures circumferential to the basin. These graben probably formed subsequent to the emplacement of Isidis Planitia since they transect craters that are relatively fresh. Directly west of Isidis Planitia are the ridged and cratered plains (pc) of Syrtis Major Planitia, with spectacular development of both bright and dark streaks. Changes in patterns of these wind streaks, caused by a veneer of light sediment continually being swept back and forth across darker volcanic rocks, doubtless give rise to the notable telescopic variations in the shape and size of Syrtis Major.

Other multi-ring basins present in these quadrangles are Cassini at 328°W, 23°N; Schiaparelli at 342°W, 40°S; Bakhuysen at 344°W, 22°S; and Huygens at 305°W, 14°S. The rugged mountains bounding the Hellas basin are visible in the south of MC 21, and the mountains bounding the Isidis basin to the south are present in the extreme upper right of that quadrangle.

MC 14, 15, 22, 23 (fig. C.10)

Cratered terrain (cu) underlies most of MC 22 and 23; plains (p and pv) are predominant in MC 14 and 15. The two terrains are separated by an erosional belt of hummocky terrain, fretted (hf).

The various landforms of the Elysium volcanic province are displayed in MC 15. These include Albor Tholus, Elysium Mons, and Orcus Patera. Knobby terrain along the eastern margin of MC 15 separates the Elysium province from the Tharsis volcanic province. Another area of former volcanic activity is present in MC 22. Tyrrhenum Patera, a relatively old structure with radial flow features emanating from a central complex caldera, is situated at 254°W, 22°S. A similar structure, Hadriacum Patera, occurs at 265°W, 30°S. Tyrrhenum Patera is situated on Hesperia Planum, a region of densely cratered plains (pc) with widespread wind streaks. The northern rim of the Hellas basin is exposed in the southwest corner of MC 22.

Degraded craters are confined primarily to cu.

Some are present in the knobby terrain to the east of the Elysium volcanic province, supporting the speculation that this is a highly eroded peninsula of cratered terrain. Fresh crater distributions are approximately similar on cu and pc. The plains of the Elysium volcanic province contain more craters than do the Tharsis plains, indicating that the Elysium volcanic activity is older.

Southern Hemisphere Lambert Maps

This set of quadrangles is shown in figures C.11–C.13. The dominant terrain type is cratered terrain. As in the northern Lambert belt, densities of fresh craters increase away from the pole. This is partly due to better resolution in pictures taken from lower elevations at lower latitudes (fig. 4.45), but is also the consequence of a debris mantle that is confined to the higher latitudes. The cratered terrain is notably mottled at higher latitudes. Traced northward this effect disappears. In part this mottling may be nothing more than a spurious artifact associated with picture enhancement of very low-contrast images, but it may also be related to the presence of sediment blankets on the surface.

MC 24, 25 (fig. C.11)

Cratered terrain in the northern part of MC 25 is highly fractured. The more prominent faults, Claritas Fossae and Thaumasia Fossae, are transected by moderately cratered plains (pm) of Solis Planum. A multi-ring basin, Lowell, is located at 81°W, 51°S. Argyre Dorsum, a mountain belt concentric to the Argyre basin further east in MC 26, is located along the east side of MC 25.

MC 26, 27 (fig. C.12)

The multi-ring Argyre basin is located in MC 26. It is surrounded by a 300-km wide rim of rugged mountains. Degraded craters are fewer in this rim than in the surrounding cratered terrain. This suggests the presence of a mantling ejecta deposit around the basin. Plains that fill the interior of the Isidis basin are young, as evidenced by the general absence of fresh craters.

The western half of the Hellas basin occurs in MC 27. Rugged, fault-bounded mountains, Hellespontus Montes, are situated to the west of the basin. As in the case of Argyre, a slight decrease in degraded craters adjacent to the basin suggests the presence of an ejecta blanket. The interior plains of Hellas are relatively young, and lack fresh craters. Several old, degraded volcanic calderas are present to the south of the Hellas basin, along with cratered plains (pc) of probable volcanic origin.

MC 28, 29 (fig. C.13)

The eastern half of Hellas occupies most of this quadrangle. Very little rim material is preserved. To the south, the rugged rim deposits are covered by cratered plains (pc), presumably volcanic deposits. An old, degraded caldera, Amphitrites Patera, is present at 300°W, 58°S. The eastern and northeastern margins of Hellas are topographically low regions occupied by plains materials (p), that are probably volcanic in origin. Several long lava channels run down into the basin. North of these lava channels, along the northern boundary of the quadrangle, is a degraded shield volcano with a large central caldera, Hadriacum Patera.

Table C.1 Legend for figures 3.2, C.2 through C.13.

POLAR UNITS

pi	**Permanent ice**	
ld	**Layered deposits**—thin (less than 100 m) continuous laminae at B-frame resolution. Appear as shallow benches alternating with steeper slopes. Probably exposed by wind erosion of layered dust deposits.	
ep	**Etched plains**—Irregular, pitted surface. Thick, unlayered deposits eroded by wind.	

VOLCANIC UNITS

v	**Volcanic constructs**—shields, domes or cones.
pv	**Volcanic plains**—few craters, lobate scarps interpreted as flow fronts.
pm	**Moderately cratered plains**—lacks volcanic features.
pc	**Cratered plains**—most densely cratered plains unit, contains ridges resembling those of the lunar maria, also contains older eroded volcanic features.

MODIFIED UNITS

hc	**Hummocky terrain, chaotic**—chaotic terrain, disrupted and tilted blocks of local surface material adjacent to channels or in closed basins.
hf	**Hummocky terrain, fretted**—marginal unit to northern scarp, located between cratered terrain and undivided plains. Large blocks near contact with cratered terrain, smaller blocks out into plains.
hk	**Hummocky terrain, knobby**—isolated regions of 10 km knobs, not clearly related to areas of chaotic or fretted hummocky terrain.
c	**Channel deposit**—smooth floors of channels, probably composed of alluvial, wind blown, or mass-wasted wall materials.
p	**Plains, undivided**—sparsely to moderately cratered at A frame resolution, B frames show irregular ridges, scarps and channels, general scoured appearance.
g	**Grooved terrain**—arcuate regions of linear mountains adjacent to Olympus Mons, 1-5 km wide and typically 100 km long, outer zone has finer texture.

ANCIENT UNITS

cu	**Cratered terrain, undivided**—densely to moderately cratered uplands, most ancient of all surfaces.
m	**Mountainous terrain**—rugged basin margin material, probably eroded basin ejecta.

Figure C.2. Geologic and crater maps for MC 1.

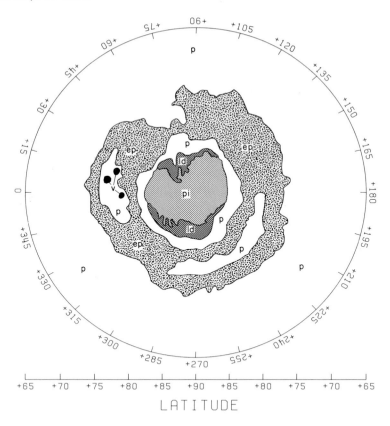

LATITUDE

MC-1, ALL

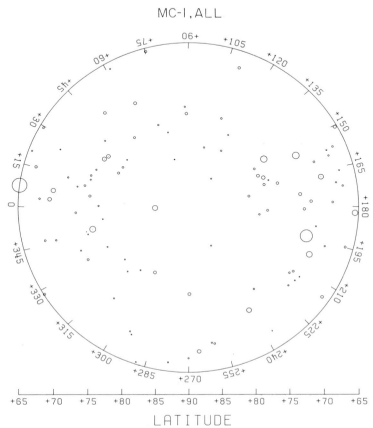

LATITUDE

MC-I, FRESH

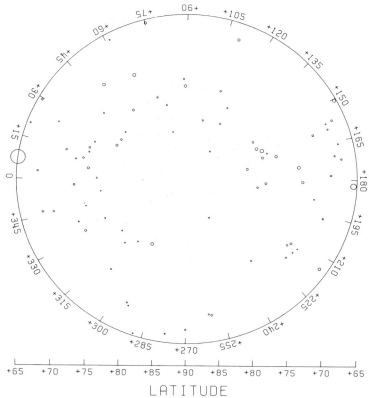

LATITUDE

MC-I, DEGRADED

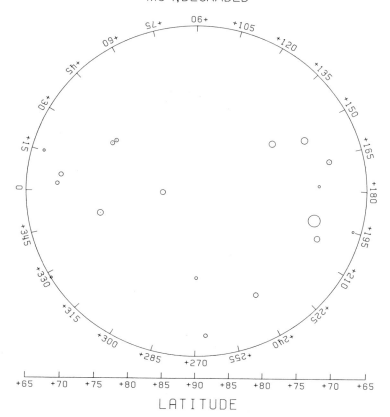

LATITUDE

Figure C.3. Geologic and crater maps for MC 30.

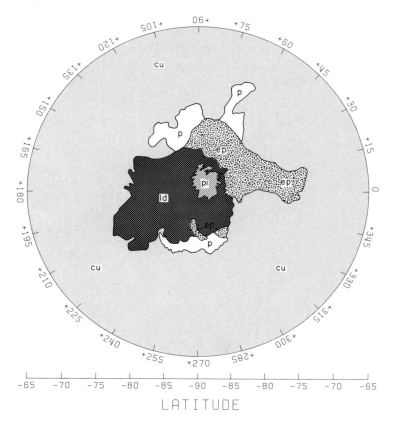

LATITUDE

MC-30, ALL

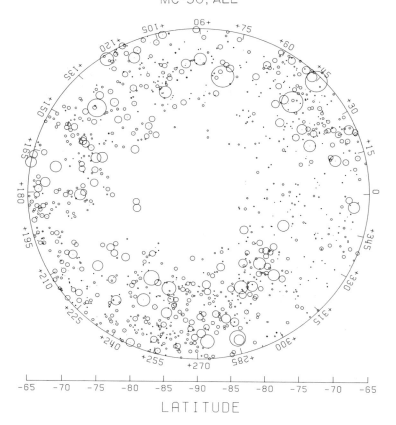

LATITUDE

MC-30, FRESH

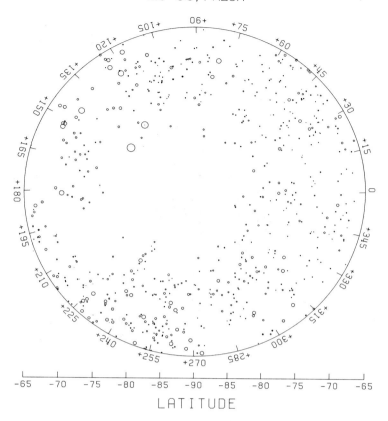

LATITUDE

MC-30, DEGRADED

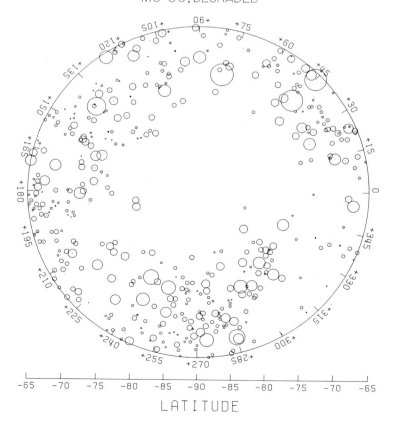

LATITUDE

Figure C.4. Geologic and crater maps for MC 2, 3.

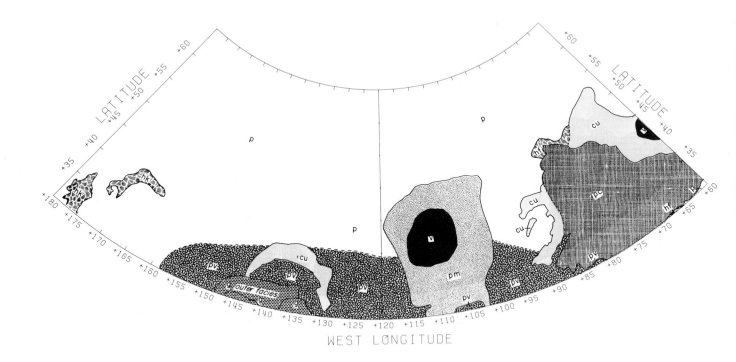

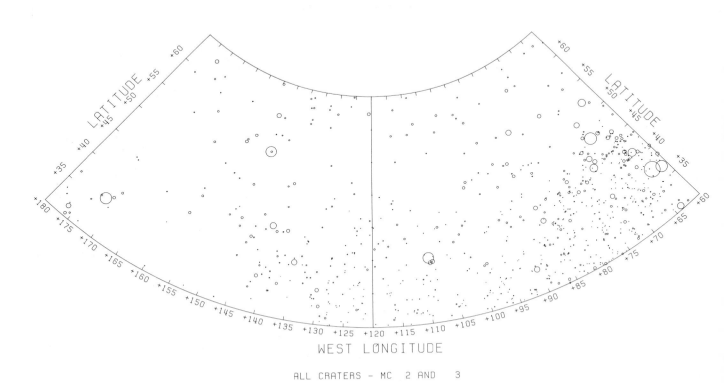

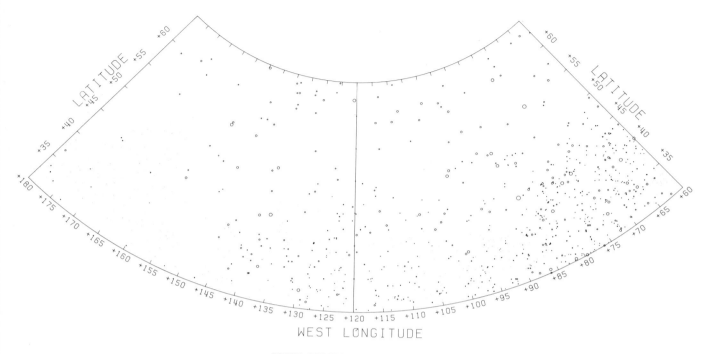

FRESH CRATERS — MC 2 AND 3

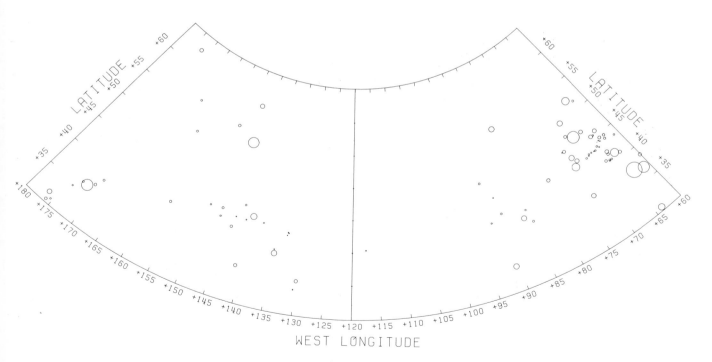

ALL DEGRADED — MC 2 AND 3

Figure C.5. Geologic and crater maps for MC 4, 5.

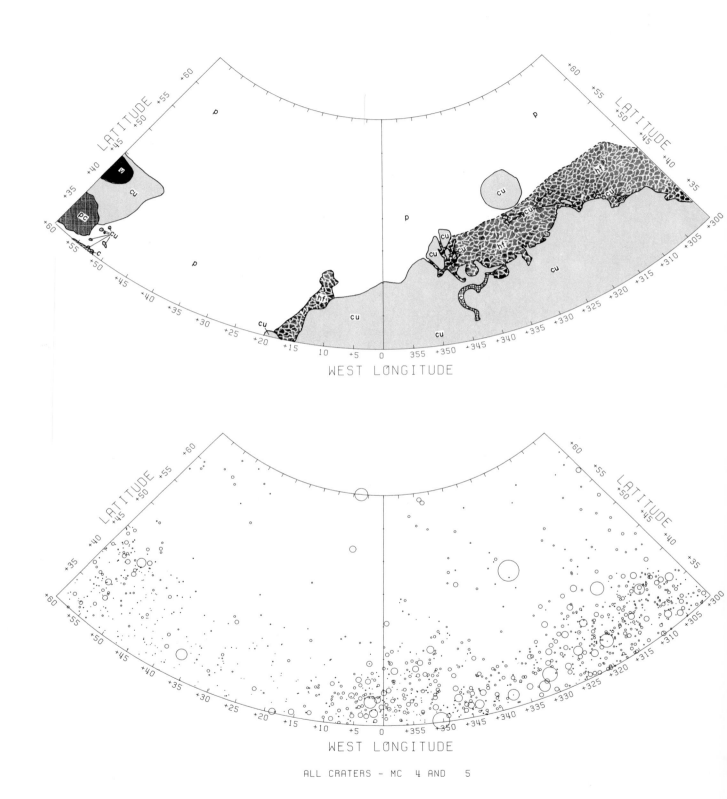

WEST LONGITUDE

WEST LONGITUDE

ALL CRATERS - MC 4 AND 5

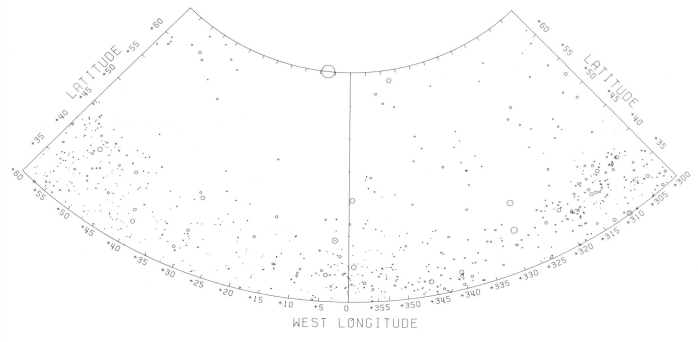

FRESH CRATERS — MC 4 AND 5

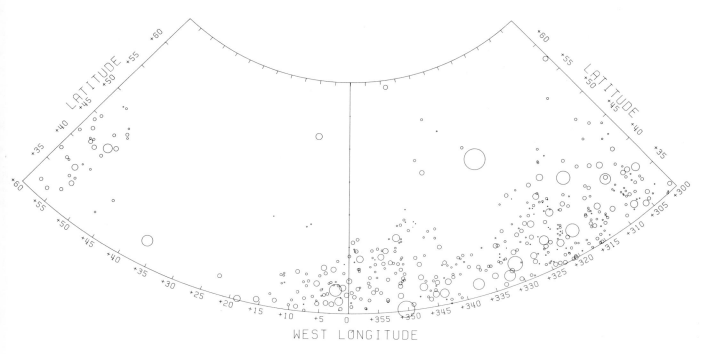

ALL DEGRADED — MC 4 AND 5

Figure C.6. Geologic and crater maps for MC 6, 7.

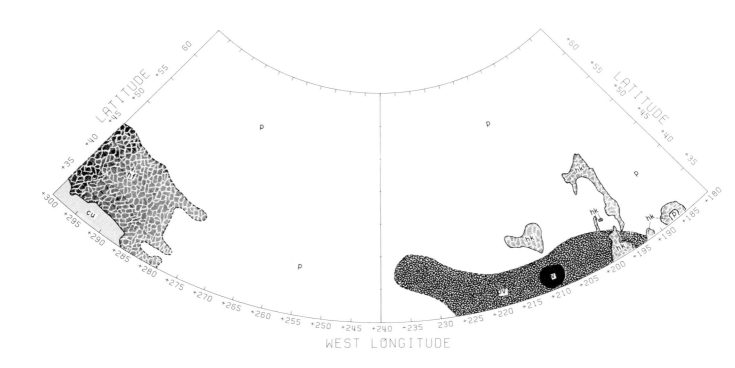

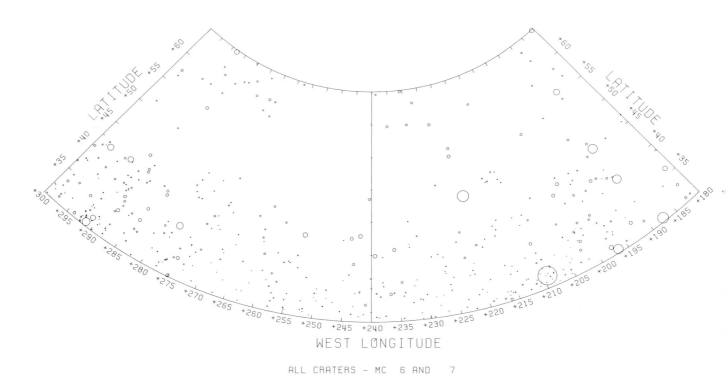

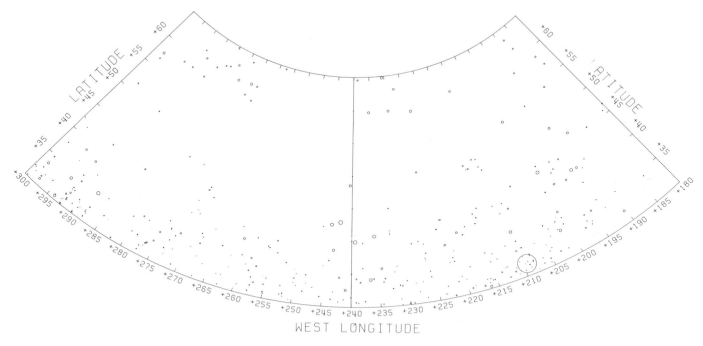

WEST LONGITUDE

FRESH CRATERS - MC 6 AND 7

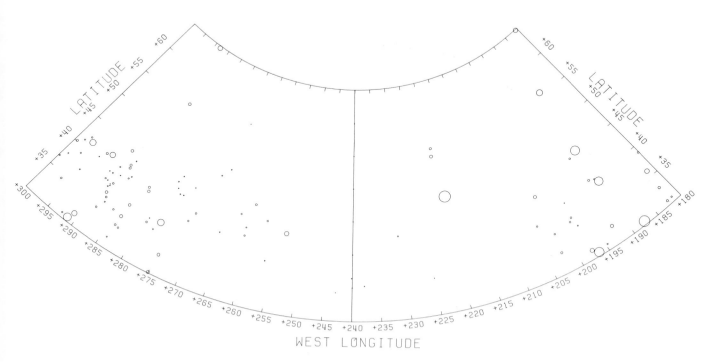

WEST LONGITUDE

ALL DEGRADED - MC 6 AND 7

Figure C.7. Geologic and crater maps for MC 8, 9, 16, 17.

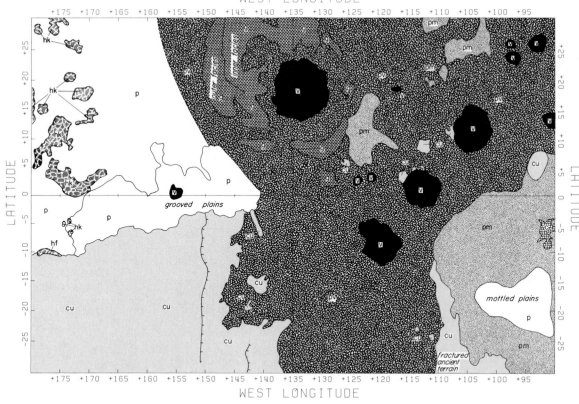

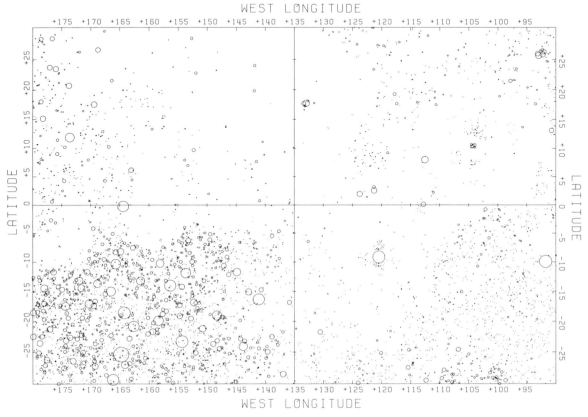

ALL FRESH CRATERS — MC 8,9,16,17

WEST LONGITUDE

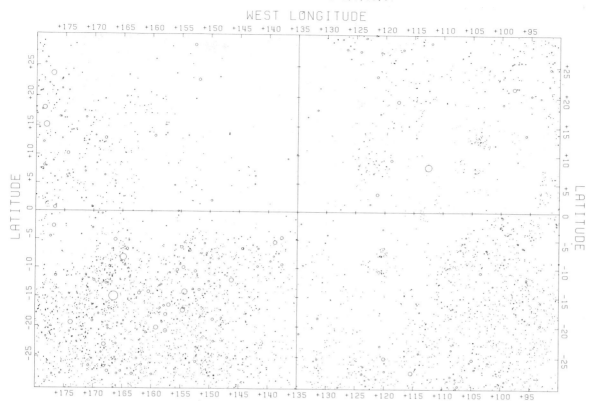

WEST LONGITUDE

ALL DEGRADED CRATERS — MC 8,9,16,17

WEST LONGITUDE

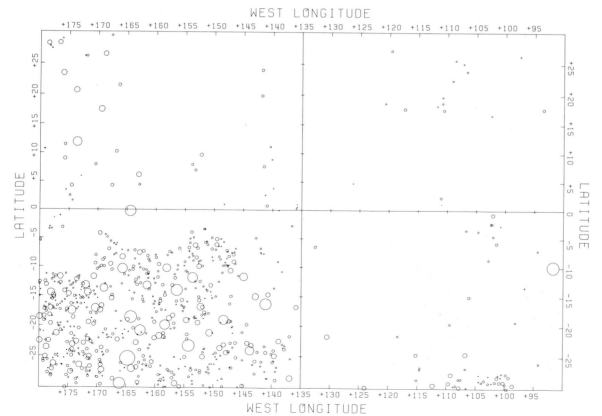

WEST LONGITUDE

Figure C.8. Geologic and crater maps for MC 10, 11, 18, 19.

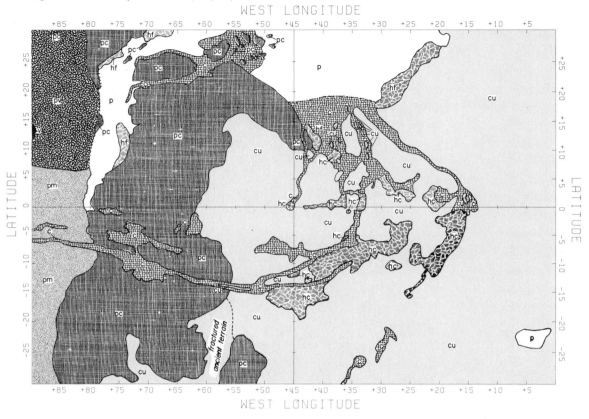

ALL CRATERS - MC 10,11,18,19

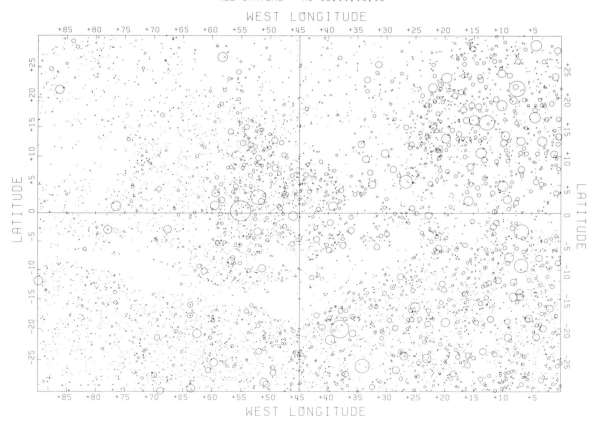

ALL DEGRADED - MC 10,11,18,19

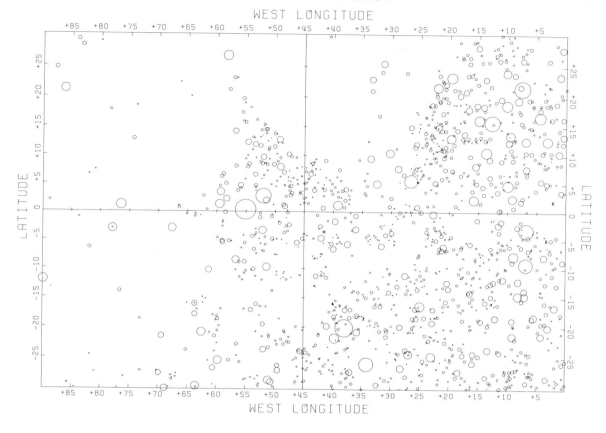

ALL FRESH CRATERS - MC 10,11,18,19

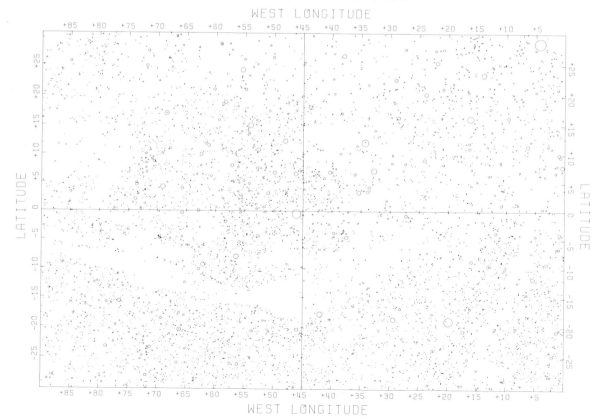

Figure C.9. Geologic and crater maps for MC 12, 13, 20, 21.

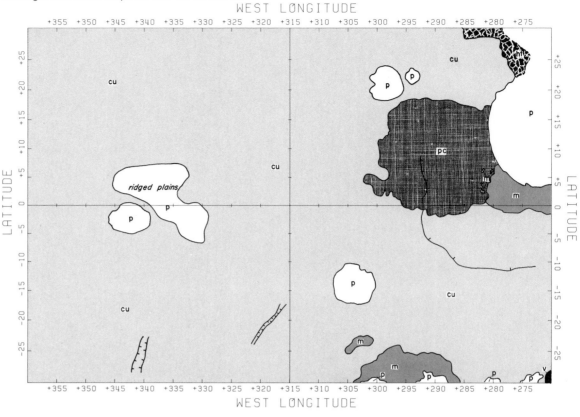

ALL FRESH CRATERS - MC 12,13,20,21

WEST LONGITUDE

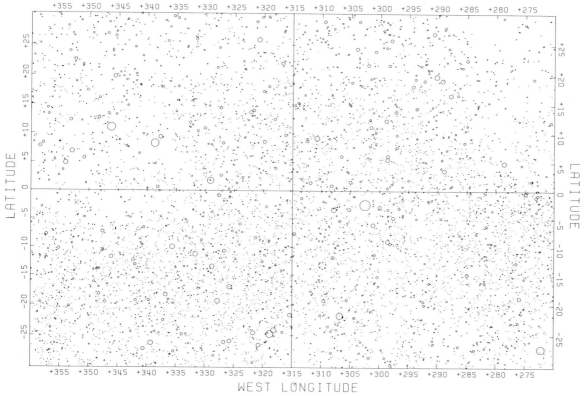

WEST LONGITUDE

ALL DEGRADED - MC12,13,20,21

WEST LONGITUDE

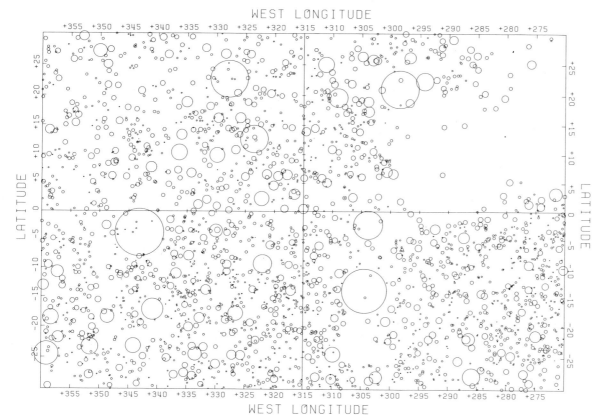

WEST LONGITUDE

Figure C.10. Geologic and crater maps for MC 14, 15, 22, 23.

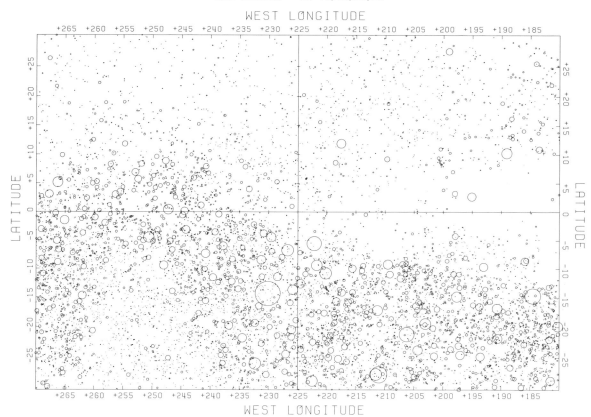

ALL FRESH CRATERS - MC 14,15,22,23

WEST LONGITUDE

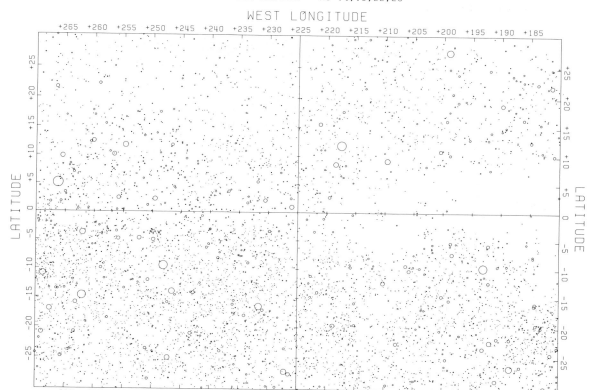

WEST LONGITUDE

ALL DEGRADED - MC 14,15,22,23

WEST LONGITUDE

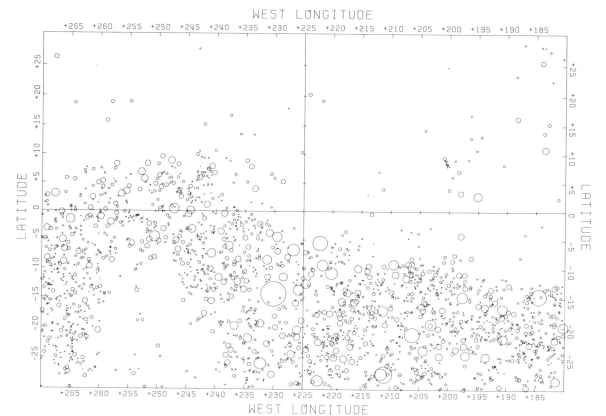

WEST LONGITUDE

Figure C.11. Geologic and crater maps for MC 24, 25.

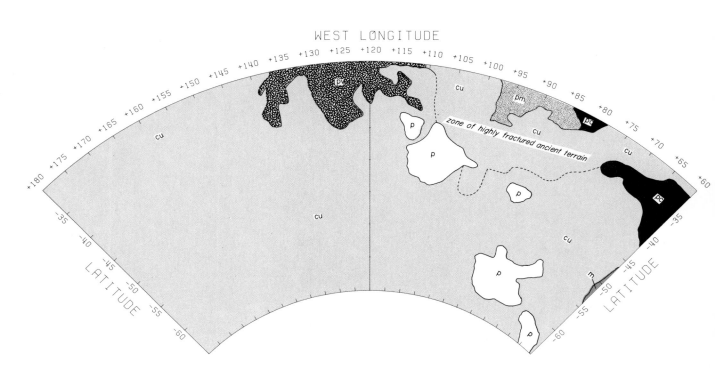

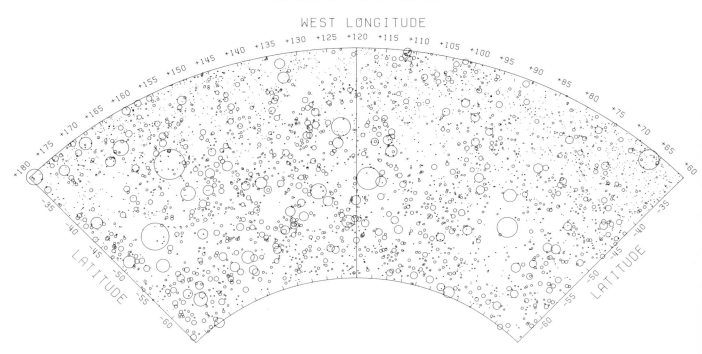

FRESH CRATERS — MC 24 AND 25

WEST LONGITUDE

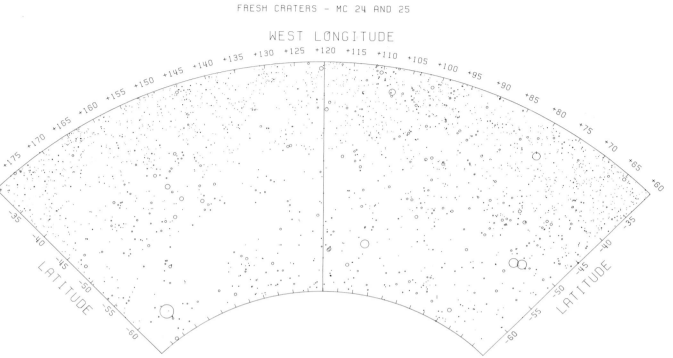

ALL DEGRADED — MC 24 AND 25

WEST LONGITUDE

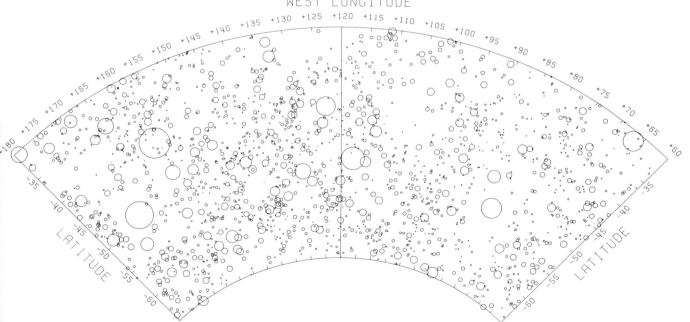

Figure C.12. Geologic and crater maps for MC 26, 27.

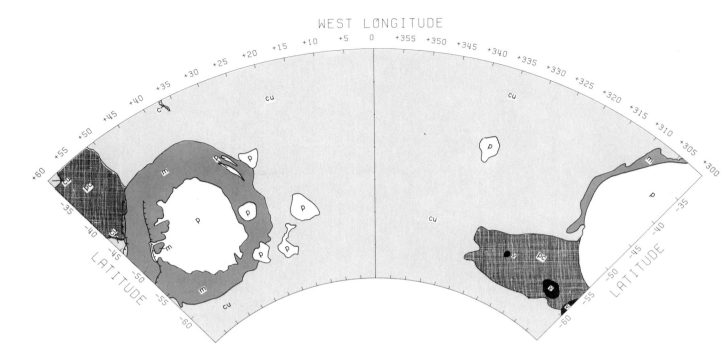

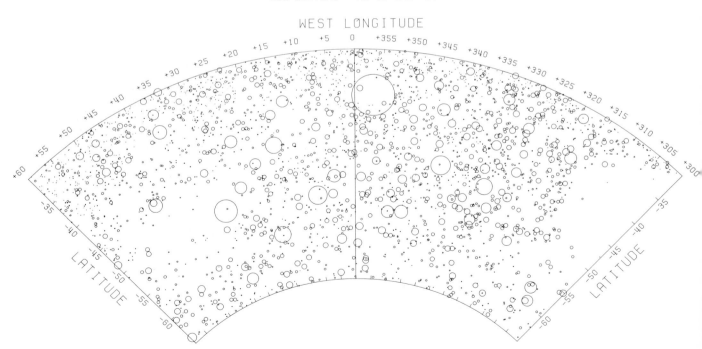

FRESH CRATERS – MC 26 AND 27

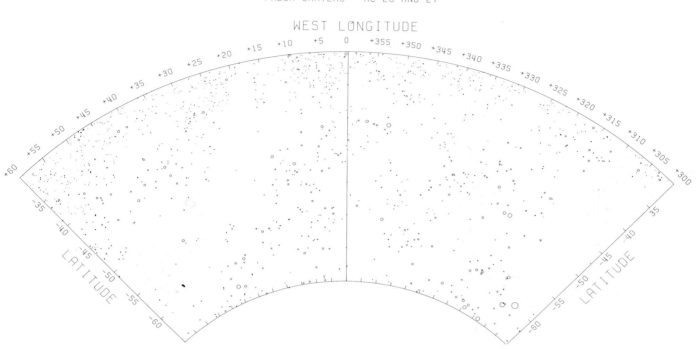

ALL DEGRADED – MC 26 AND 27

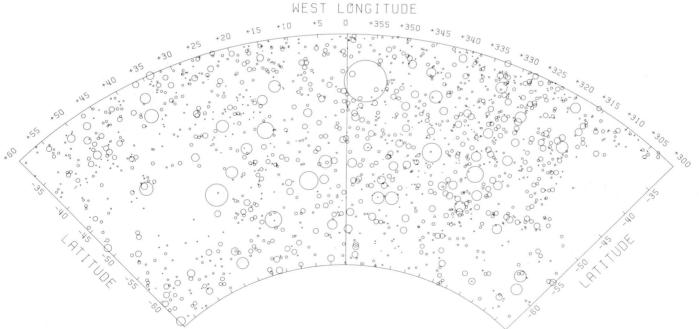

Figure C.13. Geologic and crater maps for MC 28, 29.

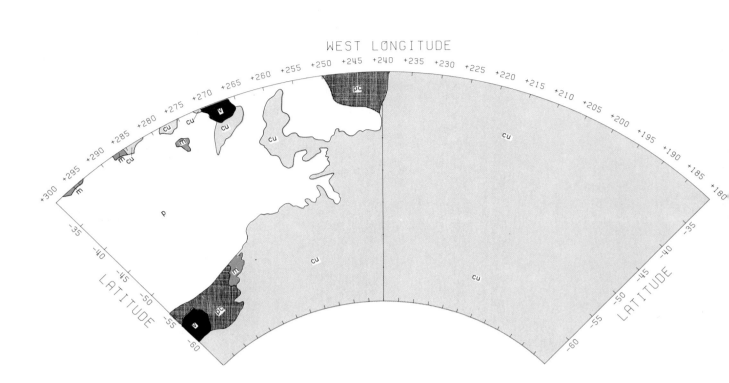

ALL DEGRADED — MC 28 AND 29

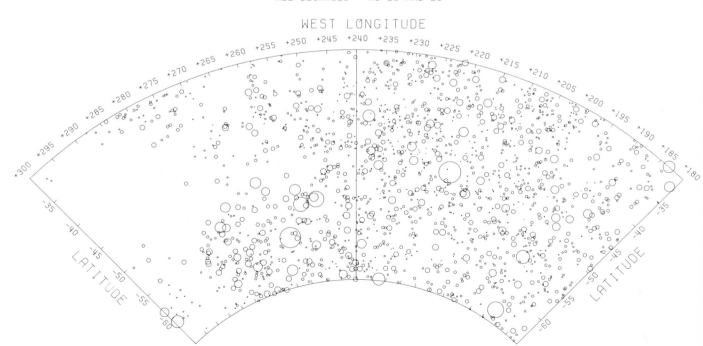

FRESH CRATERS — MC 28 AND 29

WEST LONGITUDE

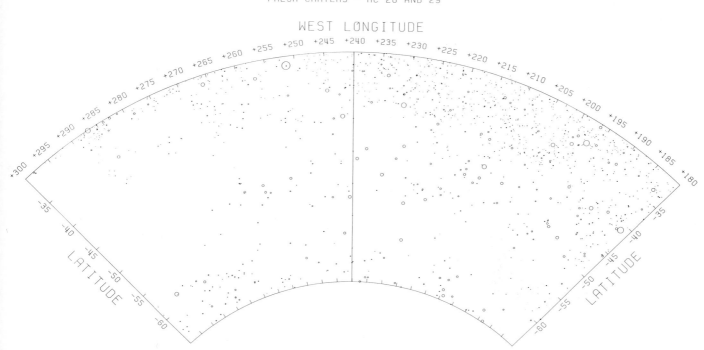

ALL CRATERS — MC 28 AND 29

WEST LONGITUDE

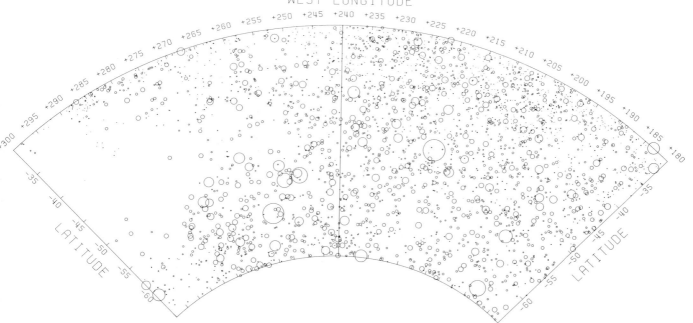

References

Adams, J. B., and T. B. McCord, 1969. Mars: interpretation of spectral reflectivity of light and dark regions: *Jour. Geophys. Res.*, v. 74, pp. 4851-4856.

Ahrens, T. J., 1972. State of mantle materials: *Tectonophysics*, v. 13, pp. 189-219.

Akimoto, S.-I., and H. Fujisawa, 1968. Olivine-spinel solid solution equilibria in the system Mg_2SiO_4-Fe_2SiO_4: *Jour. Geophys. Res.*, v. 73, pp. 1467-1479.

Anders, E., and J. R. Arnold, 1965. Age of craters on Mars: *Science*, v. 149, pp. 1494-1496.

Anderson, D. E., Jr., and C. W. Hord, 1971. Mariner 6 and 7 ultraviolet spectrometer experiment: analysis of hydrogen D. Lyman-Alpha data: *Jour. Geophys. Res.*, v. 76, p. 6666.

Anderson, D. L., 1967. Phase changes in the upper mantle: *Science*, v. 157, pp. 1165-1173.

———, 1972. Internal constitution of Mars: *Jour. Geophys. Res.*, v. 77, pp. 789-795.

———, and R. L. Kovach, 1972. The lunar interior: *Phys. of Earth Planet. Interiors*, v. 6, pp. 116-122.

———, E. S. Gaffney and P. F. Low, 1967. Frost phenomena on Mars: *Science*, v. 155, pp. 319-322.

Anderson, D. M., L. W. Gatto, and F. Ugolini, 1973. An examination of Mariner 6 and 7 imagery for evidence of permafrost terrain on Mars, *in Permafrost*: The North American contribution to the second International Conference: Nat. Acad. Sci., Washington, D.C., pp. 499-508.

Arvidson, R. E., 1972. Aeolian processes on Mars: erosive velocities, settling velocities, and yellow clouds: *Geol. Soc. America Bull.*, v. 83, pp. 1503-1508.

———, 1973. Aeolian processes on Mars: erosive velocities, settling velocities, and yellow clouds—reply: *Geol. Soc. America Bull.*, v. 84, pp. 340-350.

———, 1974a. Morphologic classification of Martian craters and some implications: *Icarus*, v. 22, pp. 264-271.

———, 1974b. Wind-blown streaks, splotches, and associated craters on Mars: statistical analysis of Mariner 9 photographs: *Icarus*, v. 21, pp. 12-27.

———, and M. Coradini, 1975. Crater diameter-frequency distributions for Martian fretted terrain: Abstracts, International Colloquium on Planetary Geology, Rome, Italy.

———, and T. A. Mutch, 1974. Sedimentary patterns in and around craters from the Pinacate volcanic field, Sonora, Mexico: some comparisons with Mars: *Geol. Soc. America Bull.*, v. 85, pp. 99-104.

———, R. Drozd, C. Hohenberg, G. Popeau, and C. Morgan, 1975a. Horizontal transport in the regolith, modification of features, and erosion rates on the lunar surface: *The Moon*, v. 13, pp. 67-79.

———, M. Coradini, A. Carusi, A. Coradini, M. Fulchignoni, C. Fedrico, R. Funiciello, M. Salomone, 1975b. Latitudinal variation of wind erosion of crater ejecta deposits on Mars: *Icarus*, in press.

Bagnold, R. A., 1941. *The Physics of Blown Sand and Desert Dunes*: Methuen and Co., 265 pp.

Baker, B. H., P. A. Mohr, and L. A. J. Williams, 1972. Geology of the eastern rift system of Africa: *Geol. Soc. America Spec. Paper 136*, 67 pp.

Baker, V. R., 1973. Paleohydrology and sedimentology of Lake Missoula flooding in eastern Washington. *Geol. Soc. America Spec. Paper 144*, 79 pp.

———, and D. J. Milton, 1974. Erosion by catastrophic floods on Mars and Earth: *Icarus*, v. 23, pp. 27-41.

Baksi, A. K. and N. D. Watkins, 1973. Volcanic production rates: Comparison of oceanic ridges, islands, and the Columbia Plateau basalts: *Science*, v. 180, pp. 493-496.

Baldwin, R. B., 1949. *The Face of the Moon*: Chicago, Univ. Chicago Press.

———, 1963. *The Measure of the Moon*: Chicago, Univ. Chicago Press.

———, 1965. Mars: an estimate of the age of its surface: *Science*, v. 149, pp. 1498-1499.

Balsamo, S. R., and J. W. Salisbury, 1973. Slope angle and frost formation on Mars: *Icarus*, v. 18, pp. 156-163.

Barker, E. S., 1967. A determination of the Martian CO_2 abundance: *Astrophys. Jour.*, v. 147, pp. 379-381.

———, 1971a. Observations of the Martian 1.2 CO_2 bands: *Planetary Atmospheres*, IAU Symposium No. 40, pp. 191-195.

———, 1971b. Variations of the Martian CO_2 abundance with Martian season: *Planetary Atmospheres*, IAU Symposium No. 40, pp. 196-202.

———, R. A. Schorn, A. Woszczyk, R. G. Tull, and S. J. Little, 1970. Mars: detection of water vapor during the southern hemisphere spring and summer season: *Science*, v. 170, pp. 1308-1310.

Barth, C. A., C. W. Hord, and A. I. Stewart, 1973. *An Atlas of Mars: Local Topography: Rpt.*, Laboratory for Atmospheric and Space Physics, Univ. of Colorado, Boulder, Colo.

Batson, R. M., 1973. Cartographic products from the Mariner 9 mission: *Jour. Geophys. Res.*, v. 78, pp. 4424-4435.

Behrman, C., G. Crozaz, R. Drozd, C. Hohenberg, C. Ralston, R. Walker, D. Yuhas, 1973. Radiation history of the Apollo 16 site (abstract), *in Lunar Science IV*, v. 1, pp. 54-56, Houston, Lunar Science Institute.

Belcher, D., J. Veverka, and C. Sagan, 1971. Mariner photography of Mars and aerial photography of Earth: some analogies: *Icarus*, v. 15, pp. 241-252.

Beloussov, V. V., 1962. *Basic Problems in Geotectonics*: New York, McGraw-Hill Co.

Belton, M. J. S., and D. M. Hunten, 1968. A search for O_2 in Mars and Venus: A probable detection of oxygen in the atmosphere of Mars: *Astrophys. Jour.*, v. 158, pp. 963-974.

———, A. L. Broadfoot, and D. M. Hunten, 1968. Abundance and temperature of CO_2 on Mars during the 1967 opposition: *Jour. Geophys. Res.*, v. 73, pp. 4795-4806.

Binder, A. B., 1966. Mariner IV: analysis of preliminary photographs: *Science*, v. 152, pp. 1053-1055.

———, and D. P. Cruikshank, 1966. Lithological and mineralogical investigation of the surface of Mars: *Icarus*, v. 5, pp. 521-525.

———, and J. C. Jones, 1972. Spectrophotometric studies of the photometric function, composition, and distribution of the surface materials of Mars: *Jour. Geophys. Res.*, v. 77, pp. 3005-3020.

———, and D. W. McCarthy, Jr., 1972. Mars: The lineament systems: *Science*, v. 176, pp. 279-281.

———, and D. R. Davis, 1973. Internal structure of Mars: *Phys. of Earth Planet. Interiors*, v. 7, pp. 477-485.

Binns, R. A., 1970. (Mg, Fe)$_2$SiO$_4$ Spinel in a meteorite: *Phys. of Earth Planet. Interiors*, v. 3, pp. 156-160.

Birch, F., 1954. The Earth's mantle: Elasticity and constitution: *American Geophys. Union Trans.*, v. 35, p. 79.

———, 1958. Differentiation of the mantle: *Geol. Soc. America Bull.*, v. 69, pp. 483-486.

Blanchet, P., 1957. Development of fracture analysis as exploration method: *American Assoc. Petrol. Geol. Bull.*, v. 41, pp. 1748-1759.

Blasius, K. R., 1973. A study of Martian topography by analytic photogrammetry: *Jour. Geophys. Res.*, v. 78, pp. 4411-4423.

Boyce, P. B., 1973. Remote sensing photometric studies of Mars in 1971: *Icarus*, v. 18, pp. 134-141.

———, and D. T. Thompson, 1972. A new look at the Martian "Violet Haze" problem. I. Syrtis Major-Arabia, 1969: *Icarus*, v. 16, pp. 291-303.

Bretz, J. H., 1923. The channeled scablands of the Columbia Plateau: *Jour. of Geology*, v. 31, pp. 617-649.

———, 1969. The Lake Missoula floods and the channeled scabland: *Jour. of Geology*, v. 77, pp. 505-543.

Briggs, F. H., and F. D. Drake, 1972. Interferometric observations of Mars at 21-cm wavelength: *Icarus*, v. 17, pp. 543-547.

Briggs, G. A., 1974. The nature of the residual Martian polar caps: *Icarus*, v. 23, pp. 167-191.

———, and C. B. Leovy, 1974. Mariner 9 observations of the Mars north polar hood: *Bull. American Met. Soc.*, v. 55, pp. 278-294.

Brinkmann, R. T., 1969. Dissociation of water vapor and evolution of oxygen in the terrestrial atmosphere: *Jour. Geophys. Res.*, v. 74, pp. 5355-5368.

———, 1971. Mars: Has nitrogen escaped?: *Science*, v. 174, pp. 943-945.

Brocoum, S. J., and I. W. D. Dalziel, 1974. The Sudbury basin, the Southern province, the Grenville front, and the Penokean orogeny: *Geol. Soc. America Bull.*, v. 85, pp. 1571-1580.

Broecker, W. S., and J. Van Donk, 1970. Insolation changes, ice volumes, and the O^{18} record in deep-sea cores: *Rev. Geophys. and Space Phys.*, v. 8, pp. 169-198.

Brouwer, D., and A. J. J. Van Woerkom, 1950. The secular variations of the orbital elements of the principal planets: *Astron. Pap. American Ephemeris Naut. Alm.*, v. 13 (part 2), pp. 81-107.

Brown, H., 1952. Rare gases and the formation of the Earth's atmosphere, pp. 258-266 *in* Kuiper, G. P., Edgerton, *The Atmospheres of the Earth and Planets*, 2nd ed.: Chicago, Univ. Chicago Press, 434 pp.

Brunsden, D., and R. Kesal, 1973. Slope development on a Mississippi River Bluff in historic time: *Jour. of Geology*, v. 81, pp. 577-597.

Bryan, W. B., 1973. Wrinkle-ridges as deformed surface crust on ponded mare lava: *Geochim. et Cosmochim. Acta*, Suppl. 4, v. 1, pp. 93-106.

Bucher, W. H., 1924. The pattern of the Earth's mobile belts: *Jour. of Geology*, v. 32, pp. 265-290.

Bullard, F. M., 1962. *Volcanoes: In History, In Theory, In Eruption*: Univ. of Texas Press, Austin, 441 pp.

Burroughs, E. R., 1917. *A Princess of Mars*, McClurg.

————, 1918. *The Gods of Mars*, McClurg.

————, 1919. *The Warlord of Mars*, McClurg.

————, 1920. *Thuvia, Maid of Mars*, McClurg.

————, 1922. *The Chessmen of Mars*, McClurg.

————, 1928. *The Master Mind of Mars*, McClurg.

————, 1931. *A Fighting Man of Mars*, Metropolitan.

————, 1936. *Swords of Mars*, Burroughs.

————, 1940. *Synthetic Men of Mars*, Burroughs.

————, 1948. *Llana of Gathol*, Burroughs.

————, 1964. *John Carter of Mars*: New York, Canaveral Press.

Cantril, H., 1940. *The Invasion from Mars*: Princeton, N.J., Princeton University Press, 228 pp.

Capen, C. F., 1971. Martian yellow clouds—past and future: *Sky and Telescope*, v. 41, pp. 2-4.

————, and L. J. Martin, 1971. The developing stages of the Martian yellow storm of 1971: *Lowell Observ. Bull.*, no. 157, v. III, pp. 211-216.

Carleton, N. P., A. Sharma, R. M. Goody, W. L. Liller, and F. L. Roesler, 1969. Measurement of the abundance of CO_2 in the Martian atmosphere: *Astrophys. Jour.*, v. 155, pp. 323-331.

————, and W. A. Traub, 1972. Detection of molecular oxygen on Mars: *Science*, v. 178, pp. 988-992.

Carr, M. H., 1973. Volcanism on Mars: *Jour. Geophys. Res.*, v. 78, pp. 4049-4062.

————, 1974a. The role of lava erosion in the formation of lunar rilles and Martian channels: *Icarus*, v. 22, pp. 1-23.

————, 1974b. Tectonism and volcanism in the Tharsis region of Mars, *Jour. Geophys. Res.*, v. 79, pp. 3943-3949.

————, H. Masursky, and R. S. Saunders, 1973. A generalized geologic map of Mars: *Jour. Geophys. Res.*, v. 78, pp. 4031-4036.

Chamberlin, T. C., 1890. The method of multiple working hypothesis: reprinted *in Science*, 1965, v. 148, pp. 754-759.

Chao, E. C. T., L. A. Soderblom, J. M. Boyce, D. Wilhelms, and C. A. Hodges, 1973. Lunar light plains deposits (Cayley Formation)—A reinterpretation of origin (abstract), *in Lunar Science IV*, Lunar Science Institute, v. 1, pp. 127-128.

Chapman, C. R., 1974. Cratering on Mars: I. Cratering and obliteration history: *Icarus*, v. 22, pp. 272-291.

————, and R. R. Haefner, 1967. A critique of methods for analysis of the diameter-frequence relation for craters with special application to the Moon: *Jour. Geophys. Res.*, v. 72, pp. 549-557.

————, J. B. Pollack, and C. Sagan, 1968. Analysis of the Mariner 4 photography of Mars: *Spec. Rep. 268*, Cambridge, Mass., Smithson. Astrophys. Observ.

————, J. B. Pollack, and C. Sagan, 1969. An analysis of the Mariner 4 cratering statistics: *Astron. Jour.*, v. 74, pp. 1039-1051.

Christensen, E. J., 1975. Martian topography derived from occultation, radar, spectral, and optical measurements: *Jour. Geophys. Res.*, v. 80, pp. 2909-2913.

Cintala, M. J., J. W. Head, and T. A. Mutch, 1975. Depth/diameter relationships for Martian and lunar craters: *American Geophys. Union Trans.*, v. 56, p. 389.

Classen, J., 1975. A large crater field recognized in central Europe: *Sky and Telescope*, v. 49, pp. 365-367.

Colton, G. W., K. A. Howard, and H. J. Moore, 1972. Mare ridges and arches in southern Oceanus Procellarum: *Apollo 16 Preliminary Science Report*, *NASA Spec. Publ. SP-315*, pp. 29.90-29.33.

Conel, J. E., 1969. Structural features relating to the origin of lunar wrinkle ridges: Jet Propulsion Laboratory, *Space Programs Summary 37-56*, v. III, pp. 58-63.

————, and D. B. Nash, 1970. Spectral reflectance and albedo of Apollo 11 lunar samples: Effects of irradiation and vitrification and comparison with telescopic observations: *Geochim. et Cosmochim. Acta*, Suppl. 1, v. 3, pp. 2013-2023.

————, F. P. Fanale, and R. J. Phillips, 1972. Physical constraints on lunar thermal history: An analysis of results from investigation of lunar samples and geophysical data: *JPL Tech. Memo 900-522*, Jet Propulsion Laboratory, Pasadena, Cal.

Conrath, B. J., 1975. Thermal structure of the Martian atmosphere during the dissipation of the dust storm of 1971: *Icarus*, v. 24, pp. 36-46.

————, R. Curran, R. Hanel, V. Kunde, W. Maguire, J. Pearl, J. Pirraglia and J. Welker, 1973. Atmospheric and surface properties of Mars obtained by infrared spectroscopy on Mariner 9: *Jour. Geophys. Res.*, v. 78, pp. 4267-4278.

Cooper, W. S., 1958. Coastal Sand Dunes of Oregon and Washington: *Geol. Soc. America, Memoir 72*, 169 pp.

Cordell, B. M., R. E. Lingenfelter, and G. Schubert, 1974. Martian cratering and central peak statistics: Mariner 9 results: *Icarus*, v. 21, pp. 448-456.

Cox, A., R. R. Doell, and G. B. Dalrymple, 1963. Geomagnetic polarity epochs and Pleistocene geochronology: *Nature*, v. 198, pp. 1049-1051.

Cruikshank, D. P., and C. A. Wood, 1972. Lunar rilles and Hawaiian volcanic features: possible analogues: *The Moon*, v. 3, pp. 412-447.

————, W. K. Hartmann, and C. A. Wood, 1973. Moon: 'Ghost' craters formed during mare filling: *The Moon*, v. 7, pp. 440-452.

Cummings, D., and G. I. Shiller, 1971. Isopach map of the Earth's crust: *Earth-Sci. Rev.*, v. 7, pp. 97-125.

Curran, R. J., B. J. Conrath, R. A. Hanel, V. G. Kunde, and J. C. Pearl, 1973. Mars: Mariner 9 spectrographic evidence for H_2O ice clouds: *Science*, v. 182, pp. 381-383.

Cutts, J. A., 1971. Martian spectral reflectivity properties from Mariner 7 observations: Ph.D. thesis, Pasadena, Cal., Calif. Institute of Technology, 92 pp.

——, 1973a. Wind erosion in the Martian polar regions: *Jour. Geophys. Res.*, v. 78, pp. 4211-4221.

——, 1973b. Nature and origin of layered deposits of the Martian polar regions: *Jour. Geophys. Res.*, v. 78, pp. 4231-4249.

——, 1974. Mariner Mars 1971 television picture catalog, v. I: *JPL Tech. Memorandum 33-585*, Jet Propulsion Laboratory, Pasadena, Cal.

——, 1975. Mars: Two possible impact structures near the South pole (Abs.) Division of Planetary Sciences, American Astronomical Society, 1975 meeting.

——, L. A. Soderblom, R. P. Sharp, B. A. Smith, and B. C. Murray, 1971. The surface of Mars, III: light and dark marking: *Jour. Geophys. Res.*, v. 76, pp. 343-356.

——, J. Veverka, and J. D. Goguen, 1972. The location of the Mountains of Mitchel and evidence for their nature in Mariner 7 pictures: *Icarus*, v. 16, pp. 528-534.

——, and R. S. U. Smith, 1973. Eolian deposits and dunes on Mars: *Jour. Geophys. Res.*, v. 78, pp. 4139-4154.

Cuzzi, J. N., and D. O. Muhleman, 1972. The microwave spectrum and nature of the subsurface of Mars: *Icarus*, v. 17, pp. 548-560.

Dalgarno, A., and M. B. McElroy, 1970. Mars: is nitrogen present?: *Science*, v. 170, pp. 167-168.

Davies, M. E., 1974. Photogrammetric measurement of Olympus Mons on Mars: *Icarus*, v. 21, pp. 230-236.

——, and D. W. G. Arthur, 1973. Martian surface coordinates: *Jour. Geophys. Res.*, v. 78, pp. 4355-4394.

Davis, W. M., 1926. Biographical memoir of Grove Karl Gilbert, 1843-1918: *Mem. Nat. Acad. Sci.*, v. 21, 303 pp.

Day, A. L., 1939. The hot spring problem: *Geol. Soc. America Bull.*, v. 50, pp. 317-336.

DeHon, R. A., 1971. Cauldron subsidence in lunar craters, Ritter and Sabine: *Jour. Geophys. Res.*, v. 76, pp. 5712-5718.

Delsemme, A. H., and D. C. Miller, 1970. Physicochemical phenomena in comets. II. Gas adsorption in the snows of the nucleus: *Planet. Space Sci.*, v. 18, pp. 717-730.

——, and A. Wenger, 1970. Physico-chemical phenomena in comets. I. Experimental study of snows in a cometary environment: *Planet. Space Sci.*, v. 18, pp. 709-715.

Dence, M. R., 1968. Shock zoning at Canadian craters: petrography and structural implications; *in* B. French and M. Short, eds., *Shock Metamorphism of Natural Materials*, Baltimore, Mono Book Corp., pp. 169-184.

——, 1972. The nature and significance of terrestrial impact features; *in Proceedings, 24th Internat. Geol. Congress, sec. 15*, pp. 17-89.

——, M. J. S. Innes, and P. B. Robertson, 1968. Recent geological and geophysical studies of Canadian craters; *in* B. French and M. Short, eds., *Shock Metamorphism of Natural Materials*, Baltimore, Mono Book Corp., pp. 339-362.

Dent, B., 1973. Glacial exhumation of impact craters on the Canadian Shield: *Geol. Soc. America Bull.*, v. 84, pp. 1667-1672.

de Vaucouleurs, G., 1954. *Physics of the Planet Mars*: London, Faber and Faber, Ltd., 365 pp.

——, 1967. A low-resolution photometric map of Mars: *Icarus*, v. 7, pp. 310-349.

DeWys, J., 1968. Sec. 3.2, p. 7, *in* C. M. Michaux and R. L. Newburn, Jr., eds., *Mars Scientific Model, Document no. JPL 606-1*, Pasadena, Cal., Jet Propulsion Lab.

Dietz, R. S., 1961. Continent and ocean basin evolution by spreading of the sea floor: *Nature*, v. 190, pp. 854-857.

——, 1964. Sudbury structure as an astrobleme: *Jour. of Geology*, v. 72, pp. 412-434.

Dohnanyi, J. S., 1972. Interplanetary objects in review: Statistics of their masses and dynamics: *Icarus*, v. 17, pp. 1-48.

Dolginov, S. S., Y. G. Yeroshenko, and L. N. Zhuzgov, 1972. The magnetic field in the very close neighborhood of Mars according to data from the Mars 2 and Mars 3 spacecraft: *Goddard Space Flight Center Rept. X-690-72-434*.

Dollfus, A., 1961a. Visual and photographic studies of the planets, at the Pic du Midi: ch. 15 *in* G. P. Kuiper and B. M. Middlehurst, eds., *Planets and Satellites*, v. III of The Solar System: Chicago, Univ. of Chicago Press.

——, 1961b. Polarization studies of the planets; ch. 9 *in* G. P. Kuiper and B. M. Middlehurst, eds., *Planets and Satellites*, v. III of The Solar System: Chicago, Univ. of Chicago Press.

Downs, G. S., R. M. Goldstein, R. R. Green, and G. A. Morris, 1971. Mars radar observations, a preliminary report: *Science*, v. 174, pp. 1324-1327.

Duxbury, T. C., 1974. Phobos: central network analysis: *Icarus*, v. 23, pp. 290-299.

Eggleton, R. E., and E. I. Smith, 1967. Preliminary geologic map of the Rümker Quadrangle of the Moon: U.S. Geol. Survey Open File Report.

——, and G. G. Schaber, 1972. Cayley formation interpreted as basin ejecta: *Apollo 16 Preliminary Science Report, NASA Spec. Publ. SP-315*, pp. 29.7-29.16.

El-Baz, F., 1972. King crater and its environs: *Apollo 16 Preliminary Science Report, NASA Spec. Publ. SP-315*, pp. 29.62-29.70.

Elsasser, W. M., 1972. Viscous stratification of the Earth and convection: *Phys. of Earth Planet. Interiors*, v. 6, pp. 198-204.

Elston, D. E., E. L. Boudette, J. P. Schafer, W. R. Muehlberger, and J. R. Sevier, 1972. Apollo 16 Field Trips: *Geotimes*, v. 17, pp. 27-30.

Elston, W., A. Laughlin, and J. Brown, 1971. Lunar near-side tectonic patterns from Orbiter IV photographs: *Jour. Geophys. Res.*, v. 76, pp. 5670-5674.

Engdahl, E. R., and E. A. Flinn, 1969. Seismic waves reflected from discontinuities within Earth's mantle: *Science*, v. 163, pp. 177-179.

Fairbridge, R. W., 1972. Planetary spin-rate and evolving cores: *Annals N.Y. Acad. Sci.*, v. 187, pp. 88-107.

Fanale, F. P., 1971a. A case for catastrophic early degassing of the Earth: *Chemical Geology*, v. 8, pp. 79-105.

————, 1971b. History of Martian volatiles: implications for organic synthesis: *Icarus*, v. 15, pp. 297-303.

————, and W. A. Cannon, 1971. Adsorption on the Martian regolith: implications for the Martian volatile budget and diurnal brightening: *Nature*, v. 230, pp. 502-504.

————, and W. A. Cannon, 1974. Exchange of adsorbed H_2O and CO_2 between the regolith and atmosphere of Mars caused by changes in surface insolation: *Jour. Geophys. Res.*, v. 79, pp. 3397-3402.

————, and D. B. Nash, 1971. Potassium-uranium systematics of Apollo 11 and Apollo 12 samples: implications for lunar material history: *Science*, v. 171, pp. 282-284.

Farmer, C. B., 1971. The strengths of H_2O lines in the 8200 A° region and their application to high dispersion spectra of Mars: *Icarus*, v. 15, pp. 190-196.

————, 1973. Liquid water on Mars: Manuscript prepared for Viking Lander Site Working Group, Hampton Va., Langley Research Center.

Ferrians, O. J., R. Kachadoorian, and G. W. Greene, 1969. Permafrost and related engineering problems in Alaska: *U.S. Geol. Survey Prof. Paper 678*, 37 pp.

Fielder, G., 1963a. On the topography of Mars: *Astron. Soc. of the Pacific, Publ.*, v. 75, pp. 75-76.

————, 1963b. Lunar Tectonics, *Geol. Soc. London Quart. Jour.*, v. 119, pp. 65-69.

————, 1965a. *Lunar Geology*: Lutterworth Press, London.

————, 1965b. Photographs of Mars taken by Mariner IV: *Nature*, v. 207, p. 1381.

Fischbacher, G. E., L. J. Martin, and W. A. Baum, 1969. Seasonal behavior of the Martian polar caps: *Astron. Soc. of the Pacific, Publ.*, v. 81, p. 538.

Flammarion, C., 1892. *La planète Mars et ses conditions d'habitabilité*: Gauthier-Villars et Fils., V. I, 1892; V. II, 1909, Paris.

Flint, R. F., 1957. *Glacial and Pleistocene Geology*: New York, Wiley.

Focas, J. H., 1961. Etude photométrique et polarimétrique des phénomènes saisonniers de la planète Mars: *Annales d'Astrophysique*, v. 24, no. 4, pp. 309-325.

Frey, H., 1974. Surface features on Mars: ground-based albedo and radar compared with Mariner 9 topography: *Jour. Geophys. Res.*, v. 79, pp. 3907-3916.

Fuller, R. E., 1927. The closing phase of a fissure eruption: *American Jour. Sci.*, 5th series, v. 14, pp. 228-230.

Gaffey, M. J., 1974. A systematic study of the spectral reflectivity characteristics of the meteorite classes with applications to the interpretation of asteroid spectra for mineralogical and petrological identifica-tion: Ph.D. thesis, Massachusetts Institute of Technology.

Gatto, L. W., and D. M. Anderson, 1975. Alaskan thermokarst terrain and a possible Martian analog: *Science*, v. 188, pp. 255-257.

Gault, D. E., 1970. Saturation and equilibrium conditions for impact cratering on the lunar surface: criteria and implications: *Radio Sci.*, v. 5, pp. 273-291.

————, J. E. Guest, J. B. Murray, D. Dzurisin, and M. C. Malin, 1975. Some comparisons of impact craters on Mercury and the Moon: *Jour. Geophys. Res.*, v. 80, pp. 2444-2460.

Gierasch, P., and C. Sagan, 1971. A preliminary assessment of Martian wind regimes: *Icarus*, v. 14, pp. 312-318.

Gifford, F. A., Jr., 1964. The Martian canals according to a purely aeolian hypothesis: *Icarus*, v. 3, pp. 130-135.

Gilluly, J., 1969. Geological perspective and the completeness of the geologic record: *Geol. Soc. America Bull.*, v. 80, pp. 2303-2312.

————, J. C. Reed, Jr., W. M. Cady, 1970. Sedimentary volumes and their significance: *Geol. Soc. America Bull.*, v. 81, pp. 353-376.

Glasstone, S., 1968. *The Book of Mars*; *NASA Spec. Publ. SP-179*.

Goettel, K. A., 1972. Partitioning of potassium between silicates and sulphide melts: experiments relevant to the Earth's core: *Phys. of Earth Planet. Interiors*, v. 6, pp. 161-166.

Goldich, S. S., and C. E. Hedge, 1974. 3,800-Myr granitic gneiss in south-western Minnesota: *Nature*, v. 252, pp. 467-468.

Golitsyn, G. S., 1973. On the Martian dust storms: *Icarus*, v. 18, pp. 113-119.

Greeley, R., 1971. Lava tubes and channels in the lunar Marius Hills: *The Moon*, v. 3, pp. 289-314.

————, 1973. Mariner 9 photographs of small volcanic structures on Mars: *Geology*, v. 1, pp. 175-180.

————, and D. E. Gault, 1970. Precision size–frequency distributions of craters for 12 selected areas of the lunar surface: *The Moon*, v. 2, pp. 10-76.

————, J. D. Iversen, J. B. Pollack, N. Udovich, and B. White, 1974. Wind tunnel simulations of light and dark streaks on Mars: *Science*, v. 183, pp. 847-849.

Griggs, D. T., 1939. A theory of mountain building: *American Jour. Sci.*, v. 237, pp. 611-650.

Grolier, M. J., G. E. Ericksen, J. F. McCauley, and E. C. Morris, 1974. The desert land forms of Peru; a preliminary photographic atlas: *U.S. Geological Survey, Interagency Rpt.*, Astrogeol. 57.

Grossman, L., 1974. Chemical fractionation in the solar nebula: *Proceedings of the Soviet-American Conference on the Cosmochemistry of the Moon and Planets*: *NASA Spec. Publ.*, in press.

Grove, T. L., D. Walker, J. Longhi, E. Stolper, and J. Hays, 1973. Petrology of rock 12002 and origin of picritic basalts at Oceanus Procellarum: *Geochim. et Cosmochim. Acta*, Suppl. 4, v. 1, pp. 995-1011.

Guest, J. E., 1971. Centers of igneous activity in the maria, *in* G. Fielder, ed., *Geology and Physics of the Moon*, pp. 41-53, Amsterdam, The Netherlands, Elsevier.

Hamilton, W. L., 1972. Tidal triggering of Moonquakes: *Science*, v. 176, pp. 1258-1259.

Hanel, R., B. Conrath, W. Hovis, V. Kunde, P. Lowman, W. Maguire, J. Pearl, J. Pirraglia, C. Prabhakara, B. Schoachman, G. Levin, P. Straat, and T. Burke, 1972. Investigation of the Martian environment by infrared spectroscopy on Mariner 9: *Icarus*, v. 17, pp. 423-442.

Hanks, T. C., and D. L. Anderson, 1969. The early thermal history of the Earth: *Phys. of Earth Planet. Interiors*, v. 2, pp. 19-29.

Hartmann, W. K., 1966a. Early lunar cratering: *Icarus*, v. 5, pp. 406-418.

———, 1966b. Martian cratering: *Icarus*, v. 5, pp. 565-576.

———, 1971a. Martian cratering II: asteroid impact history: *Icarus*, v. 15, pp. 396-409.

———, 1971b. Martian cratering III: theory of crater obliteration: *Icarus*, v. 15, pp. 410-428.

———, 1972a. Interplanetary variations in scale of crater morphology—Earth, Mars, Moon: *Icarus*, v. 17, pp. 707-713.

———, 1972b. Paleocratering of the Moon: review of post-Apollo data: *Astrophys. Space Sci.*, v. 17, pp. 48-64.

———, 1973a. Martian cratering, 4, Mariner 9 initial analysis of cratering chronology: *Jour. Geophys. Res.*, v. 78, pp. 4096-4116.

———, 1973b. Martian surface and crust: review and synthesis: *Icarus*, v. 19, pp. 550-575.

———, 1974a. Geological observations of Martian arroyos: *Jour. Geophys. Res.*, v. 79, pp. 3951-3957.

———, 1974b. Martian and terrestrial paleoclimatology: relevance of solar variability: *Icarus*, v. 22, pp. 301-311.

———, and C. A. Wood, 1971. Moon: origin and evolution of multi-ring basins: *The Moon*, v. 3, pp. 2-78.

———, and M. J. Price, 1974. Mars: clearing of the 1971 dust storm: *Icarus*, v. 21, pp. 28-34.

Hays, J. F., and D. Walker, 1975. Lunar igneous rocks and the nature of the lunar interior: *Proceedings of the Soviet-American Conference on the Cosmochemistry of the Moon and Planets*: NASA Special Publication, in press.

Head, J. W., 1973. Small-scale analogs of the Cayley Formation and Descartes Mountains in impact-associated deposits: *Apollo 16 Preliminary Science Report, NASA Spec. Publ. SP-315*, pp. 29.16-29.20.

———, 1974a. Orientale multi-ringed basin interior and implications for the petrogenesis of lunar highland samples: *The Moon*, v. 11, pp. 327-356.

———, 1974b. Lunar dark-mantle deposits: possible clues to the distribution of early mare deposits: *Geochim. et Cosmochim. Acta*, Suppl. 5, v. 1, pp. 207-222.

———, 1974c. Morphology and structure of the Taurus-Littrow Highlands (Apollo 17): evidence for their origin and evolution: *The Moon*, v. 9, pp. 355-395.

———, 1974d. Stratigraphy of the Descartes region (Apollo 16): implications for the origin of samples: *The Moon*, v. 11, pp. 77-99.

———, 1975. Processes of lunar crater degradation:

changes in style with geologic time: *The Moon*, v. 12, pp. 299-329.

———, 1976. Lunar volcanism in space and time: *Rev. Geophys. and Space Phys.*, in press.

———, and D. D. Lloyd, 1971. Near-terminator photography: *Apollo 14 Preliminary Science Report, NASA Spec. Publ. SP-272*, pp. 297-300.

———, and A. F. Goetz, 1972. Descartes region: evidence for Copernican age volcanism: *Jour. Geophys. Res.*, v. 77, pp. 1368-1374.

Heiken, G. H., D. S. McKay, and R. W. Brown, 1974. Lunar deposits of possible pyroclastic origin: *Geochim. et Cosmochim. Acta*, Suppl. 5, v. 2, pp. 1703-1718.

Herr, K. C., and G. C. Pimentel, 1969. Infrared absorptions near three microns recorded over the polar cap of Mars: *Science*, v. 166, pp. 496-499.

Hess, S. L., 1973. Martian winds and dust clouds: *Planet. Space Sci.*, v. 21, pp. 1549-1557.

Hess, H. H., 1938. Gravity anomalies and island arc structure with particular reference to the West Indies: *American Phil. Soc., Proc.*, v. 79, pp. 71-96.

———, 1962. History of ocean basins; *in* A. E. J. Engel, H. L. James, and B. F. Leonard, eds., *Petrologic studies: A volume in honor of A. F. Buddington: Geol. Soc. America*, pp. 599-620.

Hilsenrath, J., C. W. Beckett, W. S. Benedict, L. Fano, H. J. Hoge, J. F. Masi, R. L. Nuttall, Y. S. Touloukian, H. W. Wooley, 1955. *Tables of Thermal Properties of Gases*, National Bureau of Standards Circular 564, U.S. Government Printing Office.

Hodges, C. A., 1972a. Geologic map of part of Descartes region—Apollo 16: U.S. Geological Survey Geologic Atlas of the Moon, I-748, sheet 2 of 2.

———, 1972b. Descartes region: possible analogs around the Orientale basin: *Apollo 16 Preliminary Science Report, NASA Spec. Publ. SP-315*, pp. 29.20-29.33.

———, 1973. Mare ridges and lava lakes: *Apollo 17 Preliminary Science Report, NASA Spec. Publ. SP-330*, pp. 31.12-31.21.

Hodgson, R. A., 1961. Regional study of jointing in Comb Ridge, Navajo Mountain area, Arizona and Utah: *American Assoc. Petrol. Geol. Bull.*, v. 45, pp. 2-38.

Holland, H. D., 1962. Model for the evolution of the Earth's atmosphere, *in Petrologic studies: A volume in honor of A. F. Buddington: Geol. Soc. America*, pp. 447-477.

———, 1964. On the chemical evolution of the terrestrial and cytherian atmospheres: *in* P. Brancazio and A. G. W. Cameron, eds., *The origin and evolution of atmospheres and oceans*: New York, John Wiley, pp. 86-102.

Holmes, A., 1913. *The Age of the Earth*: London, Harper Bros.

———, 1931. Radioactivity and Earth movements: *Geol. Soc. Glasgow, Trans.*, v. 18, pp. 559-606.

———, 1965. *Principles of Physical Geology*, 2nd ed., New York, The Ronald Press Co.

Holmlund, G. L., 1973. Mariner Mars 1971 science data team library, final report: *JPL Report 610-243*, Jet Propulsion Laboratory, Pasadena, Cal.

Hopkins, W. 1839. Researches in physical geology: *Roy. Soc. London, Phil. Trans.*, v. 129, pp. 381-385.

Hoppin, R. A., and Palmquist, J. C., 1965. Basement influence on later deformation: the problem, techniques of investigation and examples from Bighorn Mountains: *American Assoc. Petrol. Geol. Bull.*, v. 49, pp. 993-1004.

Hord, C. W., C. A. Barth, A. I. Stewart, and A. L. Lane, 1972. Mariner 9 ultraviolet spectrometer experiment: Photometry and topography of Mars: *Icarus*, v. 17, pp. 443-456.

Houck, J. R., J. B. Pollack, C. Sagan, D. Schaack, and J. A. Decker, Jr., 1973. High altitude infrared spectroscopic evidence for bound water on Mars: *Icarus*, v. 18, pp. 470-480.

Howard, K. A., 1973. Avalanche mode of motion: Implications from lunar examples: *Science*, v. 180, pp. 1052-1055.

———, 1974. Fresh lunar impact craters: review of variations with size: *Geochim. et Cosmochim. Acta*, Suppl. 5, v. 1, pp. 61-69.

———, and B. R. Larsen, 1972. Lineaments that are artifacts of lighting: *Apollo 15 Preliminary Science Report, NASA Spec. Publ. SP-289*, pp. 25.58-25.62.

———, and H. Masursky, 1968. Geologic map of the Ptolemaeus Quadrangle of the Moon: U.S. Geol. Survey Misc. Geol. Inv. Map I-566.

———, J. W. Head, and G. A. Swann, 1972. Geology of Hadley Rille: *Geochim. et Cosmochim. Acta*, Suppl. 3, v. 1, pp. 1-14.

———, and W. R. Muehlberger, 1973. Lunar thrust faults in the Taurus-Littrow region: *Apollo 17 Preliminary Science Report, NASA Spec. Publ. SP-330*, pp. 31.22-31.25.

———, D. E. Wilhelms, and D. H. Scott, 1974. Lunar basin formation and highland stratigraphy: *Rev. Geophys. and Space Phys.*, v. 12, pp. 309-327.

Huguenin, R. L., 1974. The formation of goethite and hydrated clay minerals on Mars: *Jour. Geophys. Res.*, v. 79, pp. 3895-3905.

Hunt, G. R., L. M. Logan, and J. W. Salisbury, 1973. Mars: components of infrared spectra and the composition of the dust cloud: *Icarus*, v. 18, pp. 459-469.

Hunten, D. M., 1971. Composition and structure of planetary atmospheres: *Space Sci. Rev.*, v. 12, pp. 539-599.

Inge, J. L., and W. A. Baum, 1973. A comparison of Martian albedo features with topography: *Icarus*, v. 19, pp. 323-328.

Ingersoll, A. P., 1970. Mars: Occurrence of liquid water: *Science*, v. 168, pp. 972-973.

———, 1974. Mars: the cases against permanent CO_2 frost caps: *Jour. Geophys. Res.*, v. 79, pp. 3403-3410.

Isachsen, Y. W., 1973. Spectral geological content of ERTS-1 imagery over a variety of geological terrains in New York State: *in Proceedings, Symposium on Management and Utilization of Remote Sensing Data, Sioux Falls. American Soc. Photogrammetry.*

Jamison, D., 1965. Some speculation on the Martian canals: *Astron. Soc. of the Pacific, Publ.*, v. 77, pp. 394-395.

Johnson, T. V., and F. P. Fanale, 1973. Optical properties of carbonaceous chondrites and their relation to asteroids: *Jour. Geophys. Res.*, v. 78, pp. 8507-8518.

Johnston, D. H., T. R. McGetchin, and M. N. Toksöz, 1974. The thermal state and internal structure of Mars: *Jour. Geophys. Res.*, v. 79, pp. 3959-3971.

Jones, K. L., 1974a. Evidence for an episode of crater obliteration intermediate in Martian history: *Jour. Geophys. Res.*, v. 79, pp. 3917-3931.

———, 1974b. Martian obliterational history: Ph.D. thesis, Providence, R.I., Brown Univ.

Jones, S. E., and N. O. Cook, 1974. Color pictures of planets from black-and-white images: *Sky and Telescope*, v. 47, pp. 57-59.

Kane, J., J. Kasold, M. Suda, P. Metcalf, and S. Caccamo, 1973. Alpine glacial features of Mars: *Nature*, v. 244, pp. 20-21.

Kaplan, L. D., G. Münch, and H. Spinrad, 1964. An analysis of the spectrum of Mars: *Astrophys. Jour.*, v. 139, pp. 1-15.

———, J. Connes, and P. Connes, 1969. Carbon monoxide in the Martian atmosphere: *Astrophys. Jour.*, v. 157, pp. L187-L192.

Katterfeld, G. N., 1966. Volcanic activity on Mars (transl.): *NASA Tech. Transl.*, F-410. 16 pp.

———, and P. Hedervari, 1968. Ring-shaped and linear structures on Mars: *Astron. Zh.*, v. 45, pp. 1091-1100: transl. in *Sov. Astron.-A.J.*, v. 12, pp. 863-871.

Kaula, W. M., 1968. *An Introduction to Planetary Physics*: New York, John Wiley. 490 pp.

———, G. Schubert, and R. E. Lingenfelter, 1972. Analysis and interpretation of lunar lasar altimetry: *Geochim. et Cosmochim. Acta*, Suppl. 3, v. 3, pp. 2189-2204.

Kendall, P., and H. Briggs, 1933. The formation of rock joints and cleat in coal: *Proc. Roy. Soc. Edin.*, v. 55, p. 193.

King, J. S., and J. R. Riehle, 1974. A proposed origin of the Olympus Mons escarpment: *Icarus*, v. 23, pp. 300-317.

Kliore, A. J., D. L. Cain, G. S. Levy, V. R. Eshelman, G. Fjeldbo, and F. D. Drake, 1965. Occultation experiment: Results of the first direct measurement of Mars atmosphere and ionosphere: *Science*, v. 149, pp. 1243-1248.

———, D. L. Cain, G. Fjeldbo, B. L. Seidel, and S. I. Rasool, 1972a. Mariner 9 S-band Martian occultation experiment: Initial results on the topography and atmosphere of Mars: *Science*, v. 175, p. 313.

———, D. L. Cain, G. Fjeldbo, B. L. Seidel, M. J. Sykes, and S. I. Rasool, 1972b. The atmosphere of Mars from Mariner 9 radio occultation measurements: *Icarus*, v. 17, pp. 484-516.

———, G. Fjeldbo, B. L. Seidel, M. J. Sykes, and P. M. Woiceshyn, 1973. S-band radio occultation measurements of the atmosphere and topography of Mars with Mariner 9: Extended mission coverage of polar and intermediate latitudes: *Jour. Geophys. Res.*, v. 78, pp. 4331-4351.

Koskela, P. E., 1973. Mariner Mars 1971 television picture catalog, sequence design and picture coverage: *JPL Tech. Memo*, 33-585, v. 2. Jet Propulsion Laboratory, Pasadena, Cal.

———, M. R. Helton, L. N. Seeley, and S. J. Zawacki, 1972. Mariner Mars 1971 television picture catalog, sequence design and picture coverage: *JPL Tech. Memo*, 33-585, v. 1. Jet Propulsion Laboratory, Pasadena, Cal.

Kuenen, Ph. H., 1936. The negative isostatic anomalies in the East Indies (with experiments): *Leidsche Geol. Med.*, v. 8, pp. 169-214.

Kuiper, G. P., 1949. Survey of planetary atmospheres, chapt. 12, *Contributions from the McDonald Observatory*, no. 161, pp. 304-345.

———, 1956. Note on Dr. McLaughlin's paper: *Astron. Soc. of the Pacific, Publications*, v. 68, p. 219.

Lacroix, A., 1904. *La Montagne Pelée et ses éruptions*: Masson et Cie, Paris, 662 pp.

Lamar, D. L., 1962. Optical ellipticity and internal structure of Mars: *Icarus*, v. 1, pp. 258-265.

Lammlein, D., J. Dorman, and G. Latham, 1972. Tidal triggering of Moonquakes: *Science*, v. 176, p. 1259.

Langseth, M. G., S. J. Keihm, and J. L. Chute, 1973. Heatflow Experiment: *Apollo 17 Preliminary Science Report, NASA Spec. Publ. SP-330*, pp. 9.1-9.24.

Latham, G., M. Ewing, J. Dorman, D. Lammlein, F. Press, N. Toksöz, G. Sutton, F. Duennebier, and Y. Nakamura, 1971. Moonquakes: *Science*, v. 174, pp. 687-692.

Lederberg, J., and C. Sagan, 1962. Microenvironments for life on Mars: *Proc. Nat. Acad. Sci.*, v. 48, pp. 1473-1475.

Leighton, R. B., B. C. Murray, R. P. Sharp, J. D. Allen, and R. K. Sloan, 1965. Mariner IV photography of Mars: Initial results: *Science*, v. 149, pp. 627-630.

———, and B. C. Murray, 1966. Behavior of carbon dioxide and other volatiles on Mars: *Science*, v. 153, pp. 136-144.

———, N. H. Horowitz, B. C. Murray, R. P. Sharp, A. G. Herriman, A. T. Young, B. A. Smith, M. E. Davies, and C. G. Leovy, 1969a. Mariner 6 and 7 television pictures: Preliminary analysis: *Science*, v. 166, pp. 49-67.

———, N. H. Horowitz, B. C. Murray, R. P. Sharp, A. G. Herriman, A. T. Young, B. A. Smith, M. E. Davies, and C. G. Leovy, 1969b. Television observations from Mariners 6 and 7: *Mariner-Mars 1969, A Preliminary Report, NASA Spec. Publ. SP-225*.

Leovy, C. B., 1973. Exchange of water vapor between the atmosphere and surface of Mars: *Icarus*, v. 18, pp. 120-125.

———, and Y. Mintz, 1969. Numerical simulation of the atmospheric circulation and climate of Mars: *Jour. Astron. Sci.*, v. 26, pp. 1167-1190.

———, B. A. Smith, A. T. Young, and R. B. Leighton, 1971. Mariner Mars 1969: atmospheric results: *Jour. Geophys. Res.*, v. 76, pp. 297-312.

———, G. A. Briggs, and B. A. Smith, 1973. Mars atmosphere during the Mariner 9 extended mission: Television results: *Jour. Geophys. Res.*, v. 78, pp. 4252-4266.

Levinthal, E. C., W. B. Green, J. A. Cutts, E. D. Jahelka, R. A. Johansen, M. J. Sander, J. B. Seidman, A. T. Young, and L. A. Soderblom, 1973. Mariner 9 image processing products: *Icarus*, v. 18, pp. 75-101.

Lewis, J. S., 1972. Metal/silicate fractionation in the solar system: *Earth Planet. Sci. Letters*, v. 15, pp. 286-290.

———, 1974a. The temperature gradient in the solar nebula: *Science*, v. 186, pp. 440-443.

———, 1974b. The chemistry of the solar system: *Scientific American*, v. 230, pp. 50-65.

Loomis, A. A., 1965. Some geologic problems of Mars: *Geol. Soc. America Bull.*, v. 76, pp. 1083-1104.

Lorell, J., G. H. Born, E. J. Christensen, J. F. Fordan, P. A. Laing, W. L. Martin, W. L. Sjogren, I. I. Shapiro, R. D. Reasenberg, G. L. Slater, 1972. Mariner 9 celestial mechanics experiment: Gravity field and pole direction of Mars: *Science*, v. 175, pp. 317-320.

Lowell, P., 1895. *Mars*, Boston.

———, 1906. *Mars and Its Canals*, New York, Macmillan.

———, 1909. *Mars as the Abode of Life*, New York, Macmillan. 288 pp.

Lucchitta, B. K., 1973. Photogeology of the dark material in the Taurus-Littrow region of the moon: *Geochim. et Cosmochim. Acta*, Suppl. 4, v. 1, pp. 149-162.

———, 1974. The walls of Tithonius and Ius Chasmas, Mars: *Geol. Soc. America 1974 Ann. Mtg. Abstracts*, v. 6, no. 7, p. 851.

Lyubimova, Y. A., 1974. Heat flow and thermal history of the Moon: *Proceedings of the Soviet-American Conference on the Cosmochemistry of the Moon and Planets: NASA Spec. Publ.*, in press.

Lyttleton, R. A., 1965a. On the internal structure of the planet Mars: *Mon. Not. Roy. Astron. Soc.*, v. 129, pp. 21-39.

———, 1965b. Note on the structure of Mars. *Mon. Not. Roy. Astron. Soc.*, v. 130, pp. 95-96.

———, 1972. On the formation of planets from a solar nebula: *Mon. Not. Roy. Astron. Soc.*, v. 158, pp. 463-483.

———, 1973. The end of the iron-core age: *The Moon*, v. 7, pp. 422-439.

McCall, G. J. F., 1966. Implications of the Mariner IV photography of Mars: *Nature*, v. 211, pp. 1384-1385.

McCauley, J. F., 1967. The nature of the lunar surface as determined by systematic geologic mapping: *in* S. K. Runcorn, ed., *Mantles of the Earth and Terrestrial Planets*, New York, Interscience, pp. 431-460.

———, 1969. The domes and cones in the Marius Hills region: evidence for lunar differentiation? *American Geophys. Union Trans.*, v. 50, p. 229.

———, 1973. Mariner 9 evidence for wind erosion in the equatorial and mid-latitude regions of Mars: *Jour. Geophys. Res.*, v. 78, pp. 4123-4138.

———, and M. H. Carr, J. A. Cutts, W. K. Hartmann, H. Masursky, D. J. Milton, R. P. Sharp, and D. E. Wilhelms, 1972. Preliminary Mariner 9 Report on the Geology of Mars: *Icarus*, v. 17, pp. 289-327.

McCord, T. B., 1969. Comparison of the reflectivity and color of bright and dark regions on the surface of Mars: *Astrophys. Jour.*, v. 156, pp. 79-86.

———, and J. B. Adams, 1969. Spectral reflectivity of Mars: *Science*, v. 163, pp. 1058-1060.

———, J. H. Elias, and J. A. Westphal, 1971. Mars: the spectral albedo (0.3-2.5) of small bright and dark regions: *Icarus*, v. 14, pp. 245-251.

McElroy, M. B., 1972. Mars: an evolving atmosphere: *Science*, v. 175, pp. 443-445.

McGetchin, T. R., and J. W. Head, 1973. Lunar cinder cones: *Science*, v. 180, pp. 68-71.

———, and G. W. Ullrich, 1973. Xenoliths in maars and diatremes with inferences for the Moon, Mars, and Venus: *Jour. Geophys. Res.*, v. 78, pp. 1833-1853.

———, M. Settle, and J. W. Head, 1973. Radial thickness variation in impact crater ejecta: implications for lunar basin deposits: *Earth Planet. Sci. Letters*, v. 20, no. 2, pp. 226-236.

McGill, G. E., 1971. Attitude of fractures bounding straight and arcuate lunar rilles: *Icarus*, v. 14, pp. 53-58.

———, and D. U. Wise, 1972. Regional variations in degradation and density of Martian craters: *Jour. Geophys. Res.*, v. 77, pp. 2433-2441.

McLaughlin, D. B., 1954a. Volcanism and aeolian deposition on Mars: *Geol. Soc. America Bull.*, v. 65, pp. 715-717.

———, 1954b. Interpretation of some Martian features: *Astron. Soc. of the Pacific, Publ.*, v. 66, pp. 161-170.

———, 1954c. Wind patterns and volcanoes on Mars: *Observatory*, v. 74, pp. 166-168.

———, 1954d. Further notes on Martian features: *Astron. Soc. of the Pacific, Publ.*, v. 66, pp. 221-229.

———, 1955a. Additional evidence of volcanism on Mars: *Geol. Soc. America Bull.*, v. 66, pp. 769-772.

———, 1955b. Changes on Mars, as evidence of wind deposition and volcanism: *Astron. Jour.*, v. 60, pp. 261-270.

———, 1956a. The volcanic-aeolian hypothesis of Martian features, *Astron. Soc. of the Pacific, Publ.*, v. 68, pp. 211-218.

———, 1956b. A new theory of Mars: *Michigan Alumnus Quarterly Review*, v. 62, pp. 301-307.

Macdonald, G. A., 1972. *Volcanoes*: Englewood Cliffs, N.J., Prentice-Hall.

Malin, M. C., 1974a. Lunar red spots: possible pre-mare materials: *Earth Planet. Sci. Letters*, v. 21, pp. 331-341.

———, 1974b. Salt weathering on Mars: *Jour. Geophys. Res.*, v. 79, pp. 3888-3894.

Marcus, A. H., 1964. A stochastic model of the formation and survival of lunar craters, I: distribution of diameter of clean craters: *Icarus*, v. 4, pp. 460-472.

———, 1966a. A stochastic model of the formation and survival of lunar craters, II: approximate distribution of diameter of all observable craters: *Icarus*, v. 5, pp. 165-177.

———, 1966b. A stochastic model of the formation and survival of lunar craters, III: filling and disappearance of craters: *Icarus*, v. 5, pp. 178-189.

———, 1966c. A stochastic model of the formation and survival of lunar craters, IV: on the randomness of crater centers: *Icarus*, v. 5, pp. 190-200.

———, 1966d. A stochastic model of the formation and survival of lunar craters, V: approximate diameter distribution of primary and secondary craters: *Icarus*, v. 5, pp. 590-605.

———, 1967. A stochastic model of the formation and survival of lunar craters, VI: initial depth, distribution of depths, and lunar history: *Icarus*, v. 6, pp. 56-74.

Marov, M. Y., and G. I. Petrov, 1973. Investigations of Mars from the Soviet automatic stations Mars 2 and 3: *Icarus*, v. 19, pp. 163-179.

Martin, L. J., 1974a. The major Martian yellow storm of 1971: *Icarus*, v. 22, pp. 175-188.

———, 1974b. The major Martian dust storms of 1971 and 1973: *Icarus*, v. 23, pp. 108-115.

Masursky, H., 1973. An overview of geological results from Mariner 9: *Jour. Geophys. Res.*, v. 78, pp. 4009-4030.

———, R. M. Batson, W. Borgeson, M. H. Carr, J. McCauley, D. Milton, R. Wildey, D. Wilhelms, B. Murray, N. Horowitz, R. Leighton, R. Sharp, W. Thompson, G. Briggs, P. Chandeysson, E. Shipley, C. Sagan, J. Pollack, J. Lederberg, E. Levinthal, W. Hartmann, T. McCord, B. Smith, M. Davies, G. de Vaucouleurs, and C. Leovy, 1970. Television experiment for Mariner Mars 1971: *Icarus*, v. 12, pp. 10-45.

———, R. M. Batson, J. F. McCauley, L. A. Soderblom, R. L. Wildey, M. H. Carr, D. J. Milton, D. E. Wilhelms, B. A. Smith, T. B. Kirby, J. C. Robinson, C. B. Leovy, G. A. Briggs, A. T. Young, T. C. Duxbury, C. H. Acton, B. C. Murray, J. A. Cutts, R. P. Sharp, S. Smith, R. B. Leighton, C. Sagan, J. Veverka, M. Noland, J. Lederberg, E. Levinthal, J. B. Pollack, J. T. Moore, W. K. Hartmann, E. N. Shipley, G. de Vaucouleurs, M. F. Davies, 1972. Mariner 9 television reconnaissance of Mars and its satellites: preliminary results: *Science*, v. 175, pp. 294-305.

Matthews, R. K., 1974. *Dynamic Stratigraphy*: Englewood Cliffs, N.J., Prentice-Hall. 370 pp.

May, P. R., 1971. Pattern of Triassic-Jurassic diabasic dikes around the North Atlantic in the context of predrift position of the continents: *Geol. Soc. America Bull.*, v. 82, pp. 1285-1292.

Menard, H. W., 1964. *Marine Geology of the Pacific*: New York, McGraw-Hill. 271 pp.

———, and M. S. Smith, 1966. Hypsometry of ocean basin provinces: *Jour. Geophys. Res.*, v. 71, pp. 4305-4325.

Metz, W. D., 1974. Update on Mars: clues about the early solar system: *Science*, v. 183, pp. 187-189.

Middlehurst, B. M., 1967. An analysis of lunar events: *Rev. Geophys.*, v. 5, pp. 173-189.

Milankovitch, M., 1941. *Canon of Insolation and the Ice-age Problem*: Translated from German, 1969, U.S. Dept. Commerce and National Science Foundation, publ. TT67-51410/1 and /2, 484 pp.

Miller, S. L., 1961. The occurrence of gas hydrates in the solar system: *Proc. Nat. Acad. Sci.*, Washington, D.C., v. 47, pp. 1798-1808.

Miller, S. L., and W. D. Smythe, 1970. Carbon dioxide clathrate in the Martian ice cap: *Science*, v. 170, pp. 531-533.

Milton, D. J., 1973. Water and processes of degradation in the Martian landscape: *Jour. Geophys. Res.*, v. 78, pp. 4037-4048.

———, 1974. Carbon dioxide hydrate and floods on Mars: *Science*, v. 183, pp. 654-656.

Ming, L.-C., and W. A. Bassett, 1975. The postspinel phases in the Mg_2SiO_4-Fe_2SiO_4 system: *Science*, v. 187, pp. 66-68.

Moorbath, S., R. K. O'Nions, and R. J. Pankhurst, 1973. Early Archean age for the Isua iron formation, West Greenland: *Nature*, v. 245, p. 138.

Moore, H. J., 1964. Density of small craters on the lunar surface, *Astrogeological Studies Annual Progress Report*, Aug. 25, 1962–July 1, 1963, Part D: Studies for Space Flight Program, U.S. Geol. Survey Rept., pp. 34-51.

———, 1965. Geologic map of the Aristarchus Quadrangle of the Moon: U.S. Geol. Survey Misc. Inv. Map I-465.

———, 1967. Geologic map of the Seleucus Quadrangle of the Moon: U.S. Geol. Survey Misc. Inv. Map I-527.

———, 1971. Geologic interpretation of lunar data: *Earth Sci. Rev.*, v. 7, pp. 5-33.

———, 1972. Ranger and other impact craters photographed by Apollo 16: *Apollo 16 Preliminary Science Report, NASA Spec. Publ. SP-315*, pp. 29.45-29.50.

———, C. A. Hodges, and D. H. Scott, 1974. Multi-ringed basins—illustrated by Orientale and associated features: *Geochim. et Cosmochim. Acta*, Suppl. 5, v. 1, pp. 71-100.

Morgan, W. J., 1971. Convection plumes in the lower mantle: *Nature*, v. 230, pp. 42-43.

Morisawa, M. E., 1962. Quantitative geomorphology of some watersheds in the Appalachian Plateau: *Geol. Soc. America Bull.*, v. 73, pp. 1025-1046.

Moroz, V. I., and L. V. Ksanfomaliti, 1972. Preliminary results of astrophysical observations of Mars from Mars-3: *Icarus*, v. 17, pp. 408-422.

Morris, E. C., T. A. Mutch, and H. E. Holt, 1972. Atlas of geologic features in the dry valleys of South Victoria Land, Antarctica: *U.S.G.S., Interagency Report, Astrogeol. 52*.

———, R. M. Batson, E. A. Cernan, V. L. Freeman, M. H. Hait, H. E. Holt, K. A. Howard, E. D. Jackson, K. B. Larson, V. S. Reed, J. J. Rennilson, H. H. Schmitt, D. H. Scott, R. A. Sutton, O. Stuart-Alexander, G. A. Swann, N. J. Trask, G. E. Ulrich, H. G. Wilshire, and E. W. Wolfe, 1973. Preliminary geologic investigation of the Apollo 17 landing site: *Apollo 17 Preliminary Science Report, NASA Spec. Publ. SP-330*, 6-1-6-91.

Muehlberger, W. R., R. M. Batson, E. L. Boudette, C. M. Duker, R. E. Eggleton, D. P. Elston, A. W. England, V. L. Freeman, M. H. Hait, T. A. Hall, J. W. Head, C. A. Hodges, H. E. Holt, E. D. Jackson, J. A. Jordon, K. B. Larson, D. J. Milton, V. S. Reed, J. J. Rennilson, G. G. Schaber, J. P. Schafer, L. T. Silver, D. Stuart-Alexander, R. L. Sutton, G. A. Swann, R. L. Tyner, G. E. Ulrich, H. G. Wilshire, E. W. Wolfe, and J. W. Young, 1972. Preliminary geologic investigation of the Apollo 16 landing site: *Apollo 16 Preliminary Science Report, NASA Spec. Publ. SP-315*, 6-1-6-81.

———, L. A. Soderblom, R. P. Sharp, and J. A. Cutts, 1971. The surface of Mars, 1, cratered terrains: *Jour. Geophys. Res.*, v. 76, pp. 313-330.

———, L. A. Soderblom, J. A. Cutts, R. P. Sharp, D. J. Milton, and R. B. Leighton, 1972. Geological framework of the south polar region of Mars: *Icarus*, v. 17, pp. 328-345.

———, W. R. Ward, and S. C. Yeung, 1973. Periodic insolation variations on Mars: *Science*, v. 180, pp. 638-640.

Murray, B. C., and M. C. Malin, 1973a. Polar wandering on Mars? *Science*, v. 179, pp. 997-1000.

———, and M. C. Malin, 1973b. Polar volatiles on Mars—theory versus observation: *Science*, v. 182, pp. 437-443.

———, M. J. S. Belton, G. E. Danielson, M. E. Davies, D. E. Gault, B. Hapke, B. O'Leary, R. G. Strom, V. Suomi, N. Trask, 1974. Mercury's surface: preliminary description and interpretation from Mariner 10 pictures: *Science*, v. 185, pp. 169-179.

———, R. G. Strom, N. J. Trask, and D. E. Gault, 1975. Surface history of Mercury: implications for terrestrial planets: *Jour. Geophys. Res.*, v. 80, pp. 2508-2514.

Murray, J., 1888. On the height of the land and the depth of the ocean: *The Scottish Geographical Magazine*, v. 4, no. 1, pp. 1-41.

Murray, J. B., 1971. Sinuous rilles, in G. Fielder, ed., *Geology and Physics of the Moon*: Amsterdam, The Netherlands, Elsevier, pp. 27-29.

Mutch, T. A., 1972. *Geology of the Moon: A Stratigraphic View*: rev. ed., Princeton, N.J., Princeton Univ. Press. 391 pp.

Nakamura, Y., G. Latham, D. Lammlein, M. Ewing, F. Duennebier, and J. Dorman, 1974. Deep lunar interior inferred from recent seismic data: *Geophys. Res. Letters*, v. 1, pp. 137-140.

Nash, D. B., 1974. The relative age of escarpments in the Martian polar laminated terrain based on morphology: *Icarus*, v. 22, pp. 385-396.

Neugebauer, G., G. Münch, S. C. Chase, Jr., H. Hatzenbeler, E. Miner, and D. Schofield, 1969. Mariner 1969: preliminary results of the infrared radiometer experiment: *Science*, v. 166, pp. 98-99.

———, G. Münch, H. Kieffer, S. C. Chase, Jr., and E. Miner, 1971. Mariner 1969 infrared radiometer results: temperature and thermal properties of the Martian surface: *Astron. Jour.*, v. 76, pp. 719-728.

Noland, N., J. Veverka, and J. B. Pollack, 1973. Mariner 9 polarimetry of Phobos and Deimos: *Icarus*, v. 20, pp. 490-502.

Oberbeck, V. R., 1975. The role of ballistic erosion and sedimentation in lunar stratigraphy, *Reviews of Geophys. and Space Phys.*, v. 13, pp. 337-362.

———, and M. Aoyagi, 1972. Martian doublet craters: *Jour. Geophys. Res.*, v. 77, pp. 2419-2432.

————, and R. H. Morrison, 1973. The lunar herring-bone pattern: *Apollo 17 Preliminary Science Report, NASA Spec. Publ. SP-330*, pp. 32.15-32.29.

————, F. Horz, R. H. Morrison, and W. L. Quaide, 1973. Emplacement of the Cayley formation: *NASA Tech. Memo TM-X-62*, 302.

————, R. H. Morrison, F. Horz, W. L. Quaide, and D. E. Gault, 1974. Smooth plains and continuous deposits of craters and basins. *Geochim. et Cosmochim. Acta, Suppl.* 5, v. 1, pp. 111-136.

————, W. Quaide, and R. Arvidson, 1975. Secondary cratering on Mercury, the Moon and Mars: Abstracts, International Colloquium on Planetary Geology, Rome, Italy.

O'Connor, J. T., 1968. Mineral stability on the Martian surface: *Jour. Geophys. Res.*, v. 73, pp. 5301-5311.

Offield, T. W., and H. A. Pohn, 1970. Lunar crater morphology and relative-age determination of lunar geologic units—Part 2. Applications: U.S. Geol. Survey Paper 700-C, C163-169.

O'Leary, B. T., and D. G. Rea, 1967. Mars: influence of topography on formation of temporary bright patches: *Science*, v. 155, pp. 317-319.

Öpik, E. J., 1951. Collision probabilities with the planets: *Proc. Roy. Irish Acad.*, v. 54A, pp. 165-199.

————, 1963. Survival of comet nuclei and the asteroids: *Advances in Astronomy and Astrophysics*, v. 2, pp. 219-262.

————, 1965. Mariner IV and craters on Mars: *Irish Astron. Jour.*, v. 7, pp. 92-104.

————, 1966. The Martian surface: *Science*, v. 153, pp. 255-265.

————, 1971. Cratering and the Moon's surface: *Advances in Astronomy and Astrophysics*, v. 8, pp. 108-337.

Otterman, J., and F. E. Bronner, 1966. Martian wave of darkening: a frost phenomenon?: *Science*, v. 153, pp. 56-60.

Owen, T., and H. P. Mason, 1969. Mars: water vapor in its atmosphere: *Science*, v. 165, pp. 893-895.

Papanastassiou, D. A., and G. J. Wasserburg, 1971a. Lunar chronology and evolution from Rb-Sr studies of Apollo 11 and 12 samples: *Earth Planet. Sci. Letters*, v. 11, pp. 37-62.

————, and G. J. Wasserburg, 1971b. Rb-Sr ages of igneous rocks from the Apollo 14 mission and the age of the Fra Mauro formation: *Earth Planet. Sci. Let.*, v. 12, pp. 36-48.

————, and G. J. Wasserburg, 1972. Rb-Sr age of Luna 16 basalt and the model age of the lunar soils: *Earth Planet. Sci. Letters*, v. 13, pp. 368-374.

Patterson, C., 1956. Age of meteorites and the Earth: *Geochim. et Cosmochim. Acta*, v. 10, pp. 230-237.

Peale, S. J., 1973. Water and the Martian W cloud: *Icarus*, v. 18, pp. 497-501.

————, G. Schubert and R. E. Lingenfelter, 1975. Origin of Martian channels: clathrates and water: *Science*, v. 187, pp. 273-274.

Perls, T. A., 1973. Carbon suboxide on Mars: evidence against formation: *Icarus*, v. 20, pp. 511-512.

Pettengill, G. H., C. C. Counselman, L. P. Rainville, and I. I. Shapiro, 1969. Radar measurements of Martian topography: *Astron. Jour.*, v. 74, pp. 461-482.

————, A. E. E. Rogers, and I. I. Shapiro, 1971. Martian craters and a scarp as seen by radar: *Science*, v. 174, pp. 1321-1324.

Petterssen, S., 1969. *Introduction to Meteorology*, 3rd ed., New York, McGraw-Hill.

Phillips, R. J., and R. S. Saunders, 1975. The isostatic state of Martian topography: *Jour. Geophys. Res.*, v. 80, pp. 2893-2898.

————, J. E. Conel, E. A. Abbott, W. L. Sjogren, and J. B. Morton, 1972. Mascons: progress toward a unique solution for mass distribution: *Jour. Geophys. Res.*, v. 77, pp. 7106-7114.

————, R. S. Saunders, and J. F. Conel, 1973. Mars, crustal structure inferred from bouger gravity anomalies: *Jour. Geophys. Res.*, v. 78, pp. 4815-4820.

Pieri, D., 1975. Small channels on Mars: A preliminary assessment of distribution, morphology, age and origin: Oral presentation at Fifth Ann. Planetology Program Principal Investigators' Meeting, Pasadena, Cal., Calif. Institute of Technology.

Pieters, C., T. B. McCord, S. Zisk, and J. B. Adams, 1973. Lunar black spots and nature of Apollo 17 landing area: *Jour. Geophys. Res.*, v. 78, pp. 5867-5875.

Pike, R. J., 1968. Meteoric origin and consequent endogenetic modification of large lunar craters—a study in analytical geomorphology: Ph.D. thesis, Univ. of Michigan, Ann Arbor, Mich., 404 pp.

————, 1971. Genetic implications of the shapes of Martian and lunar craters: *Icarus*, v. 15, pp. 384-395.

————, 1972. Geometric similitude of lunar and terrestrial craters: *24th IGC, Sect. 15*, pp. 41-47.

————, 1974. Craters on Earth, Moon, and Mars: Multivariate classification and mode of origin: *Earth Planet. Sci. Letters*, v. 22, pp. 245-255.

Plummer, W. T., and R. K. Carson, 1969. Mars, is the surface colored by carbon suboxide?: *Science*, v. 166, pp. 1141-1142.

Pohn, H. A., and T. W. Offield, 1970. Lunar crater morphology and relative-age determination of lunar geologic units: Pt. 1, classification, *U.S. Geol. Surv. Prof. Paper 700-C*, pp. C153-C162.

Pollack, J. B., E. H. Greenberg, and C. Sagan, 1967. A statistical analysis of the Martian wave of darkening and related phenomena: *Planet. Space Sci.*, v. 15, pp. 817-824.

————, and C. Sagan, 1969. An analysis of Martian photometry and polarimetry: *Space Sci. Rev.*, v. 9, pp. 243-299.

————, D. Pitman, B. N. Khare, and C. Sagan, 1970a. Goethite on Mars: a laboratory study of physically and chemically bound water in ferric oxides: *Jour. Geophys. Res.*, v. 75, pp. 7480-7490.

————, R. N. Wilson, and G. G. Goles, 1970b. A reexamination of the stability of goethite on Mars: *Jour. Geophys. Res.*, v. 75, pp. 7491-7500.

————, J. Veverka, M. Noland, C. Sagan, T. C. Duxbury, C. H. Acton, Jr., G. H. Born, W. K. Hartmann, and B. A. Smith, 1973. Mariner 9 television observations of Phobos and Deimos: *Jour. Geophys. Res.*, v. 78, pp. 4313-4326.

———, O. B. Toon, and C. Sagan, 1974. Physical properties of the particles composing the great Martian dust storm of 1971: *Bull. American Astron. Soc.*, v. 6, p. 370.

Potter, D. B., 1976. Geologic atlas of Mars. Hellas quadrangle (MC-28). U.S. Geol. Survey Misc. Inv. Map, in press.

Price, N. J., 1966. *Fault and Joint Development in Brittle and Semibrittle Rocks*, New York, Pergamon Press.

Pritchard, E. B., and E. F. Harrison, 1973. *Mars exploration options in the 1977-1990 period:* AAS/AIAA Astrodynamics Conf., Vail, Colo., July 16-18, 1973.

Prospero, J. M., E. Bonatti, C. Schubert, 1970. Dust in the Caribbean atmosphere traced to an African dust storm: *Earth Planet. Sci. Letters*, v. 9, pp. 287-293.

Quaide, W. L., D. E. Gault, and R. A. Schmidt, 1965. Gravitative effects on lunar impact structures: *Annals of N.Y. Acad. Sci.*, v. 123, pp. 563-572.

———, and V. K. Oberbeck, 1968. Thickness determinations of the lunar surface layer from lunar impact craters: *Jour. Geophys. Res.*, v. 73, pp. 5247-5270.

Ramsey, W. H., 1948. On the constitution of the terrestrial planets: *Monthly Notices Roy. Astron. Soc.*, v. 108, pp. 406-413.

Rawcliffe, R. C., C. D. Bartley, F. Li, E. Gordon, and D. Carta, 1974. Meteor of August 10, 1972: *Nature*, v. 247, pp. 449-450.

Rea, D. G., B. T. O'Leary, and W. M. Sinton, 1965. The origin of the 3.58- and 3.69-micron minima in the infrared spectra: *Science*, v. 147, pp. 1286-1288.

Ringwood, A. E., 1966. Chemical evolution of the terrestrial planets: *Geochim. et Cosmochim. Acta*, v. 30, pp. 41-104.

———, 1970. Phase transformations and mantle dynamics: *Earth Planet. Sci. Letters*, v. 14, pp. 233-241.

———, and A. F. Reid, 1968. High pressure transformation of spinels I: *Earth Planet. Sci. Letters*, v. 5, pp. 245-250.

———, and A. Major, 1970. The system Mg_2SiO_4-Fe_2SiO_4 at high pressures and temperatures: *Phys. of Earth Planet. Interiors*, v. 3, pp. 89-108.

———, and S. P. Clark, 1971. Internal constitution of Mars: *Nature*, v. 234, pp. 89-92.

Ritchey, J. E., et al., 1930. The geology of Ardnamurchan, Northwest Mull and Coll: *Mem. Geol. Surv. Scotland*.

Roddy, D. J., 1968. The Flynn Creek structure, Tennessee; *in* B. M. French and N. M. Short, eds., *Shock Metamorphism of Natural Materials*, Baltimore, Mono Book Corp., pp. 291-322.

Roosen, R. G., and B. G. Marsden, 1975. Observing prospects for Halley's Comet: *Sky and Telescope*, v. 49, pp. 363-364.

Ross, H. P., 1968. A simplified mathematical model for lunar crater erosion: *Jour. Geophys. Res.*, v. 73, pp. 1343-1354.

Rubey, W. W., 1951. Geologic history of sea water: *Geol. Soc. America Bull.*, v. 62, pp. 1111-1147.

———, 1955. Development of the hydrosphere and atmosphere, with special reference to the probable composition of the early atmosphere: *Geol. Soc. America Special Paper* 62, pp. 631-650.

Runcorn, S. K., ed., 1962. *Continental Drift*, New York, Academic Press.

Safranov, V. S., 1974. Time scale of the formation of the Earth and planets and its role in their geochemical evolution: *Proceedings of the Soviet-American Conference on the Cosmochemistry of the Moon and Planets: NASA Spec. Publ.*, in press.

Sagan, C., 1971. The long winter model of Martian biology: a speculation: *Icarus*, v. 15, p. 511.

———, 1973a. Liquid carbon dioxide and the Martian polar laminas: *Jour. Geophys. Res.*, v. 78, pp. 4250-4251.

———, 1973b. Sandstorms and eolian erosion on Mars: *Jour. Geophys. Res.*, v. 78, pp. 4155-4162.

———, and J. B. Pollack, 1966. On the nature of the canals of Mars: *Nature*, v. 212, pp. 117-121.

———, and J. B. Pollack, 1967. A windblown dust model of Martian surface features and seasonal changes: *Smithsonian Astrophys. Observ. Spec. Rept. No. 255*, Cambridge, Mass.

———, and J. B. Pollack, 1968. Elevation differences on Mars: *Jour. Geophys. Res.*, v. 73, pp. 1373-1387.

———, and J. B. Pollack, 1969. Windblown dust on Mars: *Nature*, v. 223, pp. 791-794.

———, and J. Veverka, 1971. The micro-wave spectrum of Mars: an analysis: *Icarus*, v. 14, pp. 222-234.

———, E. C. Levinthal, and J. Lederberg, 1968. Contamination of Mars: *Science*, v. 159, pp. 1191-1196.

———, J. Veverka, and P. Gierasch, 1971. Observational consequences of Martian wind regimes: *Icarus*, v. 15, pp. 253-278.

———, J. Veverka, P. Fox, R. Dubisch, J. Lederberg, E. Levinthal, L. Quam, R. Tucker, J. B. Pollack, and B. A. Smith, 1972. Variable features on Mars: preliminary Mariner 9 television results: *Icarus*, v. 17, pp. 346-372.

———, O. B. Toon, and P. J. Gierasch, 1973a. Climatic change on Mars: *Science*, v. 181, pp. 1045-1049.

———, J. Veverka, P. Fox, R. Dubisch, R. French, P. Gierasch, L. Quam, J. Lederberg, E. Levinthal, R. Tucker, B. Eross, and J. B. Pollack, 1973b. Variable features on Mars, 2, Mariner 9 global results: *Jour. Geophys. Res.*, v. 78, pp. 4163-4196.

———, J. Veverka, R. Steinbacher, L. Quam, R. Tucker, and B. Eross, 1974. Variable features on Mars, IV, Pavonis Mons: *Icarus*, v. 22, pp. 24-47.

Saheki, T., 1955. Martian phenomena suggesting volcanic activity: *Sky and Telescope*, v. 14, pp. 144-146.

Salisbury, J. W., and G. R. Hunt, 1968. Martian surface materials: effect of particle size on spectral behavior: *Science*, v. 161, pp. 365-366.

————, and G. R. Hunt, 1969. Compositional implications of the spectral behavior of Martian surface: *Nature*, v. 222, pp. 132-136.

Saunders, R. S., 1973. The furrowed terrain of Mars (abstract): *Bull. American Astron. Soc.*, v. 5, p. 295.

————, 1974. Mars crustal structure (abstract): *Bull. American Astron. Soc.*, v. 6, p. 372, abs.

Schaber, G. G., 1973. Lava flows in Mare Imbrium: geologic evidence from Apollo orbital photography: *Geochim. et Cosmochim. Acta*, Suppl. 4, v. 1, pp. 73-92.

Schorn, R. A., and L. D. Gray, 1967. The Martian surface pressure: *Astrophys. Jour.*, v. 148, pp. 663-664.

————, H. Spinrad, R. C. Moore, H. J. Smith, and L. P. Giver, 1967. High-dispersion spectroscopic observations of Mars, II. The water-vapor variations: *Astrophys. Jour.*, v. 146, pp. 743-752.

————, C. B. Farmer, and S. J. Little, 1969. High-dispersion spectroscopic studies of Mars, III. Preliminary results of the 1968-1969 water-vapor studies: *Icarus*, v. 11, pp. 283-288.

Schubert, G., R. E. Lingenfelter, and S. J. Peale, 1970. The morphology, distribution, and origin of lunar sinuous rilles: *Rev. Geophys. and Space Phys.*, v. 8, pp. 199-255.

Schultz, P. H., and D. E. Gault, 1975. Seismic effects from major basin formations on the Moon and Mercury: *The Moon*, v. 12, pp. 159-177.

Schumm, S. A., 1974. Structural origin of large Martian channels: *Icarus*, v. 22, pp. 371-384.

Scott, D. H., 1972. Preliminary geologic map of the Cydonic region of Mars, candidate Viking Mars 1975 landing site 16: *Rept. to Viking Project Office*, NASA Langley Research Center, Hampton, Va.

————, 1974a. The geologic significance of some lunar gravity anomalies (abstract): in *Lunar Science V*, Lunar Science Institute, v. 2, p. 639.

————, 1974b. Figure caption, p. 210: *Mars as Viewed by Mariner 9, NASA SP-329*, 225 pp.

Seidman, J. B., W. B. Green, P. L. Jepsen, R. M. Ruiz, and T. E. Thorpe, 1973. A user's guide to the Mariner 9 television reduced data record: *JPL Tech. Memo., 33-628*, Jet Propulsion Laboratory, Pasadena, Cal.

Sengör, A. M. C., and I. C. Jones, 1975. A new interpretation of Martian tectonics with special reference to the Tharsis region: *Geol. Soc. America Abstracts*, v. 7, p. 1264.

Sharp, R. P., 1968. Surface processes modifying Martian craters: *Icarus*, v. 8, pp. 472-480.

————, 1973a. Mars: fretted and chaotic terrains: *Jour. Geophys. Res.*, v. 78, pp. 4073-4083.

————, 1973b. Mars: south polar pits and etched terrain: *Jour. Geophys. Res.*, v. 78, pp. 4221-4230.

————, 1973c. Mars: troughed terrain: *Jour. Geophys. Res.*, v. 78, pp. 4063-4072.

————, L. A. Soderblom, B. C. Murray, and J. A. Cutts, 1971a. The surface of Mars 2: uncratered terrains: *Jour. Geophys. Res.*, v. 76, pp. 331-342.

————, B. C. Murray, R. B. Leighton, L. A. Soderblom, and J. A. Cutts, 1971b. The surface of Mars, 4: south polar cap: *Jour. Geophys. Res.*, v. 76, pp. 357-368.

————, and M. C. Malin, 1975. *Channels on Mars*: Geol. Soc. America Bull., v. 86, pp. 593-609.

Sharpless, B. P., 1945. Secular accelerations in the longitudes of the satellites of Mars: *Astron. Jour.*, v. 51, pp. 165-185.

Shoemaker, E. M., 1960. Penetration mechanics of high velocity meteorites, illustrated by Meteor Crater, Arizona: *International Geol. Congress, XXI sess.*, v. 18, pp. 418-434.

————, 1966. Progress in the analysis of the fine structure and geology of the lunar surface from Ranger VIII and IX photographs: in Ranger VII and IX experiments' analyses and interpretations: *JPL Tech. Rept. No. 32-800*, Part 2, Jet Propulsion Laboratory, Pasadena, Cal., pp. 275-284.

Short, N. M., and T. E. Bunch, 1968. A worldwide inventory of features characteristic of rocks associated with presumed meteorite impact craters; *in* B. M. French and N. M. Short, eds., *Shock Metamorphism of Natural Materials*, Baltimore, Mono Book Corp., pp. 255-266.

Silver, L. T., 1971. U-Th-Pb isotopic systems in Apollo 11 and 12 regolith materials and a possible age for the Copernicus event (abstract): *American Geophys. Union Trans.*, v. 52, p. 534.

Sinton, W. M., 1957. Spectroscopic evidence for vegetation on Mars: *Astrophys. Jour.*, v. 126, pp. 231-239.

————, 1967. On the composition of Martian surface materials: *Icarus*, v. 6, pp. 222-228.

Sjogren, W. L., J. Lorell, L. Wong, N. D. Downs, 1975. Mars gravity field based on a short-arc technique: *Jour. Geophys. Res.*, v. 80, pp. 2899-2908.

Slipher, E. C., 1962. *The Photographic Story of Mars*, Cambridge, Mass., Sky Publishing Corporation.

Smith, E. I., 1974. Rümker Hills: a lunar volcanic dome complex: *The Moon*, v. 10, pp. 175-181.

————, and A. G. Sanchez, 1973. Fresh lunar craters: morphology as a function of diameter, a possible criterion for crater origin: *Modern Geology*, v. 4, pp. 51-59.

Smith, R. L., 1966. Terrestrial calderas, associated pyroclastic deposits, and possible lunar applications, *in* W. N. Hess, D. H. Menzel, and J. A. O'Keefe, eds., *The Nature of the Lunar Surface: Proc. of the 1965 IAU-NASA Symposium*, Baltimore, Md., Johns Hopkins, pp. 241-258.

Smith, S. A., and B. A. Smith, 1972. Diurnal and seasonal behavior of discrete white clouds on Mars: *Icarus*, v. 16, pp. 509-521.

Snyder, L. M., 1971. Mariner 9 TV subsystem calibration: *JPL Tech. Rept. 610-202*, Jet Propulsion Laboratory, Pasadena, Cal.

Soderblom, L. A., 1970. A model for small-impact erosion applied to the lunar surface: *Jour. Geophys. Res.*, v. 75, pp. 2655-2661.

————, 1972. The process of crater removal on the lunar maria: *Apollo 15 Preliminary Science Report, NASA Spec. Publ. SP-289*, pp. 25.87-25.91.

Soderblom, L. A., and J. M. Boyce, 1972. Relative ages of some near-side and far-side terra plains based on Apollo 16 metric photography: *Apollo 16 Preliminary Science Report, NASA SP-315*, pp. 29.3-29.6.

———, and L. A. Lebofsky, 1972. Technique for rapid determination of relative ages of lunar areas from orbital photography: *Jour. Geophys. Res.*, v. 77, pp. 279-296.

———, M. C. Malin, J. A. Cutts, and B. C. Murray, 1973a. Mariner 9 observations of the surface of Mars in the north polar region: *Jour. Geophys. Res.*, v. 78, pp. 4197-4210.

———, T. J. Kreidler, and H. Masursky, 1973b. Latitudinal distribution of a debris mantle of the Martian surface: *Jour. Geophys. Res.*, v. 78, pp. 4117-4122.

———, R. A. West, B. M. Herman, T. J. Kreidler, and C. D. Condit, 1974. Martian planetwide crater distributions: implications for geologic history and surface processes: *Icarus*, v. 22, pp. 239-263.

Spinrad, H., G. Münch, and L. D. Kaplan, 1963. The detection of water vapor on Mars: *Astrophys. Jour.*, v. 137, pp. 1319-1321.

Stearns, H. T., 1966. *Geology of the State of Hawaii*, Palo Alto, Calif., Pacific Books.

———, and G. A. Macdonald, 1946. Geology and ground-water resources on the Island of Hawaii: *Hawaii Div. Hydrog. Bull.*, v. 9.

Strain, P., and F. El-Baz, 1975. Sinuous rilles of the Harbinger Mountain region of the Moon: *Lunar Science VI*, v. 2, pp. 786-788, Lunar Science Institute.

Strom, R. G., 1964. Analysis of lunar lineaments, I: Tectonic maps of the Moon: *Comm. Lunar and Planet. Lab.*, no. 39, pp. 205-221.

———, 1972. Lunar mare ridges, rings, and volcanic ring complexes, *in* S. Runcorn and H. Urey, eds., *The Moon, Symposium 47 of the International Astronomers Union*, pp. 187-215.

———, N. J. Trask, and J. E. Guest, 1975. Tectonism and volcanism on Mercury: *Jour. Geophys. Res.*, v. 80, pp. 2478-2507.

Stuart-Alexander, D., and K. A. Howard, 1970. Lunar maria and circular basins—a review: *Icarus*, v. 12, pp. 440-456.

Surkov, Y. A., and G. A. Fedoseyev, 1974. Radioactivity of the moon, planets, and meteorites: *Proceedings of the Soviet-American Conference on the Cosmochemistry of the Moon and Planets: NASA Special Publ.*, in press.

Swann, G. A., N. G. Bailey, R. M. Batson, V. L. Freeman, M. H. Hait, J. W. Head, H. E. Holt, K. A. Howard, J. B. Irwin, K. B. Larson, W. R. Muehlberger, V. S. Reed, J. J. Rennilson, G. G. Schaber, D. R. Scott, L. T. Silver, R. L. Sutton, G. E. Ulrich, M. G. Wilshire, and E. W. Wolfe, 1972. Preliminary geologic investigation of the Apollo 15 landing site: *Apollo 15 Preliminary Science Report, NASA Spec. Publ. SP-289*, 5-1-5-112.

Takenouchi, S., and G. C. Kennedy, 1964. The binary system H_2O-CO_2 at high temperatures and pressures: *American Jour. Sci.*, v. 262, pp. 1055-1074.

———, and G. C. Kennedy, 1965. Dissociation pressure of the phase CO_2-5¾H_2O: *Jour. of Geol.*, v. 73, pp. 383-390.

Tatsumoto, M., P. D. Nunes, R. J. Knight, C. E. Hedge, and D. M. Unruh, 1973. U-Th-Pb, Rb-Sr and K measurements of two Apollo 17 samples (abstract): *American Geophys. Union Trans.*, v. 54, pp. 614-615.

Taylor, S. R., 1975. *Lunar Science: A Post-Apollo View*, Pergamon.

———, C. C. Annette, and A. L. Graham, 1969. Trace element abundances in andesites: *Contr. Mineral. and Petrol.*, v. 23, pp. 1-26.

Tera, F., D. A. Papanastassiou, and G. J. Wasserburg, 1974. Isotopic evidence for a terminal lunar cataclysm: *Earth Planet. Sci. Lett.*, v. 22, pp. 1-21.

Thompson, D. T., 1972. Brief history of the Martian "Violet Haze" problem: *Rev. Geophys. and Space Phys.*, v. 10, pp. 919-933.

———, 1973a. A new look at the Martian "Violet Haze" problem: II. "Blue Clearing" in 1969: *Icarus*, v. 18, pp. 164-170.

———, 1973b. Time variation of Martian regional contrasts: *Icarus*, v. 20, pp. 42-47.

Thorarinsson, S., 1970. The Lakagigar eruption of 1783: *Bull. Volcanologique*, ser. 2, v. 33, pp. 910-927.

Thorpe, T., 1972. Mariner 9 television imaging performance evaluation: *JPL Rept. 610-237*, v. 2, Jet Propulsion Laboratory, Pasadena, Cal.

Tjia, H. D., 1970. Lunar wrinkle ridges indicative of strike-slip faulting: *Geol. Soc. Amer. Bull.*, v. 81, pp. 3095-3100.

Toksöz, M. N., F. Press, K. Anderson, A. Dainty, G. Latham, M. Ewing, J. Dorman, D. Lammlein, G. Sutton, F. Duennebier, and Y. Nakanura, 1972. Lunar crust: structure and composition: *Science*, v. 176, pp. 1012-1016.

———, and S. C. Solomon, 1973. Thermal history and evolution of the Moon: *The Moon*, v. 7, pp. 251-278.

Tombaugh, C. W., 1966. Evidence that the dark areas on Mars are elevated mountain ranges: *Nature*, v. 209, p. 1338.

Tozer, D. C., 1972. The present thermal state of the terrestrial planets: *Phys. Earth Planet. Interiors*, v. 6, pp. 182-197.

Trask, N. J., 1970. Geologic maps of early Apollo landing sites: Suppl. to U.S. Geol. Surv. Misc. Geol. Inv. Maps I-616-I-627.

———, 1972. Geologic Map, Chryse, Viking Landing Site (Mission "A"): *Rept. to Viking Project Office*, NASA Langley Research Center, Hampton, Va.

———, and J. F. McCauley, 1972. Differentiation and volcanism in the lunar highlands: photogeologic evidence and Apollo 16 implications: *Earth Planet. Sci. Letters*, v. 14, pp. 201-206.

———, and J. E. Guest, 1975. Preliminary geologic terrain map of Mercury: *Jour. Geophys. Res.*, v. 80, pp. 2461-2477.

Tull, R. G., 1970. High-dispersion spectroscopic observations of Mars. IV. The latitudinal distribution of atmospheric water vapor: *Icarus*, v. 13, pp. 43–57.

Turekian, K. K., and S. P. Clark, Jr., 1969. Inhomogeneous accumulation of the Earth from the primitive solar nebula: *Earth Planet. Sci. Letters*, v. 6, pp. 346-348.

Turner, G., 1971. ^{40}Ar–^{39}Ar ages from the lunar maria: *Earth Planet. Sci. Letters*, v. 11, pp. 169-191.

Tyrrell, G. W., 1937. Flood basalts and fissure eruption: *Bull. Volcanologique*, ser. 2, no. 1, pp. 89-111.

Udden, J. A., 1914. Mechanical composition of clastic sediments: *Geol. Soc. America Bull.*, v. 25, pp. 655-744.

Urey, H. C., 1955. The cosmic abundances of potassium, uranium, and thorium and the heat balances of the Earth, the Moon, and Mars: *Proc. Nat. Acad. Sci.*, v. 41, pp. 127-144.

———, 1959. The atmospheres of the planets; *in Handbuch der Physik*, Berlin, Springer-Verlag, pp. 363-418.

Van Blerkom, D. J., 1971. The effect of haze on the visibility of Martian surface features: *Icarus*, v. 14, pp. 235-244.

Van Tassel, R. A., and J. W. Salisbury, 1964. The composition of the Martian surface: *Icarus*, v. 3, pp. 264-269.

Vening Meinesz, F. A., 1930. Maritime gravity surveys in the Netherlands East Indies, tentative interpretation of the results: *Ned. Akad. Wetensch, Proc.*, ser. B, v. 33, pp. 566-577.

———, 1947. Shear patterns of the Earth's crust: *American Geophys. Union Trans.*, v. 28, pp. 1-61.

Vernekar, A. D., 1972. Long-period global variations of incoming solar radiation: *Meteorol. Monograph*, v. 12.

Veverka, J., J. Goguen, and W. Liller, 1973. Multicolor polarimetric observations of the great 1971 Martian duststorm: *CRSR 550*, Laboratory for Planetary Studies, Cornell Univ., Ithaca, N.Y.

———, and C. Sagan, 1974. McLaughlin and Mars: *American Sci.*, v. 62, pp. 44-53.

———, M. Noland, C. Sagan, J. Pollack, L. Quam, R. Tucker, B. Eross, T. Duxbury, and W. Green, 1974a. A Mariner 9 atlas of the moons of Mars: *Icarus*, v. 23, pp. 206-289.

———, C. Sagan, L. Quam, R. Tucker, and B. Eross, 1974b. Variable features on Mars III: comparison of Mariner 1969 and Mariner 1971 photography: *Icarus*, v. 21, pp. 317-368.

Vine, F. J., 1966. Spreading of the ocean floor: new evidence: *Science*, v. 154, pp. 1405-1415.

———, and D. H. Matthews, 1963. Magnetic anomalies over oceanic ridges: *Nature*, v. 199, pp. 947-949.

Vogt, P. R., 1974. Volcano spacing, fractures, and thickness of the lithosphere: *Earth Planet. Sci. Letters*, v. 21, pp. 235-252.

Vortman, L. J., 1968. Craters from surface explosions and scaling laws: *Jour. Geophys. Res.*, v. 73, pp. 4621-4635.

Wade, F. A., and J. N. DeWys, 1968. Permafrost features on the Martian surface: *Icarus*, v. 9, pp. 175-185.

Ward, W. R., 1973. Large-scale variations in the obliquity of Mars: *Science*, v. 181, pp. 260-262.

———, 1974. Climatic variations on Mars, I. Astronomical theory of insolation: *Jour. Geophys. Res.*, v. 79, pp. 3375-3386.

———, B. C. Murray, and M. C. Malin, 1974. Climatic variations on Mars, II. Evolution of carbon dioxide atmosphere and polar caps: *Jour. Geophys. Res.*, v. 79, pp. 3387-3395.

Wasserburg, G. J., F. MacDonald, F. Hoyle, and W. A. Fowler, 1964. Relative contributions of U, Th, and K to heat production in the Earth: *Science*, v. 143, pp. 465-467.

———, and D. A. Papanastassiou, 1972. Age of an Apollo 15 mare basalt; lunar crust and mantle evolution: *Earth Planet. Sci. Letters*, v. 13, pp. 97-104.

———, J. C. Huneke, D. A. Papanastassiou, F. A. Podosek, and G. Turner, 1972. Age determinations on samples from the Apollo 14 landing sites: *Space Res.*, v. 12, pp. 39-42.

Waters, A. C., 1961. Stratigraphic and lithologic variations in the Columbia River basalt: *American Jour. Sci.*, v. 259, pp. 583-611.

Wegener, A., 1912. Die Entstehung der Kontinente: *Geol. Rundschau*, v. 3, pp. 276-292.

Weihaupt, J., 1974. Possible origin and probable discharges of meandering channels on the planet Mars: *Jour. Geophys. Res.*, v. 79, pp. 2073-2076.

Weill, D. F., R. A. Grieve, I. S. McCallum, and V. Bottinga, 1971. Mineralogy and petrology of lunar samples. Microprobe studies of samples 12021 and 12022 and viscosity of melts of selected lunar compositions: *Geochim. et Cosmochim. Acta*, Suppl. 2, v. 1, pp. 413-430.

Wells, H. G., 1898. *War of the Worlds*, New York, London, Harper.

Wells, R. A., 1966. An analysis of Martian clouds and their topographical relationships: *European Space Research Organization*, SN-54.

West, M., 1974. Martian volcanism: additional observations and evidence for pyroclastic activity: *Icarus*, v. 21, pp. 1-11.

Wetherill, G. W., 1972. The beginning of continental evolution: *Tectonophysics*, v. 13, pp. 31-45.

———, 1974. Problems associated with estimation of the relative impact rate on Mars and Moon: *The Moon*, v. 9, pp. 227-231.

Whitehead, A. B., 1974. The elevation of Olympus Mons from limb photography: *Icarus*, v. 22, pp. 189-196.

Wilcox, R. E., 1959. Some effects of recent volcanic ash falls, with special reference to Alaska: *U.S. Geol. Survey Bull.*, v. 1028-N, pp. 409-476.

Wilhelms, D. E., 1970. Summary of lunar stratigraphy—telescopic observations: *U.S. Geol. Survey Prof. Paper 599-F*.

———, 1973. Comparison of Martian and Lunar multi-ringed circular basins: *Jour. Geophys. Res.*, v. 78, pp. 4084-4095.

Wilhelms, D. E., 1974. Comparison of Martian and lunar geologic provinces: *Jour. Geophys. Res.*, v. 79, pp. 3933-3941.

———, and J. F. McCauley, 1971. Geologic map of the nearside of the Moon, U.S. Geol. Survey. Misc. Geol. Inv. Map I-703.

Wilkins, G. A., 1967. The determination of the mass and ablateness of Mars from the orbits of its satellites: *in* S. K. Runcorn, ed., *Mantles of the Earth and Terrestrial Planets*, London, John Wiley Co., pp. 77-84.

Wilson, R. C., E. L. Harp, M. D. Picard, and S. H. Ward, 1973. Chaotic terrain of Mars: a tectonic interpretation from Mariner 6 imagery: *Geol. Soc. America Bull.*, v. 84, pp. 741-748.

Wise, D. U., 1974. Martian fault pattern and time sequence in relation to volcanism, northern Tharsis ridge area: *Transactions, Amer. Geophys. Union*, v. 55, p. 341.

Wise, W. S., 1969. Geology and petrology of the Mt. Hood area: a study of high Cascade volcanism: *Geol. Soc. America Bull.*, v. 80, pp. 969-1006.

Witting, J., F. Narin, and C. A. Stone, 1965. Mars: Age of its craters: *Science*, v. 149, pp. 1496-1498.

Wood, G. P., 1974. Is there another major constituent in the atmosphere of Mars?: *NASA Tech. Memo.* (NASA TM-X71999).

Woolard, E. W., 1944. The secular perturbations of the satellites of Mars: *Astron. Jour.*, v. 51, pp. 33-36.

Woronow, A., 1972. Origin of the Martian chaotic terrains: *Science*, v. 178, pp. 649-650.

———, and E. A. King, Jr., 1972. Size frequency distribution of Martian craters and relative age of light and dark areas: *Science*, v. 175, pp. 755-757.

Wyllie, P. J., 1971. Role of water in magma generation and initiation of diapiric uprise in mantle: *Jour. Geophys. Res.*, v. 76, pp. 1328-1338.

———, 1973. Experimental petrology and global tectonics. A preview: *Tectonophysics*, v. 17, pp. 189-209.

Young, R. S., and D. L. DeVincenzi, 1974. From Mars with love: *Science*, v. 186, pp. 495-501.

Zellner, B. H., and R. C. Capen, 1974. Photometric properties of the Martian satellites: *Icarus*, v. 23, pp. 437-444.

Zisk, S. H., and T. Hagfors, 1970. Radar atlas of the Moon, *Final Report*, v. 2, Contract NAS 9-7830, Lincoln Laboratory.

SUPPLEMENT

A Viking View of Mars

On June 19, 1976, Viking 1 achieved orbit around Mars. Pictures from orbit revealed that the pre-planned landing site (fig. 2.20) was dangerously rough. The time of landing was delayed several weeks while pictures were acquired of other candidate landing sites. Finally an acceptable area was identified, and, on July 20, a successful landing was accomplished in the vicinity of 48°W, 22.5°N.

Viking 2 went into orbit about Mars on August 7. Again, the region originally chosen for landing (fig. 2.21) contained too many topographic hazards. For almost a month, alternate sites in the vicinity of the 45°N latitude band were studied from orbit. Among the regions photographed, the smoothest area was in the vicinity of 226°W, 48°N, a region known as Utopia Planitia. On September 3 Viking Lander 2 descended to the surface and began its scientific reconnaissance.

Here we present some of the pictures acquired by Viking during the first months of the mission. The orbiter pictures reveal spectacular details completely beyond the resolving power of Mariner cameras. The lander pictures are equally striking, transporting the viewer directly to the surface of Mars.

We are indebted to the Viking Orbiter and Lander Imaging Science Teams for permission to use these pictures. Preliminary science reports by these two teams appear in the 27 August 1976 issue of *Science* (vol. 193).

Figure S.1. The crater Arandas, 25 km in diameter and located at 15°W, 43°N. The flow-like appearance of ejecta deposits has been recorded for many craters photographed by Viking, but was not so clearly revealed in previous Mariner pictures. The sort of ejecta morphology shown in this figure and in figure S.2 is peculiar to Mars, not having been observed on the moon or Mercury. Such morphology suggests fluidization of ejected material, perhaps related to impact in a near-surface zone of permafrost. Arandas appears morphologically fresh, with a sharp rim crest. The ejecta display multiple lobes, and radial depressions scour the ejecta surface. A central peak within the crater is surrounded by material that may have slid down the crater wall during the latter stages of the impact event.

Figure S.2. The crater Yuty, approximately 18 km in diameter, located in Chryse Planitia at 34°W, 22°N. Tongues of ejecta debris morphologically resemble ground-hugging avalanche flows on Earth. Many terrestrial flows form ridges at their leading edge similar to those observed here. The smaller crater adjacent to Yuty has been partly buried by ejecta from the larger crater. However, the crater rim is surprisingly sharp, lacking the scoured texture generally caused by ballistic erosion.

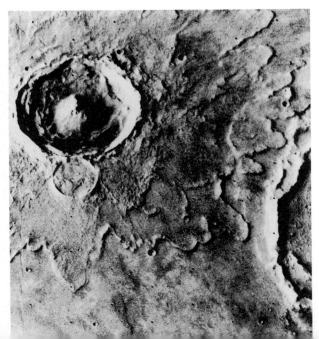

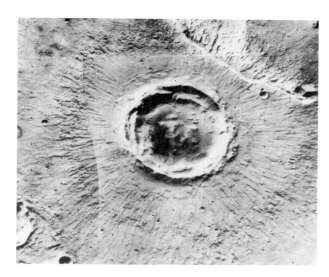

Figure S.3. A Viking mosaic showing an impact crater in the vicinity of 52°W, 23°N. The crater is about 30 km in diameter and is surrounded by radially textured ejecta similar to that observed on the moon and Mercury. The margin of Kasei Vallis is visible in the upper right.

Figure S.4. The Viking 1 orbiter recorded this view of Valles Marineris on July 3, 1976, from a range of 2000 km. The area shown is approximately 60 km by 100 km. The far wall of the canyon is about 2 km high. Layers in the wall probably indicate alternating strata of lava, ash, or wind-blown deposits. Huge landslide features are reminiscent of similar landforms on Earth. Lineations indicate the direction of flow. The large apron of debris in the center of the far wall overlaps a more degraded debris formation to the left. Permafrost melting may play a role in collapse and flow of wall material. Streaks on the canyon floor are evidence of wind action.

Figure S.5. A Viking 1 orbiter photomosaic, looking south from the equator toward the Capri Chasma region. The region is approximately 250 km by 250 km. Large blocks of material at one end of the centrally located triangular valley resemble chaotic terrain and were probably formed by melting of a subsurface ice layer with subsequent cata-strophic release of water. The lineations at the narrow end of the valley resemble fluvial flow features. Surface fractures are the result of large-scale slumping. The sinuous rille in the upper part of the picture was probably formed by volcanic activity. A field of secondary crater chains is visible in the upper right.

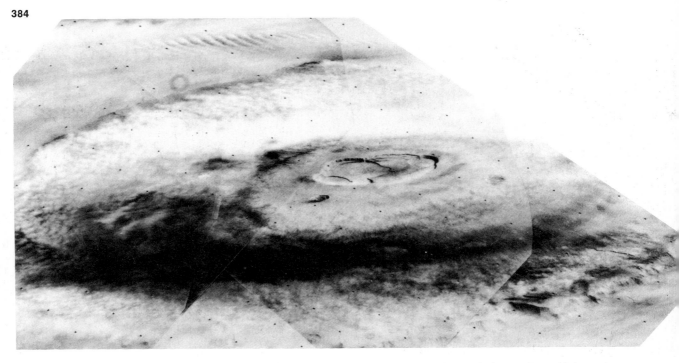

Figure S.6. An oblique view of the giant volcano Olympus Mons, approximately 600 km in diameter. A scarp that rings the volcano is visible in the upper right. The summit caldera protrudes through a cloud cover that extends up the flanks to an altitude of 19 km. The clouds, which extend several hundred kilometers beyond the mountain in the upper left, are believed to be composed of water ice, which condenses as air rises up the volcanic flanks. The clouds are a seasonal phenomenon that can be observed from Earth in the Martian afternoon during spring and summer, at which time the clouds attain their greatest development.

Figure S.7. A region in the Memnonia quadrangle, centered at 146°W, 19°S, and approximately 40 km on a side. Multiple flow fronts, probably of volcanic origin, are visible on the plains that lap up on cratered terrain. This figure, as well as figure S.8, illustrates that many of the plains that appear generally featureless at Mariner resolution, in fact contain intricate and distinctive landforms.

Figure S.8. A region on the southeastern margin of Acidalia Planitia, centered at 14°W, 41°N, and approximately 50 km on a side. This terrain, distinguished by polygonal fissures, extends for hundreds of kilometers. To the north it is covered by horizontal strata—presumably eolian deposits—extending equatorward and partly stripped away by multiple erosive events. The fissures in this picture are of unknown origin. Discounting a difference in scale, they resemble shrinkage cracks formed in a variety of terrestrial situations: lava consolidation, desiccation, and periglacial freezing and thawing.

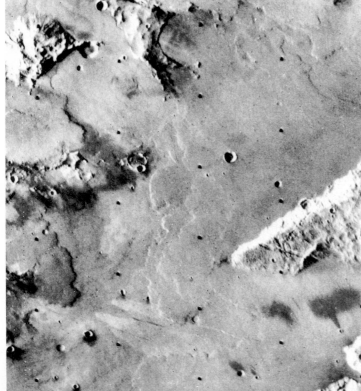

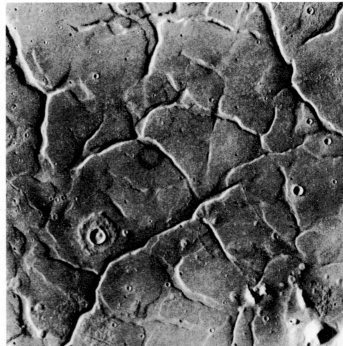

(a)
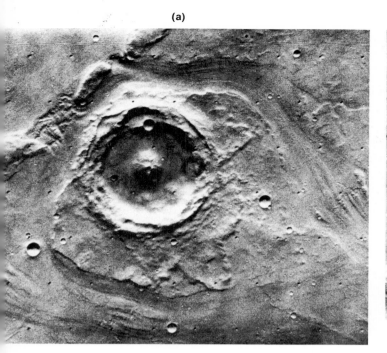

(b)
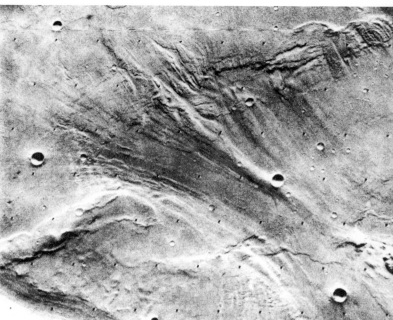

Figure S.9. This sketch map illustrates the general setting of the Viking 1 landing site, located near 48°W, 22.5°N, in Chryse Planitia. Approximately 500 km to the west lies an irregular scarp-like boundary that separates the sparsely cratered plains of Chryse from the older cratered plains of Lunae Planum (light gray). Numerous fluvial channels (Kasei Vallis, diagonal lines; and individual thin lines) flow from Lunae Planum into the Chryse basin (white), scouring the surface of the plains and modifying ejecta deposits from impact craters (dark gray). Lunae Planum also contains several prominent sinuous rilles (denoted by thick line with ball), probably volcanic in origin. They differ from structures associated with fluvial activity (thin lines) in that they lack a braided or dendritic texture. Viking pictures reveal a large number of features similar to lunar mare ridges in Chryse. It appears that volcanic plains have embayed the Lunae Planum region and that mare ridges were formed subsequently. In some cases the ridges seem to postdate the fluvial activity; in other cases the ridges have been eroded by fluvial processes.

Inset "a" shows a 15-km impact crater located about 130 km southwest of the landing site. The associated ejecta deposit has been extensively modified by fluvial activity. The crater rim was not breached and retains its original relief. Ejecta deposits in the lee of the crater remain relatively unmodified, creating a streamlined shape.

Inset "b" illustrates the interaction between fluvial activity and pre-existing ridges. The ridges, which average approximately 200 m in elevation, served to pond water upstream. Large volumes of water flowed through gaps between adjacent ridges. The size of the channel and the prominent striations illustrate the force of the erosional event.

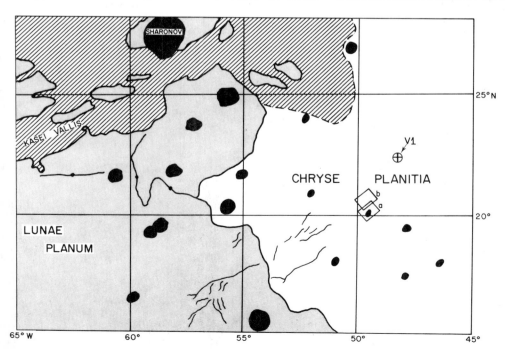

Figure S.10. This low-resolution panorama was taken by camera 2 on Viking Lander 1 approximately 10 minutes after landing, late in the Martian afternoon. The picture covers approximately 300° in azimuth. At the left side, the view is toward east-northeast; at the right side toward the northwest. The sky is brighter in the direction of the setting sun. Components of the spacecraft are visible in the nearfield; the deployed meteorology boom at the far left, the surface sampler housing just to the right, the protective housing over the spacecraft power generator on the right side, and a part of the S-band antenna in the far upper right. The nominal horizon is approximately 3 km away. Most of the surface details are within 100 m of the spacecraft. A number of hills and swales appear in the middle left proceeding toward the horizon. The surface is littered with blocks, only a few of which exceed 1 meter in size. Some of the irregularities on the skyline may be parts of raised crater rims. Areas of sediment accumulation are visible in the far-left middleground and in the region directly over the power generator housing.

Figure S.11. A high-resolution view of the Martian surface, taken with camera 2 on Viking Lander 1. The same field of view appears in the middle left of figure S.10. Several isolated dunes are visible in the midfield. An area of fractured bedrock is in the right midfield, atop a gentle ridge.

Figure S.12. A high-resolution view of the Martian surface, taken with camera 1 on Viking Lander 1. The view is toward the east, a region obscured from camera 2. Early morning lighting sharply delineates the scene. The accumulations of sediment superficially resemble terrestrial sand dunes. However, the sediment is finer grained than sand, and the "dunes" have irregular shapes. In part, they appear to be drifts, stabilized by the presence of large rocks. The linea- tions within "dunes" are internal stratification, revealed by an episode of deflation that followed the initial period of dune accretion. The meteorlogy boom is visible in the near-field. The stratified dune to the right of the boom rises about 1 m above the adjacent rock-littered surface. The large rock at far left is about 2 m across and is located about 8 m from the camera.

Figure S.13. This low-resolution panorama was taken by camera 2 on Viking Lander 2 approximately 10 minutes after landing. The format is the same as that of Figure S.10. At the left side the view is toward the northwest; at the right side toward the south. The time is Martian mid-morning with the sun toward the east. The surface is strewn with rocks out to the horizon. Most are less than 1 m across. Although fine-grained material is seen between the boulders, no sand dunes are present. This picture has been geometrically rectified to account for the fact that the spacecraft landed with an 8° tilt to the west. The horizon in the left-center is relatively featureless, indicating that it may be several kilometers distant. Toward the right (southeast) rocks are silhouetted against the skyline, indicating that the horizon is much nearer, probably because of a slight rise in that area.

Figure S.14. The approximate landing site of Viking 2, marked by an "X." The width of the area in this figure is approximately 100 km. Further south, the plains display polygonal fissures, similar to those shown in Figure S.8. However, in the landing area, a depositional mantle covers and subdues these sharp topographic features.

Figure S.15. This is the first picture taken by Viking Lander 2, minutes after touchdown. A wide variety of rocks litter a surface of finer-grained material. Boulders in the 10 to 20 cm range—some vesicular and some apparently fluted by wind—are common. Many of the pebbles have tabular or platy shapes, suggesting they may have been derived from a caliche-like surface layer, produced by upward migration of soil water that precipitated salts upon evaporation. Just as occurred with Viking Lander 1's first picture, brightness variations at the beginning of the picture scan (left edge) probably are due to dust settling after landing. A substantial amount of fine-grained material kicked up by the descent engines has accumulated in the concave interior of the footpad. The center of the image is about 1.4 m from the camera. The dimensions of the picture are about 1 m by 2 m.

Index

References to figures appear in italic type.

Library of Congress Cataloging in Publication Data

Main entry under title:

The Geology of Mars.

  Includes index.
  1. Mars (Planet)—Geology. I. Mutch, Thomas A.,
1931-
QB641.G43   559.9′23   75-30199
ISBN 0-691-08173-5